RADIATION CURING IN POLYMER SCIENCE AND TECHNOLOGY—VOLUME I

FUNDAMENTALS AND METHODS

RADIATION CURING IN POLYMER SCIENCE AND TECHNOLOGY—VOLUME I

FUNDAMENTALS AND METHODS

Edited by

J. P. FOUASSIER

Université de Haute-Alsace, Laboratoire de Photochimie Générale,
URA-CNRS No 431, ENSC Mu, 3 Rue Alfred Werner,
68093 Mulhouse Cedex, France

J. F. RABEK

Karolinska Institute, Polymer Research Group, Department of Dental Materials
and Technology, Alfred Nobels Alle 8, Box 4064, S-141 04 Huddinge, Sweden

ELSEVIER APPLIED SCIENCE
LONDON and NEW YORK

ELSEVIER SCIENCE PUBLISHERS LTD
Crown House, Linton Road, Barking, Essex IG11 8JU, England

WITH 98 TABLES AND 308 ILLUSTRATIONS

© ELSEVIER SCIENCE PUBLISHERS LTD 1993

British Library Cataloguing in Publication Data

Radiation Curing in Polymer Science and
Technology.—Vol. 1: Fundamentals and
Methods
  I. Fouassier, J. P.   II. Rabek, J. F.
547.7

ISBN 1-85166-929-9 (Volume 1)
ISBN 1-85166-939-6 (The Set)

Library of Congress Cataloging in Publication Data

Radiation curing in polymer science and technology/edited by J. P.
Fouassier, J. F. Rabek.
  p.   cm.
  Includes bibliographical references and indexes.
  Contents: v. 1. Fundamentals and methods—v. 2. Photoinitiating
systems—v. 3. Polymerisation mechanisms—v. 4. Practical
aspects and applications.
  ISBN 1-85166-929-9 (v. 1).—ISBN 1-85166-933-7 (v. 2).—ISBN
1-85166-934-5 (v. 3).—ISBN 1-85166-938-8 (v. 4)
  1. Radiation curing.   2. Polymers—Curing.   I. Fouassier, Jean
-Pierre, 1947—   I. Rabek, J. F.
TP156.C8R338   1993
668.9—dc20   92-24331
                                                            CIP

No responsibility is assumed by the Publisher for any injury and/or damage to persons or property as a matter of products liability, negligence or otherwise, or from any use or operation of any methods, products, instructions or ideas contained in the material herein.

Special regulation for readers in the USA

This publication has been registered with the Copyright Clearance Center Inc. (CCC), Salem, Massachusetts. Information can be obtained from the CCC about the conditions under which photocopies of parts of this publication may be made in the USA. All other copyright questions, including photocopying outside the USA, should be referred to the publisher.

All rights reserved. No part of this publication may be reproduced, stored in a retrieval system, or transmitted in any form or by any means, electronic, mechanical, photocopying, recording, or otherwise, without the prior written permission of the publisher.

Typeset and Printed by The Universities Press (Belfast) Ltd.

Contents

Preface vi

1. State-of-the-Art and Trends in the Radiation-Curing Market 1
    P. Dufour
2. An Introduction to Basic Principles in UV Curing 49
    J. P. Fouassier
3. Polymers in X-Ray, Electron-Beam and Ion-Beam Lithography ... 119
    L. Schlegel and W. Schnabel
4. Applications of Electron-Beam Curing 193
    J. R. Seidel
5. UV and Electron Beam Curable Pre-polymers and Diluent Monomers: Classification, Preparation and Properties 225
    N. S. Allen and M. Edge
6. Role of Grafting in UV- and EB-Curing Reactions 263
    P. A. Dworjanyn and J. L. Garnett
7. Experimental and Analytical Methods for the Investigation of Radiation Curing 329
    J. F. Rabek
8. UV-Radiation Sources and Radiation Devices for Industrial Production Processes 453
    J. F. Rabek
9. Electron-Beam Processing Machinery 503
    S. V. Nablo

Index 555

Preface

The field of Radiation Curing has grown into an important and pervasive branch of Polymer Science and is currently in full expansion throughout the world and is the basis of a considerable number of applications, not only in conventional areas, such as coatings, inks and photoresist technologies, but also in entirely new domains, such as adhesives, laser imaging and 3D-photostereolithography.

The object of the present monography is to provide a comprehensive description of the State-of-the-Art and to provide, collect and generalize on a number of available data regarding the radiation curing of polymeric materials, systems, formulations, mechanisms and kinetics. We also emphasize the viewpoint of the fundamental research conducted in this field, while giving a fully adequate treatment to applications. It strongly favours the inclusion of experimental details, apparatus and technologies. For that reason, it should be useful to all people from the polymer curing industries to learn a broad knowledge from the experts. This monography is also meant, most assuredly, for all those who are active in this field, whether students or researchers.

This monography is divided into four main parts. The first deals with the various fundamental aspects encountered in UV- and EB Curing. We purposely chose to give, in the first chapter, a bold outline of the state of the market, so that the reader might immediately have a general survey of the field. The second section is an overview of the various initiating systems, their reactivities and the mechanisms involved. The third part takes up the polymer aspect (i.e. the problems related to the organic matrix) that may be encountered in the field of Radiation Curing. The fourth part is devoted to the practical and industrial applications of radiation curing, while giving special attention to safety-related and environmental problems.

UV techniques are certainly bound to be increasingly successful in coming years, given that they appear to allow a more adequate protection of the environment. We attempt, in this monography, to assess the progress achieved in a field that is steadily on the move, taking great pains to ensure that the different chapters making up this peice of work provide the most homogeneous documentation possible.

We went to great lengths in order to secure the cooperation of the most outstanding specialists to complete this monography and would like to express their appreciation to all contributors for their efforts in the realization of this book. We are pleased to submit to the readers the State-of-the-Art in this field.

J. P. Fouassier and J. F. Rabek

'To our friends and colleagues'

List of Contributors

N. S. ALLEN
Centre for Archival Polymeric Materials, Chemistry Department, Manchester Metropolitan University, Jon Dalton Building, Chester Street, Manchester, M1 5GD, UK

P. DUFOUR
Radcure Specialties SA, Rue d'Anderlecht 33, B-1620 Drogenbos, Belgium

P. A. DWORJANYN
School of Chemistry, University of New South Wales, PO Box 1, Kensington, 2033 NSW, Australia

M. EDGE
Centre for Archival Polymeric Materials, Chemistry Department, Manchester Metropolitan University, Jon Dalton Building, Chester Street, Manchester, M1 5GD, UK

J. P. FOUASSIER
Université de Haute Alsace, Laboratoire de Photochimie Générale, URA-CNRS no 431, ENSC Mu, 3 Rue Alfred Werner, 68093 Mulhouse Cedex, France

J. L. GARNETT
School of Chemical Engineering and Industrial Chemistry, University of New South Wales, PO Box 1, Kensington, 2033 NSW, Australia

S. V. Nablo
Energy Sciences Inc., 42 Industrial Way, Wilmington, MA 01887, USA

J. F. Rabek
Karolinska Institute, Polymer Research Group, Department of Dental Materials and Technology, Alfred Nobels Alle 8, Box 4064, S-141 04 Huddinge, Sweden

L. Schlegel
Hitachi Central Research Laboratory, Kokubunji, 185 Tokyo, Japan

W. Schnabel
Hahn-Meitner Institut Berlin GmbH, Glienicker Strasse 100, D-1000 Berlin 39, Germany

J. R. Seidel
RadTech Europe, Perolles 23, CH-1700 Fribourg, Switzerland

# Chapter 1

State-of-the-Art and Trends in the Radiation-Curing Market

P. DUFOUR

Radcure Specialties SA, Rue d'Anderlecht 33, B-1620 Drogenbos, Belgium

Introduction	2
Overview of Applications	2
At home	3
In your car and at work	3
In your many varied activities	3
The Manufacturing and Distribution Chain in the Radiation-Curing Market	4
Basic raw materials	5
Monomers and oligomers	5
Formulations	5
End-users	5
Different Sources of Radiation Curing	5
Different Radiation-Curing Chemistries	7
Unsaturated polyesters in styrene (UPE)	7
Thiol-ene systems	8
Cationic systems	8
Methacrylic and acrylic systems	8
History of Radiation-Curing Technology	9
Appearance	9
Driving forces	10
Evolution	10
Market Volume Estimation	12
Evolution	12
Distribution by chemical families for 1990	14
Distribution by application field for 1990	14
Market share of radiation-curing technology	16
Coatings	16
Graphic art	16
Conclusion	17
Similarities and Differences in Europe, USA and Japan	20
Products and Suppliers	21
Products	21
Raw-material suppliers	22

    Formulators 23
    Concentration of companies 24
Graphic Art Applications 25
    Lithographic inks 25
    OPV 27
    Silk-screen printing 28
    Flexogravure 29
    Heliogravure, intaglio and ink jet 30
Coating Machines 31
    Roller coater 31
    Curtain coater 31
    Spraying 32
    Vacuum curtain application 32
    Extrusion 32
    Rotary screen 32
Coatings 32
    Wood finishing 33
    Paper upgrading 34
    Plastic coatings 35
    Metal coatings 36
    Other substrates 36
Adhesives 37
    Laminating adhesives 38
    PSA 38
    Removable PSAs 39
    Repositionable PSAs and blisters 39
Substrates and Objects 39
Limitations 40
Trends 41
    Applications to develop 41
    Future trends 43
Conclusion 46

INTRODUCTION

What is radiation curing? It is the polymerisation (in a three-dimensional network) of a chemical system which is initiated by incident radiation. The curing, that is to say the transformation from a liquid into a non-tacky solid, is very fast and occurs in less than 1 s.

The raw materials for radiation curing make up one of the fastest growing segments of the specialty chemicals industry.

This chapter will try to give an overview of the current status, the possibilities and trends in the radiation-curing market.

OVERVIEW OF APPLICATIONS

Today, we live in a world where objects are manufactured, decorated or protected by radiation curing; however, except by those individuals involved in the industry, this fact is rarely appreciated.

The consumer finds it normal to see glossy and attractive packages but does not consider how they are manufactured. Twenty four hours a day and practically all over the world (certainly, for the developed countries) we use objects coated by radiation curing.

At home
- Your alarm clock, made using printed circuit boards (PCBs), awakes you.
- The wooden bed, the poly(vinyl chloride) (PVC) cushioned floor and the sound-proof glass of your bedroom are protected or manufactured using UV or electron beam (EB) coatings.
- The toothpaste tube is decorated with UV inks and varnish.
- Many kinds of cosmetic packages are printed in UV or EB and protected with high-gloss coatings.
- Various plastic bottles are UV printed.
- The toilet seat may be protected with a UV coating resistant to all kinds of detergents.
- You use different foods to prepare your breakfast where the packages are printed with UV or EB.
- Kitchen furniture is often UV or EB coated.
- Ceramic inks on glass windows and PCBs of various kitchen equipment are manufactured using UV-radiation curing.
- PVC or parquet flooring, different furniture, doors and sometimes chairs are protected with UV or EB coatings (clear or pigmented).
- Outdoor-resistant panels are EB coated and laminated security glass is made from two glass sheets using thick (± 1 mm) UV laminating adhesive.

In your car and at work
You use several items made using radiation curing but most of the time you are not aware of it.

- Various plastic and wooden components of your car: headlamp reflectors and light covers, the black ceramic inks around the windows, and stickers are made using radiation-curing technology
- Car radios are full of PCBs using UV technology
- On the road to your office, road signs, road markings and sometimes road repairs use radiation-curing technology.
- Telephones, telefaxes, and computers use PCBs manufactured using UV technology.
- Pencils are sometimes UV coated.
- Labels are printed in UV, the adhesive and release paper are perhaps cured with UV or EB radiation.

In your many varied activities
- Razor blades and mirror-backing paint may be radiation curved.
- Most of the books and magazines you read have covers printed and protected by a glossy UV coating.

- Record sleeves, video tapes and compact discs are UV coated and magnetic audio tapes may be EB cured.
- Television cabinets are often protected with UV varnish.
- Your shoes and luggage may have been textured by means of an EB-coated paper used in casting the material.
- You pay bills with bank notes or credit cards that are totally or partially printed using UV or EB inks.
- Postcards you send, advertising panels you see along the streets, glossy covers on magazines you read in planes, in bookshops, in travel agencies, in tourist information offices are radiation cured.
- Going in for sport, you use fishing rods, skis, wooden boards in rubber boats, and crash helmets decorated and protected by UV coatings.
- In big department shops you are surrounded by colourful and glossy packages, aluminised labels and highly decorated and glossy shopping bags that were manufactured using UV or EB technology.
- You eat and drink out of packages that are often decorated with EB or UV inks and varnishes: liquid-food packages, yoghurt containers, beer cans, dry-food external packaging, etc.
- Even the printed package of the medicine that you take is probably UV cured.
- Your dentist is possibly filling your teeth using formulations cured by visible light.

It is impossible to list all the products that are on the market and use radiation curing in one or more steps of their manufacture. The above lists are certainly not complete.

The fantastic miniaturisation of electronics has been possible because of the technical possibilities offered by radiation-curing technology, but this is not part of this overview.

All our daily activities bring us into contact with radiation-cured objects and our modern world would appear old-fashioned if radiation curing were not used today.

THE MANUFACTURING AND DISTRIBUTION CHAIN IN THE RADIATION-CURING MARKET

The manufacturing and distribution chain of radiation-curing materials appears as follows:

Basic raw materials
↓
Monomers + Oligomers
↓
   Formulations ← Photoinitiators, pigments and fillers, additives
↓
     End-users ← UV oven, EB accelerators, application machines, substrates

Basic raw materials
The most important basic raw materials are polyols, anhydrides and diacids, epoxides, (meth)acrylic acids and their esters, styrene, diisocyanates.

Monomers and oligomers
Manufacturers use basic raw materials to produce the following.

- *Monomers.* Most of them are called either multifunctional monomers (MFM) or polyol polyacrylates.
- *Oligomers.* The main chemical families are polyester acrylates (PEA), epoxy acrylates (EA), urethane acrylates (UA), full acrylics (acrylated (co)polymers of (meth)acrylic esters), unsaturated polyesters diluted in styrene (maleic and fumaric unsaturations), amino acrylates (used as photoaccelerator in the photoinitiator system) and cycloaliphatic epoxies and vinyl ethers.

Formulations
Formulators manufacture ready-to-use formulations by mixing oligomers and monomers with a photoinitiator system. They often add different additives, fillers and/or pigments.

End-users
End-users utilise application machines for decorating or coating ready-to-use formulations on different substrates and UV ovens or EB accelerators for curing the liquid formulation into a solid non-tacky coating. UV or EB curing occurs in a fraction of a second.

DIFFERENT SOURCES OF RADIATION CURING

To initiate the polymerisation of the chemical formulation an incident radiation is necessary. The radiation can be ionising (e.g. EB) or electromagnetic. The electromagnetic spectrum from lower to higher energy and frequency is as follows:

Radiowaves
Microwaves
IR
Visible light
UV
X-rays
γ-rays

Radiowaves and microwaves are not powerful enough to initiate polymerisation. Curing with IR is out of the scope of this overview because it is used more as a heating source. Visible light can initiate polymerisation when using the right photoinitiator system and is used in dental fillings (mainly for safety

Table 1.1
Increase in power of UV lamps

Year	Power (W/cm)
1960	40
1970	80
1980	120
1987	160
1989	200
1990	240

reasons). UV light is the most commonly used and of course needs a photoinitiator system which will transform the absorbed light in initiating radicals or cations.

Accelerated electrons, X-rays and γ-rays are powerful enough to produce initiation by interfering with the uncured formulation. X-rays are rarely used. γ-Rays are only used in limited areas and are not included in this overview. X-rays and γ-rays require strong shielding to protect the operators and therefore are only used in very special cases.

Another source of initiating light requiring a photoinitiator is the laser (different kinds). The most known application of the laser is in stereophotolithography.

Before the beginning of the 1970s the power of UV lamps did not exceed 40 W/cm and consequently was not sufficient to overcome oxygen inhibition effectively.

Then, when the lamp manufacturers became aware of the importance of overcoming oxygen inhibition, they progressively increased the power of the UV lamps (Table 1.1).

Pulsed xenon lamps offer very high output in short bursts or pulses, which are useful for certain applications. Today, most commercial UV-curing lines are still using the medium-pressure mercury vapour lamps with outputs of 80 W/cm (200 W/in).

EB accelerators have also undergone rapid evolution. First their power requirement was reduced from several mega electron volts (MeV) to several hundred kilo electron volts (keV) (from multi-stage to single-stage acceleration), also the intensity of the beam has been increased to allow faster curing speed. The first EB gun had a point cathode producing a focused beam which was scanned across the whole width of the substrate. Then, the so-called 'electron curtain' process was commercially developed by Energy Sciences International and the Radiation Polymer Company. From a single filament it quickly developed to a multifilament cathode. In such a way the curing speed of an identical formulation could be increased. The scanning EB guns are still used and present certain advantages in special cases, but when high curing speeds are necessary guns with linear cathodes, single or preferably multiple cathodes, are now currently the state-of-the-art in EB accelerators. The number of low-voltage EB units used in production probably exceeds 120 world-wide, most being in Europe and the US. There are also about the same number of laboratory and pilot units. Each year new production lines come on stream.

It is difficult to estimate the number of UV ovens in the world. Each UV oven contains one or more UV lamps with the large width ovens often containing four or more UV lamps. The number of UV ovens probably exceeds 10 000 world-wide (this includes laboratory, pilot and production units). Each year, several hundred new ones are commissioned. The pioneers of manufacturing UV lamps and UV ovens have seen new competitors arriving in the market, certainly much more so than EB gun manufacturers. A brief market study of UV lamp and oven manufacturers during the DRUPA 1990 exhibition in Düsseldorf (Germany) indicated clearly that the number of new lines continues to increase:

- significant manufacturers reported a growth of 5–20% per year; and
- newcomers reported a growth of 25–60%.

This study clearly indicated that the market is still growing for the UV equipment suppliers. It is, therefore, possible to expect that if the number of curing units increases, radiation-curing chemicals will also continue to grow. The main disadvantage of EB against UV is the much higher capital cost, particularly at low line speed: a UV-curing oven costs maximum 100 000–130 000 $ while an EB gun costs about 1·3 million $ and in addition the EB accelerator requires inert gas purging of the curing head. However, UV and EB do not always compete with each other; on the contrary, they are often complementary. Certainly, when finishing a process is possible with UV rather than EB it will be done by UV but there are applications for which UV is impossible.

DIFFERENT RADIATION-CURING CHEMISTRIES

Four main different radiation-curing chemistries are in competition:

1. maleic–fumaric unsaturated polyester diluted in styrene;
2. thiol-ene systems;
3. cationic systems; and
4. (meth)acrylic systems.

Unsaturated polyesters in styrene (UPE)
These products are still widespread (several thousands of tonnes) mostly as wood coatings and mainly in Germany, Italy, Spain, and Eastern Europe. They are also still used in the US and Far East. They are used because of their

very low cost in comparison to the other systems, yet acrylic formulations are replacing them progressively. The main reason for this substitution is the volatility of styrene. Better surface properties and higher reactivity are obtained with acrylic-based products; higher reactivity means higher productivity. Some new formulations using maleic unsaturated polyesters but diluted in acrylated monomers are sometimes seen. Another approach increasing in popularity is to increase the reactivity of the UPE formulations by the addition of highly reactive acrylated products.

Thiol-ene systems
These systems were, at one time, intensively studied because they are not oxygen inhibited unlike the UPE and acrylic systems. Such systems are composed of mixtures of polythiol with polyallylic compounds (ideally thiol group/allylic unsaturation in a ratio of 1:1). These systems do not represent an important share of the radiation-curing market due to the odour of polythiol compounds.

Cationic systems
Mainly based on cycloaliphatic epoxies, these systems are not oxygen inhibited and have low shrinkage during curing. Being insensitive to oxygen, low photoinitiator concentrations can be used, giving lower residual odour than UV acrylic formulations which require higher concentrations of photoinitiators to overcome oxygen inhibition.

Furthermore, the low shrinkage of the cycloaliphatic epoxies is advantageous in obtaining better adhesion on difficult substrates such as metal and certain plastics. There is no published estimation of their market share on the European level, although they are unlikely to represent more than 5% of the volume of the acrylic systems at present.

Their small market share, despite all their theoretical advantages (no oxygen sensitivity and improved adhesion due to lower shrinkage), is probably due to their sensitivity towards other factors such as temperature and moisture.

Inhibition due to moisture is now well known for cationic systems and gives more unusual problems than oxygen inhibition.

Other binders appearing in the cationic systems are vinyl ether derivatives which are claimed to be highly reactive, but, up to now, initial results indicate that these products are also sensitive to moisture and shrink, substantially more than cycloaliphatic epoxies.

Cationic systems only cure under EB bombardment when they contain cationic photoinitiator. Consequently, they are not believed to be used today in EB-curing applications.

Methacrylic and acrylic systems
The most commonly used systems are formulations containing acrylic unsaturations as these are considerably more reactive than those containing methacrylic double bonds. Therefore, formulations based exclusively on

methacrylic unsaturations are used only in very special applications (e.g. dental application). In electronic applications, for example, methacrylics are often used in combination with acrylic unsaturated products in order to obtain the required combined properties after curing.

Methacrylic unsaturated products are often less irritating than their equivalent acrylic products, and after curing are thermally more resistant, harder, more abrasion-resistant but less flexible.

HISTORY OF RADIATION-CURING TECHNOLOGY

Appearance

Compared to that of most paint and coating technologies, the history of radiation-curing technology is short. The first patent was granted to Inmont in 1946 for a UV-cured ink based on UPE resins. Although theoretical discussions were published in the early 1950s, the first commercial application of this technology in coatings was in the early 1960s. This was a UV-cured particle board filler based on UPE. It began in Germany early in the 1960s and spread all over the world. The UPE systems are still widely used for wood finishing as they are cheap and reactive enough for this kind of application.

In the beginning, benzoin ether derivatives were used as photoinitiators and cured using UV lamps of 30 W/cm. Oxygen inhibition could be overcome by paraffin waxes used in the formulations and curing occurred in two steps:

1. pre-gellation using low-power lamps producing polymerisation in depth forcing the paraffin towards the surface forming a barrier film of wax; and
2. curing the surface through the thin paraffin layer using 30 W/cm mercury lamps. The line speed was 10 m/min using three to five UV lamps.

The use of radiation curing in graphic arts with line speeds of 50–400 m/min started, as soon as adequate curing equipment, reactive resins, as well as suitable photoinitiators became commercially available.

It was at the beginning of the 1970s that systems based on acrylic unsaturation became industrially used and, since that time, acrylic radiation-curing systems have enjoyed a steady growth in volume and in the number of applications. This was possible because acrylic unsaturation is the most reactive, much more than the UPE. The appearance of powerful UV lamps (80 W/cm and now up to 200 W/cm) equipped with efficient reflectors and the commercialisation of very effective photoinitiators and photoinitiator systems, specifically designed for acrylic unsaturation, played a significant part in overcoming oxygen inhibition. Consequently, in UV formulation, paraffin could be avoided and curing in ambient atmospheres became possible at high speeds. In parallel with this development of UV systems and equipment, manufacturers of EB guns introduced more powerful accelerators at lower voltages (but still enough to cure coatings from a few micrometres to a few hundred micrometres thickness). In the early 1970s, Ford in the US was the first to use an EB for curing coatings applied on interior parts of cars (dashboard). In the mid-1970s, Svedex in the Netherlands installed a line for curing doors with EB accelerators. Continuous development in equipment has brought about the single-step acceleration of electrons produced successively by 'point cathode', linear cathode and multiple cathodes.

Driving forces
During the 1960s, the threat of energy shortages and the environmental consciousness created a favourable climate for the development of radiation-curing technology.

It soon became apparent that radiation-curing technology was the only one answering the challenge of future and modern finishing. It meets the '3 E' rules fixed as a basis for each new development in the very diversified areas of coatings.

- *Energy*

- There are savings in energy achieved, because energy is just used to initiate the radical polymerisation which cures the coating.
- It is not necessary to heat the whole substrate.

- *Ecology*

- There is no pollution of the atmosphere: radiation-curing formulations do not contain volatile solvents.
- The energy used is electricity; therefore, there need be no carbon dioxide formed by burning gas or oil.

- *Economy*

- There is high productivity, because curing is done at high speed.
- There are savings in investment cost, due to compact installations.
- There are savings in materials; often, thinner coatings giving better properties are possible with radiation-curing formulations.

To these advantages can be added that final properties of the coatings are often improved, e.g. better solvent-, abrasion- and stain-resistance together with high gloss. Furthermore, acrylic chemistry offers the possibility, by adjusting the formulations, to obtain specific and diverse properties such as hardness or flexibility, gloss or matt, outdoor resistance, etc.

Evolution
In the early stages, most of the pioneers trying radiation-curing technology decided to use it only when they found an economic advantage. Rarely did industry install UV or EB finishing lines for non-polluting or energy-saving reasons. However, these two advantages were also achieved and in addition finishes with improved properties were often obtained. Progressively, industries have been using radiation-curing technology for performance reasons, to obtain coatings with properties which are difficult to achieve with more

traditional technologies. Now that regulations are becoming more and more severe towards solvent emission, decisions to use radiation-curing technology, primarily to answer ecological requirements, will become more common. Acrylic formulations were first used in lithographic printing inks on card, and as overprinting varnish to obtain a high gloss. These started rapidly in the US followed by Europe. So radiation-curing technology was first used in graphic art and on card. It has been progressively extended to other substrates. The first application in coating was wood finishing cured by EB and UV. The evolution continued and radiation-curing technology extended slowly but surely to all kinds of graphic and coating applications and now is applied on practically all substrates using almost all kinds of printing machines and coating processes.

In the beginning the pot-life stability of formulations caused problems; however, this was quite rapidly solved for monomers and oligomers containing no photoinitiators. The use of more reactive photoinitiator systems (combination of photosensitiser + photoinitiator + photoaccelerator) caused some problems due to poor pot-life stability when using specific tertiary amines as photoaccelerators. After a while, this problem was also solved.

Irritation and toxicity encountered with formulations based on acrylic unsaturation were often mentioned in the 1970s. This hindered the introduction and the development of radiation-curing technologies and a strong opposition appeared in different places mainly due to local print-trade unions. Due to the high productivity obtained with this technology, critics of radiation curing were hiding their real concerns about the possible decrease in the number of workers. But fundamentally and objectively these irritation problems were due to the lack of knowledge of the monomer manufacturers regarding the potentiality for irritation. In the very beginning when monomers were introduced, they were strongly irritant and in some cases even toxic. After the occurrence of several irritation cases, evaluations of the irritancy potential were made on different monomers and oligomers. The most toxic and irritant ones rapidly disappeared from the market and safer products were introduced. In the meantime, workers learned how to handle these products correctly and raw-material suppliers improved their manufacturing processes. Higher purity products replaced old ones which contributed towards safer formulations. Nevertheless, the danger can still come from new raw-material suppliers introducing less pure and potentially more irritating products under the pressure for supplying cheaper materials. For the 20 years that acrylic products have been used, no fatal accident has been reported.

In 1991, practically all oligomers are non-toxic (oral $LD_{50} \geq 5$ g/kg) and in the few cases of irritation, the cause is mainly monomers. Today, the most irritant commonly used monomers such as TMPTA and HDDA are replaced, when possible, by far less irritant monomers. However, the range of performance characteristics obtained with HDDA and TMPTA cannot be achieved with other monomers. By avoiding unnecessary prolonged skin contact, and with correct handling, irritation can easily be avoided.

MARKET VOLUME ESTIMATION

In this section, the author will try to give the reader a rough idea of the following:

- how big the radiation-curing market is; and
- what proportion radiation curing represents on the total graphic art and coatings market.

The estimation is done in metric tonnes (t) and not in value because the weights of binders used remain more or less the same from raw-material suppliers to end-users; and this is, of course, not the case for values.

Of course, pigments, fillers and matting agents (also solvents or water for the non-radiation-curing formulations) increase the weight of certain formulations and this is why the estimations and comparisons of radiation curing to the total of all the systems used are done in metric tonnes of resins and not of formulations.

For simplicity, the estimations are done on three main geographic areas:

NA—North America (of course, this is mainly the US with a small contribution of Canada)
EUR—Europe, but mainly Western Europe.
RW—The rest of the world (here, Japan is the most important).

Except where otherwise indicated, the volumes indicated are the acrylic binders only (pre-polymers, oligomers, monomers and photoaccelerators).

Evolution

Radiation-curing applications started in the 1960s with UPE. The acrylic binders were used in the beginning of the 1970s in the US and in Europe. Japan and RW started at the end of the 1970s. Table 1.2 gives the evolution in the three main geographic areas.

Table 1.2
Evolution of radiation-curing market volume for acrylic binders (in $\times 10^3$ t)

Year	World	NA	EUR	RW
1972	0	0	0	0
1975	1·2	1·0	0·2	0
1980	8·5	5·0	3·0	0·5
1985	23	10	9	4
1990	57	23	23	11
1995	93–100	34–37	34–37	25–26
2000	130–145	45–50	45–50	40–45

The 1970s

The growth was regular and steady in the 1970s. These were the years of the introduction of the technology, where a lot of unexpected problems were encountered. The technology had also to face resistance, as every new technology does. The 1970s were also difficult years, during which the general EUR and NA economies were stagnating, but despite this the radiation-curing technology continued to grow as steadily new applications appeared and, finally, the technology proved its specific advantages which were recognised and accepted. The NA market was at this time probably leading, due to the dynamism and openness of the US to new technologies; in the US the PVC floor market became rapidly an important area for radiation curing because the US consumer liked the high-gloss of UV finishing. EUR was more involved in graphic-art application and was careful in accepting and broadening the new technology. Japan and the RW began to use radiation-curing technology only towards the end of the decade.

From 1980 to 1990

It was during this period that radiation-curing technology became significant, being used in all the application fields where it was possible technically. The technology, which was used only in the strongly industrialised countries before 1980 (namely the US, Canada, Western Europe, and Japan), began to spread to most countries of the world. By 1990, it was used on all continents and it is only in poorly industrialised countries where it is now not used. The annual growth during this decade was around 20–23% world-wide:

17–18% in NA
21–23% in EUR
23–25% in RW

From 1990 to 1995 and 2000

It is expected that the radiation-curing market will continue to grow but the annual growth in percent will decrease, this being normal when the volumes consumed are significant. World-wide, the radiation-curing market is expected to grow at 8–10% per year based on the volume of 1990. Of course, the annual growth in NA and EUR will be less than in the RW:

1990–1995: 8–10% in NA and EUR but probably 15–18% in RW
1995–2000: about 6% in NA and EUR and 10–12% in RW

The bigger growth in RW can be explained by the following factors:

- it is starting from a lower volume;
- it will spread to new countries; and
- it will be applied in new applications already used in NA and EUR.

The overall growth world-wide will occur through new applications and the progressive replacement of older technologies. *The growth can be higher* if

Table 1.3
Radiation-curing market volume per chemical family in 1990 (acrylic binders in $\times 10^3$ t)

Chemical family	Total world-wide	NA	EUR	RW
Monomers	29·2	12·5	11·5	5·2
Photoaccelerators	2·1	0·7	1·0	0·4
Epoxy acrylates	13·8	6·5	4·5	2·8
Urethane acrylates	6·4	2·3	2·5	1·6
Polyester acrylates	5·5	1·0	3·5	1·0
Total	57·0	23·0	23·0	11·0

anti-pollution regulations become severe or if the energy becomes more expensive.

Distribution by chemical families for 1990

In 1990, the amount of acrylic binders is estimated at 57 000 t world-wide with NA and EUR being essentially at the same level of 23 000 t and the RW at 11 000 t (Table 1.3). It is worthwhile remembering that only binders based on acrylic unsaturation are considered. Neither pigments and fillers nor photo-initiators are represented; photopolymer printing plates are not included either. Most of the pre-polymers (mainly epoxy acrylates and urethane acrylates) are already diluted in small amounts of monomer. Therefore, the total volume of monomers is greater than 50% of the total volume.

Distribution by application field for 1990

In addition to the volume of acrylic resins or binders, tentative estimations are given for cationic resins and UPE (Table 1.4). As the author is not active in cationic and UPE, the figures given here can differ greatly from reality. Cationic UV seems to be small compared to acrylic UV and this fact is certain; cationic resins, mainly cycloaliphatic epoxies, are used mostly in the US and principally as a metal coating (can ends) and probably on certain plastics. UV cured UPE are used exclusively in wood finishing; they are smaller than the total UV-cured acrylics but remain dominant in wood coatings. It is expected that the growth in UPE in the next 10 years will be small in wood finishing as the UPE technology can be considered as being mature. They have an advantage, a lower price compared with acrylic, but the volatility of styrene is a major disadvantage. The application fields in radiation curing can be divided for simplicity into four areas; each area covers UV and EB curing.

Graphic art
- *Offset*: covers lithographic inks, dry offset and letterpress used to decorate various kinds of substrates.

Table 1.4
Radiation-curing market volume per application field in 1990 (acrylic binders in)

Application field	Total world-wide	NA	EUR	RW
Graphic art				
Offset	8 400	2 000	5 000	1 400
Silk-screen	3 100	1 400	1 100	600
OPV	16 700	5 200	8 500	3 000
Total	28 200	8 600	14 600	5 000
Coatings				
Wood	9 900	3 400	5 000	1 500
Plastics	8 700	6 500	1 000	1 200
Metal	1 400	400	400	600
Paper	2 100	1 000	700	400
Total	22 100	11 300	7 100	3 700
Adhesives	1 050	400	400	50
Electronics	5 650	2 500	900	2 250
Grand total acrylic binders	57 000	23 000	23 000	11 000
Cationic	1 300–1 700	1 000	200–500	100–200
UPE	24 000–30 000(?)	5 000(?)	15 000	4 000–10 000(?)

Silk-screen: all screen inks and varnishes used on all substrates except the electronic applications.

Overprinting varnish (OPV): OPVs applied by dry offset and roller coaters; some is done in flexo and gravure. Most are cured in UV and are applied on paper and board.

Coatings

Radiation-curing coatings are cured exclusively on industrial lines, mainly by UV, but EB curing is performed on a few lines each consuming large volumes.

- *Wood* including furniture, doors and parquet. Until recently, UV curing was restricted mainly to clear varnishes (glossy and matt) but now pigmented coatings are done. Wet finishing of wood is only done on a few EB-curing lines world-wide.
- *Plastic,* including coatings on rigid PVC, semi-hard PVC foils for wood lamination, rigid and flexible PVC for floor covering, and all other plastic finishing.
- *Metal*: temporary anti-corrosion protective coating, protective varnishes on metal tubes (used in metallic furniture), and protective varnishes on vacuum aluminised substrates.

- *Paper*: coatings on decorative paper for wood finishing and outdoor-resistant panels, release coating, and paper upgrading.

Adhesives

This includes mainly laminating adhesives such as plastic film lamination on printed card (UV curing), paper foil lamination on particle board (EB curing), lamination of glass sheet (UV curing) for safety and sound-proof glass, some pressure-sensitive adhesives (PSA) cured in UV and in EB, and structural adhesives (UV curing).

Electronics

This includes etch resists, solder masks and marking inks used for the manufacture of PCBs (UV curing), dry film photoresist, and conformal coatings.

Market share of radiation-curing technology

From the previous section we have an estimate of the total volume of radiation-curing monomers and oligomers and now an estimate of the total coatings will be given to illustrate the proportion that radiation curing has. Therefore, we shall use rough estimates and use several hypotheses facilitating the calculations.

Coatings

The world-wide market is 15 million t divided as follows.

- Per geographic area:

    NA—31%
    EUR—29%
    RW—40%

- Per application field:

    Building coatings—50%
    Industrial coatings—31%
    Special coatings—19%

- Wood represents 8·5% of the total coatings and plastics represent 1·15%.
- We have adopted the hypothesis that the 100% neat resin represents 30% in weight of the coating.

Graphic art

Litho + letterpress	60% resins
Screen	60% resins
Flexo	40% resins
Gravure	30% resins
OPV	40% resins

See Table 1.5 for the total volume in all systems in thousands of metric tonnes.

Table 1.5
Total volume in all systems—graphic art ($\times 10^3$ t)

Process	Total		NA		EUR		RW	
	Ink	Resins	Ink	Resins	Ink	Resins	Ink	Resins
Litho + letterpress	975	585	515	309	230	138	230	138
Flexo	250	100	125	50	85	34	40	16
Gravure	470	141	170	51	130	39	170	51
Screen	90	54	40	24	30	18	20	12
OPV	140	56	50	20	50	20	40	16
Total	1 925	936	900	454	525	249	500	233

Table 1.6
Percentage of radiation-curing resins in total coatings resins

Geographic area	Coating resins ($\times 10^3$ t)		
	All systems	Radiation cure	Radiation cure (%)
NA	1 400	11·3	0·81
EUR	1 300	7·1	0·55
RW	1 800	3·7	0·20
Total	4 500	22·1	0·49

Table 1.7
Percentage of radiation-curing resins in total resins of industrial coatings[a]

Geographic area	Coating resins used in industrial coatings (t)		
	All systems	Radiation cure	Radiation cure (%)
NA	434 000	11 300	2·60
EUR	403 000	7 100	1·76
RW	558 000	3 700	0·66
Total	1 395 000	22 100	1·58

[a] The figures are true if radiation-curing coatings are <u>only</u> used in industrial coatings.

Conclusion

What can we conclude from Tables 1.5–1.11? Even with a large margin of error in these estimations the following observations can be made.

- *Coatings*
- Radiation-curing coatings are still a very minor part of the total coatings market, i.e. about 0·5%. They are not used in 'do-it-yourself', automobile, or building coatings, and only in very few special coatings. They

have been, up to now, used only in industrial coatings, mainly in wood and plastic finishing.
- Radiation-curing coatings world-wide are less than 2% of the industrial coatings.
- Radiation-curing coatings are significant in plastic coatings only in NA and in wood coatings only in EUR (when adding acrylic + UPE).

Table 1.8
Percentage of radiation-curing resins in total resins of plastic coatings

Geographic area	Coating resins used in plastic coatings (t)		
	All systems	Radiation cure	Radiation cure (%)
NA	16 000	6 500	40·0
EUR	15 000	1 000	6·7
RW	20 000	1 200	6·0
Total	51 000	8 700	17·0

Table 1.9
Percentage of radiation-curing resins in total resins of wood coatings

Geographic area	Coating resins used in wood coatings (t)				
	All systems (t)	Acrylic (t)	Radiation cure (%)	(Acrylic + UPE) (t)	Radiation cure (%)
NA	119 000	3 900	3·3	8 900	7·5
EUR	111 000	5 000	4·5	20 000	18·0
RW	153 000	1 500	1·0	10 000	6·5
Total	383 000	10 400	2·7	38 900	10·0

Table 1.10
Resins used in graphic art ($\times 10^3$ t)

Process	Total world-wide		
	All systems	Radiation cure	Radiation cure (%)
Litho + letterpress	585	8·4	1·4
Screen	54	3·1	5·7
OPV	56	16·7	30·0
Flexo	100	—	—
Gravure	141	—	—
Total	936	28·2	3·0

Table 1.11
Resins used in graphic art (×10³ t)

Process	NA		EUR		RW			
	All systems	Radiation cure	Radiation cure (%)	All systems	Radiation cure	Radiation cure (%)	All systems	Radiation cure	Radiation cure (%)
Litho + letterpress	309	2·0	0·65	138	5·0	3·6	138	1·4	1·0
Flexo	50	—	—	34	—	—	16	—	—
Gravure	51	—	—	39	—	—	51	—	—
Screen	24	1·4	5·8	18	1·1	6·1	12	0·6	5·0
OPV	20	5·2	26·0	20	8·5	42·5	16	3·0	19·0
Total	454	8·6	1·9	249	14·6	5·9	233	5·0	2·1

Graphic art
- World-wide, radiation curing is still small in graphic art (3%), litho + letterpress (2%), and silk-screen (6%). It is only significant in OPV (30%).
- Radiation curing is not used in flexo and gravure inks because of viscosity problems, pigmentation and their impact on the speed of curing.
- Litho + letterpress inks are the most used in EUR but they represent less than 5%. One explanation is that radiation curing is not, and probably will never be used in the inks for printing newspapers because of price.
- With shares of 19% in the RW, 26% in NA and 42% in EUR, radiation curing is now significant in OPV in the three geographic areas and it approaches maturity.
- In silk-screen printing, radiation curing is between 5 and 6% in the three geographic areas, and is more if we take into account that which is used in electronics and applied by silk-screen printing.

As a conclusion, there certainly remains enough room for radiation-curing technology to grow further in coatings and graphic art.

SIMILARITIES AND DIFFERENCES IN EUROPE, USA AND JAPAN

Using the figures of Tables 1.2–1.4, several conclusions can be drawn.
- The Japanese and RW market started later and is now (in 1990) approximately half the volume of the US or EUR.
- The US uses relatively more monomers because they also use more EAs than EUR. EUR uses more PEAs; most PEAs are lower in viscosity than EA and UA, consequently, the amount of monomer can be reduced.
- Japan is still using volatile solvents in its radiation-curing formulations. This practice has the disadvantage of losing one of the key advantages of radiation-curing technology, i.e. absence of solvents, but, conversely, it allows Japanese formulators to reduce the concentration of monomers which, in turn, allows the formulator to obtain either more hardness when using highly reactive UA or PEA; or to obtain more flexible coating when using very flexible UA.
- EUR uses more radiation-curing formulations in graphic art applications and the US more in coatings.
- Offset (lithographic inks) is more widely used in EUR because EUR uses more specialised PEA specifically designed to give a good litho behaviour.
- The US is stronger in plastic coatings due to the large volume used in PVC flooring applications. Japan has successful developments on difficult plastics (probably due to the combined uses of solvents with radiation-curing binders).

- Wood finishing is well developed in EUR, certainly when UPE and acrylic are added.
- Japan and the US are stronger in electronic applications, Japan mainly with solder masks and the US with dry film resists.
- EUR has several EB lines making outdoor-resistant building panels.
- The US uses more release coatings (casting release and release against PSAs).
- Japan has an EB line for coil coating.
- Most raw-material suppliers in Japan are selling resins and, at the same time, ready-to-use formulations, whilst in EUR raw-material suppliers generally do not sell formulated products. Some important European raw-materials suppliers have formulating companies within the overall group but operating as more or less separate companies.
- In the US, some of the larger formulating companies are producing some of their own basic resins; in EUR, only a small number of formulators are doing this and it involves mainly UAs. In Korea, Taiwan and Japan the big formulators manufacture the most common EA and UA for their own use.

The radiation-curing market is tending to become more and more similar world-wide and applications developed in one geographic area are soon used in others. The US begins to use more PEA in litho inks, Europe begins to finish more plastics, Japan and the other countries of the RW begin to broaden their radiation-curing applications.

PRODUCTS AND SUPPLIERS

Products

The raw materials used in many varied formulations can be classified into different chemical families.

- Monomers
- EAs
- UAs
- PAs
- Full acrylates
- Reactive additives
- Photoaccelerators
- Photoinitiators
- UPE
- Epoxies, mainly cycloaliphatic epoxies

Furthermore, saturated oligomers diluted in monomers can be used in specific formulations and unsaturated polyesters (maleic or fumaric) diluted in acrylated monomers are also used. Finally, these chemical families can be presented as water-based, either emulsified in water, or modified to render them water dilutable.

The radiation-curing market is still benefiting from the appearance of new products and new formulations. This is an important part of the dynamism of this technology because it also encourages the development of new applications. New oligomers of all kinds are frequently introduced and occasionally new monomers also. A lot of variations are possible in the backbone of most pre-polymers (components, length of the chain) giving tremendous possibilities for new products in the future.

The large number of raw-material suppliers provokes strong competition. Therefore, each company is looking to differentiate itself from the others and to introduce new specific products on the market, some examples of which are listed below.

- Silicone acrylates for release coating against PSA
- Special PEAs for producing PSA
- Low viscous oligomers in order to reduce or eliminate the use of diluting monomers
- Chlorinated polyesters diluted in monomers
- Reactive additives
    - Flow and slip additives
    - Adhesion promotors
- Water-dilutable products

From the end of 1987 to mid-1989, there was a shortage of monomers and oligomers due principally to the shortage of basic raw materials such as acrylic acid and aliphatic isocyanates. The shortage of acrylic acid was aggravated by plant accidents. These shortages have encouraged new companies to enter the market as suppliers of oligomers and monomers and established suppliers have increased their production capacities. Consequently, the earlier shortages have not only disappeared but now there is an overabundance.

Raw-material suppliers

The main raw-material suppliers (for acrylic binders) by geographic area are given in Table 1.12. Most of these supply world-wide where necessary increasing the number of potential suppliers in each geographic area. In addition to Table 1.12 there are a number of smaller suppliers. In addition to these suppliers, there are several large formulators that produce some oligomers for in-house captive use:

NA—PPG, DeSoto (now DSM), Armstrong, Valspar, Lord, Amtico, Morton
EUR—PPG, Morton
RW—most of the big formulators (Dainippon Ink, Mitsubishi Rayon, etc.)

Attracted by the growing market, newcomers continuously enter the market generally presenting 'me too' products. These generally begin with EAs (the

Table 1.12
Main raw-material suppliers

NA	EUR	RW
Radcure Specialties	Radcure Specialties	Nippon Kayaku (Japan)
Craynor (Sartomer)	BASF	Toagosei (Japan)
Henkel	Craynor	Showa High Polymer (Japan)
CPS	Harcros	Nippon Gosei (Japan)
CL Industries	Akzo	Osaka Yuki (Japan)
	Ancomer	Shinnakamura (Japan)
	Hüls-Servo	Daicel-UCb (Japan)
	Rahn	Sunkyong-UCb (Korea)
		Eternal (Taiwan)
		Miwon (Korea)

biggest oligomer family) and monomers. As a consequence these products tend to be handled as commodities. These newcomers generally enter the market with low price levels providing strong competition. Bayer (Germany) who withdrew from the acrylic radiation-curing market, after a plant fire, is still a supplier of UPE. Most of the large wood-coating formulators produce their own UPE. Degussa and Union Carbide are suppliers of cycloaliphatic epoxies used in cationic curing. Photoinitiators are supplied by Ciba, Merck, BASF, Fratelli Lamberti, International Bio-Synthetics, Sandoz and others. Benzophenone is produced by several companies in the three geographic areas. Most of the larger raw-material suppliers of acrylic binders continuously broaden their range of products.

Formulators

In total, the number of individual formulators (inks and coatings) are at least 700 world-wide (about 300 in EUR, 300 in NA and at least 100 in the RW). The number of formulators is increasing every year and it is impossible to name them all. In addition, there have been lot of acquisitions that have occurred over the past few years and there are still newcomers entering the market. These newcomers often begin by supplying locally, often to answer the demand from one of their existing customers who has decided to use radiation-curing technology. After a while, when they feel confident in their know-how, they begin to extend their market. The result is a strong competitive supply situation in the market, encouraging new end-users to try radiation-curing technology. Table 1.13 shows a few ink formulators. They are multinationals producing in several countries and selling world-wide. These ink formulators often produce coatings too. There is a large number of significant suppliers that only produce in one country but sell to neighbouring countries and sometimes world-wide. Table 1.14 shows some of the coating formulators but multinational ones only. The same remarks can be made for the coating formulators as were made for the ink formulators. To give an idea of the growth

Table 1.13
Ink formulators—multinationals

DIC (Sun)
Coates and Lorilleux
BASF-Farben
Sicpa
Hostmann-Steinberg
Zeller + Gmelin
Casco (G-Man)
Midwest-Sericol
Avery
Croda
Toyo Ink
Tamura

in the number of formulators, in EUR their number has increased from approximately 200 in 1985 to 300 in 1989.

Concentration of companies
In the last few years, several acquisitions occurred world-wide.

- By recent reorganisation in the French chemical industry, the French group Total has now Craynor (EUR) and Sartomer (USA) who are both large suppliers to the EUR and NA markets (raw materials). Total has also in its group the combined ink manufacturer Coates and Lorilleux plus several coating formulators.
- UCB (Belgium) and Interez (USA) formed a world-wide joint venture, Radcure Specialties SA (Belgian company), dealing exclusively with the

Table 1.14
Coating formulators—multinationals

BASF Lacke
ICI Paints
Akzo Coatings (Reliance)
Morton International
Casco (Sadolin)
DuPont
Becker
Lord-Hughson
PPG
Raychem
Grace
DIC (Reichhold)
DSM
Loctite

manufacture and commercialisation of raw materials for the radiation-curing industry (oligomers and monomers). Radcure Specialties SA is now owned by UCB (Belgium).
- Several years before, Diamond-Shamrock (USA) became Henkel.

These three events mean that more than 90% of the supply of raw materials for the radiation-curing industry in NA is now in the hands of EUR companies. Acquisitions by the formulators also occurred.

- Coates and Lorilleux (already mentioned).
- DIC (Japan) acquired the world-wide ink division of Sun and recently the BASF ink division in USA (formerly Inmont).
- Sicpa (Switzerland) acquired Collie-Cooke (Australia and New Zealand) and other companies in different countries.
- AKZO (Netherlands) bought Reliance, a big formulator in the US having affiliates in EUR.

In EUR, every year there are several mergers or acquisitions of formulators occurring, this mainly being as a result of the common EUR market that will occur in 1993. These acquisitions occur because companies want to have a critical dimension before 1993. Finally, we are experiencing the development of world-wide strategies focused on the radiation-curing market. Several companies like BASF, Total (Craynor), Hüls and AKZO are completing their vertical integration manufacturing strategy:

- producing and selling basic raw materials such as acrylic acid, polyols and isocyanates;
- producing and selling radiation-curing raw materials (oligomers and monomers); and
- selling ready-to-use formulations (inks, varnishes and coatings) to end-users through their affiliates.

GRAPHIC ART APPLICATIONS

We shall now review the big application fields of radiation-curing technology. Graphic art still occupies the largest part in volume consuming almost 70% of all the acrylic binders used.

Lithographic inks
Printers have always dreamed of having an ink which is stable on the press, i.e. does not increase in viscosity due to solvent evaporation, but once on the substrate can be dried immediately. Consequently, no broadening of the dot will occur and the definition is improved. This dream is virtually achieved using UV inks because they do not contain any solvents and the binders used have very low vapour tensions. Consequently, good press stability is achieved using radiation-cured inks. Due to the high speed possible with UV inks, the ink,

once on the substrate, is exposed under the UV lamps immediately and is cured instantaneously at room temperature. In this way, the dot of ink is fixed and cannot spread. This explains why offset ink manufacturers and printers continue to have a considerable interest in UV technology. Lithographic, offset and letterpress inks are the best suited for UV technology due to the following reasons.

- High viscosity is required (therefore less diluting monomers are needed) and high reactivity is necessary (the reactivity normally decreases when the concentration in monomer increases).
- Thin layers are used; so the presence of pigments does not prevent the penetration of UV light through the thin layer of ink.
- It is possible to print several inks wet on wet, curing only once at the end; therefore it is not always necessary to dry after each ink application.
- The use of diacrylate monomers is not necessary due to the high viscosity of the inks; mainly triacrylate monomers are used reducing the potential skin irritation.
- Better gloss, print definition and quality are obtained due to reduced penetration into porous substrates because the curing of the high-viscosity ink occurs rapidly after application.
- Radiation-cured inks are used more and more on closed-surface substrates for better gloss results.

Printers progressively discovered other new advantages when using radiation-cured inks.

- A lot of problems in a print shop have always occurred due to the use of spray powder. With radiation-cured inks, spray powder is no longer necessary because the surface of the printed substrate is dried immediately and does not offset onto the back of the next card in the stacker. The inks instantaneously achieve their final properties (better abrasion and solvent resistance).
- There is lower energy consumption and no solvent emission.
- Printers can reduce their stock of finished substrates because in-line manufacture of packagings at high speed and subsequent immediate transformation after curing (cutting, folding, glueing, etc.) make it possible to produce and deliver within a day.
- Cured inks are not irritant and are non-toxic. After 20 years of using UV-cured inks not one complaint about skin irritation has been reported as a consequence of contact with UV-printed material.
- Lower odour levels are immediately reached after curing using UV inks for the decoration of external food packaging; printers do not need to wait several days before delivery as they have to do when using solvent-based inks.
- EB-cured inks give much lower residual odour levels than UV inks (due to the presence of UV photoinitiators). Therefore, EB curing is being used more and more by big printers.

Probably, more than 30 lines world-wide are printing in web offset with five or six inks applied wet-on-wet and cured in one pass. Sometimes EB overprinting varnish is applied wet-on-wet onto the five or six wet inks before EB curing. Many different applications are achieved in this way, many used for aseptic packing of liquid-food products. The appearance of extra purified binders has contributed to this development. This process allows the achievement of high gloss with almost no odour and no extractables after curing, ensuring no contamination of the food inside. Furthermore, this process is economic because high productivity is achieved due to the high-speed curing (300–500 m/min) and all operations being done in-line. Other applications with UV inks are done on plastic and metal:

- *On plastics.* UV curing is achieved at ambient temperature; consequently, no deformation of the various kinds of plastic occurs and no chemical attack occurs due to the absence of solvents. As UV inks do not contain solvents that need to be evaporated, they are ideally suited for application on impervious plastic substrates; in addition, better solvent and abrasion resistance is achieved compared to conventional inks.
- *On three- and two-piece metal cans.* Here, energy saving is most evident. Finishing in-line is possible because large thermal ovens are replaced by a small UV oven and therefore considerable time and labour are saved.

OPV

At the beginning of the 1970s, several formulators claimed that UV-cured OPV had no commercial future as they were too expensive compared to nitrocellulose varnishes. The argument 'radiation-curing formulations are too expensive' has often been heard from the beginning and is still a point when it arises into discussion about new applications. The following are often forgotten.

- Solvent-based varnish containing 20–40% dry material is compared with a UV formulation of a 100% dry material content.
- Comparisons should not be done in kilograms of products but in square metres of finished substrate, taking into account the whole finishing process.

The case of OPV in graphic art is a classic example. Printers and varnishers were rapidly convinced of the economy of UV technology. With relatively thin layers (4–8 g/m^2), OPV cured in UV gives a high gloss with smooth surfaces. In order to obtain a smooth surface with such a thin layer you need to have a precise varnishing machine. The first trials of applying UV varnish with old varnishing machines were disastrous. These old machines were working quite well with nitrocellulose varnishes at 20% dry material; after evaporation of the 80% solvent a smooth surface was obtained but the gloss remained quite low. When trying to use such a machine with UV varnish, either a thick layer was applied which made it very expensive, or when a thin layer was applied a non-smooth surface with an unacceptable finish was achieved. When varnishers were convinced of the necessity to use more precise machines, varnishing machine manufacturers rapidly improved their equipment and UV–OPV very quickly became a commercial success. Calendering, an expensive operation, is necessary to get a high gloss with nitrocellulose solvent-based varnishes. With UV, an acceptable gloss is achieved immediately and calendering is no longer necessary. The avoidance of the calendering operation made the UV technology immediately more attractive economically. UV varnishing quickly replaced some lamination applications too. Currently, the number of applications using OPV cured by UV is significant: postcards, record sleeves, book covers, and all kinds of packing materials.

UV-cured OPV are applied by many different techniques.

- *Roller coaters*. These are mainly used by specialised varnishers who apply the UV varnish over substrates already printed (principally with traditional inks).
- *Offset machines*. The possibility offered by UV to print and overprint in-line has given printers the opportunity to overprint in-house instead of by trade varnishers.

Printers are trying more and more to print with conventional offset inks and overprint with UV varnish, wet on wet. In this way they can get the benefits of conventional inks (cheaper than UV inks) with the high gloss of UV-cured OPV. This technique is in fact not so easy, as some problems can occur when applying UV varnish on wet conventional inks:

- either the UV varnish is not compatible with the wet inks, resulting in a reasonable gloss with the cured UV varnish but with poor adhesion of the UV varnish on the inks;
- or the UV varnish mixes slightly with the inks giving good adhesion but with a lower gloss after UV curing.

Some printers use an isolation layer between the wet inks and the UV varnish. They apply an emulsion which is dried with IR lamps. In this way the UV varnish gives a high-gloss finish and adheres well on the dried isolation layer.

Silk-screen printing

Silk-screen printing and overprinting started early in the 1970s quite soon after lithographic UV inks. It developed slowly and was gradually applied to more numerous applications and on a great variety of substrates. Curing is achieved using UV exclusively; EB is not used in silk-screen printing. Here, the formulators had to solve the problem of having inks with sufficient covering power but sufficiently UV transparent to achieve through-cure with UV light. With the thickness of silk-screen inks being around 15–20 μm, the concentration of pigments and the nature and the concentration of the photoinitiator system are critical. Furthermore, four-colour prints lead to several problems.

- Intercoat adhesion can be difficult to obtain when the first colour has already been cured twice.

- If four colours are superimposed at the same place, the thickness is such that it creates rigidity and brittleness. To solve the thickness problem, screens with a finer mesh are used but during the past few years an interesting development has occurred: the use of water-dilutable formulations. These theoretically have two main advantages:
  - they decrease the thickness applied;
  - the viscosity of the formulation is adjusted by using water instead of monomer (in this way, the concentration of the diluting monomers is reduced or in some cases no monomers are used at all).

The use of UV screen process printing has proliferated on both sheet-fed and rotary presses on various substrates.

- *On paper and card*: four-colour printing of large posters, labels, packaging material, etc.
- *On plastics*: on flat substrates (self-adhesive labels, etc.), and on objects such as different bottles and containers for technical and consumer items.
- *On glass*: ceramic inks contain special pigments; after curing of the inks the whole substrate is introduced into a thermal oven at 600–800°C, the cured UV binders are burned and evaporated while the special pigments are melted and fused into the glass.
- *On copper*: for the manufacture of PCBs, UV technology has rapidly been used and etch resists, solder resists and marking inks are used extensively.

The economy of UV inks and varnishes comes from the following:

- economy of energy (no thermal oven) and floor space;
- ease of handling (stable formulations with very low odour);
- stability on screen (no solvents are used, therefore no evaporation occurs and consequently, no drying on the screen occurs causing binding of the screen which has to be cleaned thoroughly before reuse);
- increased productivity (no stopping to clean the screen caused by breaks in production such as lunch or for technical reasons, and the inks can remain on the screen with no change in viscosity occurring due to the low vapour tension of the binders);
- less deformation and damage to plastic substrates; and
- better properties of the cured inks and varnishes (better solvent and abrasion resistance).

Flexogravure

This process needs lower viscosity than offset and silk-screen printing. It has to run at speeds of 100–300 m/min, which is comparable to offset and noticeably faster than silk screen. The thickness applied is about 4 μm (i.e. four times the thickness used in offset) and it is not possible to print wet on wet; therefore, each ink must be cured before applying the following one. Due to the low viscosity, the reactivity of the flexo ink is much lower than offset inks and requires more lamps to achieve full cure at high speeds. Moreover, the presence of pigments reduces the curing speed in depth. Therefore, UV flexo inks are not widely used at present.

Curing with EB, this not being affected by the presence of pigments, would make the problem of reactivity easier. The viscosity of the ink could also be reduced by increasing the temperature of the ink holder instead of diluting the ink with monomers. Curing with EB also needs irradiation after each ink application making the process expensive as an EB gun costs considerably more than a UV oven. In flexo, UV-cured OPV is more commonly used than UV inks. Applied on specific areas, different gloss levels can be achieved. Selective application of UV varnish makes the gloss higher on those areas coated and on the other areas, where there is no varnish, the gloss is much lower. It is also possible, on the same substrate, to apply a matt varnish on one flexo unit and a gloss varnish on another. In this way 'special effects' can be achieved. An important development has occurred during the last 2 years: UV flexo inks are used for the printing of labels. More and more narrow web flexo presses are being converted to UV. The inks used are 'pseudo-flexo inks' having a higher viscosity (about 1000 cP (1 Pa . s) than conventional flexo inks (about 100 cP (0·1 Pa . s). These inks are also used as modified UV letterpress inks.

Presently, 'improved print quality' seems to be the major interest rather than reduced solvent emissions. Other advantages of UV flexo inks used in narrow webs are as follows.

- Colour consistency
- Improved adhesion to films
- Less clean-up time
- Silk-screen look
- Rub and chemical resistance
- Print quality comparable to rotary letterpress
- Light fastness
- Ability to print metallic inks
- Longer repeats, etc.

This success of flexo inks is, up to now, limited to narrow web applications where it is usual to have a UV lamp after each colour. The big area of application where large volumes of inks are used is in flexible films used as food packaging. Here, neither UV nor EB are used today.

Heliogravure, intaglio and ink jet

Intaglio inks are of high viscosity and are therefore easy to prepare using radiation-curing materials. However they are applied at high coat weights and UV through curing becomes difficult; EB curing would be much easier. Limited applications exist in the printing of bank notes and other security papers.

Heliogravure requires lower viscosities than flexo and in addition thicker layers than offset and flexo are necessary. Consequently, it is difficult to achieve high reactivity in UV combined with very low application viscosities and coat weights of 4–6 μm. Printers using heliogravure inks are generally larger companies and therefore they are more able to afford the cost of installing solvent-recovery equipment.

Radiation-cured inks (UV or EB) are expensive when compared to solvent-based inks. UV curing is only used in limited amounts in OPV.

Jet inks also require low viscosities; trials have been carried out in UV and will probably be used soon. However, it is estimated that only a maximum of 10% of the jet inks could be done using UV curing.

COATING MACHINES

The coating machines are an important aspect of the success of this technology. Radiation-curing formulations do not generally contain volatile materials and therefore require precision application machines. The coating technique is often a limiting factor for the continued diversification in applications using radiation-curing technology. Coating machine manufacturers have made a lot of progress and innovation over the last 20 years to make the application of radiation-curing formulations possible on many different substrates. Coating applications were quickly developed and applied industrially. These began with wood finishing and progressively extended to other substrates. Here again, practically all application methods are used but, as we have seen with OPV applied by roller coater, more precise application machines have been developed and this has contributed to the success of radiation-curing technology in the coatings field.

Roller coater
With two or, ideally, with more rollers, thin, smooth layers can be applied on rigid and flexible substrates. Heating the rollers allows a constant temperature of the coating and decreases the viscosity without the need of using too much diluting monomer. Constant temperature avoids problems of viscosity changes occurring between winter and summer or between a cool machine at start-up and a warm machine after prolonged running. More rollers allows thinner layers to be applied. The smoothness of the coating is often obtained using other parameters of the machine: direct or reverse rollers, rollers running at different speeds, transfer via a soft roller, etc. Roller coaters are commonly employed as they allow a precise and uniform thickness of layer to be achieved.

Curtain coater
This technique is used to apply thicker layers of 50 g/m^2 or more. It is used for wood coating where a 'closed pore' finish is required. It is also ideal for substrates which have variable thicknesses and consequently this technique

does not give 'roller marks'. This technique would be ideal if thinner layers could be applied, however, as most radiation-cured coatings are 100% solids dry material, low coat weights are difficult to achieve.

Spraying
This technique generally requires a low viscosity. Once again this problem of viscosity can be solved in specific cases by warming-up the formulation before application. In most cases where spray application is used some volatile solvents are used to decrease the viscosity. Spraying can be applied on all shapes such as flat or three-dimensional objects. Instead of using solvents, it is possible to find industrial applications using emulsified formulations however the water or solvents must be evaporated first before curing with UV or EB. An alternative approach is water-dilutable formulations. A company in Germany has developed a spray gun that can apply viscosities up to 70 P (7 Pa . s); this seems ideal where a 'structured surface' is required.

Vacuum curtain application
This technique is becoming increasingly popular for application on long and narrow rigid substrates (like pieces of wood). It requires a relatively low-viscosity formulation.

Extrusion
Extrusion through a narrow die gives the same kind of application as a curtain coater but the viscosity can be much higher and thinner layers are possible. This method is used to coat a 'hot-melt' formulation. By applying at high temperatures of 80°C or more, it is possible to apply formulations that are nearly solid at room temperature and that contain almost no diluting monomer. This technique may be used more and more for the application of radiation-curing formulations in the future.

Rotary screen
Widely used in the graphic art industry but not so much in coatings, this technique may be used more and more when coatings of higher viscosities are to be applied at room temperature or when a higher thickness is necessary than can normally be applied by roller coaters.

COATINGS

Coatings based on UPE represent a large part of all radiation–curing coatings but their growth has now reached maturity. Coatings using acrylic binders, however, are gaining more of the market share each year and now represent 30% or more of all the acrylic formulations.

Wood finishing
Wood finishing is done industrially in two ways.

- *Wet finishing*: wet coatings are applied directly on the wood panel and then cured with UV or EB.
- *Dry finishing*: a foil (paper or PVC) is first finished by applying a coating cured with UV or EB, then laminated onto particle board.

Wet finishing

More and more acrylic formulations are used and are increasingly replacing UPEs. It is used for the following.

- *Door finishing*: clear open-pore finishes cured using UV or EB; pigmented coatings cured using EB are easier however, in recent years there has been a successful development using UV curing for pigmented finishes on doors.
- *Parquet flooring*: cured by UV.
- *Furniture*: open-pore and closed-pore finishes with clear lacquers cured by UV or EB.

Pigmented finishes are cured easily by EB and can now be cured by UV as well. For a few years, highly pigmented coatings up to $150 \, g/m^2$ have been cured by UV by using free isocyanate—to achieve the curing in depth—combined with a special UV-curing technique. For a long time, finishes with pigmented UV formulations were achieved by application of two to three layers, each layer not having sufficient covering power but allowing good curing in depth. By using two or three non-covering layers a coating of sufficient covering power can be achieved. Sanding in between each layer is necessary to achieve a smooth surface with good interlayer adhesion. Such finishes can be applied by roller coater or curtain coater; however, in each case the pigment concentration and the photoinitiator system must be adjusted for each specific application. Matt finishes are easier to obtain in UV than in EB. In UV, matting is obtained by utilising the phenomenon of oxygen inhibition. In EB, if too much oxygen is present, the surface does not cure properly and consequently this makes matting more difficult. When matt finishes are required by EB curing, other methods must be employed.

- Either curing through a release film applied in contact with the coating before curing, in this way during curing, the surface of the coating takes on the gloss of the film. This means that if a glossy release film is used a glossy coating is achieved and a matt finish is obtained by using a matt release film. After curing, the release film is easily removed from the cured surface.
- Or, the coating is first cured in EB in presence of air; the curing occurs in depth pushing matting agents at the surface; then the inhibited thin surface layer is cured under nitrogen with UV lamps.

Dry finishing

Clear coatings are applied on paper or PVC wood grain printed foils. On PVC foils UV curing is normally used. Glossy, satin or matt finishes are used depending on the preference the consumers have in the market. Pigmented coatings on paper are frequently done using EB curing. This kind of finishing can run at high speeds and allows the use of the same base paper with different colour finishes. Changing colour of the coating is easier than using different coloured paper and applying a protective varnish. The finished foils are laminated on wood particle board using conventional methods (pressure and temperature).

In the US there are EB lines using a combination of wet and dry methods for the finishing of particle boards in line. These start with a particle board having a good smooth and closed surface. A laminating adhesive is applied onto the board and the decorative paper foil is laminated onto the surface. Finally a protective clear coating is applied. The EB-curing process occurs in one pass, curing the laminating adhesive and the protective varnish.

Paper upgrading

Paper coating used for dry finishing of wood (described above) is an example of paper upgrading. Both UV and EB curing are used in paper upgrading but EB offers more potential. EB finishing is now used for various applications:

- protective EB varnish on paper printed with a wood imitation design;
- pigmented coating on paper before lamination onto wood;
- clear and pigmented outdoor-resistant coatings on paper and on plastic foils;
- sealing varnish of paper prior to aluminisation (also possible in UV);
- pigmented coatings to reduce the porosity of substrates used for photographic papers and mainly for printing paper (this allows excellent quality high-gloss print base to be achieved for use in high-quality packaging and advertising); and
- to obtain release properties.

Release papers look promising. There are three different kinds of release papers:

- *Caul sheet*: these give release against melamine-impregnated decorative paper foils.
- *Casting paper*: these give release against polyurethane and PVC used in the manufacture of simulated leather goods.
- *Release for labels*: these give release against PSA.

Caul and casting papers can be produced using silicone acrylate additives, the main part of the formulation being standard radiation-curable acrylics. Release systems for labels are more critical and require almost 100% silicone acrylates. Most applications are cured using EB. UV can also be used for release against labels but curing is carried out under nitrogen. Here, cationic release coatings could be a way of avoiding the use of an inert atmosphere when curing with UV. EB-cured paper upgrading, before aluminisation is more frequently used as the gloss obtained after aluminisation is better than with conventional coating methods. This market could see increased interest with the development of formulations that are alkali soluble after curing as this would allow their use for the manufacture of labels for returnable beer bottles.

Plastic coatings

PVC coatings

High-gloss coatings on PVC, for floor covering, is a big market in the US where it was developed first. These have achieved industrial success in EUR too and the market is growing in two directions:

1. coatings on rigid PVC tiles; and
2. coatings on cushioned floor (flexible PVC rolls).

In EUR, glossy finishes are not very popular; satin or matt finishes are preferred. Matt finishes are obtained a two-step in-line process under different UV lamps for each step. Up to now only UV curing is used. UV finishes on semi-rigid PVC foils for laminating onto particle boards, have already been mentioned in the section entitled 'Wood finishing'—see p. 33. PVC folding doors are also often protected with a UV varnish.

Other plastics

On PVC, good adhesion is relatively simple to achieve but this is not the case for some other plastics which are less subject to attack by the diluting monomers in the formulation: e.g. polyethylene, polypropylene, polyester or polyethylene terephthalate. Adhesion can be improved if the surface of these difficult substrates has been treated (e.g. Corona treatment or coated with an adhesion primer). Nevertheless, applications on the different plastics are becoming more and more commonplace.

- Decoration with several inks (see the section entitled 'Graphic art applications'—p. 25)
- Magnetic media using EB curing
- High-scratch-resistant coatings for car components such as mirrors, car headlamp covers, etc.
- UV varnish on car headlamp reflectors before metallisation
- Different protective or decorative coatings on panels of polycarbonate, poly(methyl methacrylate), (PMMA), polyester
- Protective varnish on skis
- Protective varnish on crash-helmets
- Protective varnish on fishing rods

Developments are progressing on different kinds of plastics. Adhesion problems are progressively overcome by the development of new binders and new formulations. Cork, used as floor- or wall-panels, although not a plastic, is also UV varnished and several lines are running in EUR.

Metal coatings

It is on metal that the energy savings offered by radiation-curing technology is greatest although good adhesion on metals is quite difficult to obtain. Radiation-cured acrylic formulations shrink during curing and this shrinkage produces tension forces in the coating which are often stronger than the adhesive forces of the coating. Radiation curable adhesion promotors have been developed that considerably improve the adhesion; however, the number of applications on metal finishing is still limited. Here again, cationic chemistry theoretically presents some advantages over acrylic chemistry. Up to now, there are no large coil coating lines using UV or EB curing although developments are progressing and a renewed interest is evident. UV applications are on steel pipes, where a temporary anti-corrosion coating is required, and are carried out industrially. Protective UV varnishes are applied on galvanised steel tubes used for the manufacture of metallic furniture. Trials and developments have been made on white base coating for three-piece cans. No industrial applications are to be found on aluminium foils. Different products can theoretically be manufactured using UV etch-resist formulations; one example is the manufacture of electrical resistors for irons, starting from a flat thin steel foil. Wire coating and protective coatings for car wheels have been achieved using EB curing.

Other substrates

Optical glassfibres

Primary and secondary coatings are applied and cured with UV. This is a high-performance application and the market will grow in accordance with the market for optical fibres. UV marking inks are also used for the marking of these coated fibres.

Glassfibre coatings

In UV, the local reinforcement of glassfibre mats is starting. The manufacture of composite materials, mainly with EB curing, will probably progress.

Security and acoustic laminated glass

These can be obtained by pouring UV formulations between two glass sheets and curing with weak TL UV lamps through the glass. This technology competes with lamination of polyvinylbutyral films which can only be done by big manufacturers because of the high investment required. This UV technology can be used by smaller window manufacturers and offers an economic and viable alternative. Other kinds of applications could derive from this specific niche in UV technology.

Manufacture of screens
Screens for screen printing can be manufactured using UV curing.

Printing plates
These already represent a large part of the market but will not be reviewed in this chapter.

Leather coatings
UV clear and pigmented coatings for leather finishing used in shoe manufacture is starting in EUR, mainly in France.

Stereo-photolithography
This technology is really a novel example of innovation through the combination of different technologies: computer-assisted design and laser curing of UV formulations.

In this way, manufacture of three-dimensional objects is possible and this process is very economic for the manufacture of prototypes. There should be a big interest in the aerospace and automotive industries.

Others
It is impossible to name all the products that are coated or decorated using radiation-curing technology: inevitably some will have been overlooked. However, it is worthwhile mentioning the following.

- Compact and video discs
- Spectacle frames
- Artificial nails
- Conformal coatings on metals, name plates, photographs, and dishes
- Holographic images
- Liquid-crystal systems
- Microporous membranes
- Dental applications
- Microencapsulating imaging systems
- Composite materials using EB, X-rays or γ-rays
- Flat batteries (obtained with UV coatings on lithium foil)
- All kinds of three-dimensional objects (chairs and stools, case goods, guitars and other musical instruments, rifle butts and gun stocks).
- Anti-fogging film
- Abrasives
- Marking inks on electronic components

ADHESIVES

This area represents a significant potential for the radiation-curing market. Radiation-curing is an ideal way of achieving hard and non-tacky coatings and therefore doubts arose about the possibility of producing adhesives, mainly PSA. However, progress never stops. What seemed impossible yesterday is sometimes possible today and this is the case with radiation-cured adhesives. Currently, this market is still limited and certainly water-based PSA will continue to hold the biggest share mainly because they are cheaper than UV- or EB-cured adhesives.

Laminating adhesives

Lamination was the first application for radiation-cured adhesives as it was far easier than for PSA. Residual tack is not always necessary and curing through one substrate is possible; consequently, oxygen inhibition is eliminated. It is possible to glue all substrates to each other and cure with UV or EB. When curing with UV, at least one substrate must be transparent to the UV light. Lamination of printed card with plastic films is now carried out on several industrial lines. Up to now, no industrial laminates produced by radiation curing have been used for food packaging. From a theoretical point of view, no good reasons can be found for this as with laminates, no direct contact with food is observed. The conventional laminating lines running today with solvent-based or water-based adhesives use low-viscosity formulations (application machines are designed for low viscosity); the solvent or water is then evaporated in a thermal oven leaving the adhesive with high tack on one substrate and then the second substrate is laminated. To come back to UV lamination, producers want to use their line with the application machine running at low viscosity. In UV, curing takes place after lamination and therefore, a lot of difficulties occur when laminating both substrates with a low viscosity non-cured adhesive.

The solution can be found in two ways:

1. by pre-curing the UV adhesive to increase the tack before laminating and finishing the curing with a second UV oven after lamination; or
2. by applying a high-viscosity UV adhesive (even applied at high temperature) then laminating and finally curing, but this means using new machines (this would not be a problem for new lines).

In the US several industrial EB-curing lines are used for the glueing of decorative paper foil onto particle boards.

PSA

Industrial lines are running in EUR and in the US but only in recent years. Achievement of PSA by radiation curing is not easy. Formulations used are often highly viscous and are normally applied using slot dies at high temperature. Curing occurs by UV or EB. The concentration of diluting diacrylate monomers is quite critical. The final tack of the cured PSA is rapidly reduced when the concentration of diacrylate monomer increases. Very good performance can be obtained with radiation-cured PSA but they are more expensive than their traditional solvent-based and water-based equivalents.

Thus, their industrial application is still limited. Recently, an interesting development has occurred, which is only possible with radiation-curing technology. Complete self-adhesive labels are produced in-line using several radiation-curing formulations: the free film supporting on one side the PSA and on the other side the decoration is made from a radiation-curable formulation; the PSA, the printing inks and overprinting varnish are all cured by radiation. Such a development makes the process viable and capitalises on the possibilities offered by radiation-curing technology. In this way, using an in-line manufacturing process of composite items the cost of radiation-cured formulations can be overcome and the total economy of the manufacturing process becomes favourable in comparison to conventional label stock production.

Removable PSA

Permanent PSA are now industrialised, therefore, it should not be too difficult technically to produce removable PSA. However, a process must offer economic viability before it can move from a technical feasibility to an industrial process. New processes have to be more economic as a whole or achieve properties that more traditional processes cannot before they are accepted industrially.

Repositionable PSA and blisters

Repositionable PSA also present radiation-curing technology with a lot of difficulties both economic and technical. 'Blisters' must be non-tacky at room temperature but rapidly stick to plastic films (polyacetate, PVC) under the influence of slight pressure and 1–3 s at 100–120°C. The technical problems encountered with repositionable PSA and blisters can be overcome with radiation-curing technology, it is only a question of time; however, the economic aspects will be more difficult to make the process viable.

SUBSTRATES AND OBJECTS

Is it possible to manufacture substrates and three-dimensional objects using radiation-curing technology? Technically, the answer is yes! But the economic aspects will again restrict the application possibilities. Radiation curing seems ideally suited to making free films but can it give a better performance and be cheaper than all the kinds of plastic films that exist today? Nevertheless, we have seen in the section entitled 'Removable PSA' that the manufacturing of a free film in-line as part of a composite item is already done industrially. Other possibilities certainly exist and it is only a question of time before these are developed (foamable films could be one application). Objects are manufactured in a limited scale by stereo-photolithography. Objects in composite materials are technically feasible and it is again a question of time. Of course, UV curing is probably very difficult in the case of pigmented substrates and composite materials but powerful EB accelerators can cure thick layers and

X-rays or γ-rays could also be used. The manufacture of substrates and three-dimensional objects is an application field where radiation-curing technology is still in its infancy. Imagination and innovation will certainly bring surprises and produce new items in the future.

LIMITATIONS

Radiation-curing technology has limitations like all other technologies. We have seen the advantages and the numerous possibilities of application of radiation-curing technology, but nevertheless it will not replace all other technologies. Even solvent-based formulations will not disappear completely.

Let us review the main limitations.

- Radiation-cured formulations are often regarded as expensive. However, if you take into account the cost per square metre of finished substrates and not only the price per kilogram of the formulation, radiation curing is often more economic than classical technologies. Further, radiation-cured formulations contain 100% dry material and should not be compared with solvent-based and water-based formulations containing from 20 to 40% dry solid. Nevertheless, the cost is an important and even a vital factor. We have seen that several applications are technically possible with radiation-curing technology but if the economy of the whole process is not correct it is not used.
- Radiation curing needs equipment that is difficult to transport, therefore the finishing and curing must occur in industrial factories.
- Thick pigmented coatings cannot be cured with UV. However, using combined curing mechanisms, thicknesses of 150 g/cm^2 or more on wood are being cured properly.
- In EB, the high investment for the electron gun makes it only viable for companies finishing several million square metres per year. High EB doses (10 Mrads (100 kGy) or more) can degrade some substrates, but most formulations, used today, are cured under nitrogen with doses below 5 Mrads (50 kGY) and often do not need more than 3 Mrads (30 kGy).
- EB curing needs an inert atmosphere. This is a challenge and an opportunity: how to overcome oxygen inhibition of acrylic formulations in EB curing. In UV, more than 95% of the formulations are cured in presence of air without any problem, but in EB it remains difficult. Nevertheless, it is possible to cure in the presence of air but higher EB doses are necessary and then, the possible degradation of the substrate has to be considered.
- Acrylated binders are irritants and some people claim that they are toxic. Generally speaking, the monomers are the most irritant and it is recommended to avoid skin contact. Oxyethylated and propoxylated polyol polyacrylates are now used on a large scale without problem.

Furthermore, all the commonly most used monomers, including HDDA and TMPTA are non-toxic as defined by oral and dermal LD_{50} values. Pre-polymers are less irritant and are also non-toxic when the oral LD_{50} is considered.
- Adhesion of radiation-curable coatings to difficult substrates such as certain plastics and metals remains one of their weak points. Where necessary, pre-treatment of the substrate can be carried out and this is widely used to improve adhesion. Adhesion promotors have been commercialised and these are effective on several metals and sometimes on plastics. Considerable efforts to provide oligomers with improved adhesion to difficult substrates are still ongoing.
- UV and EB curing is easier on flat substrates. In EB, the sides of substrates (e.g. doors) are cured efficiently. There are now many objects that are produced, clearly demonstrating that certain three-dimensional parts can be cured.
- Parts hidden from the UV or EB radiation cannot be cured; consequently, metallic structures or large complicated metallic shapes such as car bodies are not cured using radiation-curing technology. This area will remain the field of electrostatically applied powder coating or other metal treatment technologies.
- Free plastic films are more economically produced using high-molecular-weight linear polymers.
- Coating of highly porous substrates is more suited to water-based systems, mainly emulsions.
- Up to now, textile coatings remain difficult for UV or EB technology.

As with many technologies, radiation curing continues to make progress and probably some of these limitations may be overcome and provide new opportunities in the future.

TRENDS

Applications to develop
There are today a lot of applications that are possible with radiation curing but so far have only had limited application on an industrial scale or are still at pilot or laboratory stage. New applications require enhanced and often contradictory properties. It becomes increasingly difficult and requires considerable development work to optimise formulations for new applications. Radiation-curing technology offers such a broad range of final properties of the cured coating that it is theoretically possible to perform all kinds of application in 'industrial coatings'. In the modern world, it is not enough to obtain the technical properties or requirements, it must also be cost-effective, and this is a limitation. A new technology replaces an older one in an application only when the cost/performance ratio is more favourable. Either the new technology is cheaper whilst giving similar properties or it achieves a unique

performance that is not possible with the existing technology and where the cost premium is justified. These are the two main conditions for success. Nevertheless, every year new applications are developed. It is difficult to estimate today the areas in which research laboratories are probably working and developing new applications for the future. There are several applications that offer potential for the future.

- *Graphic art*
  - Flexo inks (in niche applications) and UV varnishes on top of water-based inks.
  - Litho printing cured with EB could be an alternative to replace flexo and heliogravure printing in order to avoid the solvents used in these systems.
  - Decoration with EB curing of food packaging.
  - UV, although more probably EB, curing of intaglio inks for the printing of bank notes, postage stamps and other security printing.
  - An interesting application in development is UV printing and varnishing simulating holographic effects: this can be done not only on plastic but also on paper and card.
- *Wood finishing.* UV pigmented and clear coatings to replace acid-curing systems in order to avoid solvents and emission of formaldehyde.
- *Plastic*
  - Self-supporting substrates obtained by radiation curing.
  - Anti-fogging films having a high water repellancy for several uses (window glass, bathroom mirrors, windows for refrigerated display cabinets, greenhouses, goggles, helmet shields, etc.).
  - UV-cured PVC edge-moulding is used almost exclusively on furniture but it could be applied to automotive parts.
  - Either more flexible or harder UV finishes on different plastics; the use of small concentrations of volatile solvents, still giving high solid systems, can allow a significant decrease in the concentration of diluting monomers and may provide a simple solution in achieving better adhesion together with improved flexibility and better abrasion resistance.
- *Paper*
  - Abrasive papers are an area where radiation-curing technology can bring big savings in space and energy but it will require more powerful EB accelerators (probably up to 1000–1500 kV).
  - EB paper upgrading will continue to develop new products.
- *Adhesives*
  - UV and EB PSA (permanent and removable); but only in niche applications where the cost/performance ratio brings advantages over conventional emulsion-based systems.

- Thin radiation-cured coatings can replace thicker and heavier laminating films.

- *Miscellaneous*
  - UV leather finishing (pigmented and clear).
  - UV and EB curing of composite materials.
  - Manufacture of plastic fibres for special applications.
  - Glass laminates for the manufacture of sound-proof and safety windows.
  - Decoration or coatings on various glass substrates (flat foils and bottles).
  - An application area in which it will be difficult to achieve the technical requirements is textile printing.

Future trends

As with the modern world, radiation curing is continuously evolving. One technology never replaces all others but nevertheless there are trends which are irreversible. The amount of solvents emitted into the air will have to decrease and we need cleaner water. This does not mean that all solvent-based systems will disappear but that they will be replaced where reasonably possible. Alternatively, they will have to use solvent-recovery systems (solvent recovery costs money in investment and safety) or be incinerated (burning is not an environmentally sound solution as it produces carbon dioxide and consumes precious fuel resources).

Consumers

All trends in the evolution of a technology originate from the consumer. The consumer wants a cleaner and safer world, a higher quality of life, better health and an improving standard of living (food, hygiene, communications, leisure, sport and travel).

Legislation and environment

In general, legislation for new chemicals tends to become more difficult. A few years ago, the TSCA regulation was so severe that it was practically impossible to introduce new products containing acrylic double bonds. Fortunately, it has become more realistic and it is now possible again to introduce new acrylated oligomers and pre-polymers. Nevertheless, it will become more severe against the use of products in coatings that are toxic. Fortunately, almost all the products used today in radiation curing meet the requirements of 'non-toxic' according to the oral LD_{50} value. The pressure to protect the environment is now so strong that all countries are introducing legislation preventing the pollution of air and soon it will be the same in water. This means that already today, the amount of solvent that is released into the atmosphere will have to be reduced drastically. This trend is only favourable for radiation-curing technology. Up to now, decorated and protected packaging cannot be used for direct and permanent contact with food. In EUR the regulatory authorities are now considering establishing a positive list of acrylated products for use in food packaging. Again, this can only be positive for the future of radiation technology.

Drying and curing technology
There are *not many* technologies that answer the requirements of the future:

- no pollution;
- saving energy;
- high productivity; and
- performance coatings (properties and cost).

Emulsions that are dried thermally are non-polluting with respect to air and are cheap. They are ideal for porous substrates and 'on site' painting, however, they are not ideal for saving energy, for high productivity or the properties such as gloss, scratch- and abrasion-resistance, and solvent-resistance. Powders are non-polluting for the air and water and are giving very good performance coatings on metal where they are gradually increasing their market share; however, powders consume a lot of energy and require long times to cure (6–10 minutes at 150–200°C). Radiation curing is *the only technology,* today, answering all the requirements at the same time. Radiation-curing technology will not replace all the other technologies as it has several limitations and one important limitation is the cost of the products. There already exist several applications where the technical requirements are obtained but the cost of radiation curing is too high:

- PSA (in comparison to water-based and solvent-based formulations);
- flexo inks;
- metal coatings (in comparison to powder coatings); and
- textile printing.

Radiation curing will always have difficulties on porous substrates or in application where low-viscous formulations have to be sprayed. Finally, decorative paints will remain the exclusive area for water-based systems. Radiation curing is unlikely to be displaced by conventional emulsions in OPVs as emulsions do not give the gloss levels and solvent-resistance obtained by radiation-cured coatings. The combination of different technologies will occur for the protection and decoration substrates in order to obtain the best performance of each technology, e.g. on a strong porous substrate, an emulsion will be used as primer coating and a radiation-curing formulation as a top coat. In flexo printing, the inks can be water-based and protected with a UV varnish. The use of small amounts of solvents in combination with radiation-cured formulations will open up new areas of applications and will make it possible to reduce the concentration of monomers used. This process has been used in Japan for some time and will spread to EUR and the US in some specific applications. Applications that have not been possible up to now, because the required properties could not be achieved with a single curing mechanism of the various chemistries (thermal, isocyanate, radical, cationic) will probably be solved by combining several chemistries. It is in this direction that more research and development work is carried out towards the use of the following:

- hybrid curing systems (radical + cationic); and
- dual curing (radical + thermal or alternatively thermal first, then UV or EB curing)

When the legislation becomes more severe against waste water, some habits of today will have to change drastically. Often in the case of water-based systems, waste is disposed of down the drain as it is considered to be as safe as water. When this occurs, it will again give opportunities for radiation-curing technology.

Formulators and raw-material suppliers

In the global coating and graphic arts sectors the trend is towards mergers or acquisitions of smaller companies. An additional trend is the internationalisation of businesses. In the last 15 years almost 40% of the small formulators in EUR have disappeared (probably the same is true in NA and Japan). A few large multinational companies control 50% or more of the total coatings and printing inks market. These trends will probably continue for a while. Several companies are forming joint ventures or are purchasing smaller ones in an attempt to become more competitive and to acquire a stronger buying and selling power. The global evolution of the coating and ink formulations is continuing to follow these trends. In radiation curing, we have also seen several substantial acquisitions or joint ventures over the past few years but here the number of companies involved in this technology, instead of decreasing, is increasing as a result of an increasing number of new companies entering this market. This contrary effect to the general trend is due to more and more companies becoming aware of the potentials of radiation-curing technology. They also probably consider radiation-curing technology as one of the best solutions to the challenges of the future.

Geographically

Up to a few years ago radiation curing was limited to strongly industrialised countries but it continues to spread in acceptance to all other less-developed countries of the world. Of course, in less-developed countries, radiation technology always begins with OPV or wood finishing generally using unsaturated polyester in styrene in the early stages. Also, in the beginning, ready-to-use formulations are imported from industrialised countries but eventually local formulators begin to develop their own formulations. This again contributes to the continued spreading and acceptance of radiation-curing technology in these new countries.

Products

The use of products and formulations presenting some threat due to toxicity is greatly decreasing and has practically disappeared. New products have to be non-toxic and offer as low skin irritation potential as possible. In addition, they will be more purified. Formulations after curing will have less residual odour and practically no extractable materials. Formulations will have to be as reactive as possible. But the biggest trend will probably be in the search for lower viscosity oligomers in order to reduce or, where possible, eliminate the use of diluting monomers (which are generally more irritant than the oligomers). The recent shortage in acrylic binders and the search for less irritant products are increasing the interest in alternatives to acrylic acid chemistry, consequently cationic chemistry will continue to grow, but this also has its limitations. Radiation-curable water-dilutable formulations and water-emulsified coatings will find their niche markets. Formulations based on acrylic acid chemistry will retain the greatest share of the radiation-curing market as they are the most versatile and their application fields are very wide. In a 'quality orientated society', the aesthetic appeal, appearance, environmental impact and the true quality of the coating become predominant. The consumer wants to improve the standard of living and yet becomes more price conscious. Therefore, we are living in a world where competitiveness is increasing. The two trends of 'quality' and 'price' are, in a sense, opposing each other and the right compromise is not always easily found. Therefore, the danger is for newcomers in radiation curing coming into the market with 'me too' products and cutting the price by sacrificing the 'quality level' of the products. Such a situation could lead to a bad name being given to the radiation-curing industry and in the extreme to 'accidents'.

Growth

The total coatings and inks business will probably grow with the growth in gross national products, probably 2·5–3% per annum. With the several advantages that radiation curing brings, this market is expected to grow at about 12% per annum from 1990 to 1995 and at about 6–8% per annum from 1995 to 2000. This growth could be higher if the regulation against pollution become more severe or if radiation-curing technology makes significant advances in certain areas.

CONCLUSION

Radiation-curing technology will continue to grow in existing applications. But what gives this technology a significant advantage, is its high potential for innovation. The potential innovations possible with this technology are only limited by our imagination. The true strength of radiation-curing technology will be in the development of new, currently unknown, applications which are unique and only possible with this technology and where the manufactured products will combine the excellent properties and the curing speed obtainable with radiation curing:

- production of finished packaging in one pass;
- manufacture of self-adhesive labels in-line (adhesive, substrate, printing inks and protective varnish, all being made with radiation-curable formulations); and
- objects manufactured by stereo-photolithography.

Therefore, we need the collaboration of all parties involved in the technology and need 'champions' capable of suggesting new potential applications. Every new innovation creates other new ones and it is just that, that makes this technology so exciting, so dynamic and so successful.

Chapter 2

An Introduction to the Basic Principles in UV Curing

J. P. FOUASSIER

*Université de Haute Alsace, Laboratoire de Photochimie Générale,
URA-CNRS no 431, ENSC Mu, 3 Rue Alfred Werner,
68093 Mulhouse Cedex, France*

Introduction 50
Basic Principles in Photochemistry 51
   Light 51
   Light absorption 51
   The Perrin–Jablonski's diagram 56
   Excited-state reactivity 57
   Energy transfer processes 59
Basic Photochemical Processes in Light-Induced Reactions 61
   Direct and sensitised photoinduced polymerisation 61
   Photocrosslinking processes 63
      Photoinitiated cross-linking reaction of EPDM systems 64
      Photocycloaddition reactions 64
      Thiol-ene reactions 65
   Photomodification 65
Light-Induced Polymerisation 67
   Light-curable formulations 67
      Components 67
      Problems 68
   Photoiniators and photosensitisers 73
      The reactivity of radical photoinitiators 73
      Main classes of radical photoinitiators sensitive in the UV wavelength range 77
      Photoinitiators exhibiting particular properties 85
      Synergism 96
      Visible photoinitiators 99
      Cationic photoinitiators 103
   Laser-induced polymerisation 105
   Structure–reactivity relationships 107
Conclusion 112
References 113

INTRODUCTION

Radiation-curing technologies are expanding rapidly on an industrial scale, as shown by the 100-fold increase in monomer production over the past 15 years.[1] These new technologies use light beams to start photochemical and chemical reactions in organic materials (monomers, oligomers, polymers), thus leading to the formation of a new polymeric material[2] whose final uses may be encountered in various industrial areas. Some of the most significant applications were originally related to the UV curing of coatings in the wood-finishing industry and, then, extended to the surface treatment of a large variety of substrates (plastics, metals, wires, pipes, vinyl flooring), to the coatings of optical fibres, to adhesives (laminates, sealants, bonding and pressure-sensitive systems), to dental materials and composites. They continue to develop and provide a number of economic advantages over the usual thermal operation: rapid through-cure, low energy requirements, room temperature treatment, non-polluting and solvent-free formulations, and low costs.[3,4] On the other hand, the imaging area (UV-curable inks, printing plates, high-resolution relief imaging for microcircuits in electronics, etc.) represents a large class of applications in the printing industry[5-7] and microelectronics.[8]

New applications are, nowadays, emerging in the graphic arts (e.g. UV finishing of cosmetic and external food packaging, varnishing, UV marking inks, etc.), in the coating industry (anti-corrosion coatings, protective varnishes, highly flexible and thermally resistant coatings on electric sheets, varnishes for glass cloth, etc.), in laminates (UV-laminated adhesives, glass laminates, etc.). Potential new applications are expected in leather finishing, pressure-sensitive adhesives, composites, dental materials, sealants, coatings for different glass substrates, three-dimensional curing of objects, modelling, etc. The development of high-intensity excitation sources with well-adapted emission wavelengths, as well as suitable devices for the curing of non-flat surfaces, fibres, wires, cables, cylinders, bottles or two-piece cans are likely to promote the growth of the UV-curing market in the 1990s.

Another promising area, opened up by the commercial development of various powerful lasers, is concerned with the applications of laser-induced processes in monomeric and polymeric materials[9-11] to photoimaging, microelectronics, three-dimensional machining, holographic optical elements or information recording and storage, because of the specific advantages of the laser beams, such as the coherence, the selectivity, the high energy concentration on a small surface, the short exposure times, the easy scanning of the film surface by the laser spot, etc. A fascinating field of research is becoming apparent to physicists, polymer chemists and photochemists, and thorough studies should be carried out by teams capable of solving problems lying at the frontier of chemistry and physics.

After briefly reviewing the principles used in photochemistry and underlying radiation curing, this chapter deals with the basic concepts developed in UV-curing technology. The chemistry of the materials used (with a special emphasis on photoinitiators and photosensitizers) and the main characteristics of the reactions involved are outlined. In addition, a presentation and a discussion on the photochemical reactivity of photosensitive systems are provided. Other fundamental aspects (concerned with the monomers and oligomers) are dealt with briefly, as they will be discussed in greater detail in other chapters of this book (Vol. I (Allen) and Vol. III (Takimoto & Jacobine)).

BASIC PRINCIPLES IN PHOTOCHEMISTRY

Photochemistry[12,13] is concerned with all aspects of light absorption by matter (typically from 200 to 1000 nm), which results either in a chemical change of the original molecule (photochemical processes) or a recovery of this molecule without any modification (photophysical processes).

Light

The word light denotes 'that which is perceived by the human eye', but can be more aptly referred to as optical radiations that fall into several classes: the UV (200–400 nm), the visible (400–700 nm) and the near-IR (700–1000 nm) ranges. They correspond to electromagnetic radiations exhibiting a wave-like and a corpuscular character and they travel through the vacuum with the same velocity ($c = 3 \times 10^8$ m s^{-1}). One particular radiation is characterized by its frequency (v), its wavelength (λ) or its wave number ($\bar{v}$), and transports an energy, E:

$$E = hv = h\frac{c}{\lambda} = hc\bar{v}$$

where h is Planck's constant ($6 \cdot 62 \times 10^{-34}$ J s^{-1}) which corresponds to a quantum of radiant energy or a photon. The amount of energy transported by 1 mole of photons is defined as an einstein:

$$E' = N_a E$$

where N_a is the Avogadro's number ($6 \cdot 02 \times 10^{23}$).

An electromagnetic wave displaying a single frequency is termed monochromatic. However, the waves delivered by a light source are generally polychromatic. In photochemistry, the intensity of a light beam should be defined as a number of photons, at the frequency v, in a frequency interval Δv, in unit time by unit surface. A typical spectrum of the light emitted by a high-pressure mercury lamp is shown in Fig. 2.1. Given wavelengths are selected by means of filters.[14,15]

Light absorption

A molecule can absorb a quantum of radiation and thus becomes energetically excited. The energy of the photon absorbed should correspond to the exact

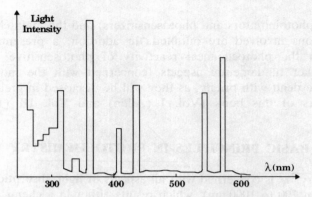

Fig. 2.1. Typical emission spectrum of a mercury lamp.

difference in energy between the initial state E^0 (ground-state) and the final state E^1 (excited state).

$$E^1 - E^0 = h\nu$$

In photochemistry, these E^0 and E^1 states are electronic states, i.e. they represent the electronic energy levels E_{el} of a molecule for a given distribution of the electrons in the molecular orbitals (e.g. non-bonding n, bonding σ, π and anti-bonding σ^*, π^*). Because of Pauli's principle, the spins of two electrons in the same orbital must be paired: as a result, there is no net spin and such a configuration corresponds to a singlet ground-state S_0). If one electron moves up to an unoccupied orbital, a singlet excited state S_1 is formed, from which a triplet excited state T_1 can be generated (the two electrons now possess parallel spins).

Generally, only a moiety of the molecule called the 'chromophoric group' or the 'chromophore' is responsible for the electronic absorption (e.g. the carbonyl group, Fig. 2.2). Promoting an electron to a higher lying energy level (or the molecule from S_0 to S_1) causes a transition. Such a transition would occur at a definite wavelength. Transitions obey selection rules which apply to spin ($S_0 \rightarrow S_n$ and $T_1 \rightarrow T_n$ transitions are allowed) and symmetry. Even when no transition is symmetry allowed, smaller terms may contribute to the perturbing potential (in the description of the light–matter interaction) and lead to transitions with much lower intensities (i.e. whose transition probability, ω, is lower).

Up to this point, electronic states have been considered for fixed nuclear geometries. In a molecule, nuclear motions result in small energy changes, depicted in terms of vibrational and rotational energy levels (E_v and E_r, respectively). The total energy of the molecule is, then, the sum $E_m = E_{el} + E_v + E_r$. Under light excitation, different photons ν_i can be absorbed, provided that

$$h\nu_i = (E_{el} + E_v + E_r)^1 - (E_{el} + E_v + E_r)^0$$

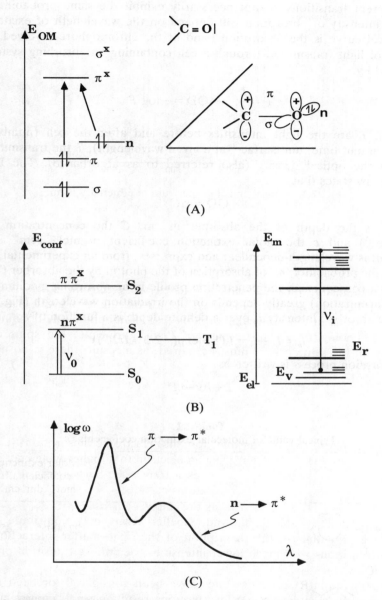

Fig. 2.2. The carbonyl chromophore: (A) molecular orbitals (OM) involved, energy levels (E_{OM}) of these OMs, $n \rightarrow \pi^*$ and $\pi \rightarrow \pi^*$ transitions under light excitation; (B) energy levels (E_{conf}) of the electronic states corresponding to a given configuration; energy levels of the molecule (E_m) when taking into account the motions of nuclei relative to one another (vibrations and rotations); (C) probability of absorption ω versus λ for an allowed transition ($\pi \rightarrow \pi^*$) and a forbidden transition ($n \rightarrow \pi^*$).

The different transitions do not necessarily exhibit the same probability, so that the intensity of absorption will depend on the wavelength of excitation: this typical curve is the absorption band of the chromophore involved. The fraction of light transmitted through a cell containing an absorbing system is expressed by

$$T = \frac{I_t}{I_0}; \quad OD = -\log T$$

where I_0, I_t are the light intensities before and after the cell (number of photons in unit time, unit surface, at a given wavelength), T the transmittance and OD the optical density (also referred to as absorbance). The Beer–Lambert law states that

$$OD = \varepsilon l C$$

where l is the depth of the absorber in cm, C the concentration in M (mol litre^{-1}) and ε the molar extinction coefficient in mol^{-1} l cm^{-1}. This coefficient is wavelength-dependent and expresses, from an experimental point of view, the probability, ω, of absorption of the photon by the absorber (Table 2.1). As a consequence, the penetration profile in an absorbing medium (at a given concentration) greatly depends on the irradiation wavelength (Fig. 2.3), since the absorbed intensity I_a over a definite depth is a function of ε:

$$I_a = I_0 - I_t = I_0(1 - \exp(-2 \cdot 3 \times l \varepsilon C))$$

The absorption process is written as

$$I + h\nu \rightarrow I^*$$

Table 2.1
Typical value of molecular extinction coefficients, ε

Structure		Molar extinction coefficients, ε mol^{-1} dm^3 cm^{-1}	
Ph–CO–Ph		22 000[a,b]	85[c,d]
(xanthone/thioxanthone structure)	X = O	200[c]	
	X = S	5 200[c]	

[a] $\lambda = 254$ nm.
[b] π–π^* transition.
[c] $\lambda = 366$ nm.
[d] $n \rightarrow \pi^*$ transition.

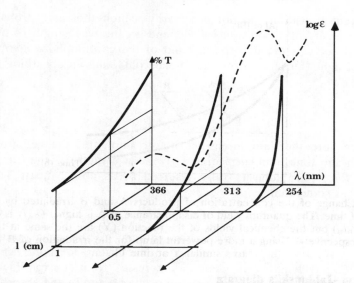

Fig. 2.3. Absorption spectrum of a molecule, $\log \varepsilon = f(\lambda)$. Fraction of light transmitted (%T) at different wavelengths as a function of the optical path (l). In this example half of the incident light is absorbed within 1 cm, 1 mm and 10 μm at 366, 313 and 254 nm, respectively.

The rate of absorption I_{abs} (or absorbed intensity that has the dimension of a number of photons absorbed in unit time by unit volume) is given by

$$I_{abs} = \frac{d}{dt}[I^*]$$

The concentration of molecules, P, formed in a particular process over the number of quanta absorbed by the system is defined as a quantum yield, ϕ. Then

$$\phi = \frac{d[P]/dt}{I_{abs}}$$

These quantum yields, ϕ, relate directly to the efficiency of a photochemical reaction, whereas the chemical yield of reaction, Y, corresponds to the amount of product, P, formed or destroyed within a period of irradiation (Fig. 2.4). If the reaction is not monophotonic with respect to the light intensity (i.e. requires more than one photon to produce one photochemical event), the following equation holds:

$$\frac{d[P]}{dt} = K I_{abs}^{\alpha} \quad (\phi = K I_{abs}^{\alpha-1})$$

and the quantum yield, ϕ, becomes intensity dependent for $\alpha \neq 1$. For $\alpha = 2$, the law is said to be biphotonic.

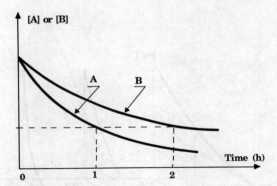

Fig. 2.4. Change of the concentration of products A and B irradiated by light, as a function of time. The quantum yield of disappearance of A is higher ($\varepsilon_A l_A$ is considered equal to $\varepsilon_B l_B$) but the chemical yields of the reaction (Y) are the same at time $t = 1$ h and 2 h, respectively. Using a more powerful lamp for the irradiation of B would lead to a similar Y at time 1 h.

The Perrin–Jablonski's diagram

Photophysical and photochemical processes occurring in an electronically excited molecule are usually represented by the Perrin–Jablonski's diagram (Fig. 2.5). Light excitation induces the $S_0 \rightarrow S_1$ transition and leads to a molecule in the first excited singlet state, S_1. Radiative and non-radiative relaxation processes arise from the S_1 state.

Fluorescence

$$S_1 \rightarrow S_0 + h\nu \quad k_r$$

Internal conversion

$$S_1 \rightarrow S_0 + kT \quad k_{IC}$$

Intersystem crossing

$$S_1 \rightarrow T_1 \quad k_{ST}$$

From the triplet state T_1, two main processes occur.

Phosphorescence:

$$T_1 \rightarrow S_0 + h\nu \quad k_p$$

Intersystem crossing

$$T_1 \rightarrow S_0 + kT \quad k'_{IC}$$

All the rate constants of these different processes involved k_r, k_{IC}, k_{ST}, k_p, k'_{IC} are first order. The lifetimes of the S_1 and T_1 states are, respectively, defined by

$$\tau_f = \frac{1}{k_r + k_{IC} + k_{ST}} \quad \tau_T = \frac{1}{k_p + k'_{IC}}$$

Introduction to the basic principles of UV curing

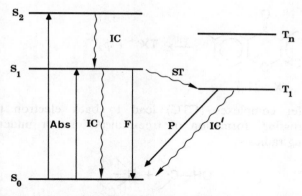

Fig. 2.5. The Perrin–Jablonski's diagram. IC—internal conversion; ST—intersystem crossing (singlet–triplet conversion); P—phosphorescence; F—fluorescence. (In the $T_1 \rightarrow S_0$ transition, IC' corresponds to intersystem crossing.)

and the quantum yields of these different processes can be evaluated, for example:

$$\phi_f = k_r \tau_f \qquad \phi_{ST} = k_{ST} \tau_f$$

Excited-state reactivity

Most photochemical processes originating from the T_1 state lead to chemical intermediates (e.g. radicals, ionic species) and then to new molecules. Well-known examples (and, in addition, useful in radiation-curing technology) include the following.

Norrish type I photoscission

Hydrogen abstraction from $n\pi^*$ state

Electron transfer

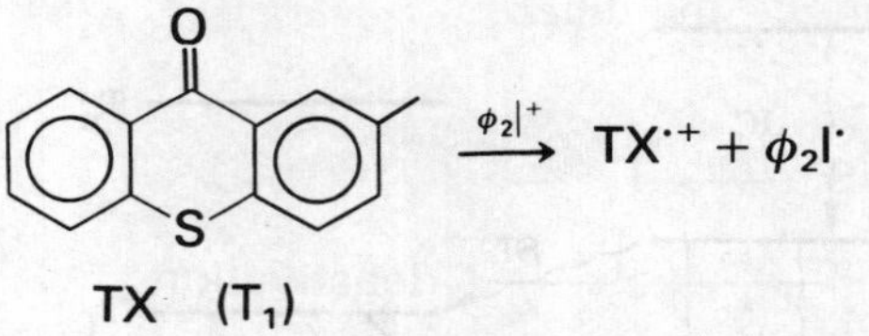

TX (T₁) →[$\phi_2 I^+$] TX·⁺ + $\phi_2 I$·

Charge transfer complexes (CTC) lead to back electron transfer after intersystem crossing, formation of free-radical ions or undergo a proton transfer yielding radicals:

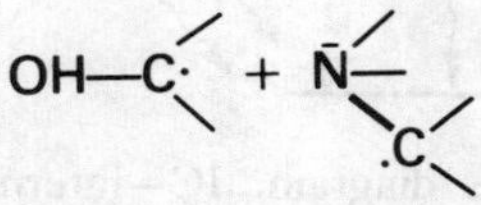

OH—C· + N—
          ·C

The occurrence of these photochemical processes shortens the triplet-state lifetime

$$\tau_T = \frac{1}{k_p + k'_{IC} + k}$$

$k = k_c$ if cleavage

$k = k_H[DH]$ if hydrogen abstraction

$k = k_e[AH]$ if electron transfer

where k_c, k_H and k_e are the rate constants of the corresponding reactions (note that k_H and k_e are bimolecular quenching rate constants, the word 'quenching' referring to a deactivation or a consumption of T_1). The dependence of τ_T on the quencher, Q, concentration (DH or AH) can be expressed by, for example:

$$\frac{1}{\tau_T} = \frac{1}{\tau_T^0} + k_q[Q]$$

with τ_T^0 being triplet-state lifetime in the absence of Q; Q = DH or AH; $k_q = k_H$ or k_e. According to the efficiency of the photochemical processes, τ_T ranges typically from a few μs to less than 1 ns (in ketones).

The quenching by molecular oxygen is a very important process of deactivation of excited states, whose mechanism is either physical of chemical. The quenching of the S_1 state results in an enhancement of the intersystem crossing yield whereas the quenching of T_1 leads to a decrease of τ_T. In a general way, the rate constant of the bimolecular process $k_q^{O_2}$ is 10^9–10^{10} mol^{-1} l s^{-1} and the oxygen concentration in an organic solvent is $[O_2] \approx 2 \times 10^{-3}$ M, so that oxygen quenching can compete efficiently only in molecules having long-lived triplet states (for many molecules, S_1 quenching is not observed because of the low value of τ_f).

The quantum yield, ϕ, of a photochemical process can be expressed as a function of the rate constants of the different reactions. Let us consider the following example:

$$S_0 \xrightarrow{h\nu} S_1 \xrightarrow{\phi_{ST}} T_1 \xrightarrow{k_c} P$$

with deactivation paths k_{IC} and $k_q[Q]$

Let us try to evaluate [P] when Q is added to the solution. The quantum yields of the formation of P in the presence (ϕ) and in the absence of (ϕ_0) of Q are given by

$$\phi_0 = \phi_{ST} \frac{k_c}{k_c + k_{IC}} \qquad \phi = \phi_{ST} \frac{k_c}{k_c + k_{IC} + k_q[Q]}$$

Then, the well-known Stern–Volmer relationship is obtained:

$$\frac{\phi_0}{\phi} = 1 + k_q \tau_T^0 [Q]$$

and a plot of ϕ_0/ϕ as a function of $[Q]$ leads to the determination of the Stern–Volmer coefficient $k_q \tau_T^0$.

Energy transfer processes

A radiationless electronic energy transfer corresponds to the passage of energy from an excited donor molecule (D) to an unexcited acceptor molecule (A), which becomes excited. Two major mechanisms occur: coulombic interactions (or dipole–dipole mechanism) between the two molecules and exchange interactions between the molecular orbitals of the two systems (collisional mechanism). If the first one can be operative even at long distance (50–100 Å (5–10 nm)), the second one requires diffusional motions of both the donor and the acceptor (Fig. 2.6). In rigid media or in polymer films, efficient energy transfer might be limited. The movement or 'hopping' of the electronic

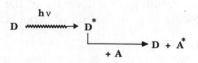

Dipole-Dipole interaction

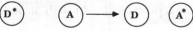

Electron exchange mechanism

Fig. 2.6. Radiationless energy transfer mechanisms.

excitation from an original excitation site to another site along a macromolecular backbone or via the solvent molecules is called energy migration. Energy transfer can take place in singlet and triplet states, such as the following.

- Singlet–singlet energy transfer:

$$^1D^* + {}^1A \longrightarrow {}^1D + {}^1A^*$$

- Triplet–triplet energy transfer:

$$^3D^* + {}^1A \longrightarrow {}^1D + {}^3A^*$$

The efficiency of the collisional process (which is mostly encountered in the systems presented here) is governed by energetics and excited-state lifetimes. First of all, an excited donor must diffuse to the acceptor within a time period in the order of its lifetime; the range, r, in which the collision can occur is a function of the viscosity of the medium: if $\tau_T = 1$ ms, $r = 15\,000$ Å ($1.5\,\mu$m) for a fluid solution (diffusion coefficient $D_c \approx 10^{-5}\,\text{cm}^2\,\text{s}^{-1}$).[12] The observed rate constant of energy transfer, k_{et}, is governed by the bimolecular rate constant, k_{diff}, for diffusion-controlled reactions (that are defined by the Debye equation or the Smoluchowski equation):

$$k_{et} = \alpha k_{diff}$$

where α is the probability of energy transfer. If every collision leads to the transfer, the process is said to be diffusion controlled ($\alpha = 1$). Secondly, an efficient energy transfer should be exothermic: the excited-state energy of the donor must be higher than that of the acceptor; an endothermic process is considerably less efficient (Fig. 2.7). Calculation of the ratio, ρ, of the energy transfer rate constants as a function of the donor–acceptor energy differences ΔE shows a 10-fold decrease when ΔE drops from $+1$ to -1.4 kcal mol^{-1}, ($+4.2$ to -5.88 kJ mol^{-1}).

The applications of energy transfer are two-fold. Quenching of an excited state by a quencher Q (which plays the role of an acceptor) may occur through energy transfer; the solution is considered as either of the following.

- A deactivation of the donor, resulting in a decrease of the efficiency of a desirable photochemical process (e.g. formation of products, P):

$$D^* \longrightarrow P$$
$$\searrow^Q \text{deactivation}$$

- A possibility of formation of a product, P, through an indirect route:

(i) $$D^* \not\longrightarrow P$$

(ii) $$A \xrightarrow{h\nu} A^* \longrightarrow P$$

(iii) $$D \xrightarrow{h\nu} D^* \longrightarrow A^* \longrightarrow P$$

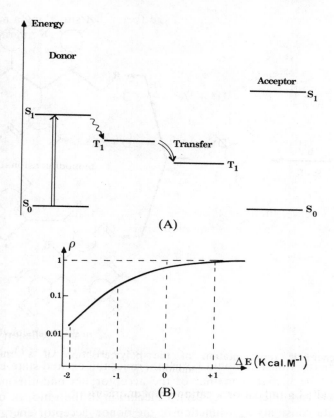

Fig. 2.7. Exothermic triplet–triplet energy transfer. (A) Energy levels of the donor and acceptor, and (B) ρ versus ΔE (see text).

In that scheme, P is formed either through the direct photolysis of A(ii), or the sensitised photolysis of A(iii), in which the energy transfer occurs after light excitation of D; D is called a sensitizer and the process a sensitization.

BASIC PHOTOCHEMICAL PROCESSES IN LIGHT-INDUCED REACTIONS

Several types of processes are generally found in light-induced reactions of organic materials, and used in the various areas of what, nowadays, is referred to as radiation curing. This section only briefly outlines some important facts, since these processes will be tackled in other chapters.

Direct and sensitised photoinduced polymerisation

Photoinduced polymerisation reactions[2,3,16–19] concern the creation of a polymer, P, through a chain reaction that has been initiated by light. According to the type of reactive species (radical or cation) formed on the monomer, M, the

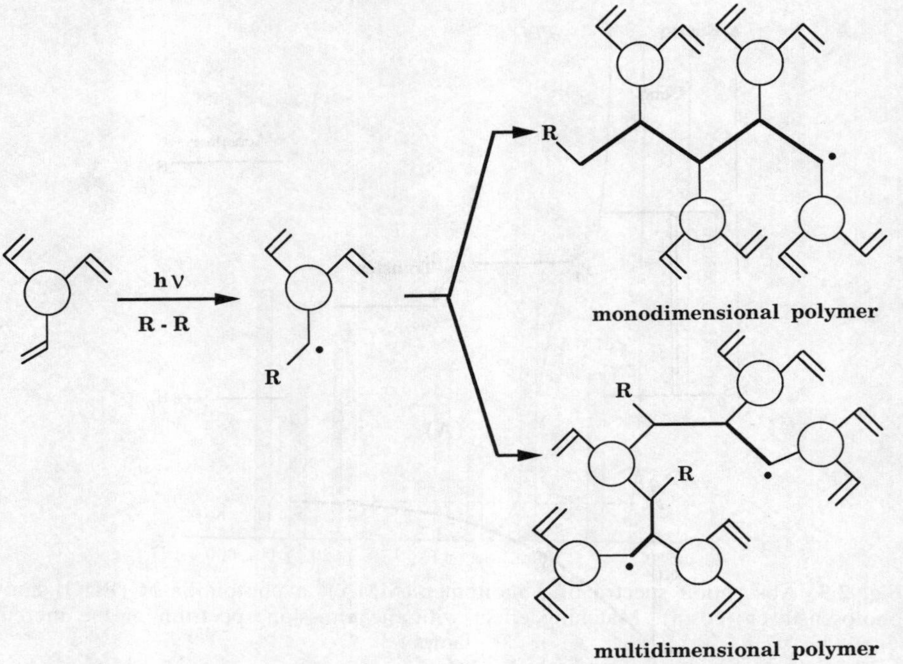

Fig. 2.8. Schematic representation of the polymerisation of a multifunctional monomer.

reaction is called a radical or a cationic polymerisation:

$$M \xrightarrow{h\nu} M^{\cdot} \rightsquigarrow (M)_n^{\cdot} \rightsquigarrow P$$

$$M \xrightarrow{h\nu} M^{+} \rightsquigarrow (M)_n^{+} \rightsquigarrow P$$

Acrylic monomers (such as methyl methacrylate MMA) or epoxides (1,2-epoxybutane 1,2-EB) exemplify this behaviour.

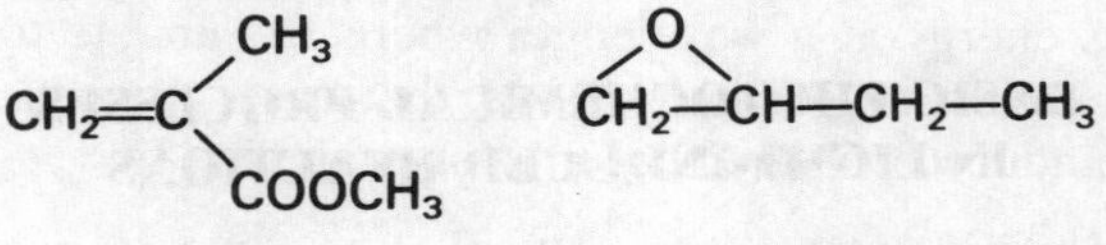

$$CH_2=C\begin{array}{c}CH_3\\COOCH_3\end{array} \qquad \underset{\diagup\ \diagdown}{\overset{O}{CH_2-CH}}-CH_2-CH_3$$

MMA   1,2-EB

As far as the chemical nature of the monomer used is concerned, an industrial formulation contains multifunctional monomers and oligomers that lead ultimately to a polymer network (Fig. 2.8).

Since the direct formation of reactive species on the monomer by light absorption:

$$M \xrightarrow{h\nu} M^{*} \longrightarrow M^{\cdot}$$

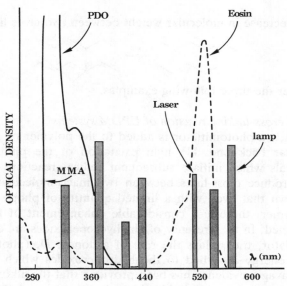

Fig. 2.9. Absorption spectra of a monomer (MMA), a photoinitiator (PDO) and a photosensitiser (eosin). Matching effect with the emission spectrum of the mercury lamp.

is not an efficient route, the initiation step of the polymerization reaction requires the presence of a photoinitiator, I, which, under light excitation, is capable of generating these reactive species (e.g. radical $R^{\cdot}$):

$$I \xrightarrow{h\nu} {}^1I^* \longrightarrow {}^3I^* \longrightarrow R^{\cdot} \xrightarrow{M} RM^{\cdot} \rightsquigarrow \rightarrow P$$

${}^1I^*$ and ${}^3I^*$ represent the S_1 and T_1 state of I, respectively. Extension of the spectral sensitivity (which corresponds to the best matching between the emission spectrum of the light source and the absorption spectrum of the formulation) can be achieved by using photosensitizers, S: their role is to absorb luminous energy at a wavelength where I is unable to operate and to transfer the excitation to *I* (Fig. 2.9). In that case, the reaction is called 'a sensitised photoinduced polymerization', for example:

$$ {}^1S \xrightarrow{h\nu} {}^1S^* \longrightarrow {}^3S^* $$
$$ \xrightarrow{I} {}^3I^* + {}^1S $$
$$ \longrightarrow R^{\cdot} $$

Photocrosslinking processes

In a general way, the photocrosslinking process[2,6,20,21] relates to the cross-linking of macromolecular chains, through a photochemical mechanism, leading to a decrease in molecular weight between two cross-links (M_c):

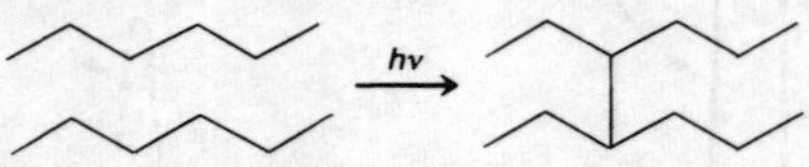

Let us consider the three following examples.

Photoinitiated cross-linking reaction of EPDM systems
In this system,[22] a photoinitiator is added to the polymer or grafted onto the macromolecular backbone. UV-light excitation of the photoinitiator yields reactive radicals which initiate subsequent radical reactions in the polymer matrix and produce cross-links between two macromolecular chains. It was generally shown that even with a limited quantity of photosensitiser grafted onto the polymer, there is a considerable enhancement of the cross-linking reaction obtained. In the presence of air, hydroperoxides are easily generated; the photoinitiator, thus, plays the role of a donor, D, (photosensitiser) and transfers the energy absorbed to the hydroperoxides which undergo radical homolytic cleavage; evidence has been provided that the alkoxy macroradicals formed are the active species and lead to ether type cross-links:

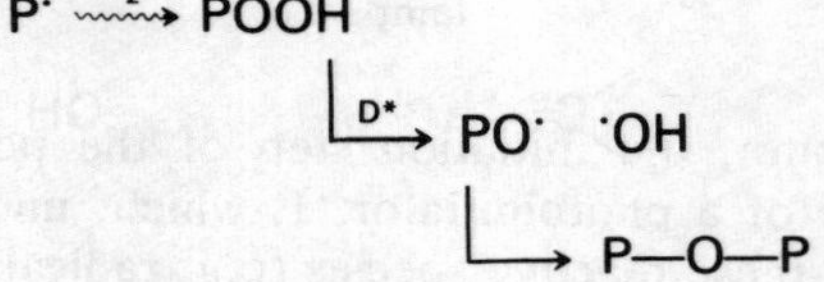

$$P^\cdot \xrightarrow{O_2} POOH$$
$$\xrightarrow{D^*} PO^\cdot \quad {}^\cdot OH$$
$$\longrightarrow P{-}O{-}P$$

Photocycloaddition reactions
This type of reaction[6,23–25] is very well known in some typical systems, such as polyvinyl-cinnamate derivatives (having the cinnamate moiety either as pendant groups or in the main polymer backbone), chalchone type compounds, bis-maleimides, anthracene derivatives, cyclised polyisoprene. For example, the reaction consists in the formation of a cyclobutane ring prepared by dimerisation of an excited chromophore with a ground-state chromophore:

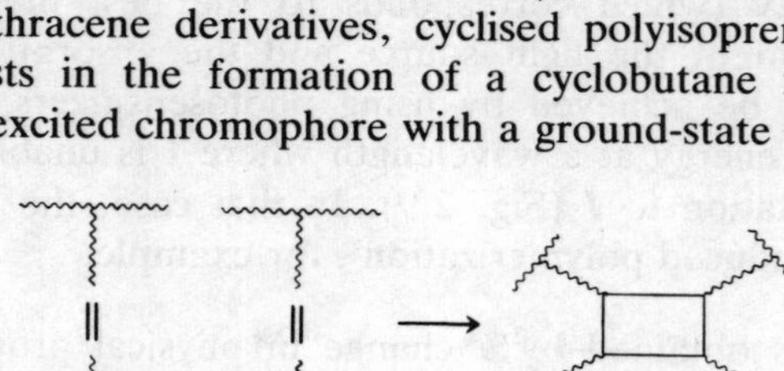

These reactions have to be sensitised in order to extend the sensitivity of the system. To overcome the difficulty of adding sensitisers, an elegant way consists in designing a polymer matrix carrying chromophoric groups exhibiting the right light absorption and an efficient photocrosslinking process:[6]

Thiol-ene reaction

This reaction corresponds to the addition of a thiol group on an ethylenic double bond in a photocrosslinkable system consisting of a polythiol plus polyene.[4,26] Cross-linking can occur either directly by irradiating under a suitable wavelength or in the presence of photoinitiators such as ketone compounds:

$$\text{>C=O} \xrightarrow{h\nu} {}^3[\text{>C=O}] \xrightarrow{RSH} \text{>C'-OH} + RS^\cdot$$

$$RS^\cdot + R'-CH=CH-R'' \longrightarrow R'-CH-\dot{C}H-R''$$
$$|$$
$$S-R$$

$$R'-CH-CH-R'' \xleftarrow{RSH}$$

Photomodification

This process[2,6,7] is obtained by a change in physical properties such as the following.

- Colour, e.g. in photochromic systems (such as spiropyrans or fulgides):

- Solubility (a high-molecular-weight polymer is transformed into a low-molecular-weight polymer):

$$\left[-CH_2-\underset{\underset{\underset{CH_3}{O}}{\underset{\|}{C=O}}}{\overset{CH_3}{\underset{|}{C}}}-CH_2-\underset{\underset{\underset{H_3C}{\overset{\|}{C}-\overset{O}{\underset{\|}{C}}-CH_3}}{\underset{\|}{\underset{N}{O}}}}{\overset{CH_3}{\underset{|}{C}}}- \right]_n \xrightarrow{h\nu} \begin{cases} \text{O—N cleavage} \\ CO_2 \text{ elimination} \\ \beta \text{ cleavage} \end{cases}$$

- Change of solubility by the modification of a functional group, e.g. in the chemistry of systems based on novolac resins and naphthalene diazoquinone sulphonates:[8]

- Solubility based on acid-catalysed depolymerisation (chemically amplified photoresists):[27]

$$Ar_3S^+ \; SbF_6^- \xrightarrow{h\nu} S_1 \rightsquigarrow H^+$$

- Ink affinity, adhesion, etc.

In all these cases, the starting polymer can form relief images on development. When such photomodification reactions are involved, the polymer is developed by preferentially dissolving the exposed areas and defined as a positive working photoresist; when using photocrosslinking, the solvent dissolves away the unexposed areas and, thus, the polymer is called a negative working photoresist (Fig. 2.10). Both systems belong to the class of photopolymers, i.e. polymers that are sensitive to light. Under high-intensity UV laser beams, the decomposition process can become a photoablation which is described as a spontaneous etching of the material surface occurring upon the absorption of a short light pulse, whose energy is greater than the ablation threshold value.[28]

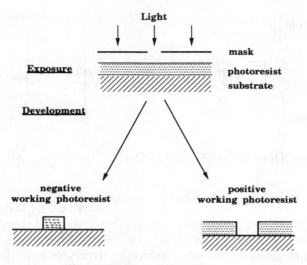

Fig. 2.10. Relief image formation in photoresist materials.

LIGHT-INDUCED POLYMERISATION

Light-curable formulations

Components

In general, during application, a photopolymerisable formulation contains four types of compounds.[3,4]

1. *A photoinitiator.* In addition, there are often several photoinitiators and photosensitisers, giving a synergistic effect (see below).
2. *A reactive diluent* whose role is to adjust the viscosity of the mixtures to an acceptable level for application; it also participates in the polymerisation reaction, for example:

$$\diagdown\!\!\!= \text{ is an acrylate function}$$

Example: $(CH_2{=}CH{-}\underset{\underset{O}{\|}}{C}{-}O{-}CH_2)_3{-}C{-}C_2H_5$

3. *An oligomer.* This is generally a multifunctional monomer which gives a cross-linking polymerisation. The backbone of the oligomers has a chemical structure which can be varied (e.g. polyester, urethane, epoxy) and confers to the polymerised material its special mechanical and physical properties: hardness, abrasion resistance, solvent resistance, elasticity, adhesion, permeability, stability under outdoor exposure, etc.,

for example:

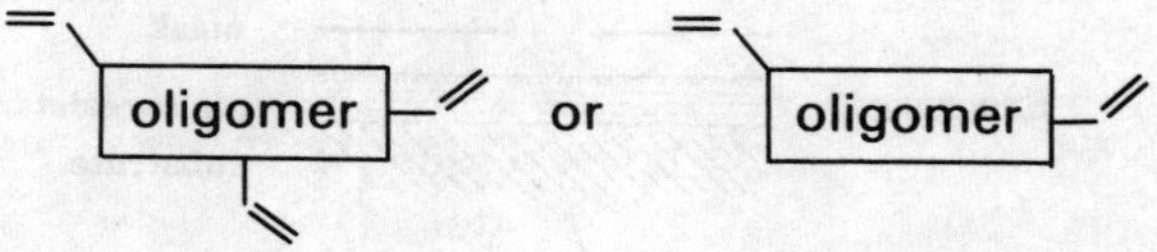

$$CH_2=\boxed{\text{oligomer}}\text{—} \quad \text{or} \quad CH_2=\boxed{\text{oligomer}}\text{—}$$

Example:

$$CH_2=\underset{|}{C}-COO\left[CH_2-\underset{\underset{OH}{|}}{CH}-CH_2-O-\underset{}{\bigcirc}-\underset{|}{C}-\underset{}{\bigcirc}-O\right]_n CH_2$$

$$-\underset{\underset{OH}{|}}{CH}-CH_2-OOC-\underset{|}{C}=CH_2$$

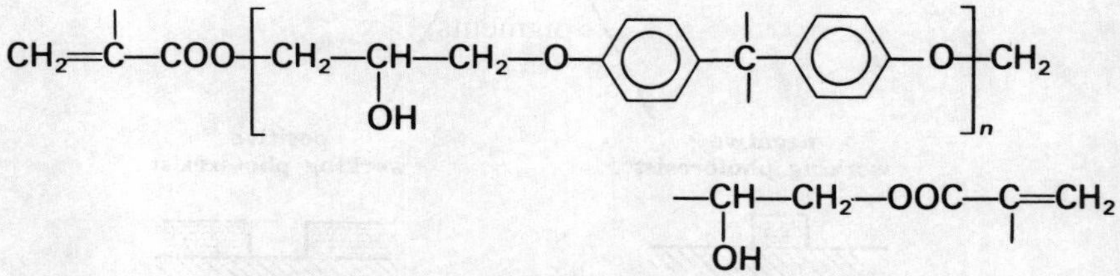

For specific applications, other monomer structures were developed, e.g. with carbamate or carbonate functional groups.[29]

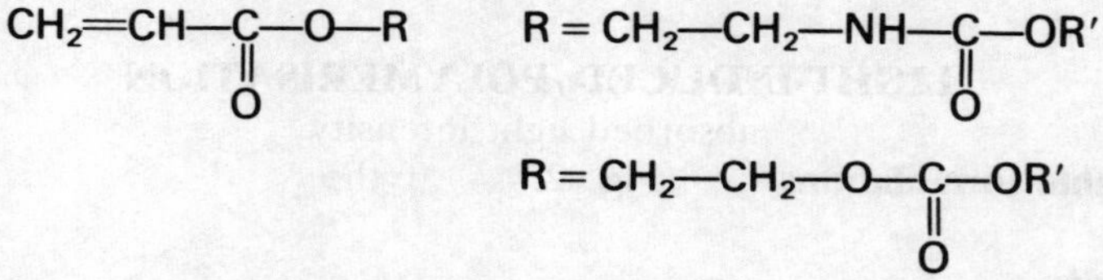

$$CH_2=CH-\underset{\underset{O}{\|}}{C}-O-R \qquad R = CH_2-CH_2-NH-\underset{\underset{O}{\|}}{C}-OR'$$

$$R = CH_2-CH_2-O-\underset{\underset{O}{\|}}{C}-OR'$$

Modifications of a given backbone were also proposed, for example:[30]

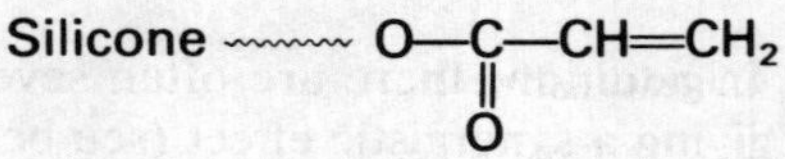

$$\text{Silicone} \sim\sim\sim O-\underset{\underset{O}{\|}}{C}-CH=CH_2$$

or

$$\text{Silicone} \sim\sim\sim CH\underset{O}{\overset{}{\diagdown}}CH_2$$

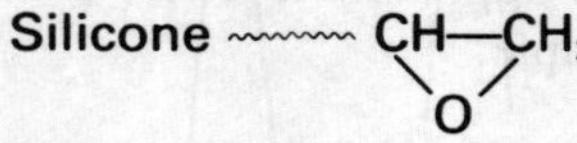

4. *Additives*. As in all industrial formulations, the role of additives is to provide special properties: fillers, pigments, stabilisers, wetting agents, anti-foam, flow aids, etc.

Depending on the application and type of formulation used, different performance levels can be attained: curing of low film thickness ($\approx$ a few μm), pigmented media, transparent media at high film thickness (>1–2 cm); short curing times ($\ll 1$ s) giving high scanning rates, etc.

Problems

Problems take place at different levels when using photopolymeric systems: from general difficulties (belonging to the field of basic research), to special difficulties (which are concerned with the use of a given system in a specific application with prescribed requirements). A formulation must, therefore, be the best possible compromise leading to optimum properties and taking into account the envisaged method of application. Furthermore, it is often difficult to consider the problems individually, because of their mutual interferences. For simplification, let us say that these interferences can be met with in very varied fields: fitted UV sources; special mechanical, physical and chemical properties; heat stability; durability; colour change; adhesion; photochemical reactivity; presence of additives and pigments, etc.

While the nature of the oligomer has a great influence on the properties of the film, the photoinitiator plays a largely decisive role on curing rate and production. Very often, studies must be done on simple media, which permit the development of models and provide evidence for elementary processes. In the light of these results, one can, thus, start studying more complicated systems. Our own experience in this field allows us to state that this is not always an easy task! As we will see, the intrinsic reactivity of a photoinitiator determines its interest. However, other factors have yet to be considered in photopolymerisation reactions: several are listed here.

Molecular absorption coefficient, ε, of the chromophoric group. This parameter determines the value of the absorbed light intensity, i.e. the required amount of energy to form the product (Fig. 2.11). If the value of ε is higher, one obviously reduces the price of the formulation, but one also eliminates the problems connected with the use of too-high concentrations.

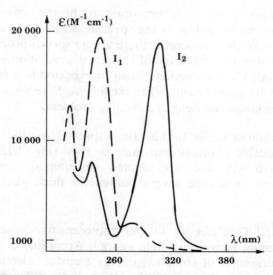

Fig. 2.11. Absorption spectra of two photoinitiators I_1 and I_2. At $\lambda = 313$ nm, $\varepsilon(I_2) \approx 50 \times \varepsilon(I_1)$. The same light absorption is, thus, achieved with $[I_1] \approx 50 \times [I_2]$.

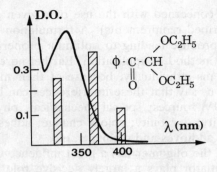

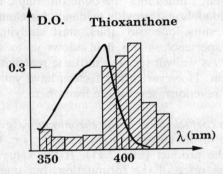

Fig. 2.12. Adaptation of absorption spectra and lamp emission spectra.

Spectral absorption range. It is important to fit the wavelength range of absorption of the photoinitiator to the spectral range of the light source (Fig. 2.12). For clear varnishes, generally, there are no special problems, because of the highly polychromatic character of the usual irradiation sources based on mercury lamps. All UV photoinitiators can be expected to meet the absorption requirements of the formulation. When media also have a specific absorbance, the use of doped lamps can help to solve this problem.

Pigmented or coloured media. In this case, a spectral 'window' has to be found, and the most suitable photoinitiator must be used (Fig. 2.13), as well as the irradiation source with the best-adapted emission spectra. To quote an instance, one can even cure several millimetre thick glass fibre-reinforced resins.

Synergistic effects. One can try to mix photosensitive systems, in order to achieve an optimum recovery of light energy. According to the system used, there can be formation of complex, energy transfer, electron transfer.[31] In certain cases, these reactions permit initiation under different radiations, e.g. visible light.

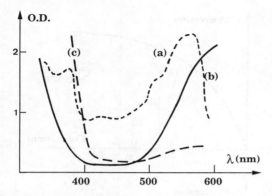

Fig. 2.13. Paint absorption spectra: (a) red, (b) blue and (c) white.

Volatility and extractability. Photoinitiators are usually low-molecular-weight compounds. They generally exhibit a high volatility and can easily be extracted during the curing process or from the cured compositions. In addition, the photolysis products obtained from the rearrangement of the radical pairs that are not consumed in the polymerization process, are generally volatile and odorous substances.

Photodegradation under outdoor exposure. The cured film, especially in the case of clear varnishes, must exhibit minimum discoloration after formation and minimum yellowing on ageing. These phenomena[32] depend on the polymer matrix, but also on the nature of the photoinitiators because of their photolysis products.

Oxygen quenching. As it is well known,[33-35] atmospheric oxygen makes the polymerization reaction of a varnish more difficult (deactivation of intermediate states, formation of oxygenated radicals), thereby slowing down the scanning rate. It also creates oxidation chemical functions (predominantly hydroperoxides, and carbonyl):

$$I \xrightarrow{h\nu} {}^1I^* \longrightarrow {}^3I^* \xrightarrow{O_2} \text{quenching}$$

$$R^\cdot \xrightarrow{M} RM^\cdot \longrightarrow RMM^\cdot$$

$$\downarrow O_2 \quad \downarrow O_2$$

$$ROO^\cdot \quad RMOO^\cdot$$

$$\text{decomposition} \longleftarrow ROOH$$

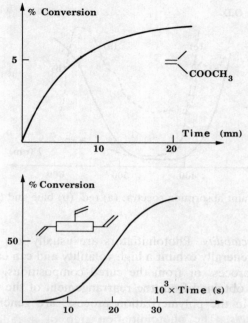

Fig. 2.14. Typical percentage conversion–time curve for a monofunctional monomer in de-aerated solution and a multifunctional monomer in aerated film.

Environment. The polymer matrix obviously plays a great part in the properties of the films, thus, produced. The important and noticeable differences of these substances, which result from the polymerisation of 'usual' monomers, have reference to the following.

- The very high polymerisation rates achieved with these media (Fig. 2.14): this is partly due to the multifunctional character of those systems, the low thickness of the films, the high initiator concentrations generally used, and the high power of the industrial irradiation sources involved.
- The formation of a relatively dense polymer network with a low average molecular weight between two cross-links, resulting from a cross-linking photopolymerisation process.
- The effect of this network on the physical and mechanical properties of the system, but also on the further stability of the coating to photochemical ageing.

Photopolymerisation reactions in other environmental conditions exhibit very typical behaviours, e.g. in micelles, microemulsions, emulsions,[36–38] and in orientated liquid crystalline acrylates.[39–40]

Interaction with stabilisers. As previously stated, oxygen also contributes significantly to the further reactions of photoageing.[41] To keep down these effects, one can add to the formulation radical traps (whose function is to interrupt the development of oxidation chain radical reactions) and/or UV absorbers (which should reduce subsequent light absorption by the medium and the sensitised decomposition of existing hydroperoxides). Investigation of the interactions between photoinitiators and photostabilisers is of prime importance.[42–44]

Light-intensity effects. Using irradiation sources such as flashlamps or laser greatly increases the complexity of the problem: very high power densities or very short light pulses become available and many new points are worth noting, notably:[45] the response of the system, the sensitivity, the non-linearity of the absorption, the multiphotonic processes, the bimolecular and the intensity dependent reactions,[46,47] etc. Whatever the problems encountered, laser-induced reactions seem to be very promising ways for specific applications, due to the advantages of laser beams.

Photoinitiators and photosensitisers

The reactivity of radical photoinitiators

The overall process. The photoinitiation step of a radical polymerisation reaction usually requires the presence of a molecule (photoinitiator) which absorbs the exciting light and leads to radical reactions onto the monomer, through processes which occur in its excited states. The photopolymerisation steps, thus, can be divided into three classes: the photochemical event which leads to the first monomer radical, the classical propagation and termination process of the reaction:

Initiation

$$I \xrightarrow{h\nu} R^{\cdot} \xrightarrow{M} RM^{\cdot}$$

Propagation

$$RM_i^{\cdot} + M \longrightarrow RM_{i+1}^{\cdot}$$

Termination

$$RM_n^{\cdot} + RM_k^{\cdot} \longrightarrow RM_{n+k}R$$
$$\longrightarrow RM_n + RM_k$$

This picture, although very schematic, shows clearly that for a given monomer, the overall efficiency of the reaction is strongly dependent on the efficiency of the photochemical processes and shows the benefits which could be derived by investigating the processes involved in the excited states of the photoinitiator. Generally, upon excitation by UV light, the photoinitiator is promoted to its first excited singlet state, and then converted into its triplet state via fast

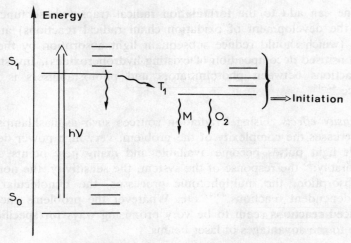

Fig. 2.15. Excited state processes.

intersystem cross-linking. This transient state yields the reactive radicals R·, which can attack a monomer molecule (M) and initiate polymerisation (Fig. 2.15). The triplet excited state may be deactivated through several processes:

- quenching by oxygen;
- non-radiative or radiative processes;
- chemical formation of products;
- reaction with other molecules;
- quenching by the monomer through a process which does not undergo any chain initiation (accordingly, this process should be regarded as a 'dead loss' pathway).

The large arrows in Fig. 2.16 show pathways which efficiently initiate the polymerisation. The efficiency of the photoinitiator can, therefore, be described by two quantum yields: the quantum yield of initiation (ϕ_i) which represents the number of starting polymer chains per photon absorbed, and the quantum yield of polymerisation (ϕ_m) which is the number of monomer units polymerised per photon absorbed. In the absence of a chain transfer reaction, ϕ_i and ϕ_m are related by the equation: $\phi_m = \phi_i \times DP_n$. In addition, the rate and degree of polymerisation are, generally, functions of four parameters: the monomer concentration [M], the optical density of the solution (OD), the incident light intensity (I_0) and the quantum yield of initiation (ϕ_i). Under conditions of low light absorption, the following expressions usually apply:

$$R_p = \frac{k_p}{k_t^{1/2}}[M]\sqrt{2\cdot 3 I_0 OD \phi_i} \qquad DP_n = \frac{k_p}{k_t^{1/2}} \frac{[M]}{\sqrt{2\cdot 3 I_0 OD \phi_i}}$$

where k_p and k_t are the rate constants of propagation and termination of the

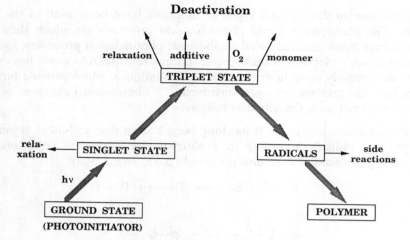

Fig. 2.16. Schematic diagram of the photoinitiation of a radical polymerisation.

polymerisation reaction. The factor ϕ_i takes into account the overall processes leading to the first monomer radicals RM$^\cdot$. It stands to reason that ϕ_i is a function of several parameters: the quantum yield of intersystem crossing (ϕ_{ST}), the rate constants of the different processes, the light intensity and the concentration of the initiator (if bimolecular reactions are involved). The analytical expression for ϕ_i is strongly dependent on the type of photoinitiator used. To gain an insight into this, the following sequence of reactions can be chosen:

$$S_0 \longrightarrow S_1 \longrightarrow T_1 \longrightarrow S_0 \qquad k_0$$
$$ \longrightarrow R^\cdot \xrightarrow{M} RM^\cdot \qquad k_\alpha, k$$
$$ \xrightarrow{M} \text{quenching} \qquad k_q$$

In this case, the quantum yield ϕ_i, is defined by

$$\phi_i = \phi_{ST}\left[\frac{k_\alpha}{k_0 + k_\alpha + k_q[M]}\right]\left[\frac{k[M]}{\Sigma k}\right]$$

where Σk stands for all the processes involved in the disappearance of R$^\cdot$. Depending on the values of the different rate constants, a large variety of cases can be expected: dependence on [M], I_0 and OD. Thus, it is apparent that the rate and the degree of polymerization are governed by three types of factors: the experimental conditions (the characteristics of the light source, the wavelength of irradiation, the number of photons emitted, the monomer and initiator concentrations, the presence of additives and oxygen, etc.), the nature of the initiator, and the nature of the monomer.

The generation of R·. Several types of initiators have been used in the last decade. The photophysical and photochemical processes on which they are founded have been generally well established: photoscission processes, hydrogen abstraction, electron transfer, etc. Extensive research work has been carried out recently using laser spectroscopy techniques, which provide further insight into the primary processes involved.[19,48] The typical behaviour of the usual photoinitiators is described as follows.

1. *Photoscission processes.* It has long been known that aryl-alkyl ketones[12] produce radicals according to a Norrish type I process (α-cleavage); other photoscissions are also observed (β- or γ-cleavage):

$$S_0 \xrightarrow{h\nu} S_1 \longrightarrow T_1 \longrightarrow R^{\cdot} + R'^{\cdot}$$

$$\phi-\underset{\underset{O}{\|}}{C}-\underset{|}{\overset{|}{C}}- \rightsquigarrow \phi-\underset{\underset{O}{\|}}{C^{\cdot}} + {}^{\cdot}\underset{|}{\overset{|}{C}}-$$

2. *Hydrogen abstraction reaction.* A ketyl type radical is generated through photoreduction of the triplet state by hydrogen donors (H donors) such as tetrahydrofuran (THF) or alcohols:

$$S_0 \xrightarrow{h\nu} S_1 \longrightarrow T_1 \xrightarrow[DH]{} K^{\cdot} + D^{\cdot}$$

$$\phi-\underset{\underset{O}{\|}}{C}-\phi + THF \longrightarrow \underset{\phi}{\overset{\phi}{\diagdown}}C^{\cdot}-OH + THF^{\cdot}$$

3. *Electron transfer.* In the presence of amines, the process of proton transfer follows an electron transfer reaction in the triplet state:

$$S_0 \xrightarrow{h\nu} S_1 \longrightarrow T_1 \xrightarrow[AH]{} [CTC] \longrightarrow A^{\cdot}$$

$$\underset{\phi}{\overset{\phi}{\diagdown}}C=O \rightleftharpoons \underset{\phi}{\overset{\phi}{\diagdown}}C^{\cdot}-O^{\ominus} \cdot \overset{\oplus}{N}\underset{CH_2-}{\diagdown}$$

$$\downarrow$$

$$\underset{\phi}{\overset{\phi}{\diagdown}}C^{\cdot}-OH + |N\underset{CH-}{\diagdown}$$

The longer the lifetime of the triplet state, the less efficient is the process of radical generation and the more efficient is the process of physical quenching by the monomer or the oxygen. In addition, this detrimental effect increases with increasing monomer concentration and, therefore, photoinitiators having long triplet lifetimes are not likely, at first view, to favour the initiation of the polymerization in a concentrated solution of monomer or in bulk.

Main classes of radical photoinitiators sensitive in the UV wavelength range
Several recent reviews have been published.[17,19,20,48,49] Most photoinitiators are based on the benzoyl chromophore (Fig. 2.17). Adequate substitution (R_1) at the *para* position and modification of the R_2 moiety has led to a large variety of compounds:

Benzophenone and thioxanthone skeleton represent the other commonly used photoinitiator structure.

Benzilketals. Substituted benzilketals are very efficient molecules,[50] capable of forming active radicals under light-excitation:[51,52]

Dialkoxyacetophenones. The photochemistry of these compounds may be explained[53,54] by a cleavage process and, to a lesser extent, by a γ-hydrogen abstraction:

Hydroxyalkylphenyl ketones. These molecules,[55] as well as the hydroxycyclohexylphenyl ketone, undergo a fast Norrish type I cleavage process:

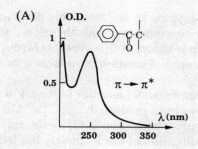
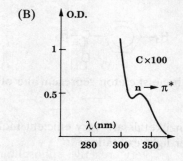
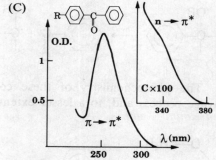
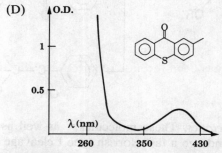

Fig. 2.17. Typical UV-absorption spectra of usual ketones: (A) and (B) aryl-alkyl ketone showing the $\pi \to \pi^*$ and $n \to \pi^*$ transitions; (C) a water-soluble benzophenone and (D) a water-soluble thioxanthone. Note, in the latter case, the high extinction coefficient around 400 nm ($\approx 5000 \text{ mol}^{-1} \text{ dm}^3 \text{ cm}^{-1}$), compared to that obtained at the maximum of the $n \to \pi^*$ transition ($\approx 100 \text{ mol}^{-1} \text{ dm}^3 \text{ cm}^{-1}$).

[Scheme: Ph–C(=O)–C(cyclohexyl)(OH) → Ph–C•(=O) + •C(cyclohexyl)(OH)]

Benzoyl oxime esters. According to the R moiety, α- or γ-cleavage may occur:[18]

[Scheme: Ph–C(=O)–C(CH₃)=N–O–C(=O)–R with hv giving either Ph–C•(=O) + •C(CH₃)=N– (α-cleavage) or Ph–C(=O)–C(CH₃)=N• + •O–C(=O)–R, which further gives Ph–C•(=O) + CH₃CN]

Benzoyl phosphine oxides. A very reactive phosphonyl radical is also formed[56–60] in this case:

[Scheme: mesityl–C(=O)–P(=O)(OR)₂ → mesityl–C•(=O) + •P(=O)(OR)₂]

Morpholino ketones and amino ketones. This new class of photoinitiators works through α-cleavage:[61]

[Scheme: Ph–C(=O)–C(R)(R)–N(morpholine) → Ph–C•(=O) + •C(R)(R)–N(morpholine)]

The newest compound has been recently reported:[62]

[Scheme: morpholine–N–C₆H₄–C(=O)–C(CH₃)(CH₃)–N(with phenyl) → morpholine–N–C₆H₄–C•(=O) + •C(CH₃)(CH₃)–N–]

Aryl-aryl sulphides. Evidence has been provided that the major photochemical process is a cleavage at the sulphur atom:[63]

Sulphoxides. These photoinitiators are built on the benzophenone and the thioxanthone skeleton. The photochemical processes involve the generation of sulphinyl radicals (Fouassier, J. P. & Lougnot, D. J. unpublished):

Sulphonyl ketones. Photoinitiators of this type[64] undergo a fast cleavage process, leading to efficient radicals; gas chromatography–mass spectrometry (GCMS) as well as laser spectroscopy data lead to the conclusion that the cleavage takes place at the β-position. Substitution on the sulphonyl moiety or α-substitution, does not change the reactivity to any significant extent.[65]

Uncleavable photoinitiators. All the above-listed photoinitiators work mostly through a cleavage process and exhibit a high intrinsic reactivity for the initiation of a polymerisation, as indicated by experiments in solution (Table 2.2).[19] However, hydrogen abstraction or electron transfer may be observed, provided that these processes are competitive with the photoscission reaction:

$$^3I \begin{array}{l} \longrightarrow \text{cleavage} \quad k_c \quad \phi_c \\ \xrightarrow{DH} \text{H-abstraction} \quad k_H[DH] \quad \phi_H \\ \xrightarrow{AH} e^- \text{ transfer} \quad k_e[AH] \quad \phi_{CT} \end{array}$$

$$\phi_{CT} = \frac{k_e[AH]}{k_c + k_e[AH] + k_H[DH]}$$

Table 2.2
Rate of polymerisation (in au) of 7 M MMA in toluene in the presence of various photoinitiators

Photoinitiator	R_p (au)
Ph–C(=O)–C(OCH₃)₂–Ph	38
Ph–C(=O)–C(CH₃)₂–OH	40
Ph–C(=O)–CH(OC₂H₅)₂	41
Ph–C(=O)–C₆H₁₀–OH (1-hydroxycyclohexyl phenyl ketone)	35
2,4,6-trimethylbenzoyl diphenylphosphine oxide	29
Ph–C(=O)–C(CH₃)=N–O–C(=O)–OC₂H₅	27
Ph–C(=O)–C(–)(–)–N(morpholino)	37
morpholino–N–C(=O)–C(–)(–)–N(Ph)(–)	19
Ph–C(=O)–C(–)(–)–SO₂–Ph	6
Ph–C(=O)–C₆H₄–S(=O)–CH₃	14

Table 2.3
Role of amine in the polymerisation of 7 M MMA in toluene

Structure	No amine ϕ_C	No amine R_p (au)	Amine ϕ_{CT}	Amine R_p (au)
Ph-C(=O)-C(OCH$_3$)$_2$-Ph	1	38	—	38
Ph-C(=O)-C(−)-SO$_2$-Ph	0.05	6	0.08	9
Ph-C(=O)-C$_6$H$_4$-S-C$_6$H$_4$-CH$_3$	0.1	9	0.4	13

This depends on the relative values of the terms k_c and $k_e[AH]$ if one evaluates, for example, the yield of charge transfer in the presence of amine AH (Table 2.3).

On the contrary, benzophenone, thioxanthone and benzil derivatives,[48] as well as other cyclic ketones such as anthrone,[66,67] dibenzosuberone,[68] xanthone,[69] ketocoumarin[70] work only through electron transfer (Table 2.4): the amine-derived radicals formed (mainly the α-amino radical) are considered as the initiating species; ketyl structures (inactive for the initiation) are simultaneously generated:

$$\text{benzophenone / thioxanthone / benzil} \xrightarrow[AH]{h\nu} A^{\cdot} + \underset{OH}{Y}$$

Except for the thioxanthones (whose lowest-lying triplet state exhibits in polar solvents a $\pi\pi^*$ or a mixed $\pi\pi^*/n\pi^*$ character),[12] an efficient photoreduction by H donors is observed.

Table 2.4
Effect of skeleton on the rate of polymerisation of 7 M MMA in toluene in the presence of 0·05 M MDEA

Structure	R_p (au)
Benzophenone (Ph-CO-Ph)	11
Ph-CO-CO-Ph	10
Chloro-thioxanthone	18
Ethoxy-xanthone (OC$_2$H$_5$)	13
Anthraquinone	10
Fluorenone	10
Dibenzosuberone	7

High rates of polymerisation, high percent conversion and excellent hardness of the coating are attained during the UV curing of a typical industrial formulation (Fig. 2.18). The effect of the photoinitiator is displayed in Fig. 2.19. The structures of the amines play a significant role in the polymerisation reaction (Table 2.5),[71] which is partly accounted for by the efficiency of (i) both the electron and the proton transfer process, and (ii) the reactivity of the amine-derived radical towards the opening of the monomer double bond.

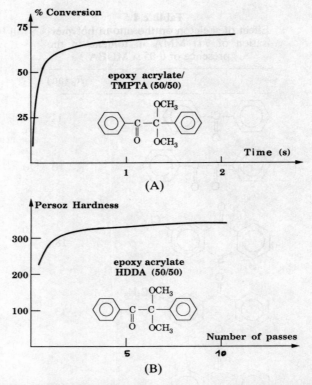

Fig. 2.18. (A) Typical percentage conversion, and (B) hardness obtained for a UV-cured coating. Higher values can be reached according to the type of formulation used.

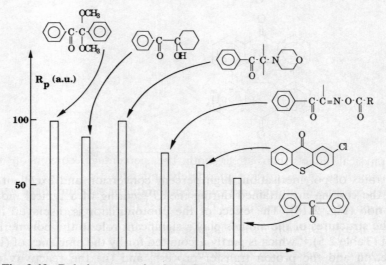

Fig. 2.19. Relative rates of polymerisation of an epoxy acrylate coating.

Table 2.5
Role of amine in the photopolymerisation of MMA (7 M) in toluene

[Thioxanthone structure with (CH)₁₁—CH₃ substituent]

Amine	Relative rate
[Ph–CH₂–NH]₂	20
Ph–CH₂–N(H)(CH₂CHOHCH₂OH)	16
(CH₃)₂N–C₆H₄–CO₂CH₂CH₂OC₄H₉	15
[Ph–CH₂–NH]₂	12
(CH₃)₂N–CH₂CH₂OH	8
Ph–CH₂–N(H)(CH(CH₃)CH₃)	2

During film experiments in aerated media, a lot of chemical, photochemical and physical effects can affect both the percentage conversion and the hardness; these effects are not adequately understood.

Usual substituent effects on a given photoinitiator structure have generally a weak influence on the reactivity (Table 2.6),[71] (Table 2.7).[53]

Photoinitiators exhibiting particular properties

UV deblockable acid-releasing systems. Different structures have been recently proposed: sulphonyl ketones,[64] oxysulphonyl ketones,[72] sulphonic acid deriva-

Table 2.6
Substituent effect in the polymerisation of 7 M MMA in toluene

Amine: (CH₃)₂N–C₆H₄–COO(C₂H₄)OC₄H₉

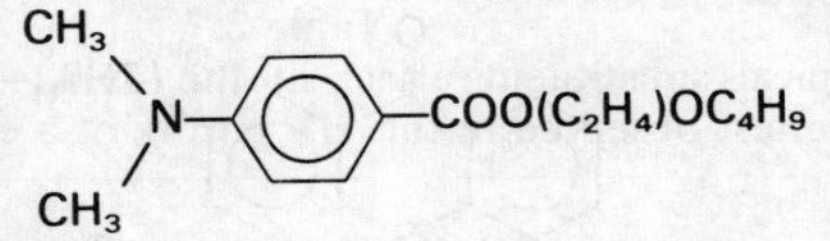

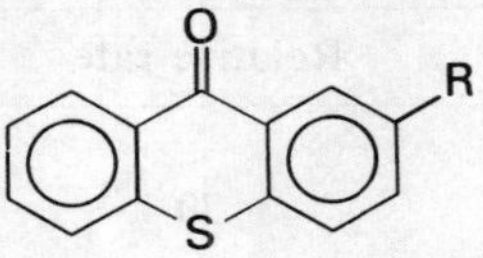

Thioxanthone with substituent R:

	Relative rate of polymerisation
R = —Cl	12
—CH(CH₃)₂	19
—C(CH₃)₃	14
—CH₃	13
—(CH)₁₁—CH₃	15
—C₆H₁₁	13
—C₆H₅	14

tives of α-hydroxymethyl benzoin:[73]

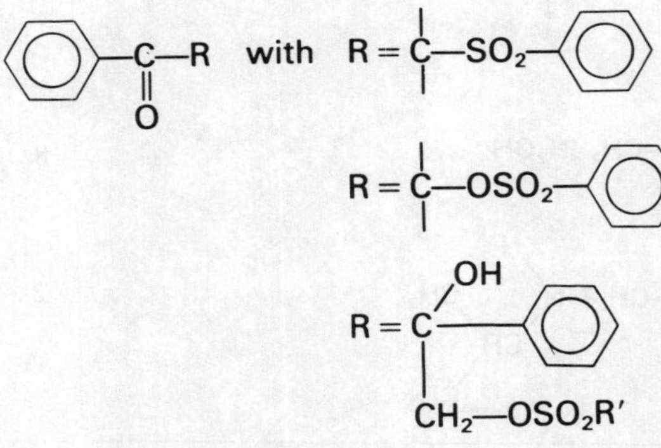

C₆H₅—C(=O)—R with R = C(–)(–)—SO₂—C₆H₅

R = C(–)(–)—OSO₂—C₆H₅

R = C(OH)(C₆H₅)—CH₂—OSO₂R′

In addition to the generation of radicals, they are able to form acid species, under light excitation, that lead to the cross-linking of aminoplast resins through a subsequent thermal reaction or to the polymerisation of hybrid systems (containing, for example, melamine derivatives and acrylic monomers). Sulphinic acids (Fouassier, J. P., unpublished) or sulphonic acids[74] are generated directly in the radical pair formed from short-lived triplet states or through hydrogen abstraction reactions with the free radicals; in oxysulphonyl ketones, an additional process involving photoreduction by H donors and acid

Table 2.7
Typical substituent effect in the dialkoxyacetophenone series, observed in the UV curing of a clear coating[53]

Substituent (R in Ph-CO-CR)	Relative rate of cure
bis(cyclohexyloxy)	126
bis(OC$_2$H$_5$)	85
bis(2-methoxyethoxy) (OCH$_2$CH$_2$OCH$_3$)$_2$	87
bis(isopropoxy)	95
bis(tert-butoxy)	76
OC$_2$H$_5$ / O-iPr (mixed)	126

elimination has been observed:[72]

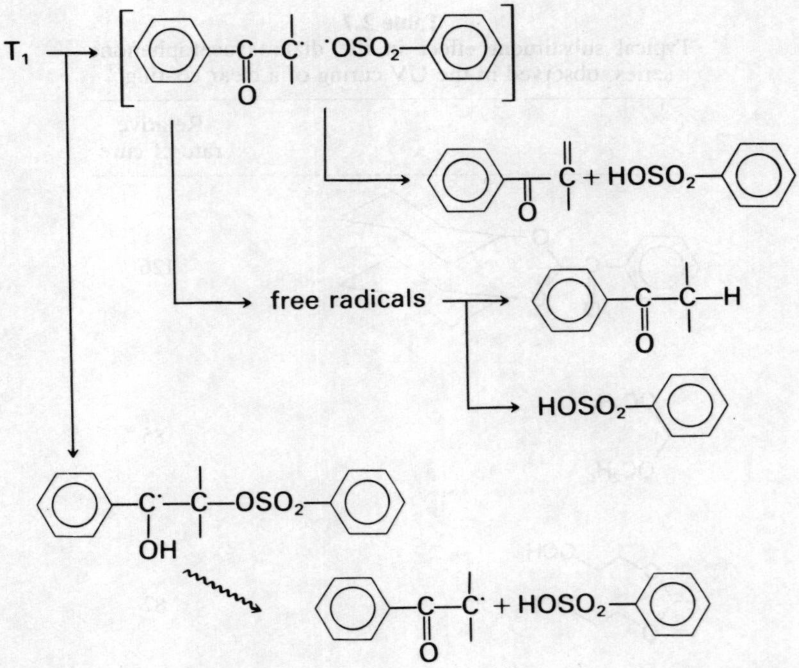

[chemical scheme showing T_1 pathway to intermediates with phenyl-C(O)-C-OSO$_2$-phenyl, producing free radicals, phenyl-C(O)-C=C + HOSO$_2$-phenyl, phenyl-C(O)-C-H, HOSO$_2$-phenyl, and phenyl-C(OH)-C-OSO$_2$-phenyl leading to phenyl-C(O)-C• + HOSO$_2$-phenyl]

Polymeric or copolymerisable photoinitiators. In order to overcome the usual problems related to the unpleasant odour caused by the volatile photolysis products (in a usual photopolymerisable film), and to decrease the extractability of a photoinitiator (e.g. in photocrosslinkable emulsions or water-borne coatings), relatively high-molecular-weight or unsaturated compounds have been described in the thioxanthone series[75] and in the hydroxylalkyl series by incorporating acrylic double bonds[76] or building-up an oligomeric structure.[77] The same has been achieved in the series of amino ketones and morpholino ketones,[61,62] benzilketals.[78,79] The behaviour of the excited states is very similar to that of the parent compounds.[80] Some examples include

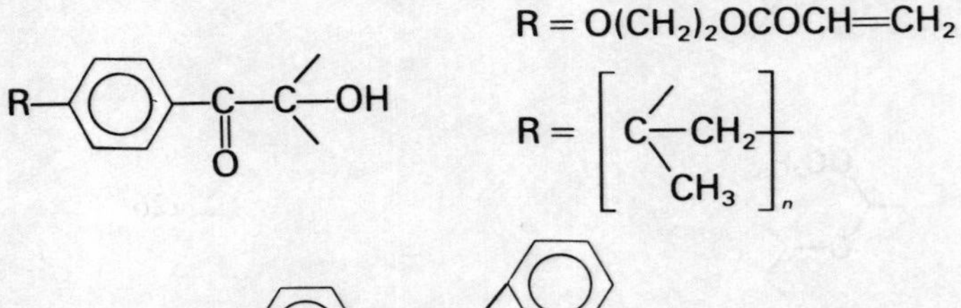

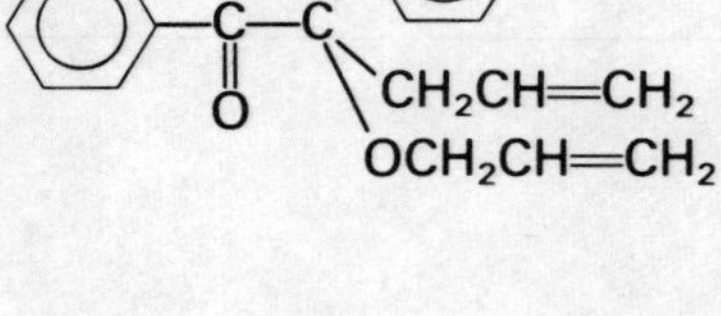

R—phenyl—C(O)—C(CH$_3$)—OH

$R = O(CH_2)_2OCOCH=CH_2$

$R = \left[\begin{matrix} C-CH_2 \\ | \\ CH_3 \end{matrix} \right]_n$

phenyl—C(O)—C(phenyl)(CH$_2$CH=CH$_2$)(OCH$_2$CH=CH$_2$)

Photoinitiators carrying long alkyl chains. Introduction of C_{10} to C_{13} alkyl chains (LC) onto the usual backbones of the photoinitiators does not change significantly the photochemical reactivity during the light exposure, but increases the compatibility in the film, for example, in

Water-soluble photoinitiators. Efficient ionic or hydrophilic compounds are suitable photoinitiators for acrylamide polymerisation,[81–83] micelle polymerisation,[84] grafting onto cellulosic materials in water,[85] water-borne coatings, aqueous dispersions, emulsions.[86] The proposed systems are mainly designed by the introduction of appropriate substituents on usual structures: thioxanthones, benzophenones, benzils,[87] hydroxyalkyl ketones[86] or newly explored skeletons such as phenacyl thiosulphate derivatives:[88]

High rates of acrylamide polymerisation in water are observed.[81–83,89] Water-borne coatings can also be efficiently cured in the presence of water-soluble photoinitiators,[86,88] although the use of hydrophobic or hydrophilic[86] or oligomeric[77] compounds seems to provide a better practical efficiency (Fig. 2.20).

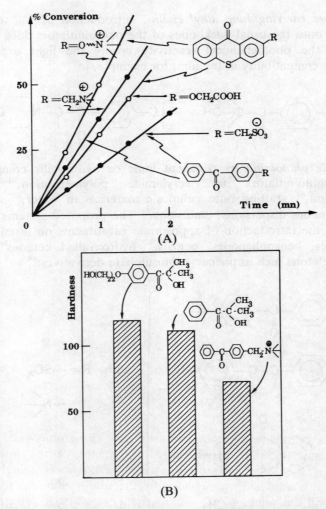

Fig. 2.20. Reactivity of water- and oil-soluble photoinitiators: polymerisation of (A) acrylamide in water,[81,82] and (B) a water-borne coating[86]: hardness of the coating as a function of the photoinitiator.

Effect of the introduction of a thioether group. The photopolymerisation of white pigmented systems has been a challenge for a long time and led to the search for suitable combinations of photosensitisers and photoinitiators. Due to the spectral window offered by titanium dioxide, the former belong generally to the thioxanthones; the basic idea to solve this problem was to lower the triplet-state energy level, E_T, of the photoinitiator, in order to enhance the energy transfer from the thioxanthone: this has been achieved for the first time by introducing a thioether group on a morpholino ketone.[61]

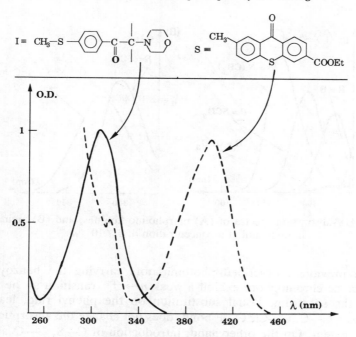

Fig. 2.21. Ground-state absorption spectra in methanol.

The presence of this group completely changes the ground-state absorption spectra, because of the charge transfer character of the lowest energy transition (Fig. 2.21) (the spectroscopic nature of the lowest-lying triplet state is probably also modified, as revealed by the totally different triplet–triplet absorption and the values of the rate constants of cleavage, electron transfer and monomer quenching) and considerably enhances the molar extinction coefficient (together with a substantial red shift of the absorption).

Photoinitiators for pigmented media. Curing of pigmented media (typically in the presence of titanium dioxide or coloured pigments) is mostly restricted to thin films and remains difficult, due to the high light absorption by the pigment, which is detrimental to the own absorption of light by the photoinitiator. Much research work has been focused on this problem and based on two ideas: either the design of red-shifted absorbing molecules acting as photoinitiators or the finding of photosensitisers able to absorb at long wavelengths and to transfer their excitation to a photoinitiator (see the next paragraph). The former idea relies on the synthesis of new or newly modified chemical structures.

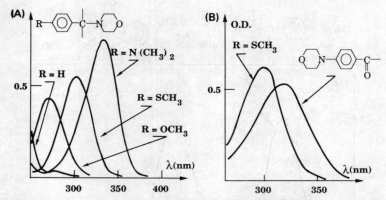

Fig. 2.22. UV-absorption spectra of (A) morpholino ketones, and (B) amino ketones in methanol at a concentration of 3×10^{-5} M.[90]

All the previously described photoinitiators carrying the benzoyl or the benzophenone chromophore exhibit a weak $n \rightarrow \pi^*$ transition in the near-UV range of the spectrum. Usual substitution on the phenyl ring, leading for example to a C—C, C—Cl, C—F bond, does not change the absorption to any significant extent. On the other hand, introduction of C—S, C—O and C—N bonds affects considerably the ground-state absorption spectrum, as shown in Fig. 2.22 for the morpholino ketone:[90]

R = H
R = OCH₃
R = N(CH₃)₂
R = SCH₃

and the amino ketone series:[62]

The same holds true when going from benzophenone to amino benzophenones or to phenyl thiobenzophenones (such as some above-mentioned aryl-aryl sulphides). Red-shifted absorption is also achieved in the thioxanthones, or by exchanging (in benzophenones or phenylalkyl ketones) the phenyl for a biphenyl (which is accompanied by the presence of a $\pi\pi^*$ ground-state). As regards the photochemical reactivity, the major problem may arise from a decrease in efficiency because of a change in the spectroscopical nature of the transient states, so that the benefit resulting from a better absorption is lost by a lack of reactivity. Table 2.8 summarises the data obtained in solution[31] and

Table 2.8
Structural effect on the excitation transfer process in the polymerisation of a white pigmented film and of 7 M MMA in a toluene solution

Structure	Relative efficiency (presence/absence of thio-xanthones)
Ph-C(=O)-C(CH₃)₂-N(morpholine)	9[a]
CH₃S-C₆H₄-C(=O)-C(CH₃)₂-N(morpholine)	8[a]
CH₃S-C₆H₄-C(=O)-C(CH₃)₂-OH	>5[a]
Ph-C(=O)-C(CH₃)₂-OH	1·6[a]
Ph-C(=O)-C₆H₄-S-Ph	3[b]
CH₃S-C₆H₄-C(=O)-C(OCH₃)₂-C₆H₄-SCH₃	1·2[b]
OH(CH₂)₂S-C₆H₄-C(=O)-C(CH₃)₂-OH	1·25[b]

[a] White pigmented film.⁹⁰
[b] 7 M MMA.³¹

film[90] polymerisation as well, and supports this remark: the intrinsic reactivity of the thiophenylmorpholino phenylamino ketone compared to that of benzilketal shows a two-fold decrease (solution experiments). However, in the curing of a pigmented film (which is conducted at a fixed concentration), this effect is largely balanced by the higher absorption of the film.[91]

Miscellaneous systems. Different systems are continuously being proposed as laboratory products, but nothing is really known about their ability to work in industrial formulations:

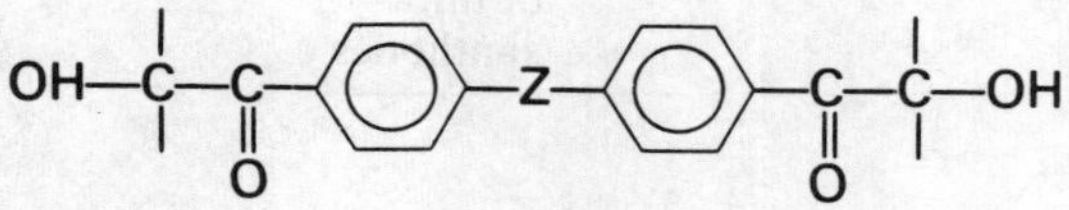

(Ref. 92)

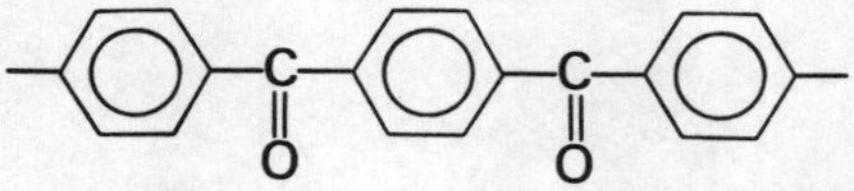

(Ref. 93)

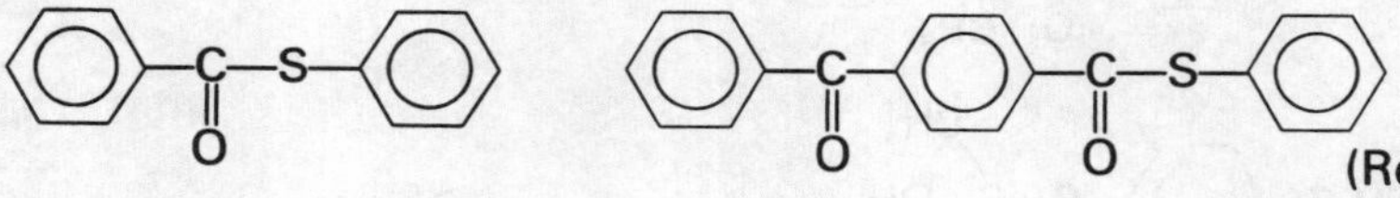

(Refs 94 and 95)

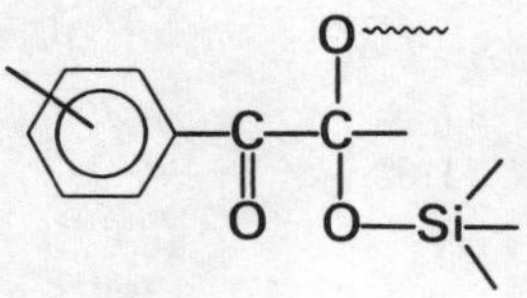

(Refs 96 and 97)

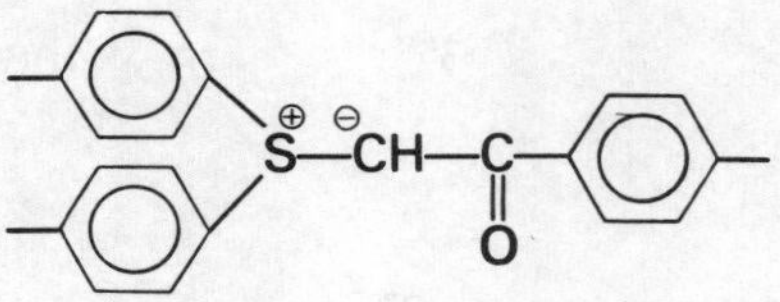

(Ref. 98)

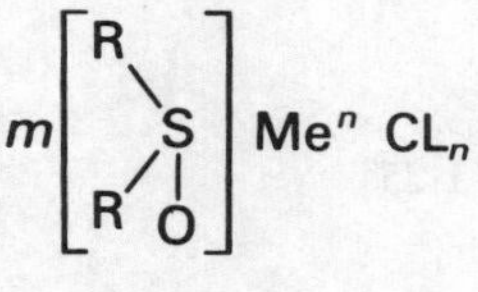

(Ref. 99)

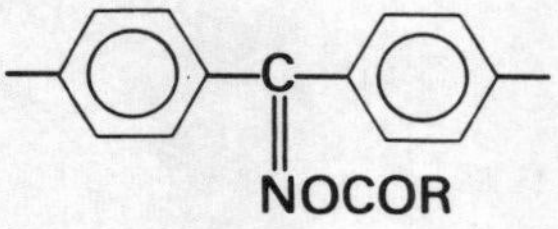

(Ref. 100)

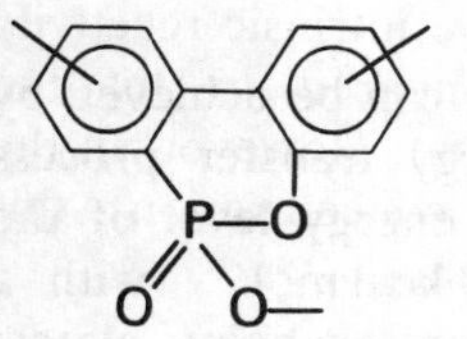

Introduction to the basic principles of UV curing 95

(Ref. 101)

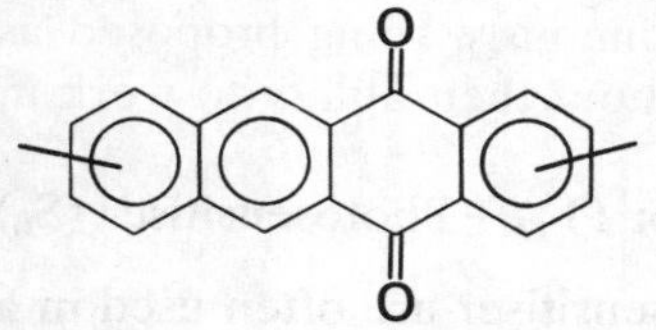

(Ref. 102)

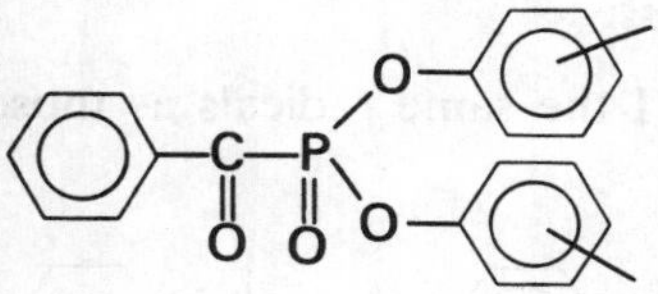

(Ref. 103)

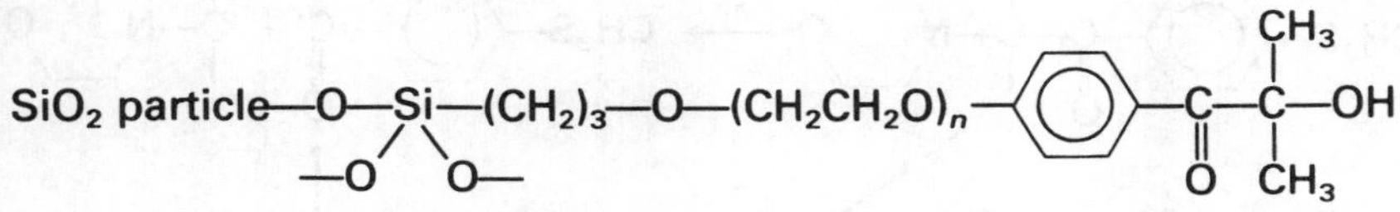

SiO_2 particle—O—Si—$(CH_2)_3$—O—$(CH_2CH_2O)_n$—⟨⟩—$\underset{O}{\underset{\|}{C}}$—$\underset{CH_3}{\underset{|}{\overset{CH_3}{\overset{|}{C}}}}$—OH

(Ref. 104)

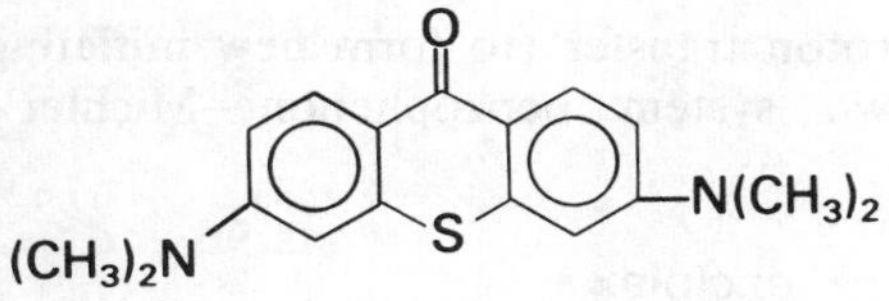

(Ref. 105)

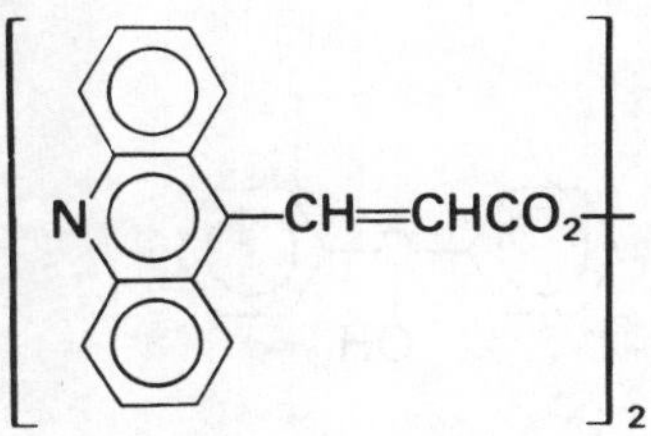

(Rel. 106)

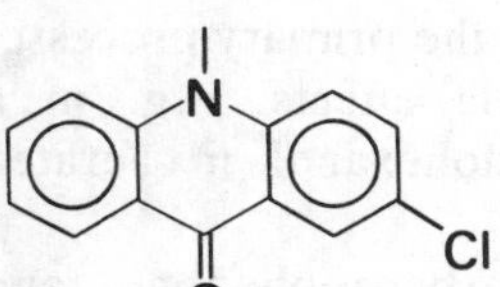

(Ref. 107)

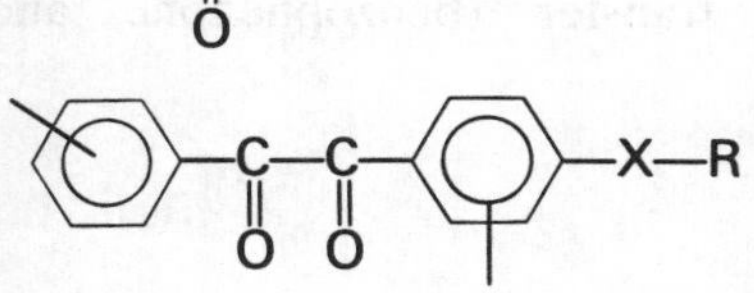

(Ref. 108)

X = S, SO, SO_2

Synergism

Extension of the spectral sensitivity of a photoinitiator, I, can be achieved by adding a photosensitiser, S. Strictly speaking, an energy transfer process should be exothermic. Such a transfer implies that the energy level of the excited donor exceeds that of the acceptor by a few kcal mol^{-1} (with a difference of 3 kcal mol^{-1} (12 kJ mol^{-1}), the energy transfer being almost diffusion controlled):

Photosensitiser (T_1) + Photoinitiator (S_0) $\longrightarrow$

Photoinitiator (T_1) + Photosensitiser (S_0)

However, the two terms photoinitiator and photosensitiser are often used in a more general sense, to denote any involved process.

1. Energy transfer (required to generate from I the same radicals as those obtained through direct excitation):[61]

2. Electron transfer followed by a proton transfer (to form new initiating radicals), e.g. in the well-known system benzophenone–Michler's ketone:[109]

3. Chemical reactions with the radicals formed during the primary processes are also found to be responsible for synergistic effects, e.g. in a combination of benzophenone and 1-benzoylcyclohexanol in aerated medium.[110]

4. Increased reactivity through energy transfer (benzophenone and trimethylbenzophenone:[111,112]

[Reaction schemes showing triplet and singlet excited state benzophenone derivatives]

The search for systems capable of working according to the process depicted in reaction 1 has been a challenge for many years. Lowering the energy level of the lowest triplet state has been clearly achieved by introducing a thioether substituent at the *para* position of the benzoyl chromophore which makes feasible the energy transfer from a thioxanthone derivative:[61]

$$S \xrightarrow{h\nu} S^* \longrightarrow CH_3S-\text{C}_6H_4-\text{C(O)}-R \longrightarrow CH_3S-\text{C}_6H_4-\overset{\cdot}{C}=O + {}^{\cdot}R$$

R = C(OCH$_3$)$_2$-C$_6$H$_5$; C(-N(morpholino))- ; C(OH)(CH$_3$)$_2$

Various experiments (laser spectroscopy,[113] GCMS,[114] CIDNP,[61] film photopolymerisation[90] support this idea but suggest that the mechanism of excitation transfer is rather complex.[115] Thus, it was shown most recently that combination of substituted thioxanthones with the methylthioethermorpholino ketone extends the photosensitivity towards the near-visible part of the spectrum (Fig. 2.21), and accelerates the curing of coatings (Fig. 2.23).

[Thioxanthone structure with substituents R$_1$, R$_2$, R$_3$, R$_4$]

$\begin{cases} R_1 = R_3 = R_4 = H \\ R_2 = CH(CH_3)_2 \end{cases}$

$\begin{cases} R_1 = R_2 = R_3 = H \\ R_4 = COO(CH_2CH_2O)_8H \end{cases}$

$\begin{cases} R_1 = CH_3 \\ R_2 = R_4 = H \\ R_3 = COOEt \end{cases}$

CH₃S—⌬—C(=O)—C(CH₃)₂—N(morpholino)

Evidence for the α-cleavage of the photoinitiator either in the absence or in the presence of thioxanthone derivatives, can be found in nuclear magnetic resonance–CIDNP spectra[61,90,115] and through GCMS,[114] which supports at least, the view of an energy transfer process. Excitation with a light filtered at $\lambda \geq 400$ nm leads to the formation of methylthiobenzaldehyde, which proceeds, in non-hydrogen-donating solvents, from a cage process in the radical pair generated after cleavage. However, complementary results[31] revealed that the excitation transfer in the presence of 2-methyl-3'-ethylester thioxanthone occurs mostly through electron transfer in toluene. The relative efficiencies of these two competitive processes (Fig. 2.24) are dependent on the stabilisation of the lowest excited state, T_1, by solvation (that governs the spectroscopic character of T_1), and on the relative positions of the triplet energy levels of the donor (the photosensitiser) and the acceptor (the photoinitiator), which determines the possibility of an energy transfer process.[115]

For the time being, the resuls obtained (Table 2.9) in a large variety of systems suggest the following trends: the presence of a thioether group at the *para* position of a benzoyl chromophore and the introduction of a morpholino moiety at the α-carbon of the carbonyl group are beneficial, as they lower E_T (and keep the reactivity at a high level) and favour the sensitization process (although a simultaneous electron transfer—which is dependent on the experimental conditions, as shown in model systems—is confirmed[115]). The polarity of the medium could play a decisive role. It may happen that the

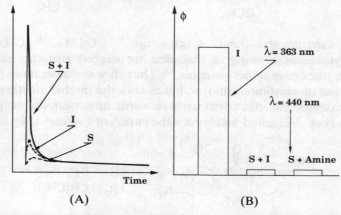

Fig. 2.23. (A) Photopolymerisation heat upon irradiation of an epoxy acrylate coating.[90] (B) Quantum yield of photopolymerisation (in au) of a PETA coating under laser lights.[114] I is 2-methyl-1-(4-methylthiophenyl)-2-morpholino-propan-1-one. S is 1-alkylester-thioxanthone in (A) and 3-methylester-2'-methyl-thioxanthone in (B).

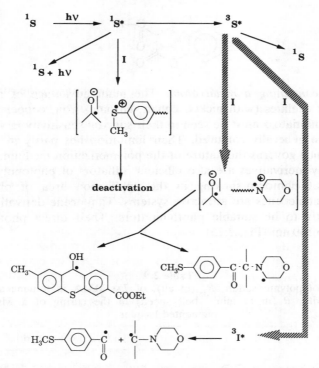

Fig. 2.24. Excited-state interaction between the photosensitiser and the photoinitiator.

reactivity of the substituted photoinitiator does not parallel that of the parent compound and decreases considerably; in that case, even with an excellent energy transfer, the practical efficiency in film experiments[90] remains low (Table 2.8).

Visible photoinitiators
It has been mentioned that the absorption of photoinitiators usually takes place in the UV part of the spectrum. Technological requirements gave rise to research endeavours directed towards the use of visible light (or near-IR) sensitive systems and the best matching between the photosensitiser (or the photoinitiator) absorption and the laser emission wavelength. Photopolymerization under visible light can be induced through different processes, involving either excitation transfer as described above, or direct production of radicals from coloured molecules. In the latter case, several systems are emerging.[116] Some examples include the following.

Ketocoumarins. Ketocoumarin structures[117] are very attractive, because of their tendency to exhibit a red-shifted ground-state absorption as a function of the R moiety and the substitution on the phenyl rings (Fig. 2.25). The electron transfer is particularly efficient.[118]

Compounds containing a metal atom. The main advantage of metal salts, complexes or chelates (with nickel, cobalt, chrome, iron, copper ions) when used as photoinitiators must be seen in their spectral sensitivity in visible light, since they are generally coloured. Their limitation lies partly in their water solubility, which governs the nature of the polymerization medium. Transition metal carbonyl derivatives are also efficient initiators of photopolymerization and agents of photocrosslinking in the application area of photographic processes. Metallocenes are attractive systems. Titanocene derivatives[119] were shown recently to be suitable photoinitiators. Their direct photosensitivity extends up to 550 nm (Fig. 2.25).

Table 2.9
Rate of polymerisation, R_p (in au), of 7 M MMA in toluene and reactivity, θ (in m min^{-1} belt speed) in the curing of a white pigmented lacquer

	R_p	θ
R = H	37	3·3[a]
OCH$_3$		2·5[a]
N(CH$_3$)$_2$		2·5[a]
SCH$_3$	26	5[a] 10[b]
	19	>170[b]
	38	— —

[a] From Ref. 90.
[b] 15 μm film.

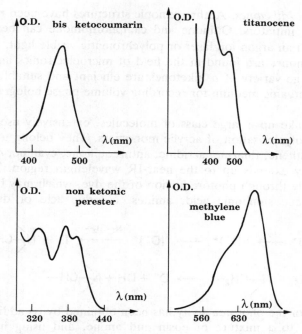

Fig. 2.25. Typical visible absorption of various coloured molecules.

Other photosensitive systems include transition-metal complexes.[120–123] (See Vol. II (Desobry)).

Non-ketonic perester structures. Such structures[124,125] are effective in vinyl polymerisation, yielding alkoxy and aryloxy free-radical pairs through homolytic decomposition.[126,127] According to the nature of the chromophore moiety (aryl groups—dyes), a strong absorption can be achieved over a wide spectral range (Fig. 2.25).

Quinones and α-diketones. Anthraquinonic structures have been recognized as polymerisation initiators. Quinone and camphorquinone can be suitable for irradiation with an argon ion laser or polychromatic visible light. Applications of naphthoquinones are found in the field of microelectronics and in curable sealants. A large variety of α-diketones are efficient and suitable photosensitive guests in organic medium for recording volume phase holograms.

Dyes. Dyes make up a large class of molecules, extensively used to start or sensitise the polymerisation of acrylic monomers. They belong to well-known structures: xanthene, thiazine, acridine, anthraquinone, cyanine, merocyanine. Photosensitivity extends up to the near-IR wavelength region. The reaction occurs primarily through photoreduction of the dye, which may be performed by using toluene sulphonic acid, amines, ascorbic acid or diketones, for example:

$$^1D^+ \xrightarrow{h\nu} {}^1(D^+)^* \longrightarrow {}^3(D^+)^* \xrightarrow{|N-CH_2-|} D^{\cdot} + {}^{\cdot}N^{\oplus}-CH_2-$$

$$D^{\cdot} + D^{\cdot} + {}^{\cdot}N^{\oplus}-CH_2- \longrightarrow D^+ + DH + \bar{N}-\dot{C}H-$$

Improvement of the photoreactivity has been obtained by (i) adding a benzoyl oxime ester[128] to a mixture of eosin and amine, and using in the photopolymerisation of glass-reinforced polyester resins in visible light or of acrylic monomers under laser beams,[129] (ii) mixing dye, amine and iodonium salt[130,131] (Table 2.10), and (iii) incorporating peroxides.[132]

Table 2.10
Reactivity of different photosensitive systems under an Ar$^+$ laser beam exposure[a] Ref. 130

	Additive	Exposure time (s)
Eosin	—	—
	PDO[b]	10
	Amine	3
	ϕ_2I^+	2
	PDO/MDEA[c]	2
	ϕ_2I^+/MDEA	0.4
Ketocoumarin	—	—
	Amine	0.5
	PDO/amine	0.5
	ϕ_2I^+	0.4
	ϕ_2I^+/Amine	0.1

[a]PETA film 30 μm; λ = 488 nm; 100 mW; ϕ = 0.6 cm; OD = 0.1.
[b]PDO—
[c]MDEA—methyldiethanol amine.

Cationic photoinitiators

A promising development of cationic photopolymerization initiated by onium salts or organo-metallic compounds seems to be taking place (excellent reviews have been recently published in this field[133–135]) because of specific advantages (e.g. excellent properties of the cured film; absence of oxygen inhibition; possible concurrent radical and cationic reaction in hybrid systems).

Diazonium salts. Epoxide ring opening with Lewis acids occurs according to the following mechanism:[136]

$$ArN_2^{\oplus}BF_4^{\ominus} \xrightarrow{h\nu} ArF + BF_3 + N_2$$

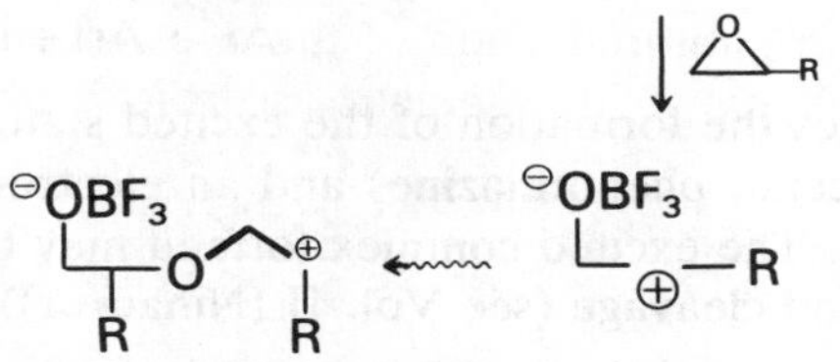

Onium salts. Among them, the best-known systems are based on the following structure: diaryliodonium, triarylsulphonium:

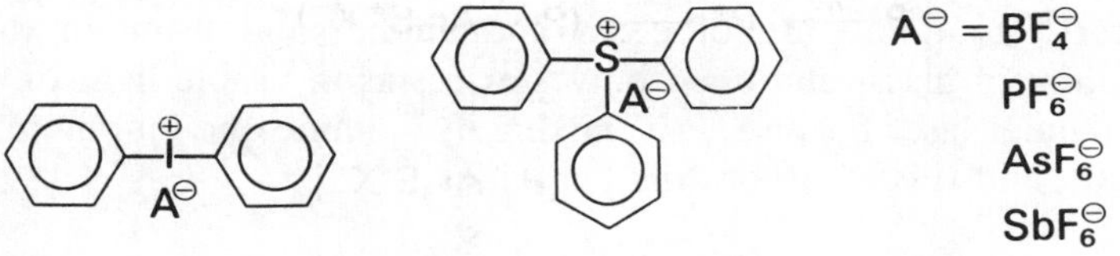

$$A^{\ominus} = BF_4^{\ominus}$$
$$PF_6^{\ominus}$$
$$AsF_6^{\ominus}$$
$$SbF_6^{\ominus}$$

A lot of other compounds have been synthesized:[133] triarylpyrylium, benzylpyridinium thiocyanate, dialkylphenacylsulphonium, dialkylhydroxphenylsulphonium, oxonium, phosphonium, nitronium, arsenium, and selenium derivatives. Manufacture of photoreactive polymeric iodonium salts for use in photocurable composition has been recently reported.[134] The synthesis and the characterisation of a series of (alkoxy phenyl) phenyl iodonium salts have been described.[135] Onium salts yield Brönsted acids under UV-light exposure:[133,137–142]

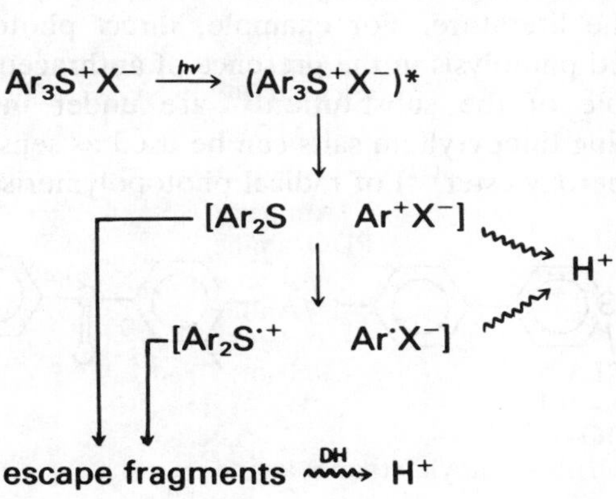

$$Ar_3S^+X^- \xrightarrow{h\nu} (Ar_3S^+X^-)^*$$

As they poorly absorb in the near-UV range, attempts have been made to develop compounds exhibiting structural modification (which leads to a higher wavelength absorption), and to add photosensitisers. According to the type of compounds, two sensitisation mechanisms are involved. The first one consists in the generation of radicals (from ketone compounds, e.g. benzophenone) which subsequently undergo electron transfer to the onium salt:

$$\text{\textbackslash C=O} \xrightarrow{h\nu} [\text{\textbackslash C=O}]^* \longrightarrow K^\cdot + D^\cdot \longrightarrow Ar_2I^\cdot + K^+(+D)X^-$$

$$\downarrow Ar_2I^+X^-$$

$$Ar^\cdot + ArI + HX(+D)$$

The second one requires the formation of the excited state of the sensitiser, P (e.g. perylene, anthracene, phenothiazine) and an electron transfer from this state to the onium salt. The excited complex formed may be modified through electron transfer or bond cleavage (see Vol. II (Nihacker)):

$$P + (Ar_3S^+X^-)^*$$
$$\uparrow$$
$$P \xrightarrow{h\nu} P^* \longrightarrow (P \cdots Ar_3S^+X^-)^*$$
$$\downarrow$$
$$[P^{\cdot+}Ar_3S^\cdot X^-]$$
$$\rightsquigarrow$$
$$H^+$$

Typical percentage conversion–time curves for the polymerisation of a cationic monomer[141] are shown in Fig. 2.26 and emphasises the role played by the nucleophilicity of the anion. Steady-state photolysis has led to a clearer understanding of the photochemistry of iodonium and sulphonium salts.[133,137,142–145] Information concerned with the processes involved and the kinetic measurements through time-resolved laser spectroscopy, is beginning to be published in the literature. For example, direct photolysis of iodonium salts,[146,147] sensitised photolysis in the presence of anthracene[148] and ketones[149] as well as the role of the substitution[150] are under investigation. Long-wavelength-absorbing thiopyrylium salts can be used as sensitisers (e.g. for the decomposition of peroxy ester[151]) of radical photopolymerisation reactions:

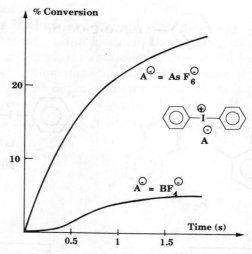

Fig. 2.26. Film photopolymerization of a commercial cationic epoxy monomer.

Organo-metallic compounds. Iron arene salts having anions with low nucleophilicity exemplify this new class of promising structures for the epoxy polymerisation.[145,152,153] Ring opening occurs probably through the loss of the arene ligand and the fixation of epoxide functions:

Laser-induced polymerization

Photopolymerization under laser beams proceeds typically just as under polychromatic light sources.[116] The real problem is concerned with the fundamental necessity of finding efficient photosensitive molecules at the laser emission wavelength. Photoinitiation has been achieved by using a large variety of lasers: nitrogen and excimer laser; ruby and YAG/Nd lasers; dye lasers; HeNe, krypton and argon ion lasers, metal vapour lasers.[154] Photosensitive systems suitable for laser-induced polymerization are classified as follows.

1. *Usual photoinitiators.* For example, those described above; working under blue[154-156] or visible[129] laser lights.
2. *Charge transfer complexes.*[157]
3. *Combination of several components.* For example, ketocoumarin/photoinitiator/iodonium salt.[130] The results shown in Fig. 2.27 and Table 2.10 demonstrate that a similar photosensitivity is obtained under blue and visible light.

106   J. P. Fouassier

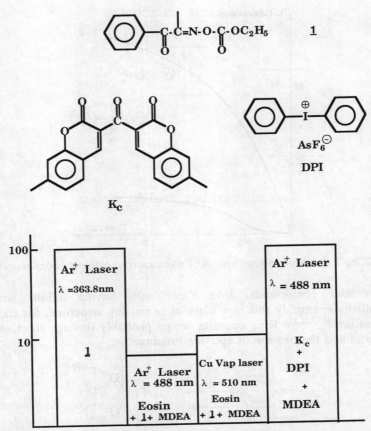

Fig. 2.27. Efficiency of various photoinitiating combinations under laser lights.[130] MDEA is methyldiethanol amine.

4. *Two photon-sensitive systems*. In specific applications such as holography, it may be beneficial to find systems working under two simultaneous irradiations of different wavelengths, so that no reaction occurs if only one wavelength is used. Two systems meeting this requirement have been proposed. One of them is called 'two photon–four level'.[158] For example, biacetyl (in a polymer matrix containing monomer) is excited by a non-coherent UV light source ($S_0 \to S_1$ transition). After intersystem crossing, the $T_1 \to T_n$ transition is induced by the red laser light beam to produce the interference pattern; the initiation of the polymerization occurs, presumably, from the upper excited triplet state. Another alternative consists in a 'two photon–two product' system, in which a ketone (1,2-dibenzoyl benzene) is formed *in situ* by specific oxidation of a starting compound (DPIBF: 2,5-diphenyl-3,4-isobenzofuran) with singlet oxygen, generated through an energy transfer between a dye (e.g. methylene blue[159] or cyanine[160]) and ground-state oxygen (Fig. 2.28).

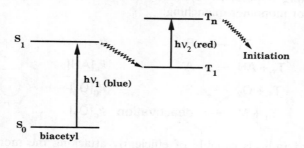

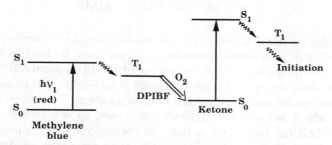

Fig. 2.28. Schematic picture of processes occurring in two photon-sensitive systems.

These last two points 3 and 4, together with the emergence of new coloured one photonsensitive systems (see Vol. III Chap. 3), will be probably developed in the future because of the expected use of laser-induced reactions in the field of microelectronics, reprography, laser imaging holography, surface treatments, three-dimensional machining, etc. Moreover, holography can also be used as a holographic spectroscopy method to investigate photochemical processes in film matrixes[161] or to monitor the extent of a polymerisation reaction.

Structure–reactivity relationships
A photosensitive molecule should obviously display quite a number of properties.

1. Its behaviour should enable a correct and useful industrial handling, as regards the problems of volatility, extractability, compatibility, odour, toxicity, storage stability, further stability of photolysis products, etc.
2. Spectroscopic properties: extinction coefficients, position of the wavelength of maximum absorption, energy level of the lowest-lying triplet state, etc.
3. An excellent reactivity of the excited states, i.e.

- efficient primary processes leading to the generation of the initiating species, making it possible to avoid an important deactivation through oxygen and/or monomer quenching:

$$T_1 \longrightarrow R^{\cdot} \qquad k_c$$
$$T_1 + AH \longrightarrow A^{\cdot} \qquad k_e[AH]$$
$$T_1 + O_2 \longrightarrow S_0 \qquad k'_q[O_2]$$
$$T_1 + M \longrightarrow \text{deactivation} \qquad k_q[O_2]$$

- production of radicals capable of efficiently attacking the monomer and thereby of starting the polymerization:

$$R^{\cdot} + M \longrightarrow RM^{\cdot} \qquad k_i[M]$$

The search for correlations between the reactivity in the excited states and the practical efficiency in film experiments represents a fascinating field of research. Unfortunately, no such endeavours have been carried through, as no quantitative study of excited-state processes is available. Up to now, such an investigation can be easily conducted only in solution,[19,48] except for a recent attempt in which holographic methods were used to win information in film matrixes.[161] Still the fact remains that the results obtained in model systems, tested in solution, represent a reasonably good approach to describe these relationships and to discuss the ability of a given photoinitiator to initiate a polymerisation in terms of the primary processes involved in the excited states. General reasonings have been recently presented in different reviews[162,163] and will be included in another chapter of this book (see Vol. II (Fouassier)). For example, Tables 2.11 and 2.12 list the yield of cleavage, ϕ_c, and electron transfer, ϕ_{CT}, calculated from the rate constants of the corresponding processes k_c, k_e, k_q. The bimolecular rate constant k_i is also shown.

$$\phi_c = \frac{k_c}{k_c + k_q[M]}$$

$$\phi_{CT} = \frac{k_e[AH]}{k_e[AH] + k_q[M] + k_0}$$

The high values of ϕ_c, ϕ_{CT} and k_i (Table 2.13) obtained for a photoinitiator are generally in line with a good practical efficiency. However, some other phenomena may affect this trend either positively or negatively, such as (i) an untimely termination of the growing polymer chain by radicals created in the early stages of the reaction (e.g. ketyl radicals[164]), (ii) by side reactions of amine radicals leading to the formation of adducts[165] in the presence of benzophenone. The scavenging effect of amines towards oxygen[166,167] and the

Table 2.11
Yield of cleavage in bulk MMA from Ref. 163

Structure	ϕ_c
Ph-C(=O)-CH(OR)-Ph (OR)	1
R-C$_6$H$_4$-C(=O)-C(CH$_3$)$_2$-OH	0·2
Ph-C(=O)-C$_6$H$_4$-S-C$_6$H$_4$-CH$_3$	0·1
CH$_3$S-C$_6$H$_4$-C(=O)-C(CH$_3$)$_2$-N(morpholino)	0·8
Ph-C(=O)-C(CH$_3$)$_2$-SO$_2$-Ph	0·05
Ph-C(=O)-CH(OC$_2$H$_5$)$_2$	0·5–1
Ph-C(=O)-C(CH$_3$)=N-O-C(=O)-OC$_2$H$_5$	≤1
(2,6-(CH$_3$)$_2$-C$_6$H$_3$)-C(=O)-P(=O)(Ph)$_2$ with additional CH$_3$	1^a

^{a}Quantum yield: 0·5.^{48}

Table 2.12
Yield of electron transfer in the presence of 0·05 M methyl diethanolamine in bulk MMA from Ref. 163.

Structure	Yield
Ph–C(=O)–Ph	0·3
thioxanthone with R substituent	0·9
Ph–C(=O)–C$_6$H$_4$–S–C$_6$H$_4$–CH$_3$	0·4
Ph–C(=O)–C(CH$_3$)$_2$–SO$_2$–Ph	0·06
Ph–C(=O)–C(=O)–Ph	1
chromanone-benzoyl compound	0·8

multiphonic cleavage of benzil[168] may also be mentioned in this respect:

$$\text{Ph–C(=O)–C(=O)–Ph} \xrightarrow{h\nu_1} T_1 \xrightarrow{h\nu_2} T_2 \longrightarrow 2\ \text{Ph–C(=O)}^\bullet$$

It is noteworthy that some typical behaviours in film experiments remain unexplained, e.g. the high reactivity of

Ph–C(=O)–C(–)(–N-morpholine)

in the presence of thioxanthones[90] (although no detectable effect was observed

Table 2.13
Rate constants of interaction of radicals with monomers from Ref. 48

Initiators	Initiating radicals	k_i (mol^{-1} l s^{-1})
Ph–C(=O)–C(OCH$_3$)$_2$–Ph	Ph–C·(=O)	10^5
Ph–C(=O)–C(CH$_3$)$_2$–OH	Ph–C·(=O)	0.9×10^5
	·C(CH$_3$)$_2$–OH	5.4×10^5
Mesityl–C(=O)–P(=O)(Ph)$_2$	–C(=O)– , >P·=O	10^5, 4.1×10^7
Ph–C(=O)–C(Ph)(OSO$_2$–Ph)	Ph–S(=O)$_2$–O·	10^{6a}
Ph–C(=O)–C(CH$_3$)$_2$–SO$_2$–Ph	Ph–S·O$_2$	10^{4a}
Mesityl–C(=O)–P(=S)(O)(Ph)	Ph$_2$P·=S	1.9×10^6

^{a} From Ref. 163.

in solution[169]) but also the high efficiency of diphenoxybenzophenone:[170]

$$\text{Ph—O—C}_6\text{H}_4\text{—C(=O)—C}_6\text{H}_4\text{—O—Ph}$$

in spite of a photochemical reactivity similar to that of benzophenone itself.[112]

CONCLUSION

This chapter was intended to give an overview of the main photoprocesses occurring in the UV curing of organic materials. While some points were purposely stressed, others were briefly reviewed, according to the main chapters of the present book. Nevertheless, this introductory paper should convince the reader that, nowadays, UV curing is really a well-established technology.

The above-mentioned ideas spawned a great deal of research work aimed at the synthesis of new efficient molecules with the desirable properties, the investigation of excited-state processes, the establishment of structure–reactivity relationships, the determination of the mechanisms involved in fast photopolymerisation of multifunctional monomers in film matrix under air and intense illumination. The finding of efficient, sensitive molecules at the particular wavelengths delivered by the lasers, e.g. in the visible or in the near-IR, or for the polymerisation of pigmented media, presents a formidable challenge. Although outstanding resistance to photoageing has already been obtained by adding UV absorbers and photooxidants, much has to be done in the future toward multicomponent systems where synergistic processes reinforce the protective effect. Special emphasis on the excited-state processes in photoinitiators and photosensitisers (as revealed by time-resolved laser spectroscopies, steady-state investigations, holographic spectroscopy in film matrix) resulted in new information useful for the design of ever more efficient systems. It is believed that, apart from a practical approach provided by the testing of many compounds, the working out of improved systems should be founded on basic research, involving picosecond, nanosecond, Raman, CIDNP or CIDEP laser photolysis. Multiphotonic spectroscopy and the investigation of processes occurring in upper excited states should again be expected to launch new research vistas in the field of photochemical reactivity and, thus, lead to the development of specific applications.

The research work required to support these new trends calls for the presence of photochemists and chemists striving for an understanding of the mechanisms involved and able to work out new efficient materials for the coming years.

REFERENCES

1. Prane, J. W. Proc. of Radcure 86, Baltimore, S.M.E. Ed., Dearborn (Michigan), 1986.
2. Rabek, J. F. *Mechanisms of Photophysical and Photochemical Reactions in Polymer: Theory and Practical Applications.* Wiley, New York, USA, 1987.
3. Pappas, S. P. (ed.) *UV-Curing: Science and Technology.* Technology Marketing Corp., Stamford, CT, USA, 1978 and 1985.
4. Roffey, C. G. *Photopolymerization of Surface Coatings.* J. Wiley, New York, USA, 1982.
5. Holman, R. & Oldring, P. (eds) *UV and EB Curing Formulation for Printing Inks, Coatings and Paints.* SITA, London, UK, 1988.
6. Reiser, A. *Photoreactive Polymers: The Science and Technology of Resists.* J. Wiley, New York, USA, 1989.
7. Jäckel, K. P. *Prog. Org. Coatings,* **16** (1988) 355.
8. Feit, E. D. & Wilkins, C. W. (eds) *Polymer Materials for Electronic Applications* (ACS Symp. Ser. 184). American Chemical Society, Washington, DC, USA, 1982.
9. Fouassier, J. P. & Rabek, J. F. (eds) *Lasers in Polymer Science and Technology: Applications.* CRC Press, Boca Raton, FL, USA, 1990.
10. Shaver, D. C., Stern, L. A., Economou, N. P. & Ehrlich, D. J. SPIE, *Lasers Microlithog.,* **774** (1987) 101.
11. Osgood, R. M. & Deutsch, T. F. *Science,* **227** (1985) 709.
12. Turro, N. J., *Modern Molecular Photochemistry.* Benjamins, New York, USA, 1978.
13. Wayne, R. P. *Principles and Applications of Photochemistry.* Oxford Science and Publications, Oxford, UK, 1988.
14. Braun, A., Maurette, M. T. & Olivers, E. *Technologie photochimique.* Presses Polytechniques Romandes, Lausanne, Switzerland, 1986.
15. Rabek, J. F. *Experimental Methods in Photochemistry and Photophysics.* J. Wiley, Chichester, UK, 1982.
16. Fouassier, J. P., *J. Chim. Phys.,* **80** (1983) 339.
17. Hageman, H. J. *Prog. Org. Coatings,* **13** (1985) 123.
18. Fouassier, J. P. *Makromol. Chem. Macromol. Symp.,* **18** (1988) 157.
19. Fouassier, J. P. In *Photopolymerization and Photoimaging Science and Technology,* ed. N. S. Allen. Elsevier Applied Science, 1989, p. 209.
20. Green, G. E., Stark, B. P. & Zahir, S. A. *J. Macromol. Sci.* (Rev. Macromol. Chem.), **C21**(2) (1981–82) 187.
21. Meier, K. & Zweifel, H., *J. Photochem.,* **35** (1986) 353.
22. Bousquet, J. A. & Fouassier, J. P., *Polym. Degrad. Stabil.,* **5** (1983) 113.
23. Reiser, A., *J. Chim. Phys.,* **77** (1980) 469.
24. Finter, J., Lohse, F. & Zweifel, H. *J. Photochem.,* **28** (1985) 175.
25. Hasagawa, M. In *Comprehensive Polymer Science.* Pergamon Press, 1989, p. 1.
26. Jacobine, A. F., Glaser, D. M. & Nakos, S. T. In *Radiation Curing of Polymeric Materials* (ACS Symp. Ser. 417), ed. C. E. Hoyle & J. F. Kinstle. American Chemical Society, Washington, DC, USA, 1990.
27. Willson, C. G. & Fechet, J. M. J. *Proc. SPIE Conf.* **771** (1987) 24.
28. Srinivasan, R. *Science,* **234** (1986) 559.
29. Chevallier, F., Chevalier, S., Decker, C. & Moussa, K. *Soc. Manuf. Eng. Techn.* (FC 87), **9** (1987) 1.
30. Trieschmann, C., Hockemeyer, F. & Preiner, G., German Patent 3708958, 1988.
31. Fouassier, J. P. & Lougnot, D. J. In *Radiation Curing of Polymeric Materials* (ACS Symp. Ser. 417), ed. C. E. Hoyle & J. F. Kinstle. American Chemical Society, Washington, DC, USA, 1990, p. 59.

32. Decker, C. & Bendaikha, T. In *Advances in the Stabilization and Controlled Degradation of Polymers* (Vol. 1), ed. A. Patsis. Wiley, 1989, p. 143.
33. Phan, X. T. *J. Radiat. Curing*, **13**(1) (1986) 11.
34. Phan, X. T. *J. Radiat. Curing*, **13**(1) (1986) 18.
35. Hageman, H. J. & Jansen, L. G. *J. Makromol. Chem.*, **189** (1988) 2781.
36. Kraeutler, B. & Turro, N. J. *Chem. Phys. Lett.*, **70** (1980) 270.
37. Kuo, P. L., Turro, N. J., Tseng, C. M., El-Aasser, M. S. & Vanderhoff, J. W. *Macromolecules*, **20** (1987) 1216.
38. Candau, F. & Sing Leong, Y. *J. Polym. Sci. Polym. Chem. Edn*, **23** (1985) 193.
39. Broer, D. J., Finkelmann, H. & Kondo, K. *Makromol. Chem.*, **189** (1988) 185.
40. Griffin, C., Hoyle, C. E. & Venkaratam, K. *SPIE Int. Proc.*, **1213** (1990).
41. Scott, G. (ed.) *Developments in Polymer Stabilization*. Elsevier Applied Science, London, UK, 1987.
42. Hult, A. & Ranby, B. *Polym. Degrad. Stabil.* **9** (1984) 1.
43. O'Connor, D. B., Scott, G. W., Coulter, D. R., Gupta, A., Webb, S. P., Yeh, S. W. & Clark, J. H. *Chem. Phys. Lett.*, **121** (1985) 417.
44. Chirinos Padron, A. J. *J. Photochem. Photobiol. A*, **49** (1989) 1.
45. Fouassier, J. P., Jacques, P., Lougnot, D. J. & Pilot, T. *Polym. Photochem.*, **5** (1984) 57.
46. Buback, M., Huckestein, B. & Leinhos, U., *Makromol. Chem. Rapid Commun.*, **8** (1987) 473.
47. McGimpsey, G. & Scaiano, J. C. *J. Amer. Chem. Soc.*, **109** (1987) 2179.
48. Schnabel, W. In *Lasers in Polymer Science and Technology: Applications*, ed. J. P. Fouassier & J. F. Rabek. CRC Press, Boca Raton, FL, USA, 1990, Vol. II, p. 95
49. Gatechair, L. R. & Wostratzky, D. In *Adhesive Chemistry*, ed. L. H. Lee. Plenum, 1984, p. 409.
50. Kirchmayer, R., Berner, G. & Rist, G. *Farbe Lack*, **3** (1980) 224.
51. Koyanagi, M., Futami, H., Mukai, M. & Yamauchi, S. *Chem. Phys. Lett.*, **154** (1989) 577.
52. Läuffer, M. & Dreeskamp, H. *J. Magnet. Reson.*, **60** (1984) 357.
53. Christensen, J. E., Jacobine, A. F. & Scanio, C. J. V., *J. Radiat. Curing*, **8**(3) (1981) 1.
54. Fouassier, J. P. & Lougnot, D. J. *J. Chem. Soc. Faraday Trans.* 1, **83** (1987) 2935.
55. Eichler, J., Herz, C. P., Naito, I. & Schnabel, W. *J. Photochem.*, **12** (1980) 225.
56. Sumiyoshi, T., Schnabel, W. & Henne, A. *J. Photochem.*, **30** (1985) 63.
57. Jacobi, M. & Henne, A. *J. Radiat. Curing*, **10** (1983) 16.
58. Baxter, J. E., Davidson, R. S., Hageman, H. J., McLauchlan, K. A. & Stevens, D. G. *J. Chem. Soc. Chem. Commun.*, (1987) 73.
59. Sumiyoshi, T., Schnabel, W. & Henne, A. *J. Photochem.*, **32** (1986) 191.
60. Majima, T. & Schnabel, W. *J. Photochem. Photobiol. A*, **50** (1989) 31.
61. Rutsch, W., Berner, G., Kirchmayer, R., Husler, R. & Rist, G. In *Organic Coatings: Science and Technology* (Vol. 8), ed. G. D. Porfitt & A. V. Patsis. Marcel Dekker, New York, USA, 1986.
62. Desobry, V., Dietliker, K., Husler, R., Misev, L., Rembold, M., Rist, G. & Rutsch, W. In *Radiation Curing of Polymeric Materials* (ACS Symp. Ser. 247), ed. C. E. Hoyle & J. F. Kinstle. American Chemical Society, Washington, DC, USA, 1989.
63. Fouassier, J. P. & Lougnot, D. J. *Polymer*, **31** (1990) 418.
64. Li Bassi, G., Cadona, L. & Broggi, F. Radcure 86, Tech. Paper 4-27, SME Ed., Dearborn, Michigan (1986).
65. Fouassier, J. P., Lougnot, D. J. & Scaiano, J. C. *Chem. Phys. Lett.*, **160** (1989) 335.
66. Redmond, R. W. & Scaiano, J. C. *J. Photochem. Photobiol. A.*, **49** (1989) 203.

67. Netto Ferreira, J. C., Murphy, W. F., Remond, J. W. & Scaiano, J. C. *J. Amer. Chem. Soc.*, **112** (1990) 4472.
68. Netto-Ferreira, J. C., Weir, D. & Scaiano, J. C., *J. Photochem. Photobiol. A*, **48** (1989) 345.
69. Scaiano, J. C. *J. Amer. Chem. Soc.*, **102** (1980) 7747.
70. Herkstroeter, W. G. & Farid, S. *J. Photochem.*, **35** (1986) 71.
71. Fouassier, J. P. *Double Liaison—Chimie des Peintures*, **356** (1985) 173/19.
72. Berner, G., Rist, G., Rutsch, W. & Kirchmayer, R. Radcure Basel, Tech. Paper FC85-446, SME Ed., Dearborn, Michigan (1985).
73. Gour, H. A., Groenenboom, C. J., Hageman, H. J., Hakwoort, G. T. M., Osterholl, P., Overeem, T., Polman, R. I. & Van Des Werf, S. *Makromol. Chem.*, **185** (1984) 1795.
74. Fouassier, J. P. & Burr, D. *Macromolecules*, **23** (1990) 3615.
75. Catalina, F., Peinado, C., Sastre, R. & Mateo, J. L. *J. Photochem. Photobiol. A*, **47** (1989) 365.
76. Bauemer, W., Koehler, H. & Ohngemach, J. Radcure 86, Tech. Paper 4-43, SME Ed., Dearborn, Michigan (1986).
77. Li Bassi, G., Cadona, L. & Broggi, F. Radcure Europe, Tech. Paper 3-15, SME Ed., Dearborn, Michigan (1987).
78. Huesler, R., Kirchmayer, R. & Rutsch, W. Euro. Patent 304886, 1989.
79. Davidson, R. S., Lewis, S. P. & Hageman, H. J. Proc. XIII IUPAC Symp. on Photochemistry, Coventry (1990).
80. Fouassier, J. P., Lougnot, D. J., Li Bassi, G. & Nicora, C. *Polym. Commun.*, **30** (1989) 245.
81. Fouassier, J. P., Lougnot, D. J., Zuchowicz, I., Green, P. N., Timpe, H. J., Kronfeld, K. P. & Muller, U. *J. Photochem.*, **36** (1987) 347.
82. Lougnot, D. J., Turck, C. & Fouassier, J. P. *Macromolecules*, **22** (1989) 108.
83. Fouassier, J. P., Burr, D. & Wieder, F. *J. Polym. Sci. Polym. Chem. Edn*, **29** (1991) 1319.
84. Fouassier, J. P. & Lougnot, D. J. *J. Appl. Polym. Sci.*, **32** (1986) 6209.
85. Bottom, R. A., Guthrie, J. T. & Green, P. N. *Polym. Photochem.*, **6** (1985) 111.
86. Koehler, M. & Ohngemach, J. Radtech 88, North America, Tech. Paper p. 150, Radtech Ind., Ed., Northbrook, Illinois, 1988.
87. Green, P. N. *Polym. Paint Colour J.*, **175** (1985) 246.
88. Li Bassi, G., Broggi, F. & Revelli, A. Radtech 88, North America, Tech. Paper p. 160, Radtech Ind. Ed., Northbrook, Illinois, 1988.
89. Allen, N. S., Catalina, F., Green, P. N. & Green, W. A. *Eur. Polym. J.*, **22** (1986) 49.
90. Dietliker, K., Rembold, M., Rist, G., Rutsch, W. & Sitek, F. Radcure Europe, Tech. Paper 3-37, SME Ed., Dearborn, Michigan, 1987.
91. Fouassier, J. P. & Ruhlmann, D. *Eur. Polym. J.* (In press).
92. Takeshi, K., Matsushima, M. & Higuchi, Y., Jap. Patent 6277346, 1987.
93. Matsushima, M. & Komai, T., Jap. Patent 6241201, 1987.
94. Tomioka, H., Takenchi, S., Kurimoto, H., Takimoto, Y., Kawabata, M. & Harada, M. *Kobunshi Ronbunshu*, **44** (1987) 729.
95. Tomioka, H., Takimoto, Y., Kawabata, M., Harada, M., Fouassier, J. P. & Ruhlmann, D. *J. Photochem. Photobiol. A*, **53** (1990) 359.
96. Fabrizio, L. F., Lin, S. O. S. & Jacobine, A. F., US Patent 4,536,265, 1985.
97. West, R., Wolff, A. R. & Peterson, D. J. *J. Radiat. Curing*, **13** (1986) 35.
98. Mikio, H., Kataoka, S., Tanaka, M., Miyagawa, T., Takenchi, H. & Tune, H. Jap. Patent 63152606, 1988.
99. Leplyanin, G. V., Battalov, E. M. & Murinov, Y. I. *Vysokomol. Soedin Ser. B*, **30** (1988) 223.
100. Ueda, Y. & Kimura, H., Jap. Patent 6124558, 1986.

101. Kleiner, H. J., Gersdorff, J. & Bastian, U., Euro. Patent 304782, 1989.
102. Fischer, W., Baumann, M., Finter, J., Kvita, V., Mayer, C. W., Rembold, M. & Roth, M., Euro. Patent 344110, 1989.
103. Venz, S., Euro. Patent 336417, 1989.
104. Koehler, M. & Ohngemach, J. In *Radiation Curing of Polymeric Materials* (ACS Symp. Ser. 417), ed. C. E. Hoyle & J. F. Kinstle. American Chemical Society, Washington, DC, USA, 1990, p. 106.
105. Timpe, H. J. *Z. Chem.*, **30**(2) (1990) 55.
106. Hakamaru, H., Minami, Y. & Kubota, N., Jap. Patent 0269463, 1990.
107. Tatsuji, A., Yasuda, T. & Kita, N., Jap. Patent 0254269, 1988.
108. Okuma, N., Hayashi, H., Minami, T. & Ohayashi, H., Jap. Patent 01253732, 1989.
109. McGinniss, V. D., Provder, T., Kuo, C. & Gallopo, A. *Macromolecules*, **11** (1978) 405.
110. Pappas, S. P. *Radiat. Phys. Chem.*, **25** (1985) 633.
111. Li Bassi, G., Luciano, C., Gaetano, A. G. & Anna, S. Radcure Basel, p. 219, SME Ed., Dearborn, Michigan, 1985.
112. Ruhlmann, D. & Fouassier, J. P. *Euro. Polym J.*, **28**(3) (1992) 287.
113. Fouassier, J. P., Lougnot, D. J., Payerne, A. & Wieder, F. *Chem. Phys. Lett.*, **135** (1987) 30.
114. Fouassier, J. P. & Burr, D. *Euro. Polym. J.*, **27** (1991) 657.
115. Fouassier, J. P., Ruhlmann, D., Rist, G., Desobry, V. & Dietliker, K. *Macromolecules*, **25** (1992) 4182.
116. Decker, C. & Fouassier, J. P. In *Lasers in Polymer Science and Technology: Applications*, ed. J. P. Fouassier & J. F. Rabek. CRC Press, Boca Raton, FL, USA, 1990, Vol. III, p. 1.
117. Williams, J. L. R., Specht, D. & Farid, S. *Polym. Eng. Sci.*, **23** (1983) 1022.
118. Wu, S. K., Zhang, J. K., Fouassier, J. P. & Burr, D., *Photogr. Sci. Photochem.*, **2** (1989) 47.
119. Angerer, H., Desobry, V., Riediker, M., Spahni, H. & Rembold, M. Proc. Radcure Asia, p. 461, 1988.
120. Serpone, N. & Jamieson, M. A. *Coordination Chem. Rev.*, **93** (1989) 87.
121. Deshpande, D. D. & Aravindakshan, P. *Polym. Photochem.*, **4** (1984) 295.
122. Kuroki, M., Aida, T. & Inoue, S. *J. Amer. Chem. Soc.*, **109** (1987) 4737.
123. Wagner, H. M. & Purbrick, M. D. *J. Photogr. Sci.*, **29**(6) (1981) 25.
124. Neckers, D. C., *J. Radiat. Curing*, **10**(2) (1983) 19.
125. Allen, N. S., Hardy, S. J., Jacobine, A. F., Glaser, D. M., Navaratnam, S. & Parsons, B. J. *J. Photochem. Photobiol. A.*, **50** (1990) 389.
126. Neckers, D. C. & Abu-Abdoun, I. I. *Macromolecules*, **17** (1984) 2468.
127. Falvey, D. E. & Schuster, G. B. *J. Amer. Chem. Soc.*, **108** (1986) 7419.
128. Wildmann, D. & Bader, S. Euro. Patent 793030784, 1980.
129. Chesneau, E. & Fouassier, J. P. *Angew. Makromol. Chem.*, **135** (1985) 41.
130. Fouassier, J. P., Chesneau, E. & Le Baccon, M. *Makromol. Chem. Rapid Commun.*, **9** (1988) 223.
131. Kawabata, M. & Takimoto, Y. *J. Photopolym. Sci. Technol.*, **3** (1990) 147.
132. Kawaki, T., Kobayashi, M., Hayashi, K., Watanabe, M. & Honda, N., Jap. Patent 63206741, 1988.
133. Crivello, J. V. In *Advances in Polymer Science* (Vol. 62), ed. T. Saegusa. Springer Verlag, Berlin, Germany, 1984, p. 32.
134. Crivello, J. V. & Lee, J. L., US Patent 4,780,511, 1987.
135. Crivello, J. V. & Lee, J. L. *J. Polym. Sci. Polym. Chem. Edn*, **27** (1989) 3951; **28** (1990) 479.
136. Scaiano, J. C., Nguyen, K. T. & Leigh, W. J. *J. Photochem.*, **24** (1984) 79.
137. Pappas, S. P. *Prog. Org. Coatings*, **13** (1985) 35.

138. Timpe, H. J. & Kronfeld, K. P. *J. Photochem. Photobiol. A*, **46** (1989) 253.
139. Timpe, H. J. & Rajendran, A. G. *Makromol. Chem. Rapid Commun.*, **9** (1988) 399.
140. Timpe, J. H. & Schikowsky, V. *J. Prakt. Chem.*, **331** (1989) 447.
141. Fouassier, J. F. & Manivannan, G. *J. Polym. Sci. Polym. Chem. Edn*, (1991).
142. Dektar, J. L. & Hacker, N. P. *J. Org. Chem.*, **55** (1990) 639.
143. McDonald, S. A. & McKean, D. R. *J. Photopolym. Sci. Technol.*, **3** (1990) 375.
144. Dektar, J. P. & Hacker, N. P. *J. Amer. Chem. Soc.*, **112** (1990) 6004.
145. Meier, K. Radcure Basel, Tech. Paper FC85-417, SME Ed., Dearborn, Michigan (1985).
146. Pappas, S. P., Pappas, B. C., Gatechair, L. R. & Lilets, J. H. *Polym. Photochem.*, **5** (1984) 1.
147. De Voe, R. J., Sahyun, M. R. V., Serpone, N. & Sharma, D. K. *Can. J. Chem.*, **65** (1987) 2342.
148. De Voe, R. J., Sahyun, M. R. V., Schmidt, E., Serpone, N. & Sharma, D. K. *Can. J. Chem.*, **66** (1988) 319.
149. Timpe, H. J., Kronfeld, K. P., Basse, B., Fouassier, J. P. & Lougnot, D. J. *J. Photochem.* (1990).
150. Fouassier, J. P., Burr, D. & Crivello, J. V. *J. Photochem. Photobiol. A*, **49** (1989) 318.
151. Goto, Y., Yamada, E., Nakayama, M. & Tokumaru, K. *J. Polym. Sci. Polym. Chem. Edn*, **26** (1988) 1671.
152. Lohse, F. & Zweifel, H. *Adv. Polym. Sci.*, **78** (1986) 62.
153. Roloff, A., Meier, K. & Riediker, M. *Pure Appl. Chem.*, **58** (1986) 1267.
154. Fouassier, J. P., Lougnot, D. J. & Pilot, T. *J. Polym. Sci. Polym. Chem. Edn*, **23** (1985) 569.
155. Decker, C. In *Radiation Curing of Polymers*, ed. D. Randell. Royal Society of Chemistry, London, UK, 1987. p. 16.
156. Hoyle, C. E., Trapp, M. & Chang, C. H. *Polym. Mater. Sci. Eng.*, **57** (1987) 579.
157. Sadhir, R. K., Smith, J. D. B. & Castle, P. M. *J. Polym. Sci. Polym. Chem. Edn*, **23** (1985) 411.
158. Brauchle, C., Wild, U. P., Burland, D. M., Bjorklund, G. C. & Alvarez, D. C. *IBM Res. Dev.*, **26** (1982) 217.
159. Lougnot, D. J., Ritzenthaler, D., Carre, C. & Fouassier, J. P. *J. Appl. Phys.*, **63** (1988) 4841.
160. Carre, C., Ritzenthaler, D., Lougnot, D. J. & Fouassier, J. P. *Optic Lett.*, **12** (1987) 646.
161. Carre, C., Lougnot, D. J. & Fouassier, J. P. *Macromolecules*, **22** (1989) 791.
162. Fouassier, J. P. In *Focus on Photophysics and Photochemistry*, ed. J. F. Rabek. CRC Press, 1989.
163. Fouassier, J. P. *Prog. Org. Coatings*, (1990).
164. Block, H., Ledwith, A. & Taylor, A. R. *Polymer*, **12** (1971) 271.
165. Wu, S. K. & Fouassier, J. P. *Chinese J. Polym. Sci.*, **8**(1) (1990).
166. Baxter, J. E., Davidson, R. S., Hageman, H. J., Hakvoort, G. T. M. & Overeem, T. *Polymer*, **29** (1988) 1575.
167. Peeters, S., Philips, M. & Loutz, J. M. Radcure Europe, p. 3-57, SME Ed., Dearborn, Michigan (1987).
168. Scaiano, J. C. & Johnston, L. J. *Pure Appl. Chem.*, **58** (1986) 1273.
169. Ruhlmann, D. & Fouassier, J. P. *Eur. Polym. J.* (In press).
170. Decker, C. & Moussa, K. *J. Polym. Sci. Polym. Lett. Edn*, **27** (1989) 347.

Chapter 3

Polymers in X-Ray, Electron-Beam and Ion-Beam Lithography

LEO SCHLEGEL
Hitachi Central Research Laboratory, Kokubunji, 185 Tokyo, Japan

&

WOLFRAM SCHNABEL
Hahn–Meitner Institut Berlin GmbH, Glienicker Strasse 100, D-1000 Berlin 39, Germany

Introduction 120
General Considerations 123
   Chemical strategies for microstructure generation 123
      Main-chain degradation and cross-linking 123
      The dissolution inhibitor principle 124
      Network formation via cross-linking agents 124
      Chemical amplification 125
         Deprotection 129
         Cross-linking 129
         Depolymerization 132
   Chemical effects of high-energy radiation of different stopping power 132
   Resist requirements 134
      Radiation sensitivity and contrast factor 134
      Resolution and pattern profile 136
      Dry-etch-resistance 136
      Stability of resist patterns 137
      Morphology of resist films 137
      Process compatibility 138
      Costs 139
   Development of microstructures 139
      Wet development 139
      Dry development 139
      Self-development and thermal development 140
   Technical aspects of lithographic techniques 140
      EB lithography 140
      X-ray lithography 141
      IB lithography 143
   Resist classification 145
      Positive versus negative resists 145
      One-layer versus multilayer resist systems 145

Generation of Microstructures 146
   Main-chain degradation and cross-linking of polymers via direct action of
   radiation 146
      Degradable polymers 146
         PMMA 146
         Copolymers containing MMA moieties 149
         Other acrylic acid-derived polymers 150
      Cross-linkable polymers 152
         Halogenated aromatic polymers 152
         Polyimides 155
         Other resist materials 157
         Inorganic resists 159
   Applications of the dissolution inhibitor principle 159
      Diazonaphthoquinone inhibitors 159
      Polymeric dissolution inhibitors 163
   Network formation via cross-linking agents 167
      Polymers suitable for liquid development 167
      Polymers suitable for dry development 168
   Applications of the chemical amplification principle 170
      Deprotection 170
      Cross-linking by condensation and polycondensation 172
      Depolymerization 178
IB Lithography 178
   General remarks 179
   Polymer resists appropriate for wet development 180
   Polymer resists appropriate for self-development 183
Acknowledgement 183
References 184
Bibliography 191

INTRODUCTION

At present, lithographic techniques are widely used to generate microstructures especially in connection with the production of electronic microdevices such as chips for computers, etc. Generally, these techniques are based on the interaction of electromagnetic or particle radiation with matter. In Fig. 3.1 the lithographic process is schematically illustrated. It has to be pointed out that direct irradiation of the substrate (mostly silicon wafers) commonly does not result in the generation of microstructures of the required quality because of low radiation chemical yields and/or insufficient penetration of the radiation into the substrate. Therefore, technically utilized process lines include an additional step: the substrate is coated with a thin layer of radiation-sensitive material in which fine line structures are primarily generated, usually in two stages: irradiation and subsequent (commonly) liquid development. Generally, the radiation-sensitive material is called a *resist* (material) because it has to be resistant to etching agents, i.e. chemicals being capable of reacting with substrate areas made open to attack by removal (development) of the irradiated resist. These steps are also illustrated in Fig. 3.1. Most of the resists that have been applied so far are polymer-based, i.e. they consist totally or

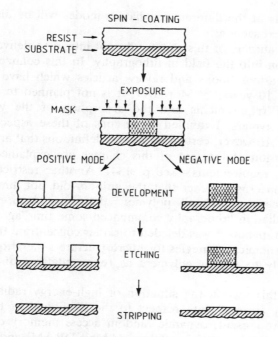

Fig. 3.1. Diagram illustrating the lithographic process.

partly of an amorphous organic polymer. This is mainly due to the fact that technically applied substrate materials, in most cases silicon wafers, can be easily coated with thin layers (usually about 1 μm thick) of amorphous polymers. Important technical requirements for appropriate resist materials comprise, apart from a significant etch resistance, good adhesion to the substrate, high radiation sensitivity and excellent line resolution. This article is devoted to polymers which can be used in high-energy radiation lithography, a technique which has been elaborated on through the last two decades quite extensively but which has been technically applied only rarely. In this respect it contrasts with photolithography (see Chapter 2) which is performed with UV light. Commonly, also photolithographic processes are operated with polymer resists. However, since the absorption modes of UV light and high-energy radiation differ appreciably, most polymers exhibit quite a different radiation sensitivity when exposed to UV light or high-energy radiation. Therefore, it seems to be appropriate to deal with the two, to some extent competing lithographic techniques, in different chapters, although they are aiming at the same goal: the generation of microstructures. Notably, the latter can be generated either by (i) focusing the radiation to selected areas of the substrate, or by (ii) irradiating the substrate through a mask. Method (i) is applicable to particle radiation (electron beam (EB) and ion beam (IB) radiation) and method (ii) to electromagnetic radiation (UV–vis light and X-rays). The technical aspects of the different irradiation modes will be discussed to some extent in the next section.

Notably, the authors of this chapter do not intend to deliver a comprehensive introduction into the field of lithography. In this connection, the reader may refer to various books and review articles which have been published during the last 10 years.[1-14] Moreover, it is not planned to discuss in detail technical resist requirements and the absorption of the various kinds of radiations in polymers. Extended discussions of these aspects can be found elsewhere.[8,11,15] However, certain terms and definitions that are needed for the full comprehension of the text of this chapter are explained in the section entitled 'Resist requirements'—see p. 134). Another restriction imposed on this chapter concerns the fact that the authors did not aim at covering all relevant publications and all polymers that have been examined since high-energy radiation lithography commenced some time ago. Because of the large number of papers it was decided to rather concentrate this review on the description of chemical strategies for microstructure generation and to refer in most cases only to typical polymers as representatives of certain polymer families.

Concluding this section the situation of high-energy radiation lithography shall be briefly assessed with respect to the likelihood of future large-scale applications. At present, dynamic random access memory devices (DRAM) with minimum features of about $0.8\ \mu m$ (4 Mbit DRAM) and of about $0.5\ \mu m$ (16 Mbit DRAM) are produced with the aid of photolithography. High numerical aperture step-and-repeat reduction projection techniques operated in conjunction with conventional mercury lamps [λ_{inc}: 436 nm (g-line) and 365 nm (i-line)] are used. It also appears that the photolithography technique in conjunction with excimer laser light sources (deep-UV lithography, $\lambda_{inc} <$ 300 nm) will be applicable to the future production of chips with features of about $0.35\ \mu m$ corresponding to 64 Mbit devices. At present (spring 1992), a prognosis on the chances of 64 Mbit devices to be fabricated with the aid of X-ray lithography cannot be made. Great difficulties particularly concerning production of defect-free X-ray masks and mask alignment and overlay accuracy still have to be overcome. However, since photolithography will probably not cope with further miniaturization new lithographic techniques operated in conjunction with high-energy radiation lithography will have to be applied. At present, only EB lithography is technically applied to some extent, mainly for the production of masks and occasionally for the production of special electronic microdevices.

It is interesting to note, that, commonly, polymers, although playing a quite essential role in the fabrication, are not contained in microelectronic devices. Moreover, it is interesting that in this kind of application quite thin polymer layers are used ($d \approx 1\ \mu m$, or less). Much thicker polymer layers (up to 500 μm) are used, however, in a novel microfabrication technique operated in conjunction with X-ray lithography (LIGA technique, German acronym for Lithografie, Galvanoformung, Abformung).[16,17] In this case films are prepared by in-situ polymerization. Patterning affords rather hard X-rays and/or multiple imaging. The process involves the following steps: primarily the polymer film is patterned by irradiation through a mask and development. Subsequently, a negative metal replica of the microstructures is produced by electroplating (galvanoforming). The metallic pattern is used as mold for the fabrication of plastic microdevices.

GENERAL CONSIDERATIONS

Chemical strategies for microstructure generation

The basic principle of microstructure generation derives from photography. Initially, a latent image is formed by subjecting an appropriate material to radiation and subsequently the latent image is developed. Regarding the generation of microstructures the interaction of radiation with the sensitive material must result in a significant change in solubility provided development is to be achieved with the aid of a liquid, i.e. by 'wet' development, which still is the state-of-the-art development technique. In the following, four methods to induce chemical changes in polymers which have been applied successfully in the past for the generation of microstructures, will be described.

Main-chain degradation and cross-linking

Solubility changes in macromolecules can be achieved readily by either main-chain scission or intermolecular cross-linking as illustrated by the reactions in eqns (1) and (2):

Main-chain scision

—M—M—M—M—M—M—M—M—M— ⟶
　　　　　　　　—M—M—M—M—M—M— + —M—M—M—　(1)

Intermolecular cross-linking:

—M—M—M—M—M—M—M—M—M—

+　　　　　　　　⟶

—M—M—M—M—M—M—M—M—M—

　　　　　—M—M—M—M—M—M—M—M—
　　　　　　　　　　　|　　　　　　　　　　(2)
　　　　　—M—M—M—M—M—M—M—M—

Main-chain scission causes a decrease in the average molar mass of polymers which commonly leads to an increase in the solubility of the polymers in a given solvent. By contrast, intermolecular cross-linking converting linear macromolecules into a three-dimensional network results in complete insolubilization if all macromolecules are integrated into the network. Actually, in most polymers high-energy radiation induces simultaneously both main-chain scission and cross-linking. However, some polymers undergo exclusively or overwhelmingly main-chain scission, such as polymethylmethacrylate, or intermolecular cross-linking, such as polystyrene. Therefore, in certain cases neat polymers can be applied as (one-component) positively or negatively acting resists provided sensitivity requirements are fulfilled.

The dissolution inhibitor principle

Polymeric systems can be made insoluble in certain liquids if blended with low-molar-mass compounds. Radiation-induced chemical alteration or decomposition of the dissolution inhibitor renders the bi- or multicomponent system soluble. Such systems act as positively working resists. Many industrially applied photolithographic processes are based on the dissolution inhibitor principle.[10] The most renowned systems functioning in this way consist of diazonaphthoquinone sulfonates **I** and **II** (2-diazo-1-oxo-1,2-dihydronaphthalene-4- or -5-sulfonic acid ester) and a phenolic resin, e.g. **III**.

I or **II** is converted into indene carboxylic acid (**IV**) and nitrogen according to the reaction in eqn (3):

$$I \xrightarrow[H_2O]{h\nu} IV + N_2 \qquad (3)$$

IV is no longer a dissolution inhibitor. It is soluble in aqueous alkali and may even promote the dissolution of the phenolic resin.

Network formation via cross-linking agents

Frequently, high-energy radiation-induced intermolecular cross-linking of linear polymers does not occur very efficiently, because of low radiation chemical yields $G(X)$. Principally, in this case the resist sensitivity can be significantly increased by blending the polymer with a compound of low molar mass that can actively participate in the cross-linking process. Such compounds can be, for example, multifunctional olefinic compounds, the C=C bonds of which readily react with free-radical sites generated at the macromolecules by bond scission. In this way, cross-linking is enhanced, since macroradicals which otherwise deactivate by other reactions are now contributing to cross-link formation. Moreover, bisazides have been reported to act as cross-linking agents in both photo- and EB lithographic processes.[3,18] The mode of action is thought to be due to reactive nitrenes that are formed after electronic excitation of the bisazide and that insert into carbon hydrogen bonds:

$$R{-}\overset{|}{\underset{|}{C}}{-}H + (N_3{-}R{-}N_3)^* + H{-}\overset{|}{\underset{|}{C}}{-}R \longrightarrow R{-}\overset{|}{\underset{|}{C}}{-}\underset{H}{N}{-}R{-}\underset{H}{N}{-}\overset{|}{\underset{|}{C}}{-}R + 2N_2 \quad (4)$$

The formation of excited bisazides, $(N_3{-}R{-}N_3)^*$, upon absorption of high-energy radiation is quite feasible if the resist system contains aromatic polymers the excited benzene rings of which are very likely to transfer excitation energy to $N_3{-}R{-}N_3$.

Chemical amplification

This fascinating method rests on the radiation-induced generation of a catalyst that is capable of inducing the chemical alteration of a major portion of the resist layer during or after the irradiation. Provided the catalyst is not consumed during the process each catalyst entity can induce the conversion of a large number of molecules in its surrounding. Chemical amplification meets extremely well the requirement of high-radiation sensitivity, which does not depend anymore on high radiation chemical yields or sophisticated development procedures. Up to now investigations and technical applications involving chemical amplification mainly concern the generation of protonic acids as catalytic species in conjunction with acid-labile polymeric matrices.[5,19–21] There are a variety of compounds acting as acid generators both upon photolysis and radiolysis. Table 3.1 shows a list of typical acid-generating compounds.[22–33]

So far, triarylsulfonium salts have played a prominent role in high-energy radiation lithography. In this case very strong acids with non-nucleophilic anions are produced:

$$Ar_3S^+MX_n \xrightarrow[RH]{rad.} Ar_2S + HMX_n^- + \text{other products} \quad (5)$$

A detailed reaction mechanism is shown in Scheme 3.1.

Recently, compounds have been described which also form bases if

Table 3.1
Acid generators

Compound	Acid	Ref.
Sulfonium and iodonium salts $Ar_3S^+X^-$, $Ar_2I^+X^-$ Ar: phenyl or substituted phenyl group X^-: BF_4^-, PF_6^-, AsF_6^-, SbF_6^-, $CF_3SO_3^-$	HX	22, 23
o-Nitrobenzyl sulfonates		24
1,2,3-Tris(methanesulfonyloxy)benzene		25
Imino sulfonates		26
4-Nitrobenzyl-9,10-ethoxyanthracene-sulfonate		27
Disulfone compounds		28

Table 3.1—continued

Compound	Acid	Ref.
Arylmethyl sulfones Ar—CH$_2$—S(=O)$_2$—CH$_2$—Ar Ar: 2-methylphenyl, 3-methoxyphenyl, naphthyl, 3,4-dichlorophenyl, 4-fluorophenyl, etc.	H$_2$SO$_4$, HO—SO—R HO—SO$_2$—R	29
α,α-Bis(arylsulfonyl)-diazomethanes R—S(=O)$_2$—C(N$_2$)—S(=O)$_2$—R R: substituted phenyl group	R—S(=O)$_2$—CH(R)—SO$_2$—OH	30
2-Aryl-4,6-bis(trichloromethyl)-1,3,5-triazines [triazine with Ar, CCl$_3$, CCl$_3$ substituents]	HCl	31
o,o'-Dibromophenols [2,6-dibromophenol with —CH—CH$_2$— substituent para to OH]	HBr	31, 32
[bisphenol A with Br at 3,3',5,5' positions and OH at 4,4']	HBr	31, 32

Table 3.1—*continued*

Compound	Acid	Ref.

1,3,5-Tris(2,3-dibromopropyl)-1,3,5-triazine-2,4,6-(1H,3H5H)trione

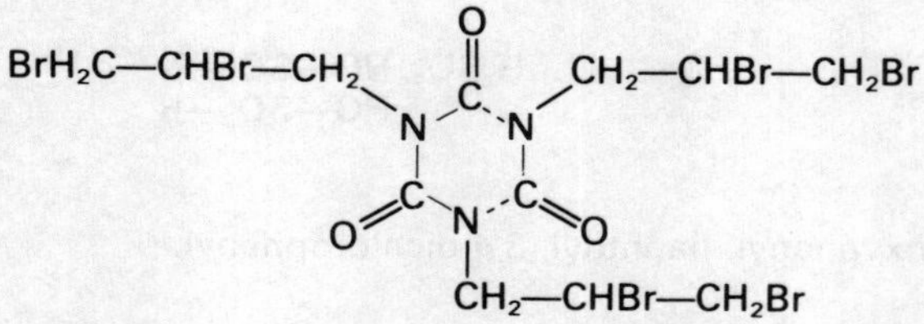

$$\text{BrH}_2\text{C—CHBr—CH}_2\text{—N} \quad \text{CH}_2\text{—CHBr—CH}_2\text{Br}$$

(triazine-trione ring structure with three CH₂—CHBr—CH₂Br substituents)

HBr   33

subjected to irradiation.[34,35] Typical examples of *base generators* are *o*-nitrobenzyl-*N*-cyclohexylcarbamate (**V**), 3,5-dimethoxycyclohexylcarbamate (**VI**) and *N*-cyclohexyl-4-methylphenyl sulfonamide (**VII**).

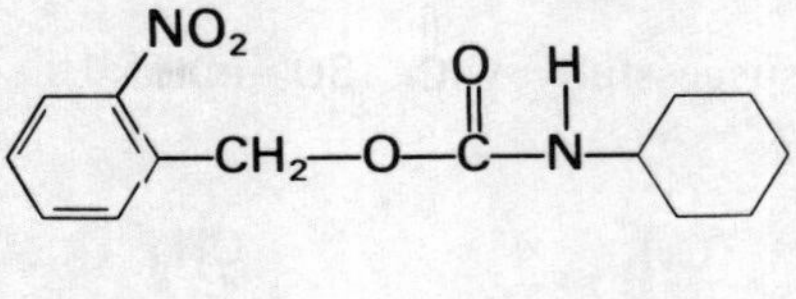

V: o-nitrobenzyl-N-cyclohexylcarbamate structure

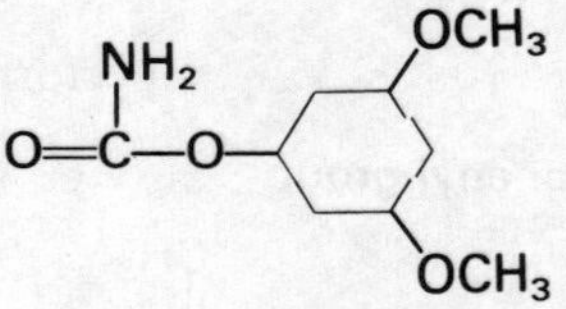

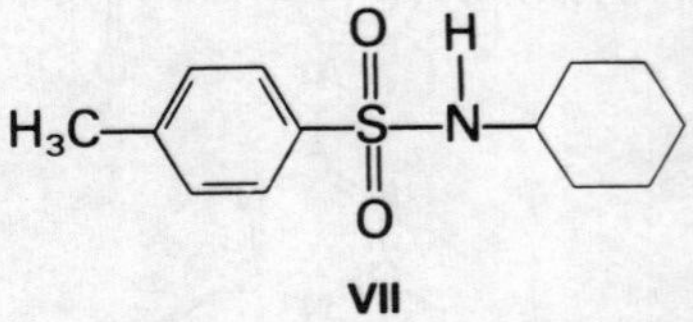

VI: 3,5-dimethoxycyclohexylcarbamate structure

VII: N-cyclohexyl-4-methylphenyl sulfonamide structure

Homolytic bond cleavage:

$$\text{Ar}_3\text{S}^+\text{MX}_n^- \xrightarrow{\text{rad.}} [\text{Ar}_3\text{S}^+\text{MX}_n^-]^* \longrightarrow \text{Ar}_2\text{S}^{\cdot+} + \text{Ar}^{\cdot} + \text{MX}_n^-$$

$$\text{Ar}_2\text{S}^{\cdot+} + \text{RH} \longrightarrow \text{Ar}_2\overset{+}{\text{S}}\text{H} + \text{R}^{\cdot}$$

$$\downarrow$$

$$\text{Ar}_2\text{S} + \text{H}^+$$

Heterolytic bond cleavage:

$$[\text{Ar}_3\text{S}^+\text{MX}_n^-]^* \longrightarrow \text{Ar}_2\text{S} + \text{Ar}^+ + \text{MX}_n^-$$

$$\text{Ar}^+ + \text{RH} \longrightarrow \text{Ar—R} + \text{H}^+$$

(Ar: phenyl or substituted phenyl group; MX_n^-: BF_4^-; PF_6^-; AsF_6^-; SbF_6^-)

Scheme 3.1. Proton generation by radiolysis or photolysis of sulfonium salts.

The radiolysis of these compounds results in the formation of amines, for example:

$$\text{o-}NO_2\text{-}C_6H_4\text{-}CH_2\text{-}O\text{-}C(=O)\text{-}NH\text{-}R \xrightarrow{rad} \text{o-}NO\text{-}C_6H_4\text{-}C(=O)\text{-}H + CO_2 + RNH_2 \quad (6)$$

Regarding microstructure generation research has been concentrated on three classes of acid-catalyzed processes during the last years: deprotection reactions, cross-linking and depolymerization. Typical examples of these processes are given below.

Deprotection. Deprotection concerns the decomposition of groups protecting functional groups attached to the polymer backbone. Upon release of the functional groups lipophilic polymers are converted into hydrophilic ones. Depending on the polarity of the developing liquid positive or negative imaging is possible.

Deprotection of poly(*t*-butoxycarbonyloxystyrene) (PBOCST) and poly(-*t*-butyl-*p*-vinyl benzoate) (PTBVB) resulting in the generation of sub-half-micrometer structures has been performed successfully with the aid of EB radiation and X-rays:[5]

PBOCST   PTBVB

Irradiated areas of these polymers are deprotected by heat treatment after irradiation in a so-called post-exposure-bake process at about 100°C, and are thus rendered soluble in aqueous alkali, for example:

$$\text{PBOCST} \xrightarrow[\Delta]{H^+} \text{H-C(CH}_2\text{)-C}_6\text{H}_4\text{-OH} + CO_2 + (CH_3)_2C=CH_2 \quad (7)$$

It turned out that one proton is capable of cleaving about 1000 *t*BOC (*t*-butoxycarbonyloxy) groups under standard processing conditions. Anisol or aqueous alkali solution served as developer for negative and positive imaging, respectively.

Cross-linking. Acid-catalyzed cross-linking (for negative imaging) can be achieved through various mechanisms: cationic polymerization, condensation reactions and electrophilic aromatic substitution.

Cationic polymerization: epoxide compounds of low molar mass of the following general structure:

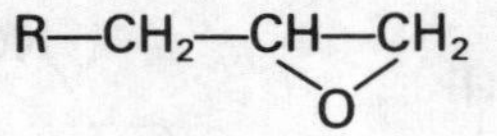

$$R-CH_2-CH-CH_2$$
$$\diagdown O \diagup$$

can be readily polymerized with the aid of Brönstedt acids. Of similar reactivity are polymers having epoxide side groups (epoxy-resins), e.g. novolak epoxides of the following structure:

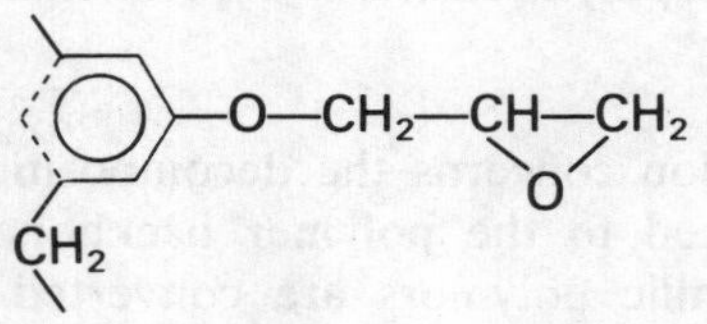

Novolak epoxide

Formulations comprised of commercially available epoxide resins and an acid generator have been employed in lithography.[21]

Crosslinking by condensation: this is achieved with systems containing a polymer (binder) with reactive sites, an acid generator and a (multifunctional) cross-linking agent. The radiation-generated acid catalyzes the reaction of the polymer with the cross-linking agent which results in a highly cross-linked insoluble network. Post-exposure baking prior to development is required to complete the condensation reaction which commonly proceeds very slowly at room temperature. Scheme 3.2 depicts the mechanism of the cross-linking of polyhydroxystyrene with the aid of a substituted melamine.

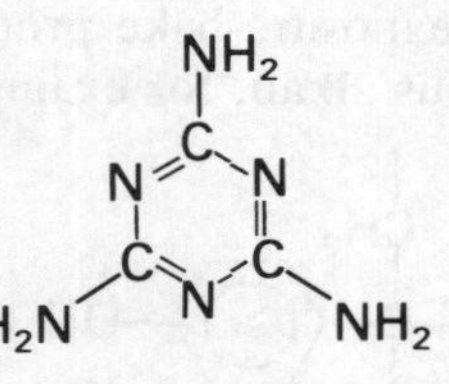

1,3,5-Triazin-2,4,6-triamine
(Melamine)

Here, the formation of the carbocation from the protonated ether group appears to be the rate-determining step. With this system sub-half-micrometer features have been generated by EB radiation.[36]

Electrophilic aromatic substitution: the aromatic rings in novolak resins of polyhydroxystyrene can undergo electrophilic substitution by a bifunctional

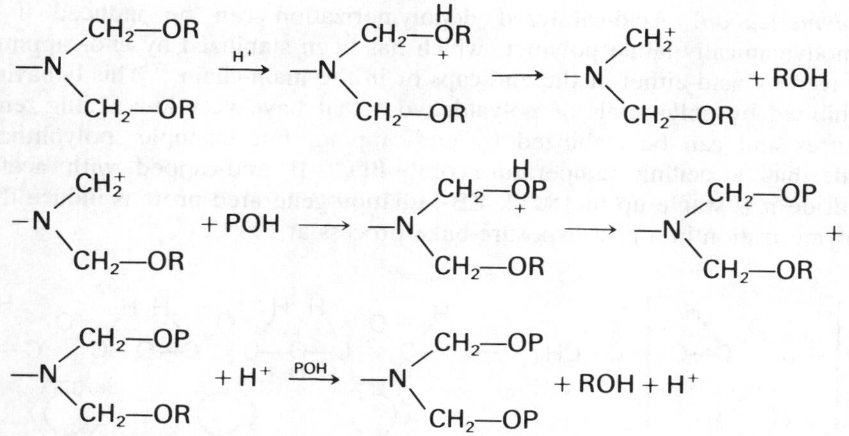

$$\begin{aligned}
-N\begin{pmatrix}CH_2-OR\\CH_2-OR\end{pmatrix} \xrightarrow{H^+} -N\begin{pmatrix}CH_2-\overset{H}{\overset{+}{O}}R\\CH_2-OR\end{pmatrix} \longrightarrow -N\begin{pmatrix}CH_2^+\\CH_2-OR\end{pmatrix} + ROH
\end{aligned}$$

$$-N\begin{pmatrix}CH_2^+\\CH_2-OR\end{pmatrix} + POH \longrightarrow -N\begin{pmatrix}CH_2-\overset{H}{\overset{+}{O}}P\\CH_2-OR\end{pmatrix} \longrightarrow -N\begin{pmatrix}CH_2-OP\\CH_2-OR\end{pmatrix} + H^+$$

$$-N\begin{pmatrix}CH_2-OP\\CH_2-OR\end{pmatrix} + H^+ \xrightarrow{POH} -N\begin{pmatrix}CH_2-OP\\CH_2-OP\end{pmatrix} + ROH + H^+$$

POH: polymer with OH groups

Scheme 3.2. Acid-catalyzed cross-linking of polymer POH (e.g. poly-*p*-hydroxystyrene) with the aid of substituted melamine.

carbocation precursor such as dibenzyl acetate. The latter forms benzyl carbocations upon reaction with protons.[21]

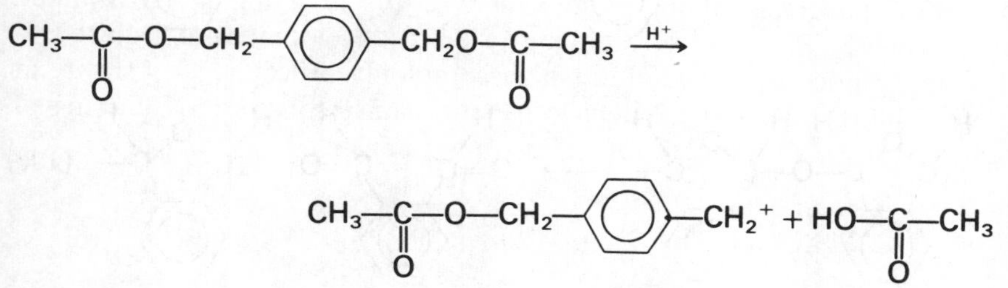

$$CH_3-\underset{O}{\underset{\|}{C}}-O-CH_2-\bigcirc-CH_2O-\underset{O}{\underset{\|}{C}}-CH_3 \xrightarrow{H^+}$$

$$CH_3-\underset{O}{\underset{\|}{C}}-O-CH_2-\bigcirc-CH_2^+ + HO-\underset{O}{\underset{\|}{C}}-CH_3 \quad (8)$$

Benzyl cations add to benzene rings of phenolic resins:

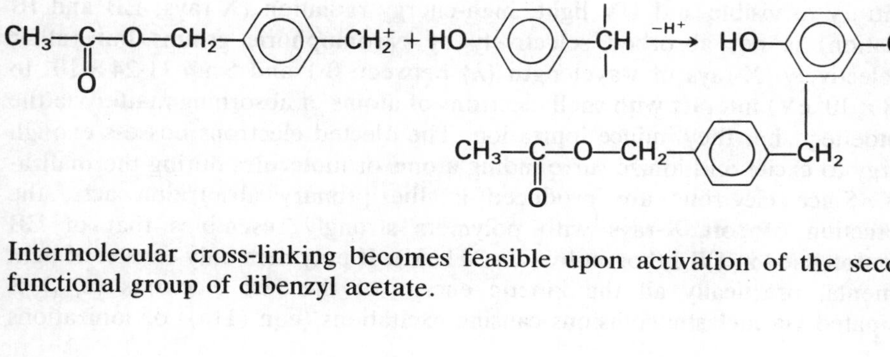

$$CH_3-\underset{O}{\underset{\|}{C}}-O-CH_2-\bigcirc-CH_2^+ + HO-\bigcirc-\overset{|}{\underset{|}{C}}H \xrightarrow{-H^+} HO-\bigcirc-\overset{|}{\underset{|}{C}}H$$
$$CH_3-\underset{O}{\underset{\|}{C}}-O-CH_2-\bigcirc-CH_2 \quad (9)$$

Intermolecular cross-linking becomes feasible upon activation of the second functional group of dibenzyl acetate.

Depolymerization. Acid-catalyzed depolymerization can be induced if a thermodynamically labile polymer, which has been stabilized by end-capping, is cleaved by acid either at the end-caps or in the main-chain.[37] This behavior is exhibited by polyacetals or polyaldehydes that have very low ceiling temperatures and can be stabilized by end-capping. For example, polyphthaldehyde has a ceiling temperature of −40°C. If end-capped with acetic anhydride it is stable up to 180°C. EB-radiation-generated protons induce the depolymerization in a post-exposure-bake process at 100°C.

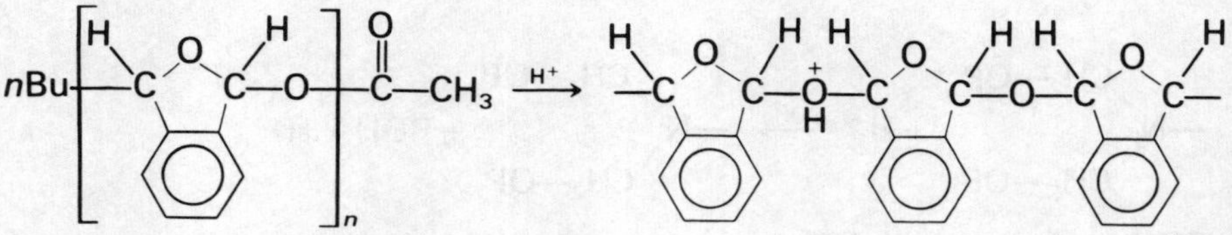

(10a)

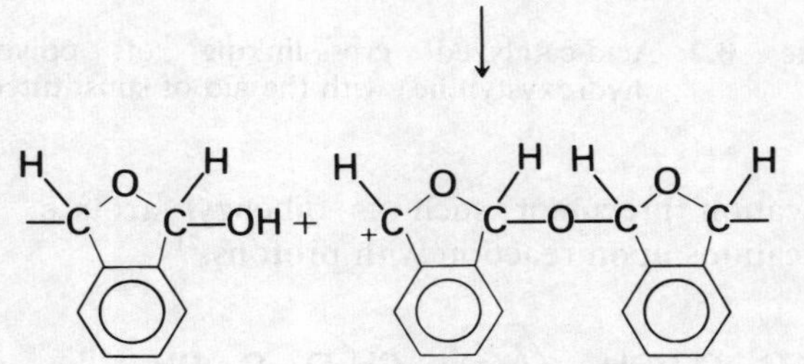

(10b)

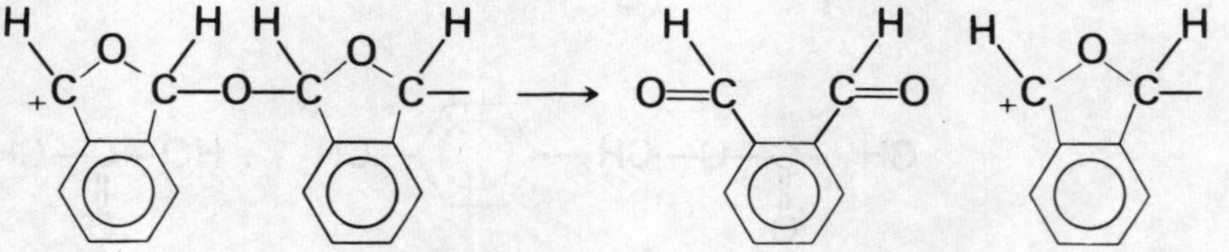

(10c)

Chemical effects of high-energy radiation of different stopping power

Contrary to visible and UV light, high-energy radiation (X-rays, EB and IB radiation) is not absorbed selectively by chromophoric groups but rather unselectively. X-rays of wavelength (λ) between 0·1 and 5 nm (1.24×10^4 to 2.48×10^2 eV) interact with shell electrons of atoms of absorbing matter via the photoeffect, i.e. they induce ionization. The ejected electrons possess enough energy to excite and ionize surrounding atoms or molecules during thermalization. Since electrons are produced in the primary absorption acts, the interaction of soft X-rays with polymers strongly resembles that of EB radiation used in EB lithography (20–30 keV). If polymers only consist of light elements, practically all the kinetic energy of the fast electrons (e^-_{fast}) is dissipated via inelastic collisions causing excitations (eqn (11a)) or ionizations (eqn (11b)). Each collision reduces the kinetic energy of e^-_{fast} by a certain amount.

$$e^-_{fast} + M \begin{cases} \xrightarrow{\text{excitation}} M^* + e^-_{fast} & (11a) \\ \xrightarrow{\text{ionization}} M^{+\cdot} + e^- + e^-_{fast} & (11b) \end{cases}$$

(M^+ and M^* denote radical cations and electronically excited molecules, respectively.) Finally, radical cations, or their decomposition products, combine with thermalized electrons. The highly excited states M^{**} formed hereby are mostly of repulsive nature, i.e. dissociation into free radicals occurs rapidly:

$$M^{+\cdot} + e^-_{therm} \longrightarrow M^{**} \longrightarrow R^1_\cdot + R^2_\cdot \qquad (12)$$

$R^1_\cdot$ and $R^2_\cdot$ denote free radicals which, together with $M^{+\cdot}$ initiate chemical reactions leading to chemical changes in the irradiated material. For further information the reader may refer to earlier reviews.[38,39]

Contrary to fast electrons, energetic ions dissipate their energy on much shorter path lengths. Ionizations and excitations occur in close sequence which results in absorption tracks of almost cylindrical shape. Principally, energetic ions can lose kinetic energy by various modes: collision with shell electrons, collision with atomic nuclei, by nuclear reactions and, at relativistic energies, also by generation of electromagnetic radiation (Cerenkov radiation). As far as IB lithography is concerned only ions of relatively low kinetic energy are of interest. At $E_{kin} < \sim 1$ MeV only collisions with shell electrons and nuclei have to be taken into account and the equation describing the total stopping power, $(-dE/dx)_{total}$ (also denoted as specific energy loss or linear energy transfer), is composed of two terms:

$$(-dE/dx)_{total} = (-dE/dx)_{el} + (-dE/dx)_{nucl} \qquad (13)$$

As can be seen from Table 3.2, where data regarding the absorption of rare

Table 3.2
Total stopping power of polystyrene and fractions of electronic and nuclear energy loss as calculated with the aid of the computer code TRIM[a] for He, Ne and Ar ions

Ion	E_0 (keV)	$(-dE/dx)_{total}$ (eV/nm)	f_{el}[b] (%)	f_{nucl}[c] (%)
He	100	140	98·6	1·4
Ne	200	370	78	22
Ar	400	700	71	29

[a] Based on Ref. 40.
[b] $100(-dE/dx)_{el}/(-dE/dx)_{total}$.
[c] $100(-dE/dx)_{nucl}/(-dE/dx)_{total}$.

gas ions in polystyrene are presented, the second term in eqn (13) is by no means negligible in these cases.

Resist requirements

The applicability of polymers as resist materials for high-energy radiation lithography depends on various parameters such as radiation sensitivity, contrast, fine line resolution, solubility in liquids, etch resistance, adhesion to substrate materials, heat resistance, etc. Table 3.3 presents requirements for some properties which are targeted for the 256 Mbit DRAM generation. In the following sub-sections several of these properties will be discussed in some detail.

Radiation sensitivity and contrast factor

Radiation sensitivity and contrast factor are defined by exposure characteristic curves as can be seen from Fig. 3.2. Here, the film thickness normalized with respect to the initial thickness is plotted as a function of the logarithm of the exposure dose, D_{exp}. In the case of positively acting resists the sensitivity, S, corresponds to the exposure dose $D_{exp}^{0 \cdot 0}$ ($S \equiv D_{exp}^{0 \cdot 0}$) at which the resist film is completely removed from the substrate during development. In the case of negatively acting resists S is mostly given by the exposure dose $D_{exp}^{0 \cdot 5}$ at which the thickness of the polymer film remaining on the substrate after development is 50% of the original film thickness ($d = 0 \cdot 5 d_0$). Occasionally, S is given by $D_{exp}^{0 \cdot 8}$ or $D_{exp}^{0 \cdot 9}$ corresponding to $d = 0 \cdot 8 d_0$ and $d = 0 \cdot 9 d_0$, respectively. Notably, D_{exp} is to be distinguished from D_{abs}, the absorbed dose denoting the energy deposited per mass or volume unit of the irradiated material. D_{exp} is the product of the radiation intensity and the time of irradiation: $D_{exp} = I \times t_{irr}$. In most cases I is readily measured and, therefore, commonly used to characterize the radiation sensitivity of polymers. If the radiation sensitivity of

Table 3.3
Requirements for important parameters of resists targeted for the 256 Mbit DRAM generation

Parameter	Specification
Sensitivity	$<2 \mu C/cm^2$ (50 keV electrons)
	$<100 mJ/cm^2$ (SOR[a] X-rays)
Contrast	>3
Resolution	$<0 \cdot 2 \mu m$
Aspect ratio	>5
Heat resistance	>130°C
Particle size	$<0 \cdot 02 \mu m$
Metal ion concentration	<100 ppb
Shelf-life time	>1 year
Film life time	>4 days

[a]SOR—synchrotron orbital radiation.

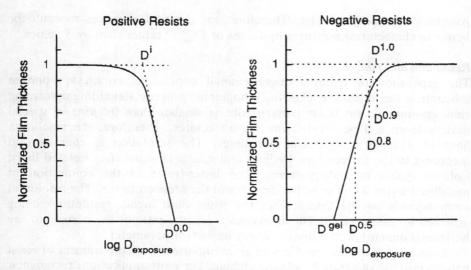

Fig. 3.2. Diagram representing exposure characteristic curves for positive and negative resists.

different polymers is to be compared it has to be taken into account that the absorption of the radiation depends on the characteristic properties of both the radiation and the absorbing material. In the case of electromagnetic radiation absorption depends, for example, on the photon energy and the elemental composition of the polymer, and in the case of particle radiation absorption depends, for example, on the stopping power $(\mathrm{d}E/\mathrm{d}x)$ of the polymer and the particle energy, E.

From the linear portions of the exposure characteristic curves the contrast factor is derived as follows:

Positively acting resists

$$\gamma = \left[\log_{10}\frac{D^{0.0}}{D^i}\right]^{-1} \tag{14}$$

Negatively acting resists

$$\gamma = \left[\log_{10}\frac{D^{1.0}}{D^{\mathrm{gel}}}\right]^{-1} \tag{15}$$

or, occasionally

$$\gamma = \left[\log_{10}\frac{D^{0.8}}{D^{\mathrm{gel}}}\right]^{-1} \tag{16}$$

It should be pointed out that data obtained from exposure characteristic curves only pertain to specific process and irradiation conditions. Moreover, the significance of S and γ values is limited. Often S is considerably lower than $D_{\mathrm{exp}}^{\mathrm{pattern}}$, the exposure dose required for the generation of the desired fine-line pattern in the polymer film. Therefore, for practical purposes it would be better to characterize polymers by values of $D_{\text{exp}}^{\text{pattern}}$ rather than by S values.

Resolution and pattern profile

The generation of patterns of very small dimensions on an appropriate substrate is the ultimate goal of the lithographic process. Regarding up-coming chip generations the finest pattern will be smaller than 0·5 µm; in special devices there will be 0·1–0·2 µm microstructures. Therefore, the resolution limit of a resist is a crucial parameter. The resolution is characterized according to the pattern type by lines and spaces of equal size, isolated lines, isolated spaces or hole patterns. The latter refer to the connection of metallized layers with the active elements of the semiconductor. The resolution limit depends on the thickness of the resist film, higher resolutions being achieved in thinner films. The *aspect ratio,* i.e. the ratio of the vertical to the horizontal dimension of a line, is a very important parameter.

The quality of the *line profile* is of great importance for the transfer of resist pattern into the substrate by plasma etching. For most applications rectangular profiles, i.e. microstructures with rectangular side walls (see Fig. 3.3(a)) are required. A tapered profile (see Fig. 3.3(b)) will result in a tapered topography as part of the resist is eroded during etching. Undercut or T-top profiles are needed for lift-off applications but also cause the formation of a tapered topography during etching as is shown by Fig. 3.3(c). Moreover, irregular profiles are detrimental for the control of the lithographic process. The latter is based on line-width determinations with the aid of a scanning electron microscope (SEM) from a viewpoint perpendicular to the wafer. In this way correct information can be obtained from rectangular profiles only.

Dry-etch-resistance

Anisotropic plasma etching is the state-of-the-art technique of pattern transfer from resist to substrate. Typical etching procedures use low-pressure plasmas based on HCF_3, CF_4, CCl_4 and SF_6. An electric field accelerates and directs ions and radical ions (the reactive species of the plasma) toward the wafer.

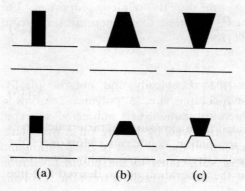

Fig. 3.3. Diagram illustrating the transfer of various line profiles into the substrate during anisotropic plasma etching.

Erosion on the surface is affected by both mechanical impact and chemical reactions of the ions. Sufficient resistance of the polymer toward the plasma is a prerequisite for pattern transfer. Generally, aromatic polymers are much more etch-resistant than aliphatic ones. Typically, phenolic resins are sufficiently resistant and poly(methyl methacrylate) (PMMA) is insufficiently resistant. Polymers that readily undergo radiation-induced main-chain scission or depolymerization (e.g. poly(olefin sulfones)) are inadequate for lithographic processes involving dry-etching.

Etching by an oxygen plasma is applied in the case of two-level resist systems which consist of a thin top layer and a thick bottom layer. Imaging of the substrate occurs by two processes. At first, the top layer is patterned conventionally. Subsequently, pattern transfer into the bottom layer is accomplished by *oxygen reactive ion etching* (O_2-RIE). The successful application of this method relies on the O_2-RIE resistance of the pattern of the top layer.

Frequently, this requirement is fulfilled by resists containing relatively heavy elements such as silicon. In this case oxide (SiO_2) is formed in the initial stage of etching which acts as a resist during the remainder of the procedure.

Stability of resist patterns
The resist pattern must retain its shape during development and subsequent process steps. Pattern stability rests essentially on heat resistance and adhesion of the polymer to the substrate. In the case of phenolic resists good adhesion to the substrate frequently is achieved by an a priori treatment of the substrate. For example, prior to spin-coating the substrate is primed in hexamethyldisilazane (HMDS) vapor.

The temperature at which resist patterns start to deform is taken as a measure for the heat resistance. Commonly, heat-resistance tests are performed by baking wafers that carry resist patterns on a hot plate. Frequently, patterns of high aspect ratio deform at higher temperatures than bulky patterns of low aspect ratio. This is probably due to the fact that fine line patterns of high specific surface transfer heat to the surroundings more easily than bulky ones which suffer from local overheating. Generally, heat resistance and glass transition temperature of a polymer are directly related properties. For practical application polymers should be resistant to temperatures up to at least 130°C.

Morphology of resist films
Resist films must be homogeneous, macroscopically and microscopically. Generally, films are prepared by spin-coating, i.e. a polymer solution is applied to the substrate (*wafer*) which subsequently is subjected to rapid spinning. Remaining solvent traces are removed by baking on a hot plate. Thickness and thickness uniformity depend on the viscosity of the solution and on the rotation speed. The addition of surfactants to the polymer solution reduces inhomogeneities in film thickness extending radially from the revolving

center (*striations*). In the case of photoresists the thickness must be controlled within a few nanometers in order to guarantee homogeneous light absorption over the whole film. This requirement is not that stringent in the case of high-energy radiation resists. However, a rather homogeneous thickness is mandatory for uniform development of microstructures. On a microscopic scale, the film homogeneity can be affected by microphase separation of resist components. This applies especially to polymer composite systems. Incompatibility of resist components may result in a rough surface of developed fine structures. The resist performance also can be affected by traces of solvent which remain in the film if pre-baking was carried out at insufficiently high temperatures.

Process compatibility
Application of a polymer as resist material in a semiconductor process line is based on additional requirements such as absence of small particles or contaminants for semiconductor devices, e.g. the content of metal ions should not exceed a few hundred ppb. Also toxicological aspects have to be considered. This concerns solvents used for spin-coating and/or development. Developers based on aqueous solutions should be preferably used. In this connection the utilization of phenolic resin-based resist systems is noticeable. In this case metal-free aqueous tetramethylammonium hydroxide solutions serve as developer. A very important issue pertains to the long-term stability of resist solutions and resist films. The shelf-life time of a solution should be at least 1 year under conventional storage conditions. The film lifetime, i.e. the stability of coated resist films, should exceed several days under ambient environment.

Resists should possess a good process stability, i.e. they should exhibit highly reproducible behavior with respect to pre-baking temperature, pre-baking time, required exposure dose, time and temperature of development. Moreover, minor parameter alterations should not affect the pattern quality. In the case of resists functioning on the basis of chemical amplification an additional baking step between exposure and development (post-exposure bake, PEB) is required, i.e. the PEB time and temperature and the time intervals between exposure and PEB and between PEB and development have to be taken into account.

Some positively working resists undergo a turnover from positive to negative mode of action at high absorbed doses. In such cases resist parts covering the alignment marks on the wafer may become insoluble upon EB irradiation because the beam is scanning these marks quite frequently. Consequently, the dose gap between film clearing and onset of cross-linking should be quite large.

A final requirement concerns full removability of the resist from the substrate after completion of the lithographic process. Film removal is accomplished by oxygen plasma stripping or by dissolution in an organic solvent. Problems may arise in the case of negative tone imaging provided the resist is converted into an insoluble gel, or in the case of resists containing silicon or other elements that form non-volatile oxides in the plasma reactor.

Costs
Regarding a semiconductor manufacturing line, the price of the resist comprises an essential part of the overall operation costs. Therefore, for being too costly, various promising materials cannot be considered for applications. Resist systems based on conventionally produced mass polymers such as phenolic resins appear to be most advantageous for applications. However, the high purity requirement in advanced resist applications even makes a special purification of raw materials for polymer synthesis necessary, which implies that a clean-room environment for resist synthesis and packaging is mandatory.

Development of microstructures

Wet development
Depending on whether a resist functions in the positive or negative mode of action either irradiated or unirradiated areas of the resist film have to be removed after exposure. This process is denoted by the term *development* (of the latent image). At present, in most cases wet development is applied, i.e. the irradiated film is treated with an appropriate liquid for a certain time. Subsequently it is rinsed in a non-solvent and dried in a nitrogen stream.

Many resists are developable in organic liquids. Drawbacks concern toxicology and in some cases distortion of microstructures due to swelling phenomena. Aqueous solutions that are favored over organic liquids for these reasons only are applicable in a few cases, especially in the development of resists based on phenolic resins. Here, a solution of tetramethylammonium hydroxide (TMAH) is used. On the laboratory scale wafers are soaked in a tank containing the developer. In production lines, the developer is either sprayed or poured onto horizontally positioned wafers.

Dry development
For dry development irradiated wafers are treated in a plasma reactor. This method is successfully applied in the case of two- or three-level resist schemes. The pattern is generated in the thin top layer by irradiation and subsequent wet development. It serves then as a mask for pattern transfer into the middle and bottom layer by plasma etching.

Recently, *top-surface-imaging* processes with an enhanced discrimination between exposed and unexposed areas have been proposed. These processes are based on the selective incorporation of relatively heavy elements, usually silicon, into exposed or unexposed areas,[41,42] by electroplating a metal film selectively on top of exposed resist parts,[43,44] or by implantation of Ga^+ ions (via IB irradiation, *vide infra* Fig. 3.7).[45] In these cases microstructures are generated without wet development.

Self-development and thermal development

Polymers having ceiling temperatures lower than room temperature depolymerize in a chain reaction provided bonds in the main-chain are cleaved. Typical polymers behaving in this way are poly(olefin sulfones) and polyaldehydes. In these cases the polymer is completely converted into low-molar-mass compounds upon irradiation. For practical applications *self-development* is undesirable because of the contamination of the exposure equipment. Preferred are, therefore, polymers that undergo *thermal development,* i.e. that depolymerize *after* irradiation on heating.

Technical aspects of lithographic techniques

EB lithography

EB fabrication of integrated circuits offers several advantages for pattern transfer: resist geometries much smaller than 1 μm can be generated and wafers can be patterned directly without a mask. Highly automated, computer-controlled EB machines are available which are used to fabricate masks for photo- and X-ray lithography and for the production of 'application specific integrated circuits' (ASIC). Moreover, special devices with critical dimensions in the range 0·2–0·5 μm such as GaAs–field effect transistors (FETs) have been produced. However, the application of EB lithography for the large-scale chip production is hampered by the fact that EB machines are slow and that their throughputs do not compete with those of optical machines. For resolutions < 100 nm, the desired pattern is 'written' into the resist layer by exposing sequentially pixel after pixel while the beam is blanked during movement. The total number of pixels per wafer can exceed 10^{11} at a pixel size of 0·1 × 0·1 μm. Improved EB machines providing variable exposure areas in the range 0·1 × 0·1 μm to 5 × 5 μm have been designed[46,47] and, at present, are used for most applications. However, the throughput is much lower than that of optical machines. Recently, techniques aiming at improving the throughput rate have been proposed such as 'electron proximity printing'[48] which uses a prefabricated stencil mask located 600 μm above the wafer, through which the latter is irradiated, or the 'cell projection technique' (see Fig. 3.4) operating with a stencil-type aperture. The demagnified image of the aperture is projected onto the wafer.[49] Other measures to improve resolution and throughput concern the increase in electron energy from currently 20–30 keV to 50–60 keV, the increase in beam current and blanking frequency and the reduction of exposure time per spot. Moreover, low ratios of exposed to unexposed areas are required for a high throughput. In this connection it is important whether positively or negatively acting resists are used. For example, the application of a negatively working resist would be advisable, if 80–90% of the chip area had to be exposed in order to generate a certain pattern in a positively working resist. Therefore, both positively and negatively

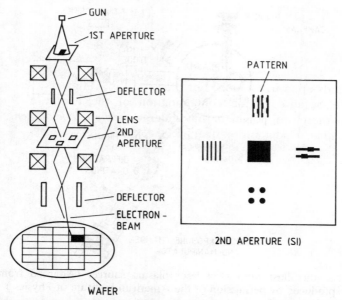

Fig. 3.4. Diagram illustrating the electron optics and the second aperture for EB cell projection lithography. (From Ref. 49, reproduced by permission of the American Institute of Physics.)

working resists with similar process and resolution characteristics should be available.

X-ray lithography
X-ray lithography is performed by exposing a wafer through a mask located in close proximity (~50 μm) to the wafer. The mask is a thin membrane consisting of material containing only light elements (mostly Si, SiN_x, SiC). It is coated with an absorber pattern consisting of heavy metals such as gold or tungsten. Considering the absorption characteristics of the mask membrane, the mask absorber and the resist material, optimum photon energies refer to the range of soft X-rays (λ: 0·2–2 nm). Appropriate available X-ray sources are electron impact, pulse plasma sources and electron storage rings. A drawback of conventional *electron impact* sources is their relatively large spot size. In order to achieve good mask fidelity on the wafer, the mask/wafer assembly must be located at a large distance (1–2 m) from the source, which implies a low X-ray intensity incident on the resist. Therefore, an economic throughput for large-scale production is not feasible. Moreover, the photon energy of X-rays produced by conventional sources is relatively high. In recent years the development of compact X-ray sources adequating industrial requirements concentrated on *plasma sources*. Among the various sources developed so far[50–52] the laser plasma and the gas plasma source initiated some

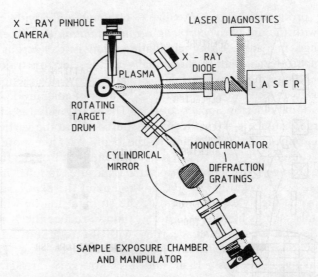

Fig. 3.5. Schematic illustration of a laser plasma source. (Adapted from Ref. 51, reproduced by permission of the American Institute of Physics.)

interest in applications for X-ray lithography. The working principle of the *laser plasma source* is depicted in Fig. 3.5. The heart is a rotating cylinder coated with gold. On focusing the beam of a pulsed high power laser (e.g. Nd:YAG or KrF excimer laser) upon the cylinder a hot plasma is formed at the surface which emits X-rays of high intensity and of the required photon energy.

Alternatively, a hot ionized *gas plasma* emitting X-rays of high intensity is formed by z-pinching a cylindrical or spherical gas volume by the electrical field generated by a short electrical pulse from a capacitor bank. The conversion efficiency of the plasma sources exceeds that of electron impact sources considerably as can be seen from Table 3.4.[53-55] Laser plasma sources are applicable for the small-scale very large-scale integration (VLSI)-chip production.

Table 3.4
Radiation sources for X-ray lithography

	Electron impact	Laser Focus	Plasma focus	Synchrotron
Soft X-ray conversion efficiency (%)	0·05[a]	8–13[b]	1·1[c]	0·1[a]

[a]Ref. 53.
[b]Ref. 54.
[c]Ref. 55.

Large-scale production can be realized, although only with considerable expenditure, with the aid of *synchrotron orbital radiation* (SOR) emitted from EB storage rings. In a storage ring, circulating relativistic electrons emit highly collimated synchrotron radiation tangentially to the electron trajectory while passing the bending magnets. The beam is of very low divergence and of an intensity, at a given wavelength, about two orders of magnitude higher than that of conventional electron impact sources. The mask/wafer assembly can be placed at a large distance (~10 m) from the source where the vertical diameter of the beam is only 1–2 cm. Scanning of the beam over the wafer (by X-ray mirrors) or scanning of the mask/wafer assembly through the beam is required. One storage ring can serve up to 20 exposure stations. As exposures are preferentially performed in ambient environment, the exposure chamber is separated from the high-vacuum system of the storage ring by windows made from light metals such as beryllium. These windows reduce the X-ray intensity significantly. Several electron storage rings with the required output characteristics are now operated in Europe, North America and Japan. The rings are very large having diameters > 15 m and are not very attractive for the production of VLSI devices on a commercial basis. Presently, compact storage rings are under development world-wide by several companies, and the development has already reached the test phase.[56]

IB lithography

Two exposure techniques for IB lithography have been developed:[10,45,57–60] the finely focused IB (FIB) exposure technique, in analogy to EB irradiation and the broad beam exposure technique in which the whole area of a mask is exposed, in analogy to X-ray irradiation. Figure 3.6 shows a schematic of a FIB source which is optimized for Ga^+ ions.[61] The ions are focused by electrostatic optics and the beam is scanned over the surface of the wafer. Ion sources like this can be operated to generate microstructures in resist layers in analogy to EBs, but they can also be used for *maskless ion implantations*. An interesting application of the ion implantation technique concerns the generation of microstructures in thick polymer layers[45,229] as can be seen from Fig. 3.7. A shallow metal layer is produced by implantation of Ga^+ ions into the resist surface. Oxygen reactive ion etching converts the metal into the oxide, Ga_2O_3, which subsequently acts as a plasma resist thus allowing the transfer of the implanted pattern into the resist. An interesting application of FIB irradiation concerns the defect repair of X-ray masks: improperly deposited metal particles can be sputtered away by an IB. Clear defects can be repaired by depositing metal spots from a metal–organic vapor under IB exposure.[62] Regarding microstructure generation in polymer resists the potential advantages of ions over electrons concern the higher focus ability. Moreover, ions scatter less than electrons in the resist material and there is no significant backscattering of ions from the substrate, but ions might damage the substrate more strongly than electrons.

The same restrictions concerning throughput exist as in the case of EB

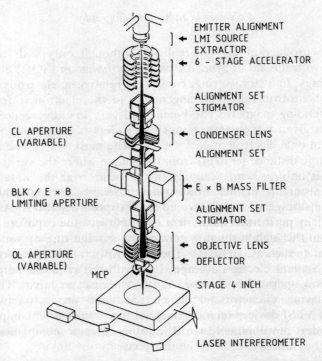

Fig. 3.6. Schematic illustration of the ion-optical system of JEOL JIBL-200S. (From Ref. 61, reproduced by permission of the American Institute of Physics.)

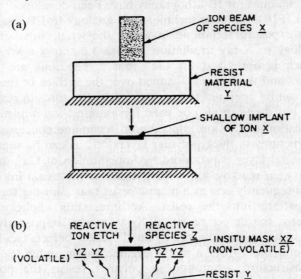

Fig. 3.7. Schematic illustrating the implantation of Ga^+ ions into the planarizing resist to produce an in-situ mask which serves to transfer the pattern into the resist by reactive ion etching. (From Ref. 45, reproduced by permission of the American Institute of Physics.)

lithography regarding applications of IB lithography for the production of microelectronic devices. IB lithography might become important for mask making for feature sizes in the nanometer range especially if X-ray lithography becomes the large-scale patterning method.

Resist classification

Positive versus negative resists

Resist materials can be classified into two categories: positively and negatively acting materials (see Fig. 3.1). Provided microstructures are obtained by liquid development, in the former case, the exposed areas become soluble in the developer liquid which results in a positive replication of the mask pattern in the resist film. In the case of negatively acting resists, the exposed parts of the film become insoluble in the developer liquid. PMMA is a typical example for positively acting resists. In this case irradiation causes main-chain scission and an increase in the free volume which results in an increased solubility. Halogen-containing polystyrene derivatives or bisazide-containing phenolic resins are typical representatives for the category of negative resists. In these cases insolubilization is caused by intermolecular cross-linking.

One-layer versus multilayer resist systems

Normally, for the production of electronic microdevices one-layer resist systems are used, i.e. the pattern generated in the resist film is directly transferred into the substrate by etching. However, the application of this technique sometimes encounters serious problems, especially if very fine structures are to be generated. These problems concern, for example, insufficient aspect ratios and line-width control over non-planar substrate surfaces. Some of these problems can be overcome by applying multilayer techniques. In the simplest case trilayer systems consisting of a thick planarizing bottom layer, a thin intermediate layer and a radiation-sensitive top layer. After imaging the top layer by exposure and development the pattern is transferred into the intermediate layer by plasma etching and subsequently into the bottom layer by RIE using an oxygen plasma. For EB lithography three-level processes have the advantage of eliminating electron backscattering from the substrate. Moreover, the line resolution can be enhanced because the top layer can be much thinner than in one-level processes. On the other hand, the performance of three-level processes is more complicated and more costly than that of one-level processes and frequently connected with a higher probability for defect generation. Therefore, tri-level systems are suitable only for the production of sophisticated devices such as for the fabrication of X-ray masks. A simplification of this process refers to a two-level system with the top layer consisting of a silicon-containing organic polymer which is both radiation-sensitive and resistant towards O_2-RIE. For a detailed description the reader may refer to Vol. IV, Chapter 12 by A. Tanaka in this book.

GENERATION OF MICROSTRUCTURES

Main-chain degradation and cross-linking of polymers via direct action of radiation

Degradable polymers

PMMA. PMMA has been the working horse as positive resist material for EB and X-ray lithography since the early days of high-energy radiation lithography and it is still the favorite resist for the small-scale fabrication of special devices.

Upon exposure to high-energy radiation PMMA undergoes bond cleavage both in the main-chain and in side groups. Main-chain scission causes a decrease in the average molar mass and side-group decomposition leads to the formation of volatile products such as CO_2 and CO.[38] Apart from main-chain scission, also the production of volatiles induces an increase in polymer solubility because volatiles cause an increase in free volume.[63] The reaction mechanism of the radiolysis of PMMA has been studied over several decades[64,65] and has been reinvestigated recently by several groups. Radicals R^1, R^2 and R^3 were detected with electron spin resonance (ESR) by electron spin echo methods. However, only the decay of R^3 was found to be correlated to main-chain scission.[66]

$$-\overset{\cdot}{C}H- \qquad -\underset{\underset{O}{\|}}{C}-OCH_2 \qquad \left[\underset{\underset{O}{\|}}{\overset{\cdot}{C}OCH_3}\right]^-$$
$$R^1 \qquad\qquad R^2 \qquad\qquad R^3$$

The elimination of oxygen from exposed PMMA surfaces was evidenced by electron energy loss spectroscopy (EELS).[67] Moreover, the influence of the kind of radiation (X-ray, EB, proton beam and deep-UV) on the radiolysis of PMMA was studied with the aid of gel permeation chromatography, Fourier transform IR (FTIR) and UV spectroscopy.[68] In all cases, it turned out that main-chain scission results in the formation of C=C bonds in the polymer backbone. The ratio of the number of decomposed side groups to the number of main-chain scissions decreases in the order γ-ray > X-ray > proton beam > EB > deep-UV.

Notably, PMMA is extraordinarily capable of yielding highly resolved patterns. Metal lines as small as 35 nm,[69] or even 10 nm,[70,71] have been fabricated as can be seen from Fig. 3.8. In the latter case a modified SEM operating at an electron energy of 350 keV was used. Fresnel zone plates fabricated with dimensions down to 75 nm[72] and 60 nm FETs fabricated with the aid of X-ray lithography are other examples demonstrating the excellent resolution capability of PMMA.[73]

Langmuir–Blodgett (LB) films made from PMMA having a thickness of 14·3 nm were successfully patterned with an EB of 10 keV. The resist thickness

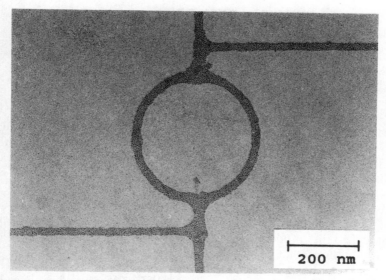

Fig. 3.8. Lift-off shadowing of PMMA patterns. The metal patterns were produced by evaporating PtPd onto developed PMMA patterns and lifting-off the metal on top of the resist by dissolving it in acetone. (From Refs 70 and 71, courtesy of Prof. Dr. A. N. Broers.)

was sufficient to etch 50-nm-deep patterns into a chromium substrate.[74] PMMA was also used to check resolution limits in the case of SOR-lithography and to examine X-ray masks with respect to Fresnel diffraction and photo-electron scattering. The results are in good agreement with those obtained by simulation programs:[75,76] a line resolution of 0·2 μm in 2-μm thick resist layers was realized.

PMMA has been applied in the laboratory-scale fabrication of a variety of electronic devices. A typical example is a prototype 64 Mbit DRAM chip with critical dimensions of 0·3 μm.[77] Here, a three-level resist system with a PMMA top layer was used. The quality of the resist pattern was improved by applying a special two-step development process. Another interesting application of PMMA resists concerns EB direct writing of metal patterns for surface acoustic wave (SAW) filter devices with critical dimensions of 85 nm.[78]

Special attention has been addressed to the fabrication of T-shaped gates for GaAs FETs. Here, a mushroom-like profile with a small foot and a wide upper part (see Fig. 3.9) is required to achieve both a fast switching time and a low resistivity of the metal line. As can be seen from Fig. 3.9, at first, trenches are formed by irradiation of the resist layer and subsequent development. Then, metal is deposited into the trenches by a vacuum metallization technique, and finally, the remaining resist layer is lifted-off by dissolution in a solvent. T-shaped resist profiles have been generated by different methods, in most cases using trilayer resist systems. A system comprised of PMMA top and

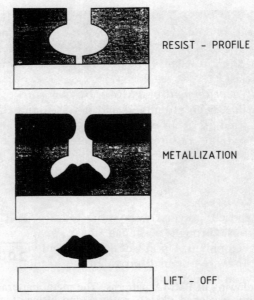

Fig. 3.9. Diagram illustrating the fabrication of T-shaped gates for GaAs field effect transistors.

bottom layers and a middle layer consisting of poly(methylmethacrylate-co-methacrylic acid) was used in combination with EB exposure.[79–81] Another trilayer system, which also was successfully applied, consists of three PMMA layers the top and the bottom layer being of high and the middle layer being of low average molar mass. In this case, SOR radiation from an electron storage ring was used for exposure.[82] Also two-level resist systems were applied, e.g. one with a PMMA top layer and poly(dimethylglutarimide) as bottom layer.[83] Alternatively, a two-level system consisting of PMMA samples of different average molar mass was examined.[84] In a modified process PMMA of high molar mass at the bottom layer and PMMA of low molar mass at the top are used. In this case, irradiation of the top layer with a broad EB and subsequent development is followed by exposure of the bottom layer to a narrow EB and an additional development step.[85] The trench profile can be modified by additional exposure of the areas next to the line center. Gate lines with a foot width of 100 nm were obtained with good reproducibility.[86]

General applications of PMMA in large-scale productions are hampered by low dry-etch-resistance and poor radiation sensitivity. EB exposure doses in the order of 100 $\mu C/cm^2$ are needed to achieve high line resolution with critical dimensions of 0·3 μm.[87] Exposure characteristic curves yield $S(EB) = $ 20–50 $\mu C/cm^2$ depending on average molar mass and development conditions. Because of the low sensitivity, numerous attempts were made to improve the radiation sensitivity of PMMA by modifications in the polymer structure, in resist composition or in process conditions. For example, the addition of 15%

ammonium perchlorate to PMMA improved the EB sensitivity from 24 $\mu C/cm^2$ to 0·8 $\mu C/cm^2$. Actually, the perchlorate does not affect the chain scission efficiency, but enhances the solubility of the polymer in the developer. Notably, addition of more than 15% perchlorate deteriorates the resolution capability of the resist.[88]

Solvent-free *self-development* of PMMA by exposure to SOR radiation at room temperature has been attempted and it was found that the decrease in resist thickness leveled-off at an exposure dose of about 10^5 mJ/cm^2. Irradiation of the resist at 180°C resulted in complete development. The sensitivity was ~1000 mJ/cm^2.[89] Incomplete self-development also was observed with plasma-polymerized PMMA, but complete positive tone development could be achieved in the case of a polymer produced by plasma polymerization of methyl methacrylate (MMA) in the presence of SF_6, tetramethyl tin, I_2 or styrene, if the exposed polymer was subjected to O_2-RIE.[90] Exposure of PMMA to electrons through a stencil mask *in vacuo* also resulted in incomplete self-development. Upon wet development vertical profile structures of 0·5 μm line width were obtained provided the irradiation was performed in the absence of oxygen (under helium).[91]

Copolymers containing MMA moieties. Poly(methyl methacrylate-*co*-tri(*n*-butyl)tin methacrylate) containing 5 wt% tin has an X-ray sensitivity, $S(X)$, of 250 mJ/cm^2. At higher tin contents the copolymer exhibits negative tone characteristics. The copolymer has a remarkably high O_2-RIE resistance.[92] $S(EB) = 0·3$ $\mu C/cm^2$ and a resolution of 0·3 μm was reported[93] for a heat-treated (250°C) poly(methyl methacrylate-*co*-*t*-butyl methacrylate) containing 68 mol% MMA. The heat-treatment converts ester side groups into acid anhydride groups according to the reaction in eqn (17):

$$\text{—CH}_2\text{—C(CH}_3\text{)(COOC(CH}_3\text{)}_3\text{)—CH}_2\text{—C(CH}_3\text{)(COOC(CH}_3\text{)}_3\text{)—} \xrightarrow{\Delta}$$
$$\text{—CH}_2\text{—C(CH}_3\text{)(CO)—CH}_2\text{—C(CH}_3\text{)(CO)—O— (anhydride)} + 2CH_2\text{=}C(CH_3)_2 + H_2O \quad (17)$$

In the case of a similar copolymer consisting of MMA and maleic anhydride moieties the dissolution characteristics were studied. The developer was selected by examining Hansen solubility parameters of solvents, non-solvents and swelling agents.[94]

It was also attempted to increase the resist sensitivity by treating exposed PMMA resist films with a reactive monomer such as methacrylic acid. Macroradicals formed by irradiation initiate the graft-copolymerization of the monomer which causes a strong difference in the solubility of exposed and unexposed areas. Although the radiation sensitivity increased, $S(EB) \approx 1\ \mu C/cm^2$, the line resolution of the system was only about 1 μm.[95] Modification of this system by ion exchange in concentrated calcium acetate solution resulted in $S(EB) \approx 2\ \mu C/cm^2$ and a high O_2-RIE-resistance of the exposed areas.[96]

Other acrylic acid-derived polymers. Apart from copolymers containing MMA many other acrylic acid-derived polymers have been synthesized in the hope of improving the radiation sensitivity and the dry-etch resistance while maintaining the high resolution capability of PMMA. Poly(methacryl anhydride) (PMAH) obtained by heat treatment of poly-*t*-butylmethacrylate can be developed in aqueous alkaline solution due to the formation of carboxylic acid during irradiation. It possesses a high sensitivity ($S(EB) = 2 \cdot 5\ \mu C/cm^2$) and satisfactory line resolution. With an organic developer, $S(EB) = 15\ \mu C/cm^2$ was much lower and the line resolution was unsatisfactory (because of swelling).[97]

Poly(dimethyl glutarimide) (PGMI), which also acts as deep UV resist[98] was applied as top layer in a bilayer system together with PMMA for the fabrication of T-shaped gates.

R: H, alkyl

Here, EB lithography was used in conjunction with deep-UV lithography. $S(EB)$ of PGMI is $\sim 100\ \mu C/cm^2$ (Refs 83 and 99).

Fluorine containing acryl-type polymers have long been known for their high EB and X-ray sensitivity.[11,100,101] However, these polymers are only applicable to processes which do not involve plasma etching because of their very poor dry-etch resistance. For example, poly(hexafluorobutyl methacrylate-*co*-glycidyl methacrylate) (tradename FBM-G) was applied for the fabrication of a *distributed feedback laser diode* (DFB-LD) by SOR lithography.[102]

Another copolymer, poly(trifluoroethyl-α-chloroacrylate-co-tetrafluoropropyl-α-chloroacrylate), consisting of the following structural repeating units was used for the fabrication of chrome masks. $S(EB) = 2\ \mu C/cm^2$ provided the developer is neat methylisobutylketone (MIBK).[103]

```
         Cl                              Cl
         |                               |
—CH₂—C—                           —CH₂—C—
         |                               |              CHF₂
    O=C—O—CH₂CF₃                   O=C—O—C—H
                                                   \
                                                    CHF₂
```

With poly(2,2,2-trifluoroethyl methacrylate) (PTFEM) it was found that the molar mass distribution affects both sensitivity and contrast.

```
         CH₃
         |
┬CH₂—C—┬
         |
    O=C—O—CH₂CF₃ ┘ₙ
```

Under optimum developing conditions the line resolution was $0.5\ \mu m$ in a $0.6\ \mu m$ thick resist layer at $D_{exp} = 8\ \mu C/cm^2$.[104] Nitrile groups containing polymers are interesting because of the improved plasma etch resistance as compared to PMMA. Films of poly(ethyl cyano acrylate), vapor polymerized on silicone wafers, possess an EB sensitivity of $12\ \mu C/cm^2$ and a high contrast.

```
         CN
         |
┬—C—CH₂—┬
         |
    O=C—O—C₂H₅ ┘ₙ
```

These films showed a higher dry-etch resistance than films obtained by spin-coating.[105] Poly(methacrylonitrile-co-methacrylic acid) having a monomer ratio MCN/MAA = 92:8 was cross-linked with tripropyleneglycol diglycidylether (TPG). This material has a high EB sensitivity ($\sim 3\ \mu C/cm^2$), but its line resolution suffers from heavy swelling in the developer (acetonitrile/toluene). Nevertheless, it could be applied in a trilayer system at a thickness of $0.3\ \mu m$.[106]

Poly(methacrylonitrile) (PMCN) has been proposed as potential EB or X-ray resist a long time ago, because of its high radiation chemical yield of main-chain scission $G(S) = 3.3$ (for γ-irradiation)[107] and its dry etch resistance being much higher than that of PMMA.

```
         CH₃
         |
┬CH₂—C—┬
         |
         CN ┘ₙ
```

However, first attempts were disappointing: $S(X)$ ($\approx 1500\,\mathrm{mJ/cm^2}$) was low when solvents such as acetonitrile or benzonitrile were used as developers.[108] However, a modified development procedure leads to a significantly better X-ray sensitivity: prior to development the polymer is 'pre-swollen', i.e. the exposed resist film is dipped into a very weak solvent, for example, methylpropylketone, which swells the polymer but is incapable of dissolving it. Actually, the exposed parts of the film are penetrated selectively by the soaking agent and transformed from the glassy state into the relaxed state. Therefore, the subsequent treatment of the film in a developer based on cyclohexanone proceeds quite fast without a significant attack on unexposed parts. In this way the X-ray sensitivity is enhanced to $S(X) \approx 200\,\mathrm{mJ/cm^2}$.[109] The reproducibility of the results is strongly determined by the *history* of the unexposed films. For example, it is essential to remove completely the casting solvent by pre-baking at 220°C. Notably, the resistance of the unirradiated parts of the PMCN film toward the swelling agent is the higher the lower the molar mass is. The penetrant was found to proceed into the exposed film with a sharp front. The dependence of the penetration velocity, v_{pen}, on the exposure dose, D, conforms to $v_{pen} \propto D^2$. This strong dose dependence is explained in terms of v_{pen} being independently influenced by both the radiation-induced free volume in the resist film and the reduced molar mass due to main-chain scissions.[110]

Cross-linkable polymers

Halogenated aromatic polymers. Halogenated polystyrene derivatives have attracted attention as negatively acting resists for electron or X-ray lithography. The introduction of halogens into aromatic hydrocarbons enhances the absorption of X-rays and also causes an increase in the radiation sensitivity because C–Cl bonds are readily cleaved which results in the formation of macroradicals and of reactive halogen radicals. The abstraction of hydrogens from the polymer by halogen radicals generates additional macroradicals, many of them deactivating by combination forming cross-links:

$$\mathrm{PX} \xrightarrow[e^-]{h\nu} \mathrm{P^{\cdot} + X^{\cdot}} \qquad (18)$$

$$\mathrm{X^{\cdot} + HP} \longrightarrow \mathrm{HX + P^{\cdot}} \qquad (19)$$

$$\mathrm{P^{\cdot} + P^{\cdot}} \longrightarrow \mathrm{P-P} \qquad (20)$$

PX: halogenated polymer; X˙: halogen atom; P˙: macroradical

Radiation-induced intermolecular cross-linking makes the polymer insoluble, i.e. irradiated resist areas are insoluble in any solvent. The halogenated polystyrene resists exhibit the following features: They possess a high dry-etch-resistance due to the high content of aromatic groups. According to Charlesby's gel formation theory, the resist sensitivity is strongly influenced by the molar mass of the polymer. The higher the average molar mass, the higher is the radiation sensitivity. Moreover, the narrower the molar mass distribution the higher is the resist contrast. On the other hand, patterns generated in resist materials of high average molar mass are difficult to develop properly because of strong swelling by the developer. Therefore, sub-micrometer features only are attainable with low-molar-mass samples or by overexposing, which leads to networks of high cross-link density. Although the extent of swelling can be influenced to some extent by properly selecting the developer liquid, swelling cannot be completely suppressed, because cross-linking has little influence on the energy of interaction between polymer and solvent. Commonly, the resolution capability depends on the feature type. Isolated resist lines can be delineated with high resolution whereas isolated spaces, or line and space patterns are not resolved due to the formation of bridges between adjacent resist lines.

Among halogenated aromatic polymers chlorinated polystyrenes have been investigated most thoroughly. In most cases these polymers were synthesized by either chloromethylation of polystyrenes or by chlorination of poly-p-methylstyrene.[111,112] In both cases the reaction proceeds unspecifically, i.e. the benzene rings of polystyrene can be chloromethylated at different positions (*para, meta, ortho*) and the chlorination of poly-p-methylstyrene can occur, apart from the benzene rings, also at the backbone in α or β position. In this way, chlorine contents corresponding to more than one chlorine atom per structural repeating unit are attainable. Notably, the modifications can be performed with polymers of narrow molar mass distribution.[113–115] Selective monochlorination of methyl groups attached to the benzene ring was achieved with the aid of sodium hypochlorite and a phase transfer catalyst. However, the molar mass distribution was broadened by this procedure.[116]

The X-ray sensitivity of these polymers, expressed in exposure dose units, strongly depends on the X-ray wavelength. It should be wavelength-independent if based on the absorbed dose.[117] The chlorine content of chlorinated poly(p-methylstyrene) also affects the resist sensitivity: the optimum sensitivity corresponds to an intermediate chlorine content. Excessive chlorination also leads to chlorine substitution at the backbone which causes, upon irradiation, chain-scission, apart from cross-linking.[117,118] A detailed investigation of the reaction mechanism by ESR measurements revealed the formation of benzyl radicals and gas chromatographic (GC) product analysis showed that hydrogen abstraction by chlorine atoms occurs mainly at the backbone in α-position. The combination of α-carbon with benzyl radicals is favored over the self-combination of α-carbon radicals and benzyl radicals. The mechanism is illustrated in Scheme 3.3.[119]

Notably, main-chain scission does not occur if a methyl group is located in *o*-position. This was concluded from results of a study on the following model compounds, that had been selectively chlorinated at specific positions: *o*-, *m*-, and *p*-methylstyrene, 2,4- and 2,5-dimethylstyrene and 2,4,6-trimethylstyrene. Patterns generated in correspondingly composed polymers have a rather high contrast.[120]

Formation of Macroradicals:

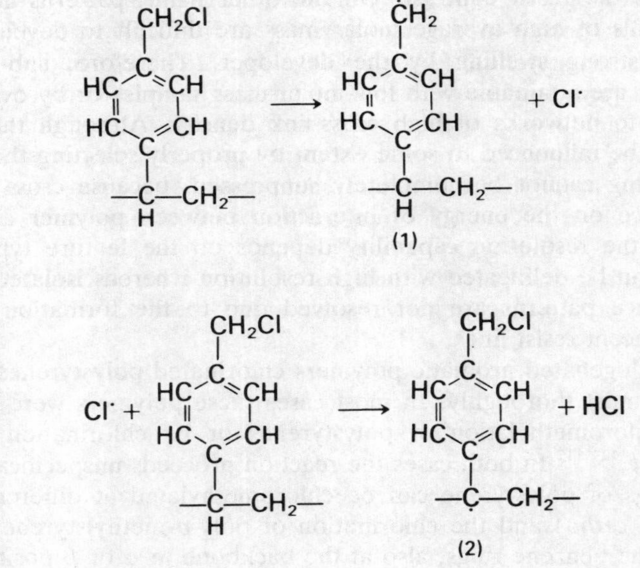

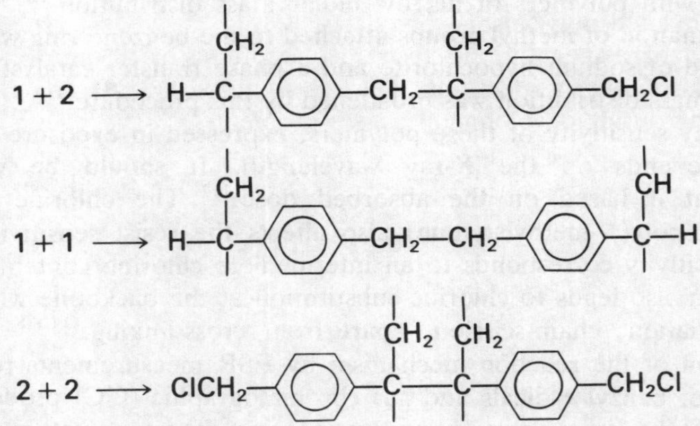

Scheme 3.3. Radiation-induced cross-linking of chloromethylstyrene-based resists.

Interestingly, several electronic devices have been fabricated using chlorine-containing polystyrene derivatives, despite the existing drawbacks due to swelling during development. Poly(chloromethylstyrene) was used for the fabrication of masks.[121] For the fabrication of III–V semiconductor devices, 40 nm structures were delineated in a 60 nm thick film of chlorinated poly-α-methylstyrene.[122] Chlorinated poly(p-methylstyrene) was applied for the production of SAW devices having critical dimensions of 0·6–1·7 μm.[123] A

100 nm gate was delineated in poly(chloromethylstyrene) with a rather high exposure dose ($D_{exp} = 130\ \mu C/cm^2$) of 40 keV EB radiation.[124] Interestingly, a 1:1 blend of poly(p-methoxystyrene), ($M_w = 110\,000$, $M_w/M_n = 1 \cdot 2$) and poly(chloromethylstyrene) ($M_w = 20\,000$, $M_w/M_n = 1 \cdot 3$) had a sensitivity, $S(EB) = 3\ \mu C/cm^2$, higher than that of the single resist components and the contrast was high inspite of the broad molar mass distribution of the blend.[125]

The swelling of poly(chloromethylstyrene) in the developer can be strongly suppressed by modifying the development process: $0 \cdot 2\ \mu m$ lines and spaces were delineated in $0 \cdot 8\ \mu m$ thick resist films by repeatedly applying the cycle: developing (0·2 s)/rinsing (5 s)/drying (5 s).[230]

Poly(trimethylgermylmethyl methacrylate-co-chloromethylstyrene) was shown to be a sensitive negatively acting EB resist, $S(EB) = 2-3\ \mu C/cm^2$, having good resistance to O_2-RIE. It was used as top layer in a two-level resist scheme.[126]

Chloromethylated poly(2-isopropenylnaphthalene) was examined with respect to its properties as negative EB resist. Isolated lines of 100 nm were fabricated with the aid of 20 keV electrons. The sensitivity was $21\ \mu C/cm^2$. Also in this case swelling during development causes problems.[127]

Polyimides. Polyimides are of importance for semiconductor technology. After curing they are quite heat resistant and, therefore, quite attractive for many applications. Many attempts have been made to utilize this class of polymers for lithographic purposes. Since after curing polyimides are insoluble in any solvent, the development of irradiated films has to be performed prior to curing. Post-development curing, on the other hand, often causes deformations of developed patterns together with limited line resolution. A conventional polyimide exposed *in vacuo* to a pulsed EB was not very sensitive: $S(EB) = 300\ \mu C/cm^2$. N-Methylpyrrolidone or aqueous alkaline solution are appropriate developers.[128] When polyimide films were doped with salts of heavy metals

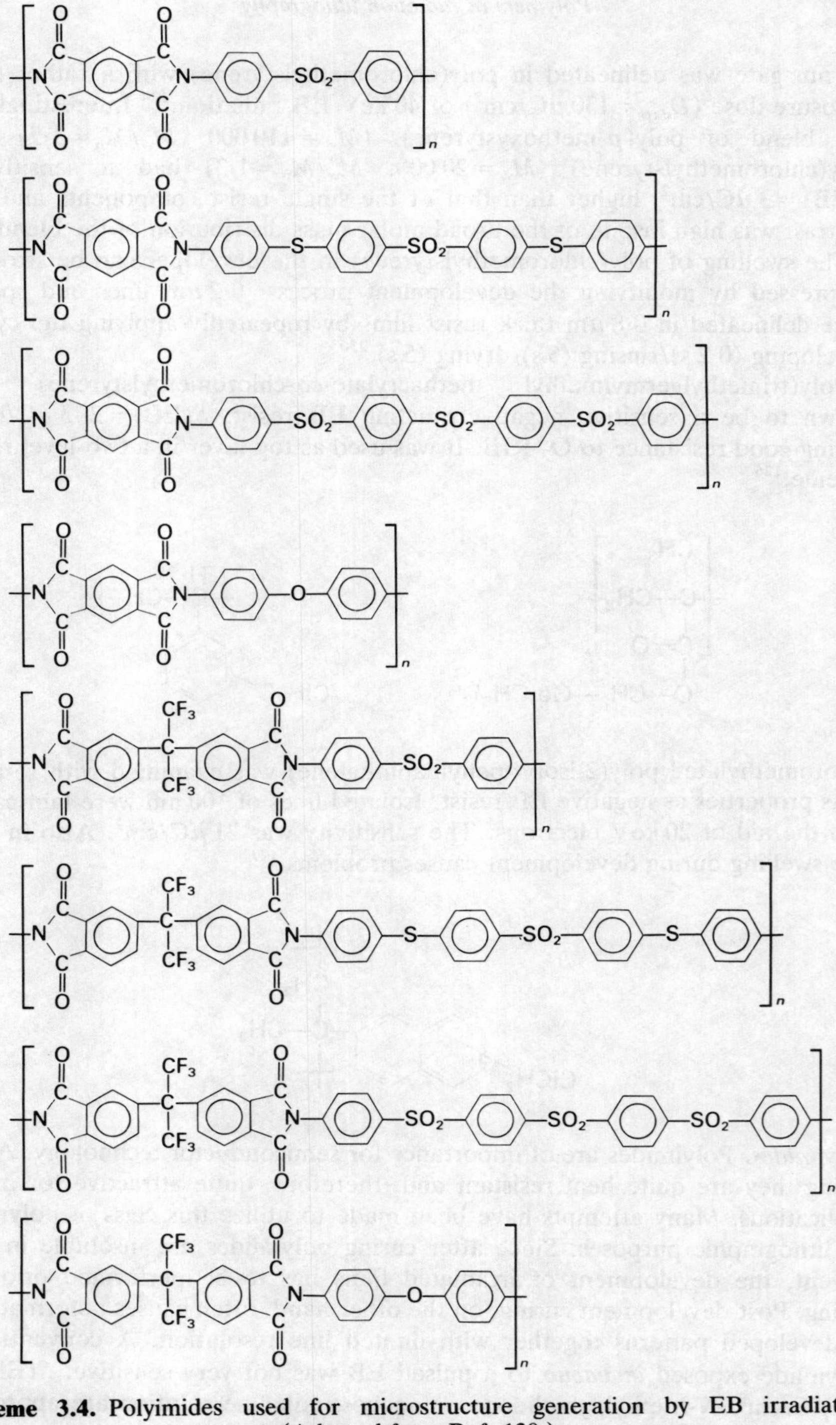

Scheme 3.4. Polyimides used for microstructure generation by EB irradiation. (Adapted from Ref. 130.)

such as silver, gold or copper, metal films were formed at the surface during curing.[129]

Polyaryline imides (for typical chemical structures see Scheme 3.4[130]) imidized prior to exposure to about 90% were quite sensitive toward EB irradiation. Optimum results pertain to $S(EB) = 1.5-2.5 \mu C/cm^2$ (20 keV electrons) and 1 μm line resolution. Upon exposure, main-chain scission occurs simultaneously with further imidization. In most cases imidization dominates over main-chain scission and, therefore, the systems exhibit negative tone behavior.[130]

Other resist materials. Polydiacetylenes with urethane side groups: poly(5,7-(bis-1,12-n-butyl-carboxy-methylene-urethane) dodecadiyne), P4BCMU, (see Scheme 3.5 for chemical structure) acted as negative resists upon irradiation with 50 keV electrons. The insolubilization of exposed areas is caused by the formation of imide type bonds in the side groups rather than by reactions of C≡C bonds in the main-chains.[131] The resolution was 80 nm.[132]

A styrene ammonium sulfonate polymer (AmPPS) of high average molar mass has an EB sensitivity $S(EB)_{0.5} = 25 \mu C/cm^2$. Water and methanol are appropriate developers. Problems of pattern displacement due to electrical

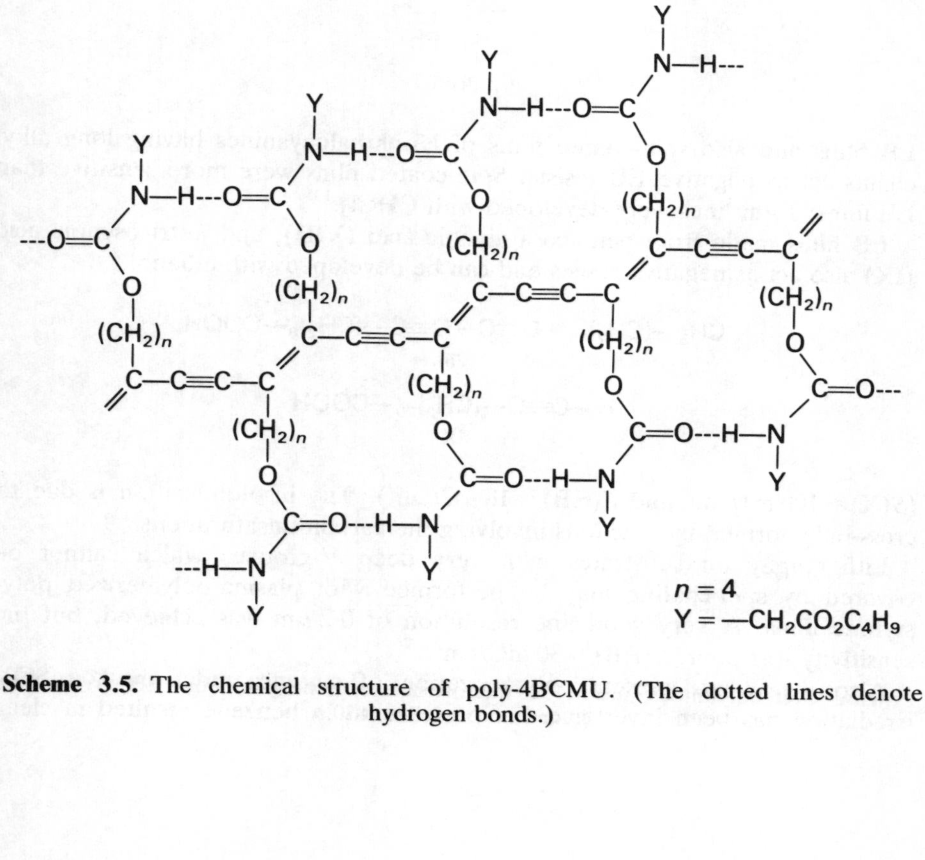

Scheme 3.5. The chemical structure of poly-4BCMU. (The dotted lines denote hydrogen bonds.)

charge effects did not occur, as the polymer has a high electrical conductivity.[133]

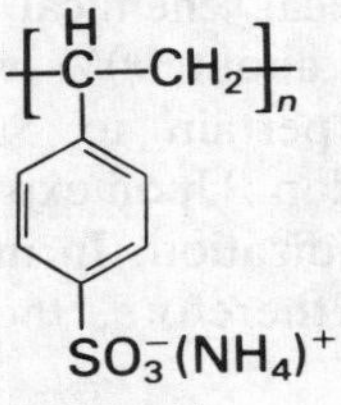

$$\left[\begin{array}{c}H\\|\\C-CH_2\\|\\\\\\\\SO_3^-(NH_4)^+\end{array}\right]_n$$

AmPPS

Poly(diphenoxyphosphazene) (PDPP) containing a sensitizer (10 wt% CBr_4) is very sensitive: $S(EB) = 0.7\ \mu C/cm^2$ and $S(X)$ 55 mJ/cm^2. Patterns can be developed in a 3:1 mixture of 2-ethoxyethyl acetate and hexane. The resist was applied as a top layer in a two-level resist system.[134]

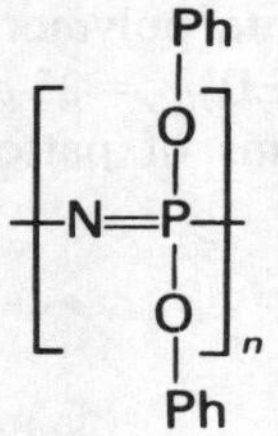

$$\left[\begin{array}{c}Ph\\|\\O\\|\\-N=P-\\|\\O\\|\\Ph\end{array}\right]_n$$

PDPP

LB films and also spin-coated films of Ni-phthalocyanines having long alkyl chains act as negative EB resists. Spin-coated films were more sensitive than LB films; 1 μm lines were developed with $CHCl_3$.[135]

LB films made from pentacosa-diynoic acid (**VIII**), and ω-tricosynoic acid (**IX**) also act as negative resists and can be developed with ethanol.

$$CH_3-(CH_2)_{11}-C\equiv C-C\equiv C-(CH_2)_8-COOH$$
VIII

$$H-C\equiv C-(CH_2)_{20}-COOH$$
IX

($S(X) \approx 100\ mJ/cm^2$ and $S(EB) \approx 10\ \mu C/cm^2$). The insolubilization is due to cross-links formed by reactions involving the carbon unsaturations.[136]

Lithography on substrates with very deep V-grooves which cannot be covered by spin-coating may be performed with plasma-polymerized polystyrene films. A very good line resolution of 0.2 μm was achieved, but the sensitivity was poor: $S(EB) \approx 30\ mC/cm^2$.[137]

Also *patternwise* plasma polymerization of styrene under masked SOR irradiation has been investigated. Development in benzene resulted in clean patterns. It turned out that the lower energy part of the SOR spectrum (35–200 Å (3·5–20 nm)) has to be applied to obtain insoluble patterns.[138]

Inorganic resists. Apart from organic polymers inorganic materials have also been assessed for lithographic applications. For example, materials based on peroxohetero(carbon)poly-tungstic acid (empirical formula: $CO_2.12WO_3.7H_2O_2.nH_2O$, $n \approx 25$) exhibited interesting properties upon doping with niobium: films spin-coated from aqueous solution and developed with sulfuric acid (pH 2) acted as negative resists. The niobium content strongly influences sensitivity and contrast. For an atomic ratio $Nb/(Nb + W) = 0.17$ the following values were reported: $S(EB) = 10 \: \mu C/cm^2$ (for 30 keV electrons) and $S(X) = 120 \: mJ/cm^2$ (Mo, L X-rays). The resist was used as top-layer in a bilayer system.[139] Metallic tungsten lines could be fabricated with this resist by reduction with hydrogen as can be seen from Fig. 3.10.[140,141]

Applications of the dissolution inhibitor principle

Diazonaphthoquinone inhibitors

Positive resists of the dissolution inhibitor type still play a prominent world-wide role in the production of microelectronic devices. Currently, diazonaphthoquinone sulfonates, DNQ, (**I** and **II**) embedded in a phenolic resin (novolak) matrix are used in conjunction with photolithography. DNQ compounds prevent the dissolution of the novolak resin in aqueous alkaline solution. Upon irradiation, DNQ is converted into soluble products with the consequence that the whole system becomes soluble in the aqueous alkaline developer. DNQ/novolak systems offer various advantages such as safe development without swelling or deformation of the microstructures. Noticeably, these systems exhibit a high dry-etch-resistance due to the high content of aromatic groups. Consequently, the question arose as to whether DNQ/novolak systems were also appropriate for high-energy radiation lithography. Actually, conventional DNQ/novolak systems, used for photolithography, are applicable both for EB and X-ray lithography. However, the sensitivity is poor: $S(EB) \approx 50 \: \mu C/cm^2$ and $S(X) \approx 1800 \: mJ/cm^2$ (SOR),[142,143] $S(X) \approx 300 \: mJ/cm^2$ (X-rays produced by laser focus).[144] Figure 3.11 shows typical exposure characteristic curves obtained at EB irradiation of an AZ-type resist,[145] and Table 3.5 presents results obtained with various diazonaphthoquinone sulfonates.[146] In this study it turned out that the 4-derivative acts more sensitively than the 5-derivative. By contrast, a higher sensitivity for the 5-derivative was reported by another group.[147] This discrepancy might be due to the fact that different novolak polymers were used, as it is well known that the dissolution inhibition efficiency depends on the mutual interaction between novolak and dissolution inhibitor.[148]

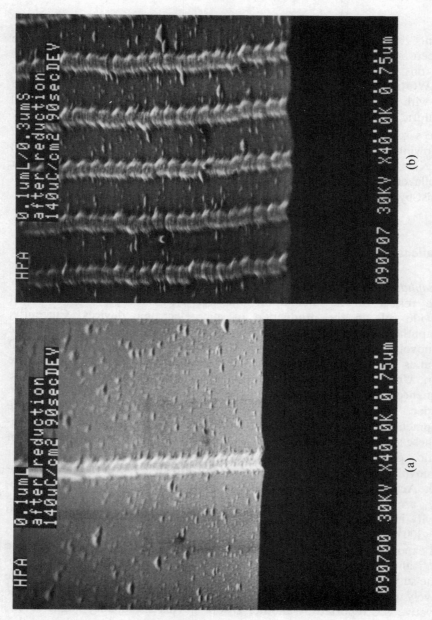

Fig. 3.10. Fine line metallic patterns generated with the aid of an inorganic resist: peroxohetero(carbon)polytungstic acid. The lines were obtained by hydrogen reduction of the developed resist at 610°C for 30 min. $D_{exp} = 140\ \mu C/cm^2$. (a) 0.1 μm line, (b) 0.1 μm lines and 0.3 μm spaces. (From Refs 140 and 141, Courtesy of Dr. T. Yoshimura.)

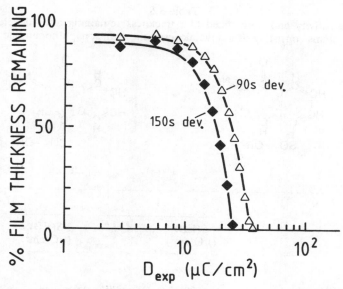

Fig. 3.11. EB exposure characteristic curve of a 0·5 μm resist layer (diazonaphthoquinone/novolak). Time of development in aqueous alkaline solution: 90 and 150 s. (From Ref. 145, reproduced by permission of the American Institute of Physics.)

The low radiation sensitivity of DNQ/novolak resists is due to the fact that high-energy radiation, in contrast to UV light, is absorbed unspecifically. There are no chromophoric groups sensitive selectively toward high-energy radiation. Therefore, apart from the desired reaction, the formation of indene carboxylic acid (**IV**) according to eqn (3), also other chemical reactions leading to new products are induced by X-rays and EB radiation. This implies that part of the absorbed energy is used to form new products and that, therefore, EB radiation or X-rays generate less indene carboxylic acid in terms of molecules per energy unit absorbed by the resist than UV light. (It should be pointed out here that, commonly, radiation chemical yields are expressed in units of molecules converted or formed per 100 eV of absorbed energy because yields related to the number of absorbed electrons or photons (quantum yield) have no meaning.) It was demonstrated that the new products can affect the dissolution rate in exposed areas.[147] For equal degrees of DNQ conversion, the increase in dissolution rate is higher for UV that for EB exposure.

Many attempts were made to improve the high-energy radiation sensitivity and the contrast of microstructures. Rather high sensitivities, obtained with a non-specified phenolic resin/DNQ system have been reported: $S(EB) = 16\,\mu C/cm^2$ (10 keV (Ref. 149) or 20 keV (Ref. 150) electrons) and $S(X) = 100\,mJ/cm^2$ (soft X-rays from a laser plasma source, $\lambda \approx 1\cdot4$ nm). Notably, the latter value corresponds to 20–30% conversion of DNQ contained in the resist.[151–153] The addition of N-heterocyclic compounds, such as imidazole, to a

Table 3.5
Radiation sensitivity and normalized film thickness remaining after development for composite systems consisting of a novolak resin and a diazonaphthoquinone derivative[a]

R	I		II	
	$S(EB)$ ($\mu C/cm^2$)	d/d_0[b]	$S(EB)$ ($\mu C/cm^2$)	d/d_0[b]
—C$_6$H$_4$—CH$_3$ (para)	38	0.70	48	0.55
—C$_6$H$_4$—CH$_3$ (meta, H$_3$C)	37	0.85	55	0.65
—C$_6$H$_4$—CH$_3$ (ortho)	40	0.85	65	0.65
—C$_6$H$_4$—C(O)—C$_6$H$_5$	—	—	40	0.80
—C$_6$H$_4$—C(CH$_3$)$_2$—C$_6$H$_5$	35	0.90	50	0.75
—C$_6$H$_4$—CH(CH$_3$)$_2$	35	0.90	55	0.70

[a] From Ref. 146, reproduced by permission of the American Institute of Physics.
[b] d and d_0: film thickness after and before development, respectively.

novolak/DNQ resist resulted in a rather high sensitivity: $S(EB) = 2\ \mu C/cm^2$. Imidazole forms a charge transfer complex with DNQ, as evidenced by UV–vis spectroscopy. The latter decomposes more readily than uncomplexed DNQ. Notably, imidazole in excess caused a diminution in contrast and line resolution.[154]

Improvements in contrast and line profile were brought about by modifying the development process: the development was interrupted by a rinse with water and subsequent drying.[145] Treatment of the resist with the developer solution prior to EB exposure also resulted in better contrast.[155]

Noteworthy are also attempts aiming at better contrast values by optimizing developer conditions (temperature and alkali concentration). The contrast became better by lowering the alkali concentration and increasing the temperature.[156]

Polymeric dissolution inhibitors

Certain polymers that can be blended with phenol–formaldehyde or cresol–formaldehyde resins act as dissolution inhibitors for the resins. Some of these polymers became prominent for high-energy lithography because of their capability of undergoing radiation-induced depolymerization, i.e. decomposition into compounds of low molar mass and high volatility. Typical examples are poly(olefin sulfones) and polyaldehydes with the following structural repeating units:

$$\left[-CH_2-\underset{R^2}{\overset{R^1}{C}}-\underset{O}{\overset{O}{S}}- \right]_n \qquad \left[-\underset{R}{\overset{H}{C}}-O- \right]_n$$

Poly(olefin sulfone)   Polyaldehyde

Since its ceiling temperature is lower than room temperature poly(2-methyl-1-pentene sulfone) (PMPS) depolymerizes spontaneously at room temperature under X-ray or EB irradiation. Once active centers in the main-chain are generated, 2-methylpentene and sulfur dioxide are formed. Because oxygen strongly retards depolymerization a free-radical mechanism is assumed for this process.[157] Initially, main-chain cleavage occurs:

$$-CH_2-\underset{CH_2CH_2CH_3}{\overset{CH_3}{\underset{|}{\overset{|}{C}}}}-\underset{O}{\overset{O}{S}}- \xrightarrow{rad.} -CH_2-\underset{CH_2CH_2CH_3}{\overset{CH_3}{\underset{|}{\overset{|}{C}}}}\cdot \ + \ \cdot\underset{O}{\overset{O}{S}}- \qquad (21)$$

The resulting macroradicals react with oxygen or initiate depolymerization according to the reactions in eqns (22a) and (22b), respectively:

$$-CH_2-\underset{\underset{CH_2CH_2CH_3}{|}}{\overset{\overset{CH_3}{|}}{C}}-\underset{\underset{O}{\|}}{\overset{\overset{O}{\|}}{S}}-CH_2-\underset{\underset{CH_2CH_2CH}{|}}{\overset{\overset{CH_3}{|}}{C^{\cdot}}} + \xrightarrow{O_2} \begin{array}{l} -CH_2-\underset{\underset{CH_2CH_2CH_3}{|}}{\overset{\overset{CH_3}{|}}{C}}-O-O^{\cdot} \quad (22a) \\ \\ \longrightarrow CH_2=\underset{\underset{CH_2CH_2CH_3}{|}}{\overset{\overset{CH_3}{|}}{C}} + SO_2 \quad (22b) \end{array}$$

Poly(olefin sulfones) not decomposing spontaneously under irradiation depolymerize upon post-exposure heating. Therefore, commonly neither liquid nor dry development procedures are needed for feature transformation. In a recent study it was demonstrated that microstructures in PMPS are self-developing upon irradiation of the polymer with a focused EB ($E > 20$ eV).[158]

After the introduction of novolak/PMPS blends to EB lithography more than 10 years ago[159] numerous studies aiming at optimum properties have been performed. Good compatibility of the two polymeric phases has been attained by selecting appropriate novolak resins, by using as a dissolution inhibitor a terpolymer composed of 2-methyl-1-pentene, methylallylether and sulfur dioxide[160,161] and by applying a proper solvent for film casting (isoamylacetate).[162] In the latter case $S(EB) = 3 \mu C/cm^2$ at $E = 15$ keV and $5 \mu C/cm^2$ at $E = 30$ keV was found.[163] $S(EB)$ and $S(X)$ decreased drastically when the resist was exposed in the presence of oxygen. This problem was overcome by coating the resist surface with an oxygen-blocking polyvinyl alcohol film ($1 \mu m$).[164] Contrast and profile of the resist are improved by interrupted development as in the case of novolak/DNQ systems.[165] Typical microstructures obtained in this way are shown in Fig. 3.12.[165] Alternatively, a good contrast is obtained by soaking the irradiated resist film in chlorobenzene for some minutes. In this way, novolak oligomers are extracted from the surface which increases the resistance of the unexposed areas of the film towards the aqueous alkaline developer.[166] A change in $S(X)$ from 500 mJ/cm^2 to 150 mJ/cm^2 was brought about by coating the film surface with an extra layer of PMPS. Post-exposure baking totally removed the top layer at exposed areas by self-development, while it protected the unexposed parts during wet development.[167] $S(X) = 0.42$ mJ/cm^2 was reported for soft X-ray exposure ($\lambda = 1.2–1.6$ nm) of this resist system.[157] At very high exposure doses ($D = 500$ mJ/cm^2) the mode of action converted from positive to negative tone behavior. In the case of a special novolak polymer turn-over from positive tone to negative tone action occurs at $D_{exp} > 2000 \mu C/cm^2$.[168]

Notably, a novolak/PMPS resist was applied for the fabrication of T-shaped gates, using a special half-tone X-ray mask.[169]

The utilization of systems based on the application of both the *acid-catalyzed*

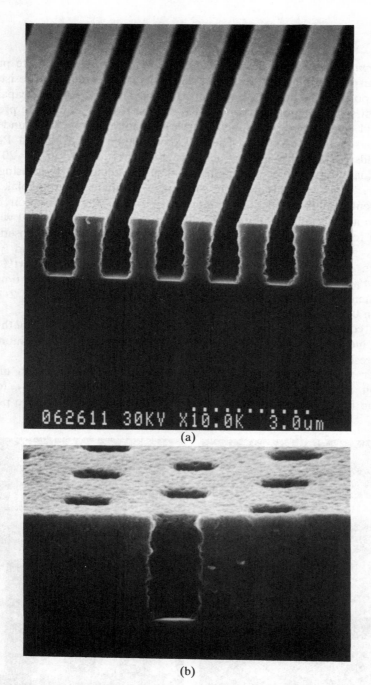

Fig. 3.12. Microstructure pattern generated in novolak containing poly(2-methyl-1-pentene sulfone) by EB radiation. (a) 0·7 μm lines and spaces, $d_{film} = 1·6\,\mu m$, $D_{exp} = 40\,\mu C/cm^2$. Development in 12 steps at 10 s. (b) 0·4 μm holes, $d_{film} = 1·5\,\mu m$, $D_{exp} = 120\,\mu C/cm^2$. Development in eight steps at 15 s. (From Ref. 165, reproduced by permission of the American Institute of Physics.)

chemical amplification and the *dissolution inhibition principle* is quite promising with respect to enhanced radiation sensitivity of phenolic resin resists. For example, polyphthalaldehydes end-capped by acylation (PPA) are capable of acting as dissolution inhibitors for novolak resins. If irradiated in the presence of an acid generator (preferably triaryl sulfonium salts) PPA undergoes spontaneous depolymerization. Thermal development of irradiated PPA at 160°C yields good quality microstructures ($S(EB) = 1 \,\mu C/cm^2$ at $E = 20$ keV). The residual non-volatile sulfonium salt can be removed by rinsing with 2-propanol.[5] PPA is highly miscible with cresol–formaldehyde novolak resins and efficiently inhibits, at concentrations <10 wt%, the dissolution of the novolak in aqueous alkali solution. Positive tone images are provided with EB irradiated formulations by post-exposure baking when PPA in exposed areas undergoes acid-catalyzed depolymerization. $S(EB) = 1 \,\mu C/cm^2$ and $\gamma = 6 \cdot 3$ have been reported[5] for a novolak formulation containing 5·7 wt% PPA and 8·5 wt% 4-thiophenoxyphenyl(diphenyl)sulfonium hexafluoroantimonate. Development in aqueous potassium hydroxide solution for at least 2·45 min resulted in 0·5 μm line/space patterns.

In this connection it is noted that also a copolymer of 3-trimethylsilylpropanal and 3-phenylpropanal functions as polymeric dissolution inhibitor for phenolic resins and can be used for EB and X-ray imaging.[170]

Another resist system based on both the chemical amplification and the dissolution inhibition principle has been commercialized recently for the application in X-ray and EB lithography.[32,171] Typical microstructure patterns

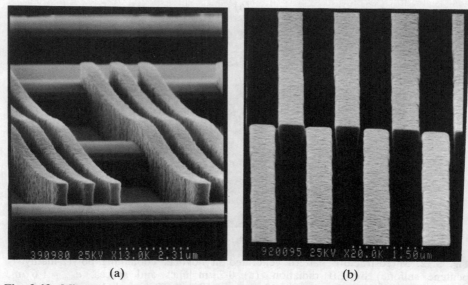

(a)   (b)

Fig. 3.13. Microstructure patterns generated in resist film AZ-PF 514 ($d_{film} \approx 1 \cdot 1 \,\mu m$) by (a) X-rays (SOR) and by (b) 30 keV EB radiation. (a) Line width: 0·7 μm. (b) Width of lines and spaces: 0·7 μm. (Courtesy of Dr G. Pawlowski and Mr. W. Meier, Hoechst AG.)

are shown in Fig. 3.13. The detailed composition of the resist has not yet been made public. The resist is said to contain, apart from a binder polymer, e.g. a novolak resin, an acetal (AC) or an orthoester compound (OE) as dissolution inhibitors, and trichloromethyltriazines or tetrabromobisphenol A (see Table 3.1) as acid generators.[172,173]

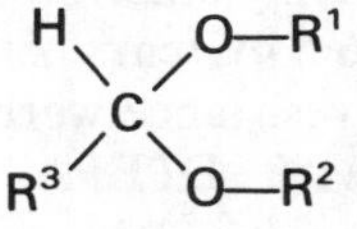

AC 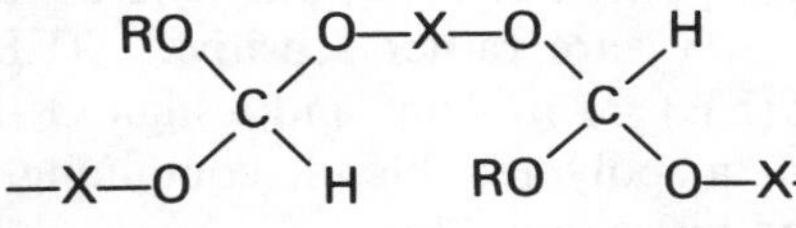   OE

Very satisfactory sensitivities were reported: $S(X) = 30\,mJ/cm^2$ for SOR[171] and 10–33 mJ/cm^2 for soft X-rays;[151,174] $S(EB) = 9\,\mu C/cm^2$ at $E = 50\,keV$ for a film thickness $d = 200\,nm$.[175] Various process parameters such as post-exposure bake conditions have been examined for this resist system.[176] From a computer-simulation study it was concluded that the rough surface of developed microstructures is attributable to the statistical distribution in which the X-ray photons are absorbed by the resist (shot noise).[177]

Network formation via cross-linking agents

There have been many attempts to enhance the sensitivity of negatively acting polymer resists with the aid of cross-linking agents. Some of the more recent investigations which were aiming at utilizing commercially or otherwise cheaply available polymers will be reviewed here. Since in the case of negative resists liquid development of microstructures is often difficult because of swelling of the unexposed resist parts it seems appropriate to arrange the different studies according to the mode of development. Actually, both liquid and dry development techniques have been applied.

Polymers suitable for liquid development

A five to 20-fold increase in $S(EB)$ was observed upon the addition of 20 wt% trimethylolpropanetrimethacrylate (TPTM) or dipentaerythritylpentaacrylate (DPEPA) to poly(chloromethylstyrene) or to a commercial terpolymer composed of vinyl chloride, vinyl acetate and maleic acid.[178]

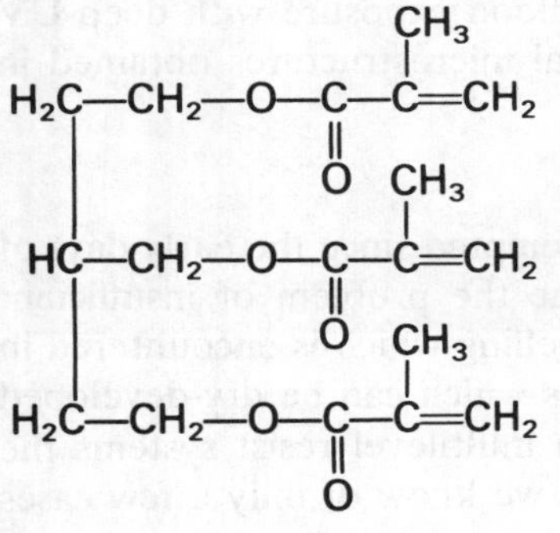

TPTM 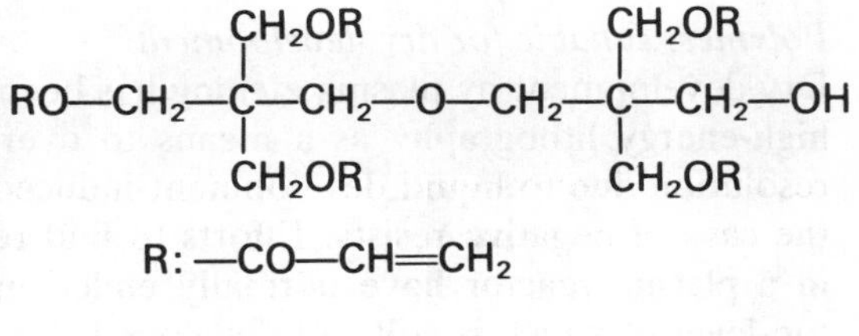   DPEPA

Problems concerning post-exposure hardening *in vacuo* and swelling during development were reported. Swelling was not important when DPEPA was used to cross-link PMMA.[179] A rather high sensitivity $S(EB)_{50} = 4\ \mu C/cm^2$ was found. An expected deterioration of the contrast due to main-chain scission of PMMA was not observed. Commercially available maleic anhydride copolymers containing either styrene, ethylene or methylvinylether cross-linked with DPEPA (20 wt%) are rather sensitive: $S(EB) \approx 6.5\ \mu C/cm^2$. A much higher sensitivity, $S(EB) = 1\ \mu C/cm^2$ and a high O_2-RIE resistance were measured in the case of a polymer blend containing 20 wt% DPEPA and 10 wt% bis(tri-*n*-butyl tin) oxide.[180]

Organic liquids are used for development of the systems dealt with so far in this section. However, it should be pointed out that there are also systems which are developable in aqueous solutions. In this connection the reader's attention is drawn to the system poly(*p*-vinyl phenol)/3,3'-diazidodiphenyl sulfone, the latter compound acting as cross-linking agent.[181]

Poly(*p*-vinyl phenol)
(Poly(*p*-hydroxystyrene))

3,3'-Diazidodiphenyl sulfone

This system performs as a negative tone resist, if developed in an aqueous solution containing 1·19 wt% TMAH. In this case there are no problems due to swelling. At higher TMAH concentration (2·5 wt%) the exposed areas also become soluble. The dissolution rate of the resin strongly depends on the average molar mass of the polymer. Notably, the exposure dose needed to render the exposed areas insoluble is much lower than the gel dose of the polymer. This resist which is sensitive to deep-UV as well as to EB radiation and X-rays has been extensively tested. Some results are reported here: $S(EB) = 12\ \mu C/cm^2$ at $E = 20\ keV$,[87] $10\ \mu C/cm^2$ at $E = 30\ keV$,[162] and $50\ \mu C/cm^2$ at $E = 50\ keV$.[182] The resolution limit is at about 0·25 μm. Contrast and sensitivity were improved by an additional flood exposure with deep-UV light that made the resist surface hard.[183] Typical microstructures obtained in this way are shown in Fig. 3.14.

Polymers suitable for dry development
Dry development by plasma etching has been attempted since the early days of high-energy lithography as a means to overcome the problem of insufficient resolution due to liquid development-induced swelling which is encountered in the case of negative resists. Efforts to find resists which can be dry-developed in a plasma reactor have normally ended up in multilevel resist systems the top-level of which is still wet-developed. So far, we know of only a few cases

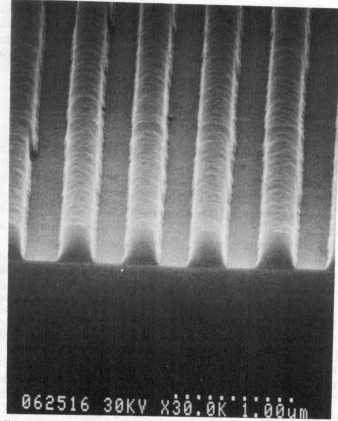

Fig. 3.14. Microstructure patterns (0·3 μm lines) generated in novolak containing bis-3,3'-azido diphenyl sulfone by EB radiation. (From Ref. 183, reproduced by permission of the American Institute of Physics.)

where radiation alters the surface of the resist film so much that plasma etching results in a significant discrimination between exposed and unexposed areas. One example is poly(methylisopropenyl ketone) (PMIPK) employed in conjunction with 2,6-bis(4'-azidobenzylidene)-4-methylcyclohexanone-1 (**X**) as a cross-linking agent.

The resist was cross-linked by X-ray exposure. Post-baking occurred at 250°C for 60 s and the development was performed in an oxygen plasma at 105°C (100% O_2, 0·8 torr (106·4 Pa), gas flow: 200 standard cm^3/min). In this way vertical resist profiles were obtained, but the sensitivity was poor: $S(X) = 1380$ mJ/cm^2.[184]

Applications of the chemical amplification principle

Deprotection

The utilization of radiation-induced chain reactions or catalyzed reactions has brought about a breakthrough with respect to the required very high radiation sensitivity of resists. Since single radiation-induced elementary acts can initiate the conversion of a large number of molecules the term *chemical amplification* is used for this kind of processes. In a general way the chemical amplification principle has been treated in the section entitled 'Chemical amplification'—see p. 125, and resist systems functioning according to this principle have been referred to occasionally in the text (for example, the section entitled 'Polymeric dissolution inhibitors'—p. 163). At this stage, we would like to describe in a concealed mode further examples which are especially demonstrating the strong power of radiation-induced acid-catalyzed or base-catalyzed reactions for the generation of microstructures.

After the acid-catalyzed deprotection of poly(*t*-butoxycarbonyloxystyrene) (PBOCST) and poly(*t*-butyl-*p*-vinyl benzoate) (PTBVB) has been proposed,[37] similar work was performed in various laboratories. Deprotection, as described by eqn (7), and analogous reactions are catalyzed by protons which are formed in the radiolysis of *acid generators* (see Table 3.1). In the case of PBOCST and PTBVB onium salts such as triphenylsulfonium hexafluoroarsenate ($Ph_3S^+AsF_6^-$) can serve as acid generators. The strong change in polarity allows either positive or negative tone imaging depending on the polarity of the developing liquid. With organic liquids negative imaging without swelling of the microstructures has been demonstrated.[37] With the aid of negative tone imaging, 18 nm wide lines were delineated in a 54 nm thick resist film coated on a Si_3N_4 membrane with a special EB writer operated at 50 kV. Positive tone imaging yielded a line resolution of ~40 nm.[185] The sensitivity of PBOCST containing 15 wt% $Ph_3S^+SbF_6^-$ is high: $S(EB) = 2-3\ \mu C/cm^2$ at $E = 20$ keV (Ref. 161) and $S(X) = 13$ mJ/cm^2 (SOR).[186] An investigation of the effect of pre-baking temperature on the line profile revealed that stronger undercut profiles are formed at lower temperatures.[155] In the case of PTBVB a higher sensitivity was measured (in negative tone imaging): $S(EB)_{0.5} = 0.5\ \mu C/cm^2$. The stability of the developed microstructures proved to be better than in the case of PBOCST.[161] Statistical copolymers of *t*BOC-protected *p*-hydroxystyrene and sulfur dioxide (**XI**) (monomer ratio 2:1 and 3:1) exhibit a rather high thermal stability. They were applied as positively acting resists in conjunction with onium salts ($Ar_3S^+AsF_6^-$ or $Ar_3S^+SbF_6^-$) and 2,6-dinitrobenzyl tosylate (**XII**) as acid generators.

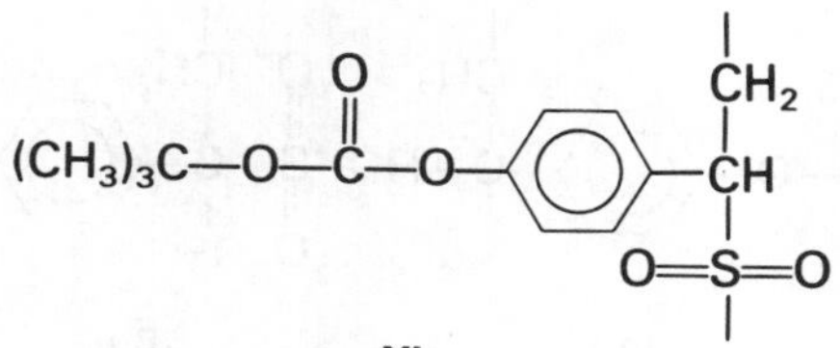

XI: (CH₃)₃C—O—C(=O)—O—C₆H₄—CH(CH₂)—S(=O)₂— (structure shown)

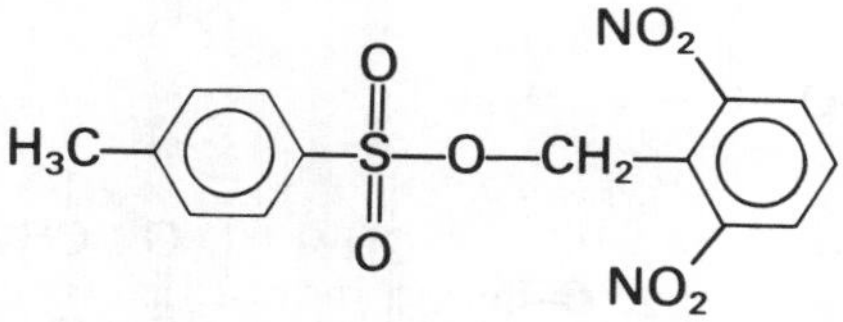

XII: H₃C—C₆H₄—S(=O)₂—O—CH₂—C₆H₃(NO₂)₂ (structure shown)

The following sensitivity values were reported: $S(EB) = 90\ \mu C/cm^2$ and $S(X) = 10\ mJ/cm^2$ (soft X-rays). The resolution was 0·25 μm.[187] Notably, the copolymer was found rather sensitive toward EB radiation or X-rays even in the absence of acid generators.[29] This behavior has been explained in terms of acid formation resulting from a cleavage of C–S bonds leading to sulfinyl radicals. Eventually, the latter are converted into sulfinic acid. Moreover, the radiolysis product sulfur dioxide was suggested to form sulfurous acid. This mechanism is illustrated in Scheme 3.6.

An interesting application of the deprotection chemistry concerns triaryl sulfonium salts acting both as acid generators and dissolution inhibitors if combined with phenolic polymers such as poly(*p*-vinyl phenol).[188] A typical sulfonium salt of this kind is tris(4-*tert*-butoxycarbonyloxyphenyl)sulfonium triflate (**XIII**):

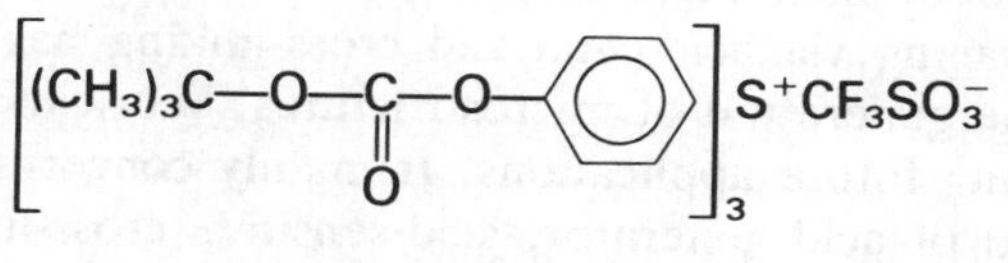

$$[(CH_3)_3C-O-C(=O)-O-C_6H_4-]_3 S^+ CF_3SO_3^-$$

XIII

Irradiation of this salt by UV light or X-rays results in the formation of CF_3SO_3H which catalyzes the decomposition of the *tert*-butylcarbonate groups in the absence of water already at room temperature. In the presence of water the acid dissociates. H_3O^+ ions, thus formed, only catalyze the deprotection significantly at elevated temperatures ($T > 70°C$).[189]

Since technical applications under water-free conditions are not feasible post-exposure baking is mandatory for fine-line generation. Resist formulations composed of poly(*p*-vinyl phenol) and 4-*tert*-butoxycarbonyloxyphenyl sulfonium salts have been examined successfully in X-ray tests using SOR: $S(X) = 50–100\ mJ/cm^2$. The resolution is very good: 0·2 μm structures have been delineated.

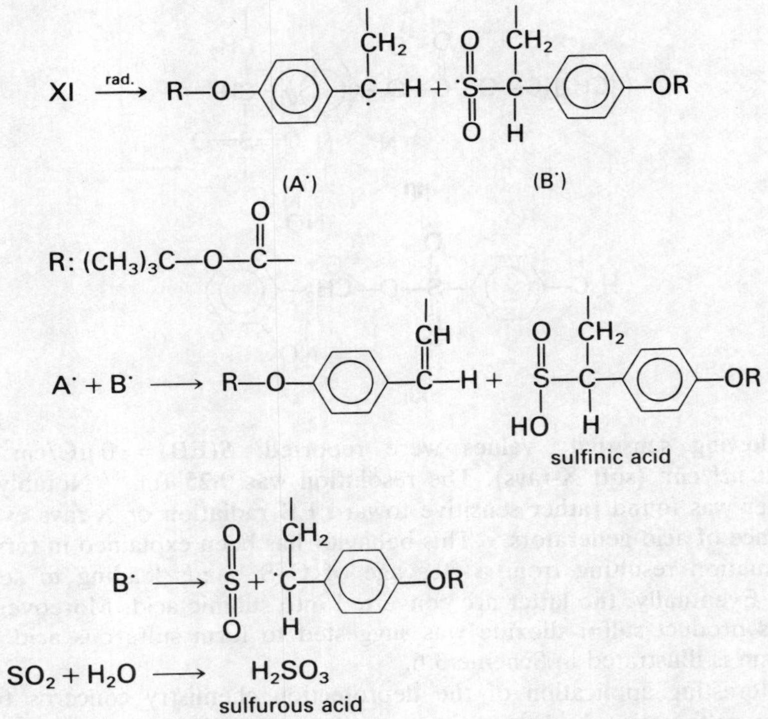

Scheme 3.6. Mechanism illustrating the formation of sulfinic acid and sulfurous acid during the radiolysis of copolymer **XI**. (Adapted from Ref. 29.)

Cross-linking by condensation and polycondensation
Negative tone imaging via acid-catalyzed cross-linking has turned out to be very useful for the generation of microstructures. Work in this field is highly promising regarding future applications. It mainly concerns three-component systems composed of acid generator, acid-sensitive cross-linker and phenolic resin. Cured formulations are developable in aqueous alkaline solution and provide high resolution and process stability. Their behavior resembles that of bisazide/phenolic resin systems described above, however, the radiation sensitivity is higher than in that case. In most cases acid-catalyzed cross-linking proceeds via condensation and in some cases by cationic polymerization.

The first resist of this type was composed of a novolak resin, hexamethoxymethylmelamine (HMMM) (cross-linker) and a halogen-containing compound (acid generator).[190] The curing mechanism is rather simple: hydrogen halide is formed by the radiolysis of the acid generator. During subsequent heating, the melamine compound reacts with the phenolic hydroxyl groups via a condensation reaction. It involves the formation of ether linkages at the hydroxyl site of the phenol with HMMM and the release of methanol. The acid-catalyzed cross-linking of poly(*p*-vinyl phenol) with the aid of HMMM is illustrated in

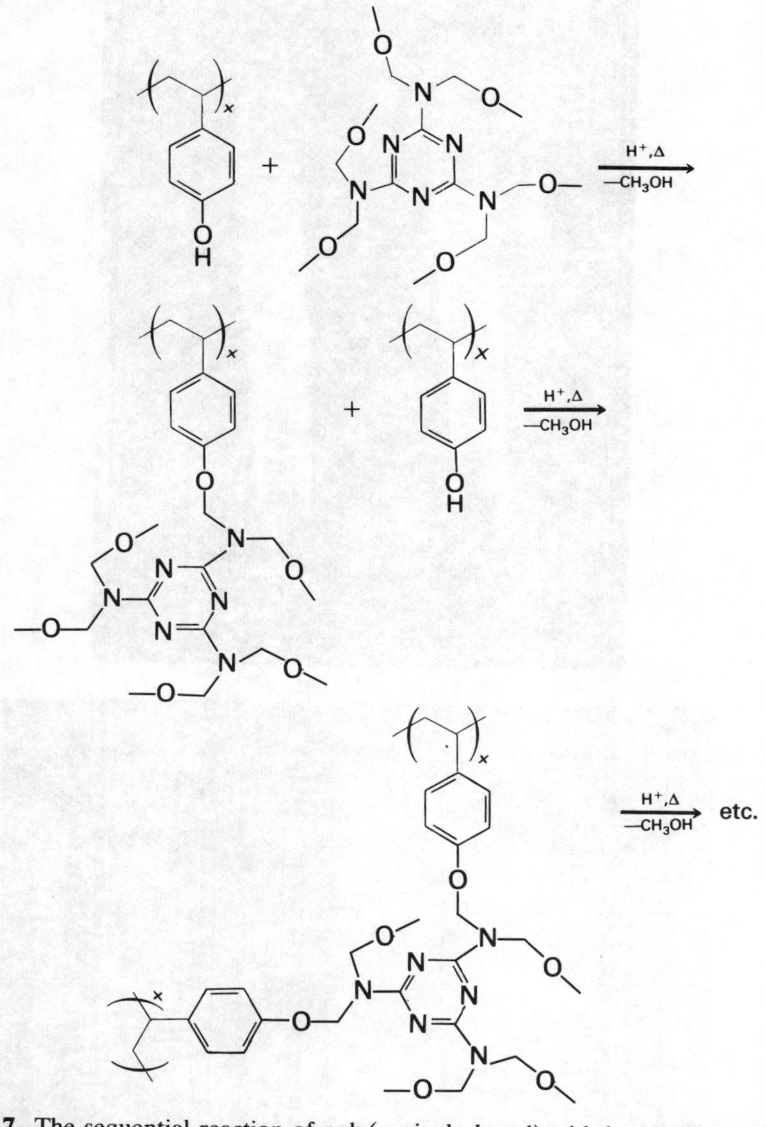

Scheme 3.7. The sequential reaction of poly(*p*-vinyl phenol) with hexamethoxymethyl melamine upon acid generation and thermal curing. (Adapted from Ref. 191.)

Scheme 3.7.[191] FTIR measurements revealed that the cross-linking reaction occurs to about 20% already during the exposure to EB radiation. Further cross-linking during post-exposure baking is completed within 1 min.[192] The decrease in solubility of the polymer in the exposed areas is caused by an increase in the average molar mass and by blocking of the base-labile phenolic

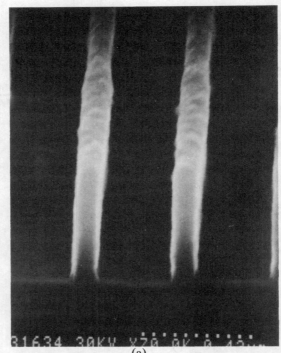

(a)

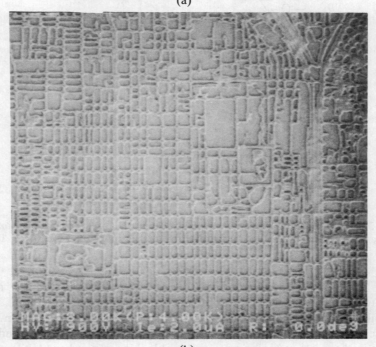

(b)

hydroxyl groups by the cross-linker.[191] The following sensitivity values were reported for novolak-based resist formulations: $S(X) = 5-25\,mJ/cm^2$ (soft X-rays),[151] $S(X) = 20-50\,mJ/cm^2$ (X-rays emitted from a Pd-anode)[190] and $S(EB) = 4-5\,\mu C/cm^2$.[36,193] Various attempts[182,194–197] to optimize this resist system with respect to process parameters resulted in an even higher sensitivity: $S(EB) = 1\cdot5\,\mu C/cm^2$, and a significantly good resolution: 65 nm in a 570 nm thick film. Problems concerning the shelf-life time were traced back to dark reactions of the cross-linker with the resin and were overcome by using a modified cross-linker.[36] Typical microstructures produced with the aid of a commercialized version of this resist (SAL-601, Shipley) are shown in Fig. 3.15. This resist was successfully applied for the fabrication of $0\cdot3\,\mu m$ devices by an EB direct write technique using a three-level system and an anti-charging layer[198] and, moreover, for the fabrication of a prototype 64 Mbit DRAM chip, also in a three-level configuration.[77]

A resist applicable for X-ray and EB lithography and based on acid-catalyzed cross-linking of phenolic resins with the aid of melamine derivatives has also been developed by Hoechst.[172,175,199,200] It has been applied for the fabrication of X-ray masks using a single layer system which was exposed to the beam of 50 keV electrons.[201] Because of the rather large film thickness needed in this case the resolution was limited to about $0\cdot5\,\mu m$. This implies that three-level resist systems have to be used for the fabrication of masks with smaller features.

A resist composed of poly(p-vinyl phenol), alkoxymethyl benzoguanamine (cross-linker) and trihalomethylarylsulfone (acid generator) was reported to be quite sensitive $(S(EB) = 1\cdot1\,\mu C/cm^2)$ and to have a better exposure linearity and a longer shelf-life time than resists containing hexamethoxymethyl melamine (cross-linker) and a triazine derivative (acid generator).[202]

Alkoxymethyl benzoguanamine

Fig. 3.15. Microstructure patterns generated in resist SAL 601 (Shipley) by EB radiation. (a) $0\cdot1\,\mu m$ lines ($D_{exp} = 15\,\mu C/cm^2$, $E = 30\,keV$, PEB = 7 min at 105°C, development: 5 min in aq. TMAH, $1\cdot19\,wt\%$). (b) Reproduction of the map of Kyoto (in part; scale $1:1\cdot5 \times 10^8$) with the aid of an EB writer of the variable-shaped beam-type that can produce rectangular items ranging from $0\cdot1 \times 0\cdot1\,\mu m$ to $5 \times 5\,\mu m$. The resist is composed of a novolak polymer containing a cross-linking agent (melamine derivative) and an acid (hydrogen halide) generator. (Courtesy of Yoshimura, T., Hitachi Ltd.)

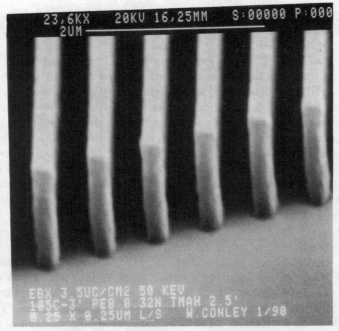

Fig. 3.16. Microstructure patterns (0·25 μm lines and spaces) generated in poly(p-vinyl phenol) containing urea/formaldehyde oligomer (cross-linking agent) and sulfonium salt (acid generator) by EB radiation. (From Ref. 204, reproduced by permission of the Society of Photo-Optical Instrumentation Engineers (SPIE).)

A two-component system composed of an epoxy-novolak, functioning both as matrix polymer and cross-linker, and $Ph_3S^+SbF_6^-$ was tested for its applicability as EB resist. As the matrix polymer is not soluble in aqueous alkaline solution, organic solvents such as methylisobutyl ketone, bis(2-methoxyethyl)ether (diglyme) or propanediol-monomethylether acetate (PGMEA) were used as developers. Swelling was minimized by using an epoxy-novolak resin of rather low molar mass. This system proved to be very sensitive: $S(EB) = 0·8 \, \mu C/cm^2$ at $E = 25$ keV and $S(X) = 15\text{--}25 \, mJ/cm^2$.[203]

Figure 3.16[204] shows microstructures generated by X-rays in a resist formulation containing poly(p-vinyl phenol) as matrix polymer, a urea-formaldehyde oligomer (cross-linker) and sulfonium salt (acid generator). The microstructures were developed in aqueous alkaline solution. The characteristics of this system are as follows: $S(X) = 90 \, mJ/cm^2$ for hard X-rays and $22 \, mJ/cm^2$ for soft X-rays; line resolution: $0·2 \, \mu m$ at $1·2 \, \mu m$ film thickness.[204] This system was found to be much more stable than a similar one containing epoxy resin instead of the urea-formaldehyde resin as cross-linking agent.[205]

Positive tone imaging via acid-catalyzed cross-linking has been reported recently.[34,35] The formulation contains, apart from the matrix polymer and an acid-sensitive cross-linker, a *radiation-sensitive base generator* and, in addition, a heat-sensitive acid generator. Irradiation results in the formation of base.

Post-exposure baking decomposes the acid generator in all parts of the resist film. However, cross-linking takes place only in unexposed areas because otherwise the acid is neutralized by the base formed during radiolysis of the base generator. The advantage of this method over the 'classical' method consists in the fact that the highly cross-linked polymer matrix formed in unexposed areas is of high thermal stability and high resistance toward overdevelopment. Among various base generators examined o-nitrobenzyl-N-cyclohexylcarbamate (**V**) and 3,5-dimethoxycyclohexylcarbamate (**VI**), provide the highest radiation sensitivity. 2-Nitrobenzyltosylate was used as thermal acid generator. Resist characteristics pertaining to pre-baking at 80°C, and post-exposure baking at 120°C are $S(X) = 560 \, mJ/cm^2$ at $d_{film} = 1 \, \mu m$; $S(EB) = 17.5 \, \mu C/cm^2$ at $E = 20 \, keV$, resolution $= 0.3 \, \mu m$ at $d_{film} = 0.5 \, \mu m$.

In another case the action of a radiation-generated base was also utilized for negative tone imaging. Here, the formulation was composed of poly(p-vinyl phenol), a cross-linking agent of the azide type[181] and a carboxylic acid, e.g. cinnamic acid (**XIV**), 3-methylindenecarboxylic acid (**XV**) or poly(p-vinyl cinnamic acid) (**XVI**).

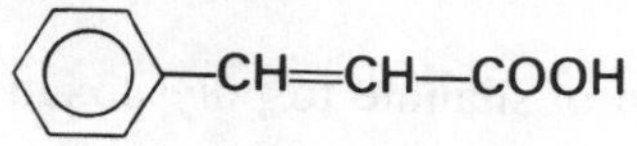

$$\text{Ph—CH=CH—COOH}$$

XIV

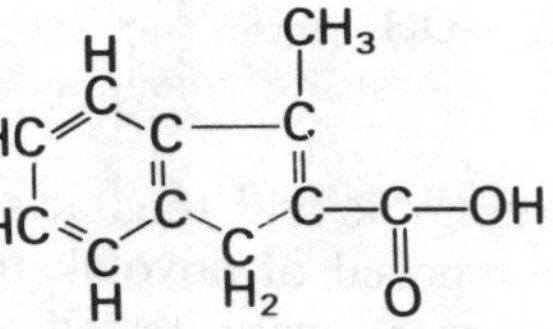

XV

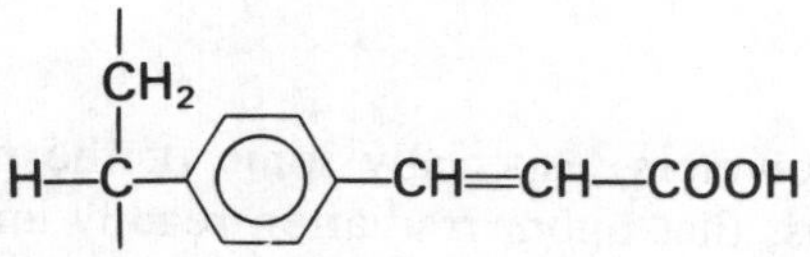

XVI

Radiolysis of the azide causes cross-linking of the polymer and in addition leads to the formation of a primary amine. The mechanism is shown in Scheme 3.8. During post-exposure baking the amine catalyzes the decarboxylation of the carboxylic acid. The dissolution inhibitor thus formed causes an additional decrease in solubility of the material in exposed areas. By the addition of carboxylic acid, the resist sensitivity was improved from $S(EB) = 15 \, \mu C/cm^2$ to $4.1 \, \mu C/cm^2$.[206]

An imaging method also using novolak resins but otherwise differing appreciably in its chemistry from the methods discussed so far is based on the

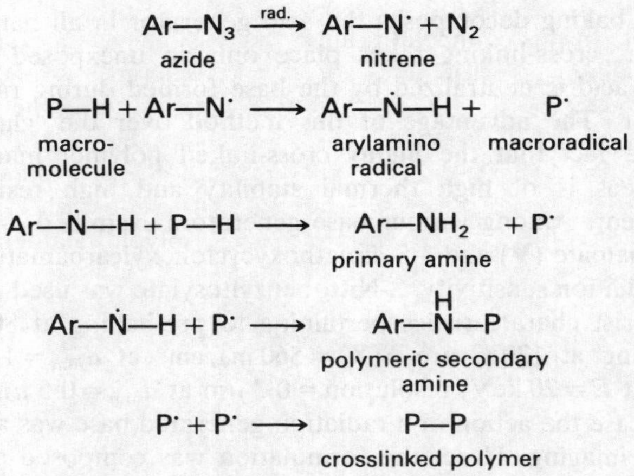

$$Ar\text{—}N_3 \xrightarrow{rad.} Ar\text{—}N^{\cdot} + N_2$$
$$\text{azide} \qquad\qquad \text{nitrene}$$

$$P\text{—}H + Ar\text{—}N^{\cdot} \longrightarrow Ar\text{—}\dot{N}\text{—}H + P^{\cdot}$$
$$\text{macro-} \qquad\qquad\qquad \text{arylamino} \quad \text{macroradical}$$
$$\text{molecule} \qquad\qquad\qquad \text{radical}$$

$$Ar\text{—}\dot{N}\text{—}H + P\text{—}H \longrightarrow Ar\text{—}NH_2 + P^{\cdot}$$
$$\text{primary amine}$$

$$Ar\text{—}\dot{N}\text{—}H + P^{\cdot} \longrightarrow Ar\text{—}\underset{\text{polymeric secondary}}{\underset{\text{amine}}{N\text{—}P}}$$

$$P^{\cdot} + P^{\cdot} \longrightarrow P\text{—}P$$
$$\text{crosslinked polymer}$$

Scheme 3.8. Radiation-induced amine formation and cross-linking of macromolecules in a system containing both azide and polymer.

acid-catalyzed condensation of silanole to polysiloxane:

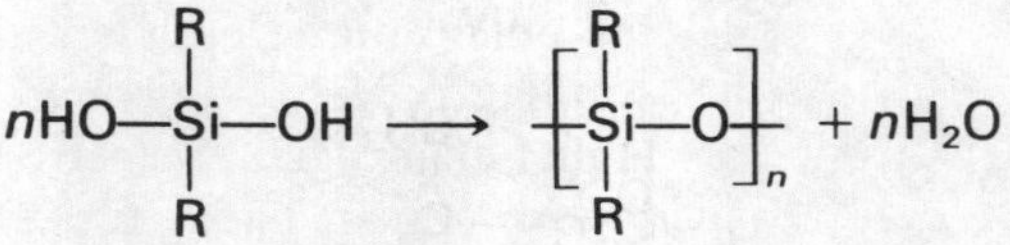

$$n\text{HO}\text{—}\underset{R}{\overset{R}{\text{Si}}}\text{—OH} \longrightarrow \left[\underset{R}{\overset{R}{\text{Si}}}\text{—O}\right]_n + n\text{H}_2\text{O} \qquad (23)$$

Polysiloxane is quite hydrophobic and thus acts as a dissolution inhibitor for the resin. Formulations composed of novolak resin, diphenylsilane diole and sulfonium salt (acid generator) were tested: $S(EB) = 0.8\ \mu C/cm^2$ at $E = 30\ keV$.[207,208]

Depolymerization

There are certain polymers, especially some of the poly(olefin sulfone) and polyaldehyde families, that upon irradiation readily undergo bond cleavage in the main-chain. The free radicals generated in this way can give rise to depolymerization, i.e. decomposition of the macromolecules into small volatile molecules via a chain reaction. This is illustrated for the case of poly(2-methyl-1-pentene sulfone) by the reactions (21), (22a) and (22b). Depending on the chemical nature of the polymer the chain reaction can propagate at room temperature or at higher temperature. In the latter case irradiation and depolymerization can be performed in separate steps, i.e. the depolymerization is carried out during post-exposure baking which involves the generation of microstructures and, therefore, is denoted *thermal development*. Apart from radiation-induced free-radical formation depolymerization can also be initiated by reactive species that are formed during radiolysis. The latter case corresponds, for example, to the generation of acid by the radiolysis of an appropriate compound being a component of the resist formulation, i.e. an acid generator. In this connection the most prominent system to be referred to is polyphthalaldehyde/sulfonium salt.[5] PPA undergoes spontaneous depolymerization if irradiated in the presence of a triphenyl sulfonium hexafluoroantimonate ($Ph_3S^+SbF_6^-$). Thermal development of irradiated PPA at 160°C yields good quality microstructures ($S(EB) = 1\ \mu C/cm^2$ at $E = 20$ keV). The residual non-volatile sulfonium salt can be removed by rinsing with 2-propanol.[5] Also PPA derivatives such as 4-chloro- or bromo-polyphthalaldehyde have been tested. Rapid depolymerization in exposed resist areas was found to occur at 100°C. The exposure doses needed for full conversion was less than $1\ \mu C/cm^2$.[209,210]

IB LITHOGRAPHY

General remarks

Currently, IB lithography is investigated intensely in various laboratories in the world. It seems that IB technology will become important for the generation of microstructures in the nanometer range, and will probably help to realize quantum effect devices. The focused IB sources presently available (see Fig. 3.6) permit ions to be accelerated to kinetic energies up to a few hundred keV. At these relatively low energies the penetration depth is rather low (see Table 3.6).[211–214] In other words, the stopping power of polymeric materials against the particles is high, i.e. much higher than against electrons of equal kinetic energy. The ions lose their energy on almost linear trajectories and there is very little emission of secondary electrons with ranges exceeding 10 nm. Moreover, there is no serious backscattering from the substrate. Therefore,

Table 3.6
Interaction of ions with PMMA (stopping power, projected range of ions and radiation-sensitivity values)

Ion	$E_0{}^a$ (keV)	dE/dx^b (eV/nm)	Projected ion range^c (nm)	$S(IB)$ (10^{13} ions/cm^2)	Ref.
He$^+$	83	170	700	1–2	211
Be^{2+}	200	300	864	2	212
Be^{2+}	100–300	223–370	506–1 150	1·5	213
Si^{2+}	200	490	382	5	212
Si^{2+}	100–300	486–513	185–575	3	213
Ga$^+$	50	1 199	49	3·8	214

^{a}Kinetic energy of ions incident on the film.
^{b}Stopping power of PMMA ($\rho = 1\cdot3$ g/cm^3) at the entrance of ions into the film, calculated with the aid of computer code TRIM 91 based on Ref. 40.
^{c}Calculated with the aid of computer code TRIM 91 based on Ref. 40.

microstructures of excellent straight walls at high contrast can be produced in polymeric resist films with IB radiation.

An interesting feature of IB lithography at high absorbed doses concerns polymer ablation which involves the decomposition of the polymer into volatile compounds. IB-induced ablation corresponds to self-development processes initiated by EB radiation or X-rays in certain polymers, e.g. poly(olefin sulfones). However, IB ablation differs from those processes by the fact that even polymers that do not decompose by a chain-reaction mechanism are ablated. However, apart from certain polymers that are especially prone to IB ablation at high absorbed doses, lithographic imaging at lower absorbed doses based on liquid development is possible with various polymers that also perform quite well in X-ray and EB lithography. Commonly, the mode of action of lithographic performance (positive or negative) of a polymer is equal for all kinds of radiation. Therefore, this section will only deal with a few typical polymers that have been applied as resists in IB lithography.

Polymer resists appropriate for wet development

The IB resist properties of PMMA have been examined rather intensely. For example, 200 keV Be^{2+} ions produced microstructure patterns of very high resolution (0·2 μm, lines and spaces) at $d_{film} = 0·7$ μm and at an exposure dose $D_{exp} = 2 \times 10^{11}$ ions/cm^2. Si^{2+} ions did not completely penetrate the film in this case and only produced rather broad grooves. The combined action of both IBs at the same place resulted in a groove with a narrow footprint and a wide upper part, which was utilized for the fabrication of T-shaped gates.[212] Sensitivity values obtained with various ion beams are compiled in Table 3.6. Notably, 50 nm lines were produced in a 60 nm thick PMMA film which was exposed to 50 keV Ga^+ ions with the aid of the FIB technique.[214] Very fine structures can also be made by ion projection lithography (irradiation through a stencil mask): 200 nm lines were generated in PMMA ($d_{film} = 350$ nm) with H^+ or He^+ ions.[211] Interestingly, PMMA behaves as a negative resist when exposed to very high doses of Ga^+ ions.[215]

If expressed in terms of an exposure dose the radiation sensitivity of polymers toward IBs is always greater than that toward EBs: $S(IB) < S(EB)$. Typical examples are presented in Fig. 3.17,[215,216] where exposure curves for two negatively acting polymers, novolak and poly(trimethylsilylstyrene$_{90}$-co-chloromethylstyrene$_{10}$), **XVII**, are shown.

XVII

Similar results were also obtained for positively acting resists. The strongly

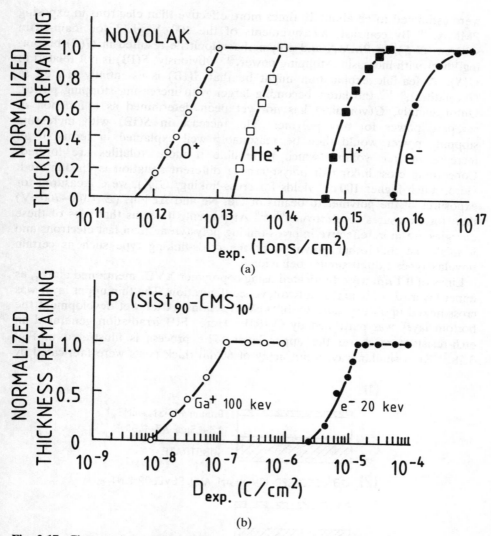

Fig. 3.17. Characteristic curves of (a) novolak and (b) poly(trimethylsilylstyrene$_{90}$-co-chloromethylstyrene$_{10}$) exposed to electrons and various ions as indicated in the graphs. ((a) Adapted from Ref. 216, (b) adapted from Ref. 215.)

differing S-values obtained in this way seem to indicate that ions are more effective than electrons, and moreover, that heavy ions are more effective than light ones. Conclusions arrived at in this way can be quite erroneous since effectivity can only be compared in terms of absolute sensitivity values which must be based on absorbed doses. However, almost all published sensitivity data are based on exposure doses (radiation fluences), that are expressed in units of (intensity × irradiation time). Based on the absorbed dose, O^+ ions were estimated to be about 10 times more effective than electrons in exposing PMMA.[216] By contrast, measurements of the 100 eV yield for main-chain scission, $G(S)$, in PMMA and polymethacrylonitrile resulted in $G(S)$ becoming lower with increasing stopping power.[217] Obviously, $S(IB)$ is not related to $G(S)$. A feasible explanation might be that $S(IB)$ is essentially related to $G(\text{volatiles})$,[63,110] the latter becoming larger with increasing stopping power. Unfortunately, $G(\text{volatiles})$ has not yet been determined as a function of stopping power for any polymer. The increase in $S(IB)$ with increasing stopping power would then be reasonably well explained in terms of an increase in free volume which is feasible if more volatiles are formed. Concerning cross-linking of polystyrene a different situation is encountered. Here, much higher 100 eV yields for cross-linking, $G(X)$, were measured for exposures of the polymer to beams of He, Ne and Ar ions ($E = 100-700$ eV) than for exposures to electrons.[107,218] At present, it seems that ions of these energies are more effective in cross-linking polystyrene than fast electrons and it might be that other polymers of the cross-linking type such as certain novolak types exhibit similar behavior.

Lines of 0·1 μm were fabricated using copolymer **XVII**, mentioned above, as upper layer ($d = 150$ nm) in a two-level configuration. The thin upper layer was cross-linked upon exposure to 100 keV Ga$^+$ ions. After wet development the bottom layer was patterned by O$_2$-RIE. Here, FIB irradiation generated an etch-resistant mask on the upper layer. The process is illustrated in Fig. 3.18.[215] In a similar way, a dot array of 60 nm thick posts were fabricated by

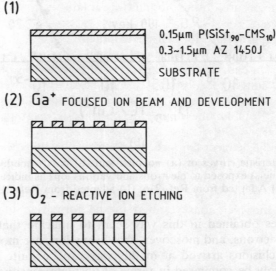

Fig. 3.18. Diagram of the processing sequence for a two-level resist system consisting of poly(trimethylsilylstyrene$_{90}$-*co*-chloromethylstyrene$_{10}$) as top layer and a novolak bottom layer. (From Ref. 215, reproduced by permission of the American Institute of Physics.)

patterning a 40 nm thick film of polyphenylsilsesquioxane (**XVIII**), a soluble ladder type polymer, with 50 keV Ga$^+$ ions.[219]

$$\begin{array}{c}
\text{Ph} \quad \text{Ph} \quad \text{Ph} \quad \text{Ph} \\
| \quad | \quad | \quad | \\
\text{HO—Si—O—Si—O—Si—O—Si—O}\sim \\
| \quad | \quad | \quad | \\
\text{O} \quad \text{O} \quad \text{O} \quad \text{O} \\
| \quad | \quad | \quad | \\
\sim\text{O—Si—O—Si—O—Si—O—Si—O—H} \\
| \quad | \quad | \quad | \\
\text{Ph} \quad \text{Ph} \quad \text{Ph} \quad \text{Ph}
\end{array}$$

XVIII

A few details concerning other polymers may be reported in the following: chloromethylated polystyrene became insoluble due to cross-linking on exposure to 260 keV Be^{2+} ions: $S(\text{IB}) = 3 \times 10^{12}$ ions/cm^2.[213] With the aid of the ion projection technique (He$^+$ ions, $D_{\text{exp}} = 1\cdot5 \times 10^{15}$ ions/cm^2) polyimide films ($d_{\text{film}} = 1\cdot2$ μm) were hardened. The exposed film areas were capable of withstanding photoablation by the light beam of an ArF excimer laser.[211] LB films of ω-tricosenoic acid (**XIX**) consisting of 75 layers ($d_{\text{total}} = 200$ nm) were exposed to Be^{2+}, Si^{2+} and Au^{2+} ions ($E_{\text{kin}} = 100$ keV).

$$\text{H}_2\text{C}=\text{CH—(CH}_2)_{20}\text{—COOH}$$
XIX

The highest sensitivity, $S(\text{IB}) = 2\cdot1 \times 10^{12}$ ions/cm^2, was observed with Au^{2+} ions. At higher exposure doses, the film was ablated.[220]

Polymer resists appropriate for self-development

Nitrocellulose was found to be most suitable for ablative microstructure generation[221,222] On exposure to 50 keV Ar$^+$ or 200 keV Si^{2+} ions this polymer is disintegrated and the exposed areas are ablated. However, rather high exposure doses (1×10^{14} ions/cm^2) were needed for clean development of microstructures. Notably, the remaining polymer is resistant to conventional plasma etching.

Systematic studies concerning process control, film thickness, nature of ions and kinetic energy of the ions were performed.[223–227] It turned out that the sputtering rate decreased with increasing film thickness and that Ar$^+$ ions were more appropriate than N$^+$, Ne$^+$, Kr$^+$ or Xe$^+$ ions. Irradiation of nitrocellulose films with N$^+$ and Ne$^+$ ions resulted in the formation of a carbonaceous residue.

ACKNOWLEDGEMENT

The authors gratefully acknowledge the assistance of Dr S. Klaumünzer of the Nuclear Physics Department of the Hahn–Meitner Institut. He performed the stopping power calculations.

REFERENCES

1. Bendig, J. & Timpe, H. J. In *Technical Applications of Photochemistry*, ed. H. Böttcher. Deutscher Verlag für Grundstoffindustrie, Leipzig, Germany, 1991, p. 172–252.
2. Davidson, T. (ed.) *Polymers in Electronics* (ACS Symp. Ser. 242). American Chemical Society, Washington, DC, USA, 1984.
3. DeForest, W. S. *Photoresists*. McGraw Hill, New York, USA, 1975.
4. Feit, E. D. & Wilkins, Jr, C. W. (eds) *Polymer Materials for Electronic Applications* (ACS Symp. Ser. 184). American Chemical Society, Washington, DC, USA, 1982.
5. Ito, H. In *Radiation Effects on Polymers* (ACS Symp. Ser. 475), ed. R. L. Clough & S. W. Shalaby. American Chemical Society, Washington DC, USA, 1991, p. 326–42.
6. Lingnau, J., Dammel, R. & Theis, J. *Solid State Technol.*, **32**(9) (1989) 105; **32** (10) (1989) 107.
7. McGillis, D. A. In *VLSI Technology*, ed. S. M. Sze. McGraw-Hill, Auckland, New Zealand, 1983, Ch. 7.
8. Moreau, W. M. *Semiconductor Lithography—Principles, Practices and Materials*. Plenum Press, New York, USA, 1989.
9. Pethrick, R. A. *Prog. Rubb. Plast. Technol.*, **3** (1987) 11.
10. Reiser, A. *Photoreactive Polymers, The Science and Technology of Resists*. J. Wiley, New York, USA, 1989.
11. Schnabel, W. & Sotobayashi, H. *Prog. Polym. Sci.*, **9** (1983) 297.
12. Tabata, Y., Mita, I., Nonogaki, S., Horie, K. & Tagawa, S. (eds) *Polymers for Microelectronics—Science and Technology*. Kodansha, Tokyo, Japan, 1990.
13. Thompson, L. F., Willson, C. G. & Fréchet, J. M. (eds) *Materials for Microlithography, Radiation Sensitive Polymers* (ACS Symp. Ser. 266). American Chemical Society, Washington, DC, USA, 1984.
14. Ueda, M. & Ito, H. *Yuki Gosei Kagaku*, **49** (1991) 437.
15. Thompson, L. F., Willson, C. G. & Bowden, M. J. (eds) *Introduction to Microlithography, Theory, Materials and Processing* (ACS Symp. Ser. 219). American Chemical Society, Washington, DC, USA, 1983.
16. Becker, E. W., Ehrfeld, W., Hagmann, P., Maner, A. & Münchmeyer, D. *Microelectron. Eng.*, **4** (1986) 35.
17. Ehrfeld, W., Bley, P., Götz, F., Mohr, J., Münchmeyer, D. & Schelb, W. *J. Vac. Sci. Technol.*, **B6** (1988) 178.
18. Jagt, C. J. & Sevriens, A. P. G. *Polym. Eng. Sci.*, **9** (1980) 297.
19. Ito, H. & Willson, C. G. *Polym. Eng. Sci.*, **23** (1983) 1012.
20. Pawlowski, G., Dammel, R., Przybilla, K.-J., Röschert, H. & Spiess, W. *J. Photopolym. Sci. Technol.*, **4** (1991) 389.
21. Reichmanis, E., Houlihan, F. M., Nalamasu, O. & Neenan, T. X. *Chem. Mater.*, **3** (1991) 394.
22. Crivello, J. V. & Lam, J. H. W. *Macromolecules*, **10** (1977) 1307.
23. Crivello, J. V., Lee, J. L. & Coulon, D. A. *Makromol. Chem. Macromol. Symp.*, **13/14** (1988) 145.
24. Neenan, T. X., Houlihan, F. N., Reichmanis, E., Kometani, J. M., Bachman, B. J. & Thompson, L. F. *Macromolecules*, **23** (1990) 145.
25. Schlegel, L., Ueno, T., Shiraishi, H., Hayashi, N. & Iwayanagi, T. *Chem. Mater.*, **2** (1990) 299.
26. Shirai, M., Katsuta, N., Tsunooka, M. & Tanaka, M. *Makromol. Chem.*, **190** (1989) 2099.
27. Yamaoka, T., Omote, T., Adachi, H., Kikuchi, N., Watanabe, Y. & Shirosaki, T. *J. Photopolym. Sci. Technol.*, **3** (1990) 275.

28. Aoai, T., Aotani, Y., Umehara, A. & Kokubo, T. *J. Photopolym. Sci. Technol.*, **3** (1990) 389.
29. Novembre, A. E., Tai, W. W., Kometani, J. M., Hanson, J. E., Nalamasu, O., Taylor, G. N., Reichmanis, E. & Thompson, L. F. *Proc. SPIE*, **1466** (1991) 89.
30. Pawlowski, G., Dammel, R., Lindley, C. R., Merrem, H.-J., Röschert, H. & Lingnau, J. *Proc. SPIE*, **1262** (1990) 16.
31. Buhr, G., Dammel, R. & Lindley, C. R. *Proc. Polym. Mater. Sci. Eng.*, **61** (1989) 269.
32. Dössel, K.-F. Euro. Patent A0312751, 1989.
33. Calabrese, G. S., Lamola, A. A., Sinta, R., Thackeray, J. W. & Berry, A. K. In *Polymers for Microelectronics*, eds Y. Tabata, I. Mita, S. Nonogaki, K. Horie & S. Tagawa. Kodansha, Tokyo, Japan, 1990, p. 435.
34. Graziano, K. A., Thompson, S. D. & Winkle, M. R. *Proc. SPIE*, **1466** (1991) 75.
35. Winkle, M. R. & Graziano, K. A. *J. Photopolym. Sci. Technol.*, **3** (1990) 419.
36. Liu, H., deGrandpre, M. & Feely, W. E. *J. Vac. Sci. Technol.*, **B6** (1988) 379.
37. Ito, H. & Willson, C. G. In *Polymers in Electronics* (ACS Symp. Ser. 242) ed. T. Davidson. American Chemical Society, Washington, DC, USA, 1984, p. 11.
38. Schnabel, W. In *Aspects of Degradation and Stabilization of Polymers*, ed. H. H. G. Jellinek. Elsevier, Amsterdam, The Netherlands, 1978.
39. Schnabel, W. *Polymer Degradation, Principles and Application*. Hanser, Munich, Germany, 1981.
40. Ziegler, J. F., Biersack, J. P. & Littmark, U. In *The Stopping Power and Ranges of Ions in Matter* (Vol. 1), ed. J. F. Ziegler. Pergamon Press, New York, USA, 1985, p. 109.
41. Pierrat, C., Tedesco, S., Vinet, F., Lerme, M. & Dal'Totto, B. *J. Vac. Sci. Technol.*, **B7** (1989) 1782.
42. Roland, B., Lombaerts, R., Jakus, C. & Coopmanns, F. *Proc. SPIE*, **771** (1987) 69.
43. Abali, L. N., Bobbio, S. M., Bohland, J. F., Calabrese, G. S., Gulla, M., Pavelchek, E. K. & Sricharoenchaikit, P. *Microelectron. Eng.*, **13** (1991) 93.
44. Calvert, J. M., Dulcey, C. S., Peckerar, M. C., Schnur, J. M., George Jr, J. H., Calabrese, G. S. & Sricharoenchaikit, P. *Solid State Technol.*, **34** (1991) 77.
45. Venkatesan, T., Taylor, G. N., Wagner, A., Wilkens, B. & Barr, D. *J. Vac. Sci. Technol.*, **19** (1981) 1379.
46. Pfeiffer, H. *J. Vac. Sci. Technol.*, **15** (1978) 887.
47. Pfeiffer, H. *IEEE Trans. Electron. Devices*, **ED-26** (1979) 663.
48. Meissner, K. & Haug, W. *J. Vac. Sci. Technol.*, **B7** (1989) 1443.
49. Nakayama, Y., Okazaki, S., Saitou, N. & Wakabayashi, H. *J. Vac. Sci. Technol.*, **B8** (1990) 1836.
50. Eberle, J., Holz, C., Lebert, R., Neff, W., Richter, F. & Noll, R. *Phys. Bl.*, **45** (1989) 333.
51. Kubiak, G. D., Outka, D. A., Rohlfing, C. M., Zeigler, J. M., Windt, D. L. & Waskiewicz, W. K. *J. Vac. Sci. Technol.*, **B8** (1990) 1643.
52. Maldonado, J. R. *J. Electron. Mater.*, **19** (1990) 699.
53. Pearlman, J. S. & Riordan, J. C. *Proc. SPIE*, **537** (1985) 102.
54. Chaker, M., Pepin, H., Bareau, V., Lafontaine, B., Toubhans, I. Fabbro, R. & Faral, B. *J. Appl. Phys.*, **63** (1988) 892.
55. Kato, T., Ochiai, I., Watanabe, Y. & Murayama, S. *J. Vac. Sci. Technol.*, **B6** (1988) 196.
56. Murphy, J. B. *Proc. SPIE*, **1263** (1990) 116.
57. Aihara, R., Sawaragi, Morimoto, H. & Kato, T. *Proc. SPIE*, **623** (1986) 196.
58. Brown, W. L., Venkatesan, T. & Wagner, A. *Nucl. Instr. Meth.*, **191** (1981) 157.
59. Parker, N. W., Robinson, W. P. & Snyder, J. M. *Proc. SPIE*, **632** (1986) 76.
60. Stengl, G., Löschner, H. & Muray, J. J. *Solid State Technol.*, **29** (1986) 119.

61. Aihara, R., Sawaragi, H., Morimoto, H., Hosono, K., Sasaki, Y., Kato, T. & Hasselshearer, M. *J. Vac. Sci. Technol.*, **B6** (1988) 245.
62. Petzold, H.-C., Burghause, H., Putzar, R., Weigmann, U., Economou, N. P. & Stern, L. A. *Proc. SPIE*, **1089** (1989) 45.
63. Ouano, A. C. In *Polymers in Electronics* (ACS Symp. Ser. 242), ed. T. Davidson. American Chemical Society, Washington, DC, USA, 1984, p. 79.
64. Chapiro, A. *Radiation Chemistry of Polymeric Systems*. Interscience, New York, USA, 1962.
65. Charlesby, A. *Atomic Radiation and Polymers*. Pergamon Press, Oxford, UK, 1960.
66. Yoshida, H. & Ishikawa, T. In *Polymers for Microelectronics*, ed. Y. Tabata, I. Mita, S. Nonogaki, K. Horie & S. Tagawa. Tokyo, Japan, 1990, p. 83.
67. Martinez, J.-P., Camon, H., Kihn, Y., Sevely, D. & Balladore, J. L. *J. Vac. Sci. Technol.*, **B6** (1988) 2254.
68. Choi, J. O., Moore, J. A., Corelli, J. C., Silverman, J. P. & Bakhray, H. *J. Vac. Sci. Technol.*, **B6** (1988) 2286.
69. Rooks, M. J., Wind, S., McEuen, P. & Prober, D. *J. Vac. Sci. Technol.*, **B5** (1987) 318.
70. Broers, A. N., Timbs, A. E. & Koch, R. *Proc. Microcircuit Eng.*, **88** (1988) 187.
71. Broers, A. N. In *Proc. Microprocess Conf. 1989*, JJAP Series 3, p. 43.
72. Vladimirski, Y., Kern, D., Chang, T. H. P., Attwood, D., Ade, H., Kirz, J., McNulty, I., Rarback, H. & Shu, D. *J. Vac. Sci. Technol.*, **B6** (1988) 311.
73. Chou, S. Y., Smith, H. I. & Antoniadis, D. H. *J. Vac. Sci. Technol.*, **B4** (1986) 253.
74. Kuan, S. W. J., Pease, R. F. W. & Frank, C. W. In *Polymers for Microelectronics*, ed. Y. Tabata, I. Mita, S. Nonogaki, K. Horie & S. Tagawa. Kodansha, Tokyo, Japan, 1990, p. 169.
75. Betz, H., Heinrich, K., Heuberger, A., Huber, H.-L. & Oertel, H. *J. Vac. Sci. Technol.*, **B4** (1986) 248.
76. Suzuki, M., Kaneko, T. & Saitoh, Y. *J. Vac. Sci. Technol.*, **B7** (1989) 47.
77. Murai, F., Nakayama, Y., Sakama, I., Kaga, T., Nakagome, Y., Kawamoto, Y. & Okazaki, S. *Jap. J. Appl. Phys.*, **29** (1990) 2590.
78. Yamanouchi, K., Meguro, T. & Cho, Y. Digest of Papers, 1st Microprocess Conf., p. 68.
79. Hülsmann, A. & Kaufel, G. *Jap. J. Appl. Phys.*, **29** (1990) 2317.
80. Tan, Z. C. H. & Silverman, S. E. *Proc. SPIE*, **1263** (1990) 217.
81. Wolfstädter, B., Colquohoun, A. & Wegner, K. Proc. Intern. Symp. Microprocess Conf., JJAP Series 3, 1989, p. 337.
82. Müller, K. H., Stelter, Th., Pouse, F. & Weidlich, H. Proc. Microcircuit Engineering 1986, p. 239.
83. Takenaka, H. & Todokoro, Y. *Proc. SPIE*, **1089** (1989) 132.
84. Reichmanis, E., McDonald, S. A. M. & Iwayanagi, T. (eds) *Polymers in Microlithography, Materials and Processes* (ACS Symp. Ser. 412). American Chemical Society, Washington, DC, USA, 1989.
85. Samoto, N., Makino, Y., Onoda, K. & Mizuki, E. *J. Vac. Sci. Technol.*, **B8** (1990) 1335.
86. Tiberio, R. C., Limber, J. M., Galvin, G. J. & Wolf, E. D. *Proc. SPIE*, **1089** (1989) 125.
87. Coyne, R. D. & Otto, O. W. *Proc. SPIE*, **773** (1987) 183.
88. Hamada, Y., Yoshikazu, T., Yamazaki, S., Kuroki, K. & Kuwano, Y. *J. Electrochem. Soc.*, **135** (1988) 2606.
89. Yamada, H., Hori, M., Morita, S. & Hattori, S. *J. Electrochem. Soc.*, **135** (1988) 966.

90. Hori, M., Yamada, H., Yoneda, T., Morita, S. & Hattori, S. *J. Electrochem. Soc.*, **134** (1987) 707.
91. Krishnaswamy, J., Li, L., Collins, G. J., Hiraoka, H. & Cado, M. A. *J. Vac. Sci. Technol.*, **B8** (1990) 39.
92. Tanaka, Y., Horibe, H., Kubota, S., Koezuka, H., Yoshioka, N., Aoyama, S., Watakabe, Y. & Maezawa, H. *Jap. J. Appl. Phys.*, **29** (1990) 2638.
93. Saeki, H. *J. Electrochem. Soc.*, **134** (1987) 1194.
94. Parsonage, E. E., Peppas, N. A. & Lee, P. I. *J. Vac. Sci. Technol.*, **B5** (1987) 538.
95. Kim, S. Y., Choi, J., Moore, J. A., Corelli, J. C., Steckel, J. & Randall, J. N. *J. Vac. Sci. Technol.*, **B4** (1986) 403.
96. Choi, J. O., Kim, S. Y., Moore, J. A., Corelli, J. C. & Steckel, A. J. *J. Vac. Sci. Technol.*, **B5** (1987) 382.
97. Miller, L. J., Brault, R. G., Granger, D. D., Jensen, J. E., Ast, C. I. van & Lewis, M. M. *J. Vac. Sci. Technol.*, **B7** (1989) 68.
98. de Grandpre, M. P., Vidusek, D. A. & Legenza, M. W. *Proc. SPIE*, **539** (1985) 103.
99. Takenaka, H., Watanabe, H., Todokoro, Y. & Inoue, M. *Jap. J. Appl. Phys.*, **29** (1990) 2879.
100. Kakuchi, M., Sugawara, S., Murase, K. & Matsuyama, K. *J. Electrochem. Soc.*, **124** (1977) 1648.
101. Kakuchi, M., Sugawara, S. & Sukegawa, K. *Rev. Electr. Commun. Lab.*, **27** (1979) 1113.
102. Nishida, T., Makao, M., Tamamura, T., Ozawa, A., Saito, Y., Nishimura, K. & Yoshihara, H. Proc. Microprocess. Conf., JJAP Series 3, 1989, p. 130.
103. Kataoka, M. & Tokunaga, A. In *Polymers for Microelectronics*. ed. Y. Tabata, I. Mita, S. Nonogaki, K. Horie & S. Tagawa. Kodansha, Tokyo, Japan, 1990, p. 327.
104. Delaire, J. A., Lagade, M., Broussoux, D. & Dubois, J. C. *J. Vac. Sci. Technol.*, **B8** (1990) 33.
105. Kelly, W. M., Doyle, A., Noonan, E., Woods, J. & Rooney, J. *Proc. SPIE*, **923** (1988) 224.
106. Suzuki, M., Ohnishi, Z. & Furuta, A. *J. Electrochem. Soc.*, **132** (1985) 1390.
107. Helbert, J. N., Poindexter, E. H., Stahl, G. A., Chen, C.-Y. & Pittman, C. U. *J. Polym. Sci. Polym. Chem. Edn*, **17** (1979) 49.
108. Helbert, J. N., Cook, C. F., Chen, C.-Y. & Pittman, C. U. *J. Electrochem. Soc.*, **126** (1979) 694.
109. Schlegel, L. & Schnabel, W. *J. Vac. Sci. Technol.*, **B6** (1988) 82.
110. Schlegel, L. & Schnabel, W. *J. Appl. Polym. Sci.*, **41** (1990) 1797.
111. Imamura, S., Tamamura, T., Harada, K. & Sugawara, S. *J. Appl. Polym. Sci.*, **27** (1982) 937.
112. Imamura, S., Tamamura, T., Sukegawa, K., Kogura, O. & Sugawara, S. *J. Electrochem. Soc.*, **131** (1984) 1122.
113. Hartney, M. A., Tarascon, R. G. & Novembre, A. E. *J. Vac. Sci. Technol.* **B3** (1985) 360.
114. Kamoshida, Y., Koshiba, M., Yoshimoto, H., Horita, Y. & Harada, K. *J. Vac. Sci. Technol.*, **B1** (1983) 1156.
115. Tarascon, R. G., Hartney, M. A. & Bowden, M. J. In *Materials for Microlithography*, ed. L. F. Thompson, C. G. Willson & J. M. J. Fréchet. American Chemical Society, Washington, DC, USA, 1984, p. 361.
116. Jones, R. G. & Matsubayashi, Y. *Polymer*, **31** (1990) 1519.
117. Yoshioka, N., Suzuki, Y., Ishio, N. & Yamazaki, T. *J. Vac. Sci. Technol.*, **B5** (1987) 546.

118. Saeki, H., Shigetomi, A., Watakabe, Y. & Kato, T. *J. Electrochem. Soc.*, **133** (1986) 1236.
119. Tanigaki, K., Suzuki, M., Saotome, Y., Ohnishi, Y. & Tateishi, K. *J. Electrochem. Soc.*, **132** (1985) 1678.
120. Jones, R. G., Matsubayashi, Y., Tate, P. M. & Brambley, D. R. *J. Electrochem. Soc.*, **137** (1990) 2820.
121. Saeki, H., Shigetomi, A. & Watanabe, Y. *J. Electrochem. Soc.*, **134** (1987) 3134.
122. Maile, B. E., Forchel, A., Germann, R., Menschig, A., Meier, H. P. & Grützmacher, D. *J. Vac. Sci. Technol.*, **B6** (1988) 2308.
123. Yoshioka, N., Sakai, A., Morimoto, H., Hosono, K. & Watanabe, Y. *J. Vac. Sci. Technol.*, **B7** (1989) 1688.
124. Wada, H., Hattori, K., Nishimura, E., Ikeda, N., Ikenaga, O., Kusokabe, H., Tamamushi, S., Kato, Y., Yoshihara, R. & Takigawa, T., Digest of Papers, 3rd Microprocess Conf. 1990, p. 54.
125. Tanigaki, K., Suzuki, M. & Ohnishi, Y. *J. Electrochem. Soc.*, **133** (1986) 977.
126. Mixon, D. A. & Novembre, A. E. *J. Vac. Sci. Technol.*, **B7** (1989) 1723.
127. Atoda, N., Doi, H. & Kokubun, K. *J. Vac. Sci. Technol.*, **B4** (1986) 386.
128. Krishnaswamy, J., Li, L., Collins, G., Hiraoka, H. & Caolo, M. A. *Proc. SPIE*, **773** (1987) 159.
129. Krishnaswamy, J., Li, L., Collins, G., Hiraoka, H. & Caolo, M. A. *Proc. SPIE*, **923** (1988) 258.
130. Chien, J. C. W. & Gong, B. M. *Polym. Eng. Sci.*, **29** (1989) 937.
131. Colton, R. J., Marrian, C. R. K., Snow, A. & Dilella, D. *J. Vac. Sci. Technol.*, **B5** (1987) 1353.
132. Dobisz, E. A., Marrian, C. R. K. & Colton, R. J. *J. Vac. Sci. Technol.*, **B8** (1990) 1754.
133. Watanabe, H. & Todokoro, Y. Digest of Papers, 1st Microprocess Conf., 1989, p. 172.
134. Hiraoka, H. & Chiong, K. N. *J. Vac. Sci. Technol.*, **B5** (1987) 386.
135. Fujiki, M., Tabei, H. & Inamura, S. *Jap. J. Appl. Phys.*, **26** (1987) 1224.
136. Ogawa, K. *Jap. J. Appl. Phys.*, **27** (1988) 855.
137. Fong, F.-O., Cuo, C., Wolfe, J. C. & Randall, J. N. *J. Vac. Sci. Technol.*, **B6** (1988) 375.
138. Hayakawa, T., Tashiro, T., Yamada, H., Morita, S., Hattori, S., Ohashi, H. & Shobatake, K. In *Polymers for Microelectronics*, ed. Y. Tabata, I. Mita, S. Nonogaki, K. Horie & S. Tagawa. Kondansha, Tokyo, Japan, 1990, p. 245.
139. Kudo, T., Ishikawa, A., Okamoto, H., Miyauchi, K., Murai, F., Mochiji, K. & Umezaki, H. *J. Electrochem. Soc.*, **134** (1987) 2607.
140. Yoshimura, T., Ishikawa, A., Okamoto, H., Miyazaki, H., Sawada, A., Tanimoto, T. & Okazaki, S. *Microelectron. Eng.*, **13** (1991) 97.
141. Yoshimura, T., Ishikawa, A., Okamoto, H., Miyazaki, H., Sawada, A., Tanimoto, T. & Okazaki, S. *Microelectron. Eng.*, **14** (1991) 149.
142. Huber, H. L., Betz, H., Heuberger, A. & Pongratz, S. *Microelectron. Eng.*, **84** (1984) 325.
143. Redaelli, R., Wells, G. M., Cerrina, F., Crapella, S. & Vento, G. *Microelectron. Eng.*, **6** (1987) 519.
144. Peters, D. W., Drumheller, J. P., Frankel, R. D., Kaplan, A. S., Preston, S. M. & Tomes, D. N. *Proc. SPIE*, **923** (1988) 28.
145. Chiong, K. G., Petrillo, K., Hohn, F. J. & Wilson, A. D. *J. Vac. Sci. Technol.*, **B6** (1988) 2238.
146. Tanigaki, K. *J. Vac. Sci. Technol.*, **B6** (1988) 91.
147. Jayaraman, T. V., Tadros, S., Beauchemin, B., Blakeney, A. J. & Greene, N. N. *Proc. SPIE*, **1089** (1989) 323.

148. Schlegel, L., Ueno, T., Shiraishi, H., Hayashi, N., Hesp, S. & Iwayanagi, T. *Jap. J. Appl. Phys.*, **28** (1989) 2114.
149. Mitchell, J. & Walker, D. M. *Proc. SPIE*, **773** (1987) 150.
150. Tang, P. P. *Proc. SPIE*, **923** (1988) 151.
151. Peters, D. W., Tomes, D., Preston, S. & Grant, R. Proc. KTI Microelectron. Seminar, San Diego, CA, 1988, p. 361.
152. Peters, D. W., Tomes, D. N., Grant, R. A. & West, R. J. *Proc. SPIE*, **1089** (1989) 179.
153. Peters, D. W. & West, R. J. *J. Vac. Sci. Technol.*, **B7** (1989) 1751.
154. Goncher, G. M., Lyngdal, J. W. & Lamer, G. L. *J. Vac. Sci. Technol.*, **B6** (1988) 384.
155. Chiong, K. G., Rothwell, M. B., Wind, S., Buckignano, J., Hohn, F. J. & Kvitek, R. *J. Vac. Sci. Technol.*, **B7** (1989) 1771.
156. Atwood, D. K., Timko, A. G., Kostelak, R. L. & Resnick, D. R. *Proc. SPIE*, **923** (1988) 141.
157. Mochiji, K., Soda, Y. & Kimura, T. *J. Vac. Sci. Technol.*, **B6** (1988) 858.
158. Ishii, K. & Matsuda, T. *Jap. J. Appl. Phys.*, **29** (1990) 2212.
159. Bowden, M. J., Thompson, L. F., Fahrenholtz, S. R. & Doerries, E. M. *J. Electrochem. Soc.*, **128** (1981) 1304.
160. Ito, H., Pedersen, L. A., MacDonald, S. A., Cheng, Y. Y., Lyerla, J. R. & Willson, C. G. In *Photopolymers, Principles—Processes and Materials*. Ellenville, NY, USA, 1985, p. 127.
161. Ito, H., Pederson, L. A., Chiong, K. N., Sonchik, S. & Tsai, C. *Proc. SPIE*, **1086** (1989) 11.
162. Shiraishi, H., Isobe, A., Murai, F. & Nonogaki, S. In *Polymers in Electronics* (ACS Symp. Ser. 242), ed. T. Davidson. American Chemical Society, Washington, DC, USA, 1984, p. 167.
163. Okazaki, S., Murai, F., Suga, O., Shiraishi, H. & Koibuchi, S. *J. Vac. Sci. Technol.*, **B5** (1987) 402.
164. Mochiji, K., Soda, Y. & Kimura, T. *J. Electrochem. Soc.*, **133** (1986) 147.
165. Yoshimura, T., Murai, F., Shiraishi, H. & Okazaki, S. *J. Vac. Sci. Technol.*, **B6** (1988) 2249.
166. Mitani, K., Okazaki, S., Murai, F. & Shiraishi, H. *J. Electrochem. Soc.*, **135** (1988) 1014.
167. Ogawa, T., Mochiji, K., Shiraishi, H. & Kimura, T. *J. Vac. Sci. Technol.*, **B7** (1989) 1684.
168. Iida, Y. & Tanigaki, K. *J. Vac. Sci. Technol.*, **B4** (1986) 394.
169. Yoshioka, N., Fujino, T., Morimoto, H., Watanabe, Y. & Abe, H. *J. Vac. Sci. Technol.*, **B8** (1990) 1535.
170. Nate, K., Inoue, T., Yokono, H. & Hatada, K. *J. Appl. Polym. Sci.*, **35** (1988) 913.
171. Dössel, K.-F., Huber, H. L. & Oertel, H. *Microelectron. Eng.*, **5** (1986) 97.
172. Lingnau, J., Dammel, R., Lindley, C. R., Pawlowski, G., Scheunemann, U. & Theis, J. In *Polymers for Microelectronics*, ed. Y. Tabata, I. Mita, S. Nonogaki, K. Horie & S. Tagawa. Kodansha, Tokyo, Japan, 1990, p. 445.
173. Lingnau, J., Dammel, R. & Theis, J. *Polym. Eng. Sci.*, **29** (1989) 874.
174. Bijkerk, F., vanDorssen, G. E., van der Wiel, M. J., Dammel, R. & Lingnau, J. *Microelectron. Eng.*, **9** (1989) 121.
175. Pongratz, S., Demmeler, R., Ehrlich, C., Kohlmann, K., Reimer, K., Dammel, R., Hessemer, W., Lingnau, J., Scheunemann, U. & Theis, J. *Proc. SPIE*, **1089** (1989) 303.
176. Ballhorn, R. U., Dammel, R., David, H. H., Eckes, Ch., Fricke-Damm, A., Kreuer, K., Pawlowski, G. & Przybilla, K. *Microelectron. Eng.*, **13** (1991) 73.

177. Oertel, H., Weiss, M., Chlebek, J., Huber, H. L., Dammel, R., Lindley, C. R., Lignau, J. & Theis, J. *Proc. SPIE*, **1089** (1989) 283.
178. Namaste, Y. M. N., Obendorf, S. K. & Rodriguez, F. *J. Vac. Sci. Technol.*, **B6** (1988) 2245.
179. Namaste, Y. M. N., Obendorf, S. K., Dems, B. C. & Rodriguez, F. *Proc. SPIE*, **1089** (1989) 339.
180. Malhotra, S., Deans, B. C., Namaste, Y. M. N., Rodriguez, F. & Obendorf, S. K. *Proc. SPIE*, **1263** (1990) 187.
181. Iwayanagi, T., Kohashi, T., Nonogaki, S., Matsuzawa, T., Douta, K. & Yamazawa, H. *IEEE Trans. Electron. Devices*, **28** (1981) 1306.
182. Chiong, K. C., Wind, S. & Seeger, D. *J. Vac. Sci. Technol.*, **B8** (1990) 1447.
183. Suga, O., Aoki, E., Okazaki, S., Murai, F., Shiraishi, H. & Nonogaki, S. *J. Vac. Sci. Technol.*, **B6** (1988) 366.
184. Tsuda, M., Oikawa, S., Yabuta, M., Yokota, A., Nakane, H., Atoda, N. & Hoh, K. *J. Vac. Sci. Technol.*, **B4** (1986) 256.
185. Umbach, C. P., Broers, A. N., Willson, C. G., Koch, R. & Laibowitz, R. B. *J. Vac. Sci. Technol.*, **B6** (1988) 319.
186. Seligson, D., Ito, H. & Willson, C. G. *J. Vac. Sci. Technol.* **B6** (1988) 2268.
187. Novembre, A. E., Tai, W. W., Nalamasu, O., Kometani, J. M., Houlihan, F. M., Neenan, T. X. & Reichmanis, E. *Polym. Prepr.*, **31** (1990) 379.
188. Schwalm, R. *Proc. Amer. Chem. Soc. Div. Polym. Mater. Sci. Eng.*, **61** (1989) 278.
189. Schwalm, R., Bug, R., Dai, G. S., Fritz, P. M., Reinhardt, M., Schneider, S. & Schnabel, W. *J. Chem. Soc. Perkin Trans. 2*, (1991) 1803.
190. Tai, E., Fay, B., Stein, C. M. & Feely, W. E. *Proc. SPIE*, **773** (1987) 132.
191. Thackeray, W., Orsula, G. W., Rajaratnam, M. M., Sinta, R., Herr, D. & Pavelcheck, E. *Proc. SPIE*, **1466** (1991) 39.
192. Tam, N. N., Ferguson, R. A., Titus, A., Hutchinson, J. M., Spence, C. A. & Neureuther, A. R. *J. Vac. Sci. Technol.*, **B8** (1990) 1470.
193. de Grandpre, M., Graziano, K., Thompson, S. D., Liu, H. & Blum, L. *Proc. SPIE*, **923** (1988) 158.
194. Bernstein, G. H., Liu, W. P., Khawaja, Y. N., Kozicki, M. N., Ferry, K. D. & Blum, L. *J. Vac. Sci. Technol.*, **B6** (1988) 2298.
195. Fedynyshyn, T. H., Cronin, M. F., Pol, L. C. & Kondek, C. *J. Vac. Sci. Technol.*, **B8** (1990) 1454.
196. Fukuda, H. & Okazaki, S. In *Proc. Microprocess. Conf. 89*, JJAP Series 3, ed. S. Namba & T. Kitayama. Tokyo, 1989, p. 161.
197. Seligson, D., Das, S., Gaw, H. & Pianetta, P. *J. Vac. Sci. Technol.*, **B6** (1988) 2303.
198. Moriizumi, K., Takeuchi, S., Fujino, T., Aoyama, S., Yoneda, M., Morimoto, H. & Watakabe, Y. In *Proc. Microprocess. Conf. 90*, ed. S. Namba & T. Kitayama. JJAP Series 4, Tokyo, 1990, p. 52.
199. Dammel, R., Dössel, K.-F., Lingnau, J., Theis, J., Huber, H., Oertel, H. & Trube, J. *Microelectron. Eng.*, **9** (1989) 575.
200. Eckes, Ch., Pawlowski, G., Przybilla, K., Meier, W., Madore, M. & Dammel, R. *Proc. SPIE*, **1466** (1991) 394.
201. Ehrlich, C., Demmeler, R., Goepel, U., Pongratz, S., Reimer, K., Dammel, R., Lingnau, J. & Theis, J. In *Proc. Microprocess. Conf.* 1989, ed. S. Namba & T. Kitayama. Tokyo, 1989, JJAP Series 3, p. 68.
202. Koyanagi, H., Umeda, S., Fukunaga, S., Kitaori, T. & Nagasawa, K. *Proc. SPIE*, **1466** (1991) 346.
203. Stewart, K. J., Hatzakis, M., Shaw, J. M., Seeger, D. E. & Neumann, E. *J. Vac. Sci. Technol.*, **B7** (1989) 1734.
204. Conley, W., Dundatschek, R., Gelorme, J., Horvat, J., Martino, R., Murphy, E.,

Petrosky, A., Spinillo, G., Stewart, K., Wilborg, R. & Wood, R. *Proc. SPIE*, **1466** (1991) 53.
205. Conley, W., Moreau, W., Perreault, S., Spinillo, G., Wood, R., Gelorme, J. & Martino, R. *Proc. SPIE*, **1262** (1990) 49.
206. Aoki, E., Shiraishi, H., Hashimoto, M. & Hayashi, N. *Proc. SPIE*, **1089** (1989) 334.
207. Shiraishi, H., Fukuma, E., Hayashi, N., Ueno, T., Tadano, K. & Iwayanagi, T. *J. Photopolym. Sci. Technol.*, **3** (1990) 385.
208. Shiraishi, H., Fukuma, E., Hayashi, N., Tadano, K. & Ueno, T. *Chem. Mater.*, **3** (1991) 62.
209. Ito, H., Ueda, M. & Schwalm, R. *J. Vac. Sci. Technol.*, **B6** (1988) 2259.
210. Ito, H., Pederson, L. A., MacDonald, S. A., Cheng, Y. Y., Lyerla, J. R. & Willson, C. G. *J. Electrochem. Soc.*, **135** (1988) 1504.
211. Stangl, G., Cekan, E., Jakob, S., Fallmann, W., Paschke, F., Buchmann, L. M., Müller, K.-P., Csepregi, L., Heuberger, A., Hammel, E., Traher, C., Löschner, H. & Stengl, G. *Microelectron. Eng.*, **9** (1989) 289.
212. Morimoto, H., Onoda, H., Kato, T., Sasaki, Y., Saitoh, K. & Kato, T. *J. Vac. Sci. Technol.*, **B4** (1986) 205.
213. Ochiai, Y., Kojima, Y. & Matsui, S. *J. Vac. Sci. Technol.*, **B6** (1988) 1055.
214. Kubena, R. L., Stratton, F. P., Ward, J. W., Atkinson, G. M. & Joyce, R. J. *J. Vac. Sci. Technol.*, **B7** (1989) 1798.
215. Matsui, S., Mori, K., Saigo, K., Shiokawa, T., Toyoda, K. & Namba, S. *J. Vac. Sci. Technol.*, **B4** (1986) 845.
216. Hall, T. M., Wagner, A. & Thompson, L. F. *J. Vac. Sci. Technol.*, **16** (1979) 1889.
217. Schnabel, W. & Klaumünzer, S. *Radiat. Phys. Chem.*, *Int. J. Radiat. Appl. Instr. Part C* **37** (1991) 131.
218. Aoki, Y., Kouchi, N., Shibata, H., Tagawara, S., Tabata, Y. & Imamura, S. *Nucl. Instr. Meth. Phys. Res.*, **B33** (1988) 799.
219. Kubena, R. L., Joyce, R. J., Ward, J. W., Garvin, H. L., Stratton, F. P. & Brault, R. G. *J. Vac. Si. Technol.*, **B6** (1988) 353.
220. Shiokawa, T., Kim, P. H., Toyoda, K., Namba, S., Suzuki, M. & Matsui, S. *J. Vac. Sci. Technol.*, **B6** (1988) 993.
221. Geis, M. W., Randall, J. N., Mountain, R. W., Woodhouse, J. D., Bromley, E. I., Astoljki, D. K. & Economou, N. P. *J. Vac. Sci. Technol.*, **B3** (1985) 343.
222. Harakawa, K., Yasuoka, Y., Gamo, K. & Namba, S. *J. Vac. Sci. Technol.*, **B4** (1986) 355.
223. Kaneko, H., Yasuoka, Y., Gamo, K. & Namba, S. *J. Vac. Sci. Technol.*, **B6** (1988) 982.
224. Kaneko, H., Yasuoka, Y., Gamo, K. & Namba, S. *Jap. J. Appl. Phys.*, **27** (1988) 1764.
225. Kaneko, H., Yasuoka, Y., Gamo, K. & Namba, S. *Jap. J. Appl. Phys.*, **28** (1989) 716.
226. Kaneko, H., Yasuoka, Y., Gamo, K. & Namba, S. *Jap. J. Appl. Phys.*, **28** (1989) 1113.
227. Yasuoka, Y., Kaneko, K., Gamo, K. & Namba, S. *Microelectron. Eng.*, **9** (1989) 543.

BIBLIOGRAPHY

Rishton, S. A., Schmid, H., Kern, D. P., Luhn, H. E., Chang, T. H. P., Sai-Halasz, G. A., Wordeman, M. R. & Gainin, E. *J. Vac. Sci. Technol.*, **B6** (1988) 1836.
Ito, H. & Schwalm, R. *J. Electrochem. Soc.*, **136** (1989) 241.

Houlihan, F. M., Shugard, A., Gooden, R. & Reichmanis, E. *Macromolecules*, **21** (1988) 2001.

Kato, T., Morimoto, H. & Nakata, H. *Proc. SPIE*, **537** (1985) 189.

Maruyama, T., Chijimatsu, T. & Kobayashi, K. *Microelectron. Eng.*, **13** (1991) 201.

Ueno, T., Shiraishi, H., Schlegel, L., Hayashi, N. & Iwayanagi, T. In *Polymers for Microelectronics*, ed. Y. Tabata, I. Mita, S. Nonogaki, K. Horie & S. Tagawa. Kodansha, Tokyo, Japan, 1990, p. 413.

Chapter 4

Applications of Electron-Beam Curing

JÜRG R. SEIDEL

RadTech Europe, Perolles 23, CH-1700 Fribourg, Switzerland

Introduction 193
EBC Applications on Paper and Board 194
   Silicone release coatings (on paper) 199
   Known industrial products 201
Printing with EBC 201
   On sheets 201
   On webs 202
   Intaglio printing 204
Converting of Films and Other Plastic Objects 204
   Protective films 205
   Protective and functional coatings 206
   Coating and EBC of three-dimensional plastic objects 207
   EBC of fibre-reinforced objects 208
Wood-Based and Building Materials 209
   EB processing of wood–cement panels 212
   Radiation-curing activities in the former USSR 214
EBC of Coatings on Metals 216
   Finishing of steel sheets with EBC in Japan 218
   Panels for tunnel walls 218
   Varnishing of aluminium foil 219
   Wire enamelling 219
   Other applications of EBC 220
Conclusions 221
References 222

INTRODUCTION

Interest in electron-beam curing (EBC) is documented and proven by a large number of publications, many conferences organised since the 1970s, development and patent activities. Success of this technology can best be confirmed by the number of installations sold. The four suppliers: two in the US, one in Europe and one in Japan, were able to sell 100 lines up to 1982. Since then the trend has increased—not always at a profitable rate—to an estimated 300 EB accelerators in the low-energy field up to the end of the 1980s.[1]

Environmental pressure (avoiding solvents) and saving of energy and production space will further promote EBC despite relatively high investment costs. Working near room temperature, creating new products, surface functions—only possible with this high-speed converting technique—are other advantages. There are several ways to describe and comment on the ever-growing field of EBC applications: we could show the historical development, arrange them in the order of volume or success, analyse them by geographical origin—because differences have resulted from continent to continent, or from one macroeconomical system to another.

The best way seems to describe them as a function of the substrate or product converted, coated and cured by EB accelerators. Within this arrangement we shall take into consideration the volume and the ease of access to the process for the newcomer. We shall look at flexible substrates first, followed by rigid, some three-dimensional products and finally look at others, which cannot easily be classified into the large categories.

EBC APPLICATIONS ON PAPER AND BOARD

Gloss is a major result when substrates are coated or varnished with acrylic formulations and cured or polymerised by EB or UV radiation. The effect is stronger with EB than with UV. One explanation for this phenomenon is that a film coated from smooth rolls has no time to flow out and adapt to the surface structure of a reasonably flat substrate, because it will be 'frozen' and hardened in a fraction of a second by the EB at normal industrial web speeds (100 m/min and more). It will rather reflect the surface structure of the coating roll, in special cases that of a polished drum or a glossy transfer film,[2,3] than the surface roughness of the substrate. Another explanation is that a solvent-free system will penetrate less in the substrate and will not show cratering from solvent evaporation encountered in a thermal drying process.

It is rather exceptional that an EB line is used to cure a clear varnish, because this can also be done with UV curing. But there are at least two companies producing *pigmented high-gloss coatings* on paper and board. Traditionally, coloured papers are obtained by gravure or flexo printing, followed by an off-line varnishing operation to obtain a protective layer with a certain gloss. It is a surprising fact that one can obtain colour and gloss, but also abrasion and, if necessary, chemical resistance in one pass with EBC. Such products are used for luxury packaging, resulting in a certain sales-appeal or they may be embossed and be used for bookbinding.

Varnishes with a viscosity of 300–1000 mPa . s are used in several plants as a *base-coat for metallisation* on clay-coated or starch-sized label paper.[4] These conditions are necessary to have the flattest base and to avoid wicking (i.e. penetration) of the varnish into the paper. The aim is to obtain, in combination with metallisation of aluminium in vacuum, a mirror-like surface, which could never be realised with a laminated aluminium foil.

'No paper dry-out' is often claimed as an advantage of EBC compared to oven drying; in vacuum-metallisation this created problems, because a paper with more than 4% relative humidity releases so much water vapour that the vacuum collapses in a very short time. New vacuum equipment had to be designed by an Italian manufacturer, with more powerful pumps and a subdivision into three chambers with different vacuum levels. In the US, pigmented base coatings were suggested; they had the advantage of hiding the paper's structure with their opacity, but they resulted in a more silky gloss and in certain cases they become completely milky in vacuum, perhaps due to evaporation of an insufficiently cured monomer.

This technique is also used by the largest beer-label printer in Europe, who adjusts the gloss by the applied coating weight: $2\,g/m^2$ results in a dull metallic finish and allows rapid label wash-off in the bottle-cleaning process. A coating weight of $5-6\,g/m^2$ results in a mirror-like metal or ink finish when overprinted. Acid groups are added to the varnish for easier removal of the labels. In Australia, one of the most efficiently used EB accelerators is used to pre- and top-coat label paper and board at a speed of up to 400 m/min and a width of 165 cm. (Fig. 4.1).

After several years of development work, EBC has been introduced to *transfer metallisation* to gain time, quality and productivity. An aluminium-metallised PET or PP film is laminated to paper or board, exceptionally to a textile substrate or non-woven. After EBC through the film, the laminate can be immediately split into a high-gloss product and a demetallised carrier film. The number of reuses/remetallisations of this film is an important factor in the economy of this process. The PP film normally fails after four to six times due to degradation, the PET film after six to 10 times due to mechanical problems or due to diminishing transfer. Users of the traditional transfer process—either solvent-based or with humidity curing—claim to reach more than 20 transfers, but the latter have to wait 2–3 days until their adhesive is cured. Another disadvantage is that the still soft adhesive + aluminium layer may be 'grained' by the surface roughness of the paper or board backside, when stored for curing in the roll (Fig. 4.2).

With the development of the *selective metal-transfer process*[5,6] the largest US gift-card producer has found another unique application. After printing the substrate with several colours by gravure, an adhesive pattern is added on the last printing station before laminating the web to a metallised film. After instant EBC through the film, thereby avoiding the problem of oxygen inhibition of the curing, the two webs are split and metallic patterns are selectively transferred on the printed product. A typical example produced are colourful parrots covered with, or caught in, golden cages. This process is in

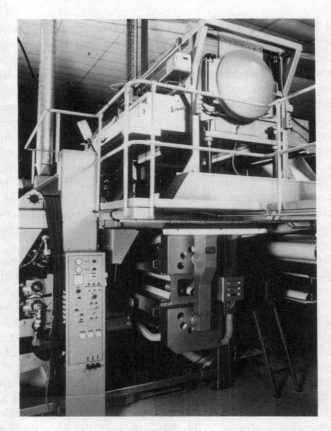

Fig. 4.1. Low-energy electron accelerator (200 kV) with four parallel linear cathodes, supplying 40 kGy at 400 m/min for the curing of base- and top-coats for vacuum metallisation of label and gift-wrap paper and board (ESI/Polytype/Vacubrite, Australia).

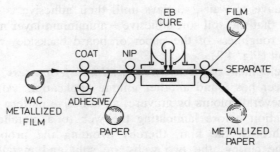

Fig. 4.2. Sketch of the EB-Aluglas transfer metallisation process with curing through the laminated film (Energy Sciences Inc.).

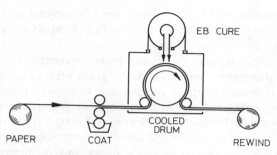

Fig. 4.3. Curing of a pre-coated layer against a polished or structured (optionally cooled) drum/cylinder (ESI).

competition with hot-stamping foils; it may be of advantage for high-speed in-line production.

Curing against a drum (Fig. 4.3) is only possible with EB and could be one of its most attractive applications, if there were not certain limitations. A polished drum produces a high-gloss surface on the substrate cured against it, as long as the coating regularly and totally releases from the drum surface. This would become a continuous transfer process without any need for a transfer film, and theoretically no inert gas is necessary because the oxygen-sensitive coating is protected by the substrate. The aim of many paper converters is the production of a high-gloss, heat-resistant, printable paper of the Kromekoat-type. A Finnish paper group is attempting to solve hidden problems to achieve this aim. To reach art-quality printability the acrylate content should be reduced and the pore-free surface should become absorbent for inks to offer good adhesion as feasible with traditional hot-calendering processes. To improve printability the coating should be silicone free. Pigmented coatings have, on the other hand, a relatively strong tendency to adhere to the metallic surface, probably because of the presence of metal-oxide groups.

Similar attempts are made by certain producers of photographic base-papers to reach a surface which is smoother but chemically as neutral as the now generally introduced PE-coated base-papers.[7,8]

For one particular application, smooth or rather structured drums have become the base of a successful, patented process[9,10] for the structuring of high-quality *casting paper* for the production of artificial leather. EBC coatings with or without silicone additions, cured against a drum without inhibition problems become reusable bases, resistant to hot DMF solvent. They offer fine, natural looking structures to the industrial product for shoe and bag manufacturing. Here, the drawback exists only for the other producers who have no access to the patented process. Embossing of certain EBC coatings could be a second-choice alternative, but nobody wants this poorer alternative.

Separation sheets (i.e. caul stock) used in the chipboard or particle-board press operation to avoid transfer of glue from one panel to the other, are another, less sophisticated EBC process, used by several producers in the US and Europe. A low-cost, silicone-free coating is applied on to a relatively heavy paper, to allow multiple reuse.

Staying with chipboard panels, we can mention another EB process used for the curing of a *protective varnish* on wood grain printed on uni *decor papers,* which will be laminated to these inexpensive furniture elements. Contrary to the gloss effects easily obtained with EBC, these papers should be as matt as real wood, at least in Central and Northern Europe. It is relatively easy to reach a silk-gloss level of around 40 Gardner with solvent-free formulations, but reaching 10–20 Gardner with more fillers and pigments has resulted in 'uncoatable' pastes because of a too-sharp rise in viscosity. Finer silicates or PE waxes have been tried as alternative matting agents, but without reaching the aim of obtaining matt varnishes which can be coated at high web speeds (>50 m/min).

Lamination of paper or lamination to board are potential processes in competition to traditional lamination with relatively inexpensive water-based adhesives. No investment solely for this field is known, but there is always the possibility to produce on existing equipment. Special properties must however justify the use of EBC.

Lamination of high-gloss papers, of aluminium foil to board are feasible, but most are done traditionally. Lamination of transparent films to board are more and more often done by UV curing for book covers, luxury packaging and publicity. *Playing-card stock* lamination is one of these examples, where EBC brings at least two advantages: a black high-gloss coating based on urethane acrylates with adhesive properties can also be used for lamination. It gives protection against transparency thanks to the black pigmentation, and because of curing at room temperature the tendency to curl was reduced with less wetting and loss of humidity in the drying oven. These playing cards with a good 'snap' are used in large quantities in gambling casinos.

Impregnation of paper is described for several applications, but not yet used in many cases. The major problem of EBC impregnants is the difficulty to reach low-enough viscosities, to allow a quick penetration and to avoid a too-high 'add-on': too high a quantity results in excessive cost and maybe in unwanted stiffening. The chances of these processes will increase in the near future: it has been suggested to coat, for example, filter paper or drafting paper,[11,12] and store it for hours or days in the roll, allowing better impregnation by hygroscopic effects. Finally, the through-impregnated paper can be cured at relatively high speeds, because it will no longer be controlled by the wicking speed of the chemical. Filter papers produced with EBC would present several advantages:

- energy and space saving, less pollution;
- traditional impregnation results in embrittlement and yellowing (with EBC, paper stays white and can be folded after curing); and
- filters become more resistant against solvents and gasoline.

Silicone release coatings (on paper)

These are used mainly as label backing paper, but also in many applications to protect an adhesive layer before its use, these products are of high interest and potential, but they are vigorously defended by the suppliers of today's or alternative silicone-curing methods. We are talking about wide-web production (2 m and more), at 300 m/min and more, resulting in at least 10 million m^2/year.

Solvent-based silicones were the first, but under ecological pressure their users looked for alternatives. Water-based systems are a logical replacement and can be coated on the existing coating equipment and dried in the same ovens with less danger and pollution.

Solvent-free systems were a third alternative defended with high-pressure methods. Both solvent-based and solvent-free systems result in silicone papers with too much humidity loss during thermal curing; they must be rehumidified. This may be feasible for one-sided coatings, but very difficult for products with two-sided siliconisation. A lot of development work has been done to reduce curing temperatures. When converters look at EB processing, they are told: 'Wait, we soon will have room-temperature curing. Why buy an expensive accelerator?'

As for high-gloss paper, the better the substrate surface is prepared in the paper-mill, the less solvent-free silicones are necessary to obtain a good release coating. It is important that base-papers are not only flat and smooth, but they must be pore-free, because solvent-free or water-free silicones will have no time to coat or fill the pores. In these places, adhesive coated against the siliconised substrate will penetrate into the release paper like roots or carrots: the release forces will go up to unacceptable values. Producers of glassine paper have learned to improve their products by sizing, improved calendaring, and by the use of colourants which do not change under EBC.

Siliconisation with EBC has a very high potential, but only very few lines are in operation. One of the handicaps is that the silicone-acrylate from the major and most advanced supplier[13] is relatively sensitive to air-inhibition leaving smeary spots with uncertain release values or a risk of transfer to the adhesive. Values below 100 ppm oxygen are suggested, preferably on the very coating surface, where the measurement of the inerting efficiency is practically impossible. High inert-gas consumption has been the major reason for slow introduction of this process, despite efforts by the equipment builders with improved treatment-zone design to reduce this consumption. UV curing, also improved with inerting, is more easily installed, especially on smaller 'do-it-yourself' equipment[14] used by label makers. Cationic UV curing in air, recently demonstrated and described,[15,16] may become a real, simple competition to EBC.

For a limited time, *hot-melt packaging paper* was produced on a toll line. Folded by home-workers to containers shown in Fig. 4.4, this release paper with an exceptionally high coating weight of 3 g/m^2, compared to 1 g/m^2 for label paper, had to offer easy handling and hot-melt transfer to a slot coater.

Fig. 4.4. Siliconised, EBC and folded containers for easier handling of hot-melts (ESI toll line).

Siliconisation of films with EBC brings more advantages and is used more often for industrial applications. There is a clear advantage for all heat-sensitive substrates like LDPE, which stand the heat of drying tunnels, and even UV only under limited working conditions (lower speed and temperatures, longer drying path). A line for *protective films* of self-adhesive roofing material was commissioned recently. Such an EB equipment offers the

Fig. 4.5. Compact electrocurtain processor for the curing of siliconised OPP film, supported by a cooled drum (ESI/NOPI/Beiersdorf).

additional advantage that the film can be cross-linked and becomes stronger and more heat-resistant.

Known industrial products

A siliconised OPP film, used by a major European promoter of radiation curing (TESASEAL, Beiersdorf, Hamburg, Germany) to protect a reusable closure strip.[17,18] During curing, the film is supported by a cooled cylinder in an elegant and compact EB processor, shown in Fig. 4.5. The use of OPP film for EBC, is somewhat surprising, because its standard makes are known to become brittle after EB exposure, because of the formation of long-living radicals which combine with oxygen and interrupt the polymeric chain, a degradation process which may take weeks to become measurable. In working with selected copolymers and/or oxygen inhibitors it became possible to obtain the necessary resistance to EB degradation.

In the US, a high-volume toll production of siliconised HDPE, which will be cut to separation sheets used in the production of printed circuit-boards, is known.[19]

PRINTING WITH EBC

On sheets

Paper and board represent the largest volume of 'offset' printed packaging materials. Most of this volume is printed on sheets, with three to eight colours, wet on wet or with interstation curing, very often with UV curing. Only two exceptions are known, where sheets are printed with EBC as the last operation: the very first application of EB in printing in the US,[20] mainly used for record jackets. After proof of a feasibility period the line was changed to a web operation with cutting to sheets after curing.

The other sheet offset printing equipment is used in the Tokyo area (Mitsumura Printing) for the printing on OPP sheets for school material and luxury-packaging folding boxes. As advantage for this unique application one cites the fact, that compared to UV curing one finds better adhesion and abrasion-resistance of the inks with EBC due to a certain grafting effect to the substrate. EBC inks do not require a protective varnish for gloss and abrasion resistance.

The major reason that there are not more sheet applications is the inerting and shielding requirements which we shall discuss with the help of Fig. 4.6. The sheets cannot be transported with clamps, because they would require too big openings with shielding and inerting problems; they cannot be moved on polymer belts because of degradation risks. They are transported on a metallic conveyor with a vacuum hold-down system, but vacuum means loss of costly and necessary inert gas, unless a sophisticated recirculation system for inert gas is provided. Circles 1–7 are blowers which extract or move nitrogen from one zone to the other, in some cases through heat-exchangers for cooling and

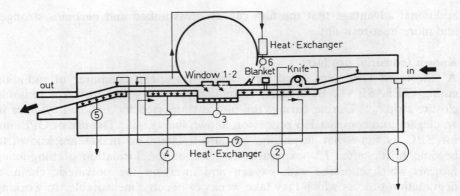

Fig. 4.6. Offset printing line for OPP sheets with EB curing. Sketch illustrates how handling and inerting problems were solved (R15/ESI/Mitsumura Printing, Japan).

filtering. The possible web speed of about 50 m/min may be acceptable for a special product with high added value, but for a possible introduction with European sheet printers a speed of at least 100 1-m sheets/min would be required.

On webs

Offset-litho printing has been well introduced in modern high-quality and high-speed printing of newspapers. No special curing is provided, because the solvent-based inks are absorbed by the porous paper or dried in short IR zones.

For printing of packaging materials two other solvent-based processes are mainly used: gravure for high-quality and flexo for high-volume, less-demanding products. Examples are as follows: flexo with one to two colours for milk packaging, gravure with four and more colours for fruit-juice, frozen-food or cigarette packaging, the latter often followed by an off-line varnishing operation providing gloss and abrasion-resistance to the packaging material.

Thanks to the initiative of a large producer of packaging equipment and complexes, thanks to pressure on press builders and chemical suppliers, it became possible to prove in the early to mid-1980s that EBC could become an important contribution and market for the curing of offset inks and varnishes (TetraPak, Westvaco, Int. Paper, Akerlund & Rausing, etc.).[21] The paste-like viscosity of these inks allows with an adapted offset technique to print wet-on-wet, without interstation curing, necessary or preferable when printing on less or non-absorbent substrates. The original idea was to replace expensive 'gravure' board by cheaper PE-coated board, the saving would pay the investment in a relatively short time. At speeds of 200–250 m/min offset inks were printed on the PE surface, most often varnished, and then cured in a compact EB processor (Fig. 4.7) with a dose of 20–30 kGy. In a later

Fig. 4.7. A typical 175 kV EB processor installed after four colour and an optional varnishing station on (PE-coated) board or folding-box carton. (ESI/Chambon/TetraPak).

development, the relatively expensive gloss and barrier varnish was replaced with an extruded PE coating and the printing was done on the board surface again. This compromise eliminated two problems encountered from time to time with the irradiation of the PE layer: a reduction of its heat-sealability (due to cross-linking?) and occasional development of odour due to PE additives.

After comparison with UV curing, EBC was selected for its higher safety and degree of cure, reduced risks of odour and taint because of formulations which function without photoinitiators and amine co-initiators. The curing speed is not influenced by pigment absorption and it is possible to cure through protective layers or laminates if necessary. The ever-growing requirements from this pioneer were met by a limited number of chemical suppliers and ink makers, which continue to improve their raw materials by reducing odour, impurities, etc., by distillation, filtration and creation of new molecules. The industry has profited in other fields from these developments of improved and safer chemicals.

The number of EBC web-offset printing lines world-wide is estimated at 35–40 in early 1991. The pioneering company has been followed by other press builders, printers and packaging companies. Some of them sworn-in for gravure printing, start to change: they consider using offset-varnishing in-line with EBC, they consider applying it to cigarette packaging, after an almost exclusive use for fruit-juice packaging. One of the press builders has even bought an EB pilot line for development and promotional activities.

Intaglio printing
At this time, most other printing techniques offer little chance for the introduction of EBC. Either their volume is too small or the inks are of such low viscosity that they need solvents or water for dilution, and interstation curing after each colour; interstation curing EB accelerators are much too expensive. Dry-offset—a compromise of flexo and offset printing, maybe rotary screen printing may become occasional candidates for EBC. With higher pigmentation and higher coating weights, UV curing may reach its limits of penetration and safe cure at high speeds. This is particularly true for intaglio printing, a field where pilot tests have been done in several countries (US, UK, Denmark, Switzerland and Japan). It has been proven that ink profiles of up to 100 μm can be safely cured with EB, with the advantage that no humidity is removed from the security paper and curl can be avoided. There is, however, a major problem: *banknote paper* is most often made of cellulose linter, a material very sensitive to degradation/scission of cellulose molecules at a dose as low as 20 kGy. Projects for Dutch cheques, US stamps and even the 'new US dollar' are ready, but the ink specialists have first to reduce the necessary dose to avoid an embrittlement of the banknote paper which can be observed in a reduction of the possible number of foldings which drops from about 4000 to about 2000 at the dose needed today. It is expected that sooner or later banknotes will become even safer thanks to an EBC 'fingerprint' and an intaglio structure which cannot be duplicated with laser copying machines.[22]

CONVERTING OF FILMS AND OTHER PLASTIC OBJECTS

A huge application of EB is the *cross-linking* of polymer (films) to obtain heat-resistance, solvent-resistance, shrink memory and many more interesting properties, but this subject is treated in detail in several other publications.[23–25]

Grafting on films (and fibres) is getting closer to the curing of coatings, because a (monomer) layer is cured on a substrate, whereby the change of its surface properties becomes the major effect looked for. Many ideas have been tried in laboratories and published, but only a few have found an industrial application. In one known case a producer of self-adhesive (insulation) tapes has found a way to render the soft PVC tape resistant to gasoline and solvents by a grafting operation. This is of particular merit, because this material often suffers from EB radiation and discolours or becomes brittle.

In another case a large glass converter describes a method to obtain an anti-misting effect on a PUR-film by grafting, with the aim to laminate this film to vehicle windshield inner sides to reduce the blinding effect of high relative humidity.[26]

One of the first users of linear cathode accelerators in the US have built their success on the regular development of *new products* not feasible with other techniques.[27,28] They have specialised in the protection, decoration and metallisation of films. Two recent developments based on a combination of coating and grafting are worth mentioning.

Adhesion of radiation-curable coatings on treated and untreated polymer surfaces is often rather difficult; without solvents, wetting is more difficult and most coatings have a tendency to shrink when EB cured. Vinyl-amino-methoxy primers not only provide *good adhesion* after EB grafting, but they may also become an intermediate to add other functions to the film surface.[29] This intermediate layer helps to further improve the properties of a *static control coating* used, for example, on LDPE as a packaging material for sensitive electronic components. The surface resistivity of $2 \times 10\,E^{10}$ ohms per square realised, falls nicely into the static dissipative range and allows the production of a transparent packaging, not feasible with the usual carbon-black additions.

More details on grafting and a scientific background will be given in the chapters on grafting and converting of textiles by EB.

Protective films

As for the release films described in the previous section, the production of protective films is evaluated and developed in several places, because of the evident advantage that heat-sensitive substrates, like LPDE, can be used. While the Europeans are still studying and looking for the right adhesives and a method for high-viscosity low-weight coatings, a Japanese company has developed their own high-quality product. Adhesives and coating method will be described below,[30] because they are a typical example where EBC can be used for a new product and process with a number of advantages. The development of the special adhesives was described earlier,[31] the clever coating method is shown in the patent drawing (Fig. 4.8).

A heated roller coater ⑤ has been selected, despite the risk of working air-bubbles into the high-viscosity adhesive. The coated adhesive film is laminated ③ against the same substrate on its first pass through the coater. This gives a smoothing effect and protects the reactive chemicals from inhibition

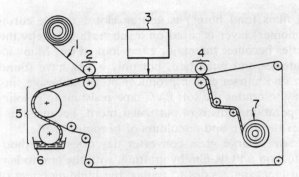

Fig. 4.8. Patent drawing for the production of peelable, protective PSA film, with double pass of the substrate (LDPE or HDPE) under the beam for inhibition protection and additional smoothening (Hitachi Chemical Co., Japan).

when they pass through the process zone ③. Having participated in the process development, we have never seen such quality requirements: the number of miniature bubbles cm^{-2} was checked under the microscope and had to be very limited to make the product acceptable. It was shown that polished steel plates, protected with films containing tiny bubbles, resulted in black corrosion spots on the steel!

Protective and functional coatings
A producer of *telephone credit cards* needed a scratch-resistant coating on one side of the tinted PVC cards. A logical way would be to have it varnished by a toll converter with a UV-curable coating. But the UV curing created too much heat with deformation of the web, a cooled UV system did not completely cure and gave poorer adhesion. Since then the substrate for the cards is toll varnished on an EB pilot line. To obtain sufficient adhesion a high-Corona treatment had to be applied, a relatively slow reverse gravure technique only was giving acceptable results. The treatment dose had to be limited, otherwise the colour darkened with a negative influence on the holographic reading of the card.

In school and seminar rooms, *erasable white boards* for felt pens are being found more often. A German producer coats PVC films with a white EBC coating which is not attacked by special solvent-based felt inks. In the US a similar self-adhesive, paper-based product 'White Magic' is toll manufactured and EB cured with a similar coating giving the desired wipability to its surface.[29]

Another new process and product is a PE film, coated with a thin *rare earth layer* and protected with an *EB-cured top-coat*.[32] It will be inserted into X-ray image cassettes, and has the function of enhancing the sensitivity of X-ray films.

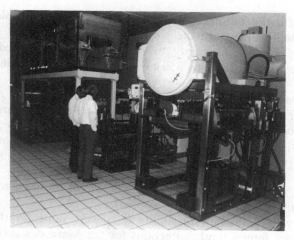

Fig. 4.9. Special clean-room EB processor for the curing and cross-linking of a PETP floppy disk substrate against a polished and cooled drum (Ref. 33/ESI/Design Magnetics, USA).

Magnetic coatings are applied as a solvent-based slurry of metal oxides most often to PET carriers: audio-, video-tapes, floppy and computer disks, etc. One of the aims—to get away from solvents—is practically not feasible. A system without solvent cannot carry a high enough number of magnetic particles and the 'viscosity' becomes too high for coatability. For high-density products a particle orientation step takes place after evaporation of part of the solvent. After EB pilot activities in several large companies on tapes, further development was concentrated on floppy disk substrates which could be treated as webs and because they promised a higher added value. Such activities were pursued in California and in Japan.[33] After evaporation of most of the solvent in an explosion-proof area, the floppy web was introduced in a high-performance EB processor shown in Fig. 4.9. The web is calendered against a cooled polished drum, and the PUR-acrylate-based magnetic coating is cured/cross-linked through the web. Such products offer, besides an excellent flatness and abrasion-resistance, an improved signal-to-noise ratio.

In normal working procedures without EB, the calendaring is done after thermal curing which results in higher wear of the calendaring rolls and in a reduced flattening effect because the coating is already hard and quite abrasive.

In Japan the EB treatment is used for increasing the abrasion resistance of the magnetic coating on *floppy disks* by a post-crosslinking effect.[34]

Coating and EBC of three-dimensional plastic objects

It started in the mid-1960s at the Ford Motor Co. in the US: decoration and EBC of full-size dashboards and comes up from time-to-time in Europe thanks to the initiatives from a German scanning-electron accelerator builder in cooperation with the automotive industry.[35] The major problem with large three-dimensional objects are the size, shielding and inerting of the process zone, as well as the uncertainty to reach the entire coated surface with low-energy electrons limited to 350 kV to maintain a self-shielded, compact equipment.

At Ford, one-colour, heat-sensitive *dashboards* could be painted to the ordered colour and cured to a relatively elastic coating in a limited space and time. They worked with several scanners developed in-house, used also for other projects, and exported to Europe for several wood-coating lines. The inert gas was produced by combustion of natural gas. Some time later, when the moulding techniques allowed plastic parts in different colours to be obtained, but after a fatal accident when a worker walked into the inerted bunker and suffocated, the line was stopped.

Almost 20 years later new feasibility tests were done for the German car industry, e.g. for lateral mud protections for the Mercedes cars. In place of a continuous system with a labyrinth and a large inert-gas consumption, a vacuum treatment box, equipped with a scanner was used for batch-type treatment. Coatings with satisfactory results, including good salt-spray tests were found, but the pilot equipment has not yet been transformed into a production line.

EBC of fibre-reinforced objects

At *Radtech '90*, a Russian Electrical Institute described in relatively precise terms the development and pilot production of glassfibre-reinforced copper-clad *prepregs*.[36] The patent-like text shows ways of solving potential problems on the way to an industrial process. A short summary of this interesting technique is best understood with the help of Fig. 4.10.

One or two glassfibre webs are unwound from rolls 4 and 5; they are impregnated with about 50% add-on of an epoxy-based reactive resin–monomer mixture. Between squeeze rolls 9, the product thickness can be controlled; from 11, for example, a copper foil can be laminated to the impregnated web. It is possible to introduce a (siliconised) PET film from 3, which will further smoothen the surface, protect the product from inhibition during irradiation (2, 13) and act as protector or separation layer in the roll. The described product has only a thickness of up to 0·2 mm and can be treated with a low-energy accelerator. The pilot-line in the Institute is equipped with a 700 kV scanner; it can therefore also treat/cure other and thicker products.

The newest and largest EBC project of acrylic resin *impregnated carbon and glassfibre composites* and structures will start to produce in 1991 in France.[32] Compared to the other processes described, all is spectacular in this unique application: only a high-energy Linac accelerator, exceptionally even X-rays, can penetrate 25 or 300 mm thick structures for aerospace applications. The concrete-shielded irradiation vault can handle workpieces of 4 m in diameter and up to 10 m length. After careful technical and economical studies Aerospatiale has proven, that the curing can be done 10 times faster than in a

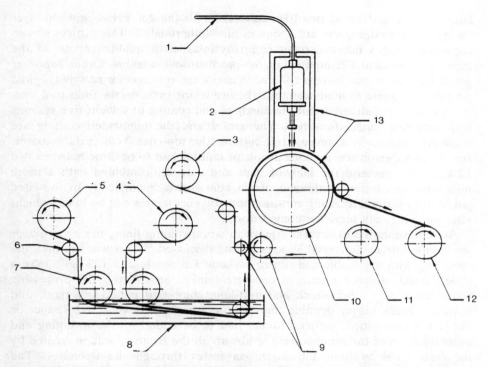

Fig. 4.10. Sketch of a production line of impregnated glassfibre/copper laminates with EBC used as bases for printed circuit boards.[36]

thermal system, thermal deformations of jigs are limited, and performance and strength of these objects is improved. Government sponsorship may have helped this costly project, but it will become a reference for other, similar applications in the future.

WOOD-BASED AND BUILDING MATERIALS

This is the oldest field of industrial EBC application. It started in Europe in the 1970s: several lines with conrete bunkers for shielding and scanning accelerators from Ford, later from High-Voltage Engineering, both of American origin. After an apparent saturation or lack of innovation, new lines with new wood-converting ideas appeared in different locations in the US. Only recently, new European companies were persuaded, or discovered that protecting or decoration of wood-based products (mainly panels) or building materials could still be an advantageous process for a growing market, where UV finishing would not satisfy because of depth of penetration, opacity or pigmentation.

The best documented *finishing and coating line for door panels* with UV lamps for the sealer and two EB accelerators for the decorative and top-layer curing is still in operation at Svedex in the Netherlands.[37] This project was so important, that a national paint company installed a smaller version of the same handling and EB equipment for product development. Other suppliers got at least their own laboraotry unit to check the coatings for reactivity, gloss control, sensitivity to inhibition, etc., before trying them on the industrial line. A lot had to be learned about the handling and coating of solvent-free systems with unusually high viscosities. The sealer and the pigmented coating are normally applied by a reverse-roll system, the top-coat by a curtain coater. Interlayer adhesion was a major problem: sanding had to be done between two EBC layers, because the lack of heat and solvent, combined with a more important and instant shrinkage, made adhesion more difficult. In the cited publication many interesting consumption and cost figures can be found which, after updating, can serve for comparison.

As representative of several American wood-coating lines, the *decor-paper lamination* line of Universal Woods will be discussed.[38] This was the first line operating with a self-shielded, linear-cathode EB accelerator (250 kV, 165 cm wide), which did not require a concrete bunker around the EB processing zone. Since up to 2-cm-thick and 4-m-long chipboard panels are rigid, and require a much bigger opening than a processor for a flexible web (paper or film), a lead-shielded, inclined tunnel had to be added on the incoming and outgoing sides of the process zone to absorb all the Bremsstrahlung created by the beam and to limit the inert gas losses through the openings. The installation shown in Fig. 4.11 has a total length of about 12 m; this is still considerably less than the space which would be taken by a line with thermal curing estimated at 50–60 m.

Fig. 4.11. Chipboard coating, laminating and EB curing line, showing inclined transport and shielding tunnel (2 × 6 m) with short, central EB processor (250 kV, 90 mA, 165 cm; ESI/Universal Woods, USA).

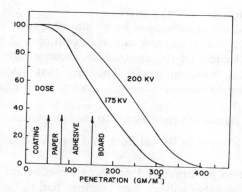

Fig. 4.12. Dosimeter curves for the evaluation of the needed acceleration voltage to give sufficient curing to the four layers: varnish, impregnated decor paper, adhesive and adhesive penetrated into the panel: 200 kV gives 80% of surface dose on panel surface.

On this line, the above-described coating and curing methods can be performed, but it is the decor-paper lamination, which shows that EBC can be particularly interesting. After sanding, the chipboard panels receive an adhesive layer of about 90 g/m², on which a decor-paper is laminated and cut. The laminate passes under one (two) curtain-coaters which deposit about 30 g/m² of a protective varnish. The printed decor-paper is very porous, it is rapidly impregnated by the adhesive and the varnish to saturation, then only the three layers are cured at once in a single pass under an EB of 30–50 kGy resulting in a very resistant 'fibre-reinforced' surface, which meets several properties of hot laminated panels. The dose–penetration chart in Fig. 4.12 shows that the beam efficiency drops as a function of coating weight for this interesting multilayer product. Acceleration voltage must be selected to guarantee curing of the lowest layer. This continuous, room temperature process compares very favourably with traditional lamination: application of varnish of the decor-paper and a separation sheet (i.e. caul stock, also obtained by EBC), then stacked and laminated with heat in a press. Larger producers may today also work with very costly and complex continuous presses, developed in the meantime.

Other companies in the US have become active and produce even larger amounts of such coated and laminated chipboard or fibre-board panels. One of them has received a special award at *RadTech '88*, for their activity and a very informative presentation including economical figures.[39,40]

After these success stories in the US, more motivation for new developments was felt in Europe in the late 1980s. For years an attempt has been made to decorate asbestos–cement panels with outdoor resistant coatings. A breakthrough was never obtained, and then asbestos was banned. Several replacement materials were developed: wood–cement panels or fibre-reinforcement with glass or organic fibres. Two initiatives and several years of hard work led to two new, very interesting product lines.

EB processing of wood–cement panels

Most EB projects are best developed by a team of several specialised suppliers with the future user: substrate, raw material, coating and handling equipment suppliers, and the builder of the electron accelerator with a responsibility for the shielding of the handling zone. The fact that several West-European suppliers were going to make this new line work in Hungary was a particular challenge.[41]

There are some significant differences to the above-described panel lines:

- the panels may be up to 40 mm thick and must be coated and cured on the sides;
- the finish had to be structured and weather-resistant (therefore, a spray method and sprayable solvent-free coatings had to be developed for both sides with 100 g/m^2 base coat and 100 g/m^2 top coat);
- both for a regular surface structure and cure the panel speed had to be precise;
- transportation and shielding were realised with two Ω-shaped tunnels and horizontal conveyors, instead of the inclined tunnel (Universal Woods) or the V-shaped conveyor (Svedex); and
- the inert atmosphere is obtained with carbon dioxide.

Figure 4.13 shows two views of one of the two EB processing zones with details described above. The accelerators are of the scanning type (250 kV, 150 mA). The belt speed can reach 30 m/min, one of the reasons to choose

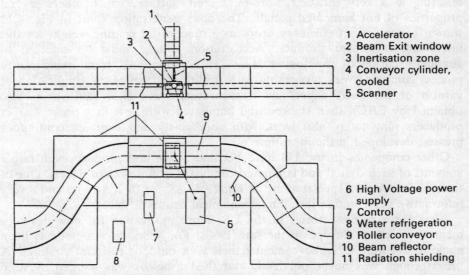

1 Accelerator
2 Beam Exit window
3 Inertisation zone
4 Conveyor cylinder, cooled
5 Scanner
6 High Voltage power supply
7 Control
8 Water refrigeration
9 Roller conveyor
10 Beam reflector
11 Radiation shielding

Fig. 4.13. One of two Ω-shaped EB processing zones for the curing of coatings on wood-cement panels of 3·2 × 1, 25 × 0·04 m (Polymer-Physik/Taubert, D/Falco, Szombathely, H; R32).

Fig. 4.14. Building panels with weather-resistant EBC coatings produced on the line shown in Fig. 4.13, used for private and public housing (Düsseldorf area, R32).

EBC rather than other curing methods. Figure 4.14 shows building panels from Falco(H) exported back to Germany and used on private or public buildings.

After having spent thousands of man-hours on such a project, it is hard for the European team involved to observe that the next two wood coating projects will be installed by competitive American working groups. On the other hand, it is an advantage to all the suppliers, if after at least 10 years of persuasion and hesitation additional lines using EBC are used in the building products industries.

In Italy, the country with the largest number of UV curing equipments installed for wood-finishing processes, the widest EB panel processor (2·6 m) will be installed in 1991 for the *pigmented gloss finishing of thin fibre-board and chipboard panels* up to 40 mm thick. The selection of EB is somewhat surprising, because UV finishing has made big progress both for pigmented and thick coatings. It may be justified by the safer cure of coatings penetrated into the fibre-composite and the lower heat load from EBC. Thin panels risk deformation by unilateral wetting—even from penetration of monomers!—or by shrinkage during polymerisation. It is hoped that the latter can be avoided by using less reactive, cheaper chemicals; but these will call for a higher dose, resulting in higher unilateral heating. Time will tell, if EBC for this record size installation was a good choice.

In Sweden, a country with enormous wood resources, a planar beam EB accelerator was installed in 1991 for the curing of coloured finishes at different gloss levels on *MD fibre-board panels,* used as indoor and outdoor building panels. This investment took about 8 years to be accepted by this plant.

EB accelerators do not always get bigger and bigger. In the 1980s, EB projects have been undertaken to cure coatings on pencils (Switzerland),[42] and on wooden profiles for picture frames, etc. (UK). In both cases a laboratory unit (175 kV, 15 cm, 10 mA) would have been sufficient for the pencils with rotation under the beam, for the profiles with a double coating and curing pass. In both cases the projects were not realised, because the coating with a die did not give the desired regularity, and even laboratory units were considered too expensive compared to available thermal-curing or UV-curing technology. An additional problem for the pencils was that the chemical suppliers were not able to present an FDA approval for their products.

Now the first application of this type has been started in late 1990 in Sweden: *small wooden base-boards* (Zargen) are vacuum-coated and cured from two sides with two 'baby-scanners' (150 kV, 20 cm, 30 mA). As it happens most often in the realisation of EB projects, the accelerators give the requested results, but problems have to be solved with the coating viscosity, application technique and recycling.[43]

The four above examples are unique applications in wood related areas. It is hoped that they will be available as positive references for other potential users, if possible accelerated with a know-how transfer licence, unless the licensee believes that finding the right solution himself is less expensive.

Developing countries are making important efforts to create an adapted version of EBC finishing of wood-based products. The *Shanghai* Applied Research Institute reported in 1988 about an EB pilot line for *wood panel finishing* with an annual capacity of 260 000 m^2 of up to 90-cm-wide panels for furniture elements.[44] Except for the curtain coater, all equipment elements, inerting with nitrogen, the coating and the ICT type 300 kV accelerator have been developed in China. The many experiences and results reported lead to the belief that a transfer to industrial production lines seems feasible.

Malaysia, the largest Asian exporter of wood, makes considerable efforts to avoid buying back converted products from other countries at a high price.[45] Imported radiation-curable chemicals are also too expensive, therefore studies are made with accelerators and help sponsored by the United Nations and Japan. Dimeric acid resins based on the plentiful palm oil were synthesised and mixed with monomers. Cured at 40 kGy they resulted in dry, but too soft coatings.

Radiation curing activities in the former USSR

Thanks to a recent review presented at *RadTech Asia '91*,[46] we can describe parallel activities in these countries, where several institutes have developed medium to high energy (scanning) EB accelerators for some years. Cross-linking and research are the major activities, but curing of coatings, mainly on wood-based products, has been industrialized in the last 3–5 years.

In these locations *wood elements for TV sets* are produced with the following typical procedure: application of a relatively thick primer (0·3–0·5 mm!), and a curtain-coated top layer. The panels of $2 \times 0.55 \times 0.018$ m move at a maximum speed of 10 m/min in a nitrogen atmosphere (0·1% oxygen) under three accelerators with stepwise increasing dose levels from 10 to 150 kGy. The total annual production of one plant reached 600 000 TV sets in a two-shift operation.

An interesting detail is, that in one plant inerting is avoided by 'covering' the coating by a PA film. It is not clear if the film is laminated to the panels or if it is peeled after curing, maybe at the end of the total manufacturing cycle.

One of the few advantages of a state-governed industry may be, that a promising idea can be installed in several places, allowing some economies and rapid know-how transfer without competitive barriers. That may have happened for certain *wood-coating* lines, but even the official reports indicate that there is not always agreement about the number and type of accelerators, inerting, chemistry to be used. For one wood-panel line a polyether + styrene chemistry is used which is slow, but much cheaper than the formulations used in the Western world. In one case they work in air 'with a lot of ventilation', in another, air curing is followed by a surface curing step with UV (a dual cure). This latter technique is applied in a *paper to asbestos–cement coating* and *lamination* line near Moscow, with an annual production of 200 000 m^2.

The Russian report on EBC lines closes with a short description of the curing of *coatings on ceramic tiles* and the planned construction of three plants for the *coating of gypsum* board. All these activities prove that also—or even more actively—EBC will be applied in the Eastern countries for its space and energy savings, but probably in the first place to *create improved surfaces* not feasible with thermal curing processes.

EBC of building materials and other rigid or three-dimensional objects seems to be more closely related to countries and companies which have a well developed engineering culture (Japan, D, DO, the former USSR) and which can, if necessary, apply medium-energy scanning accelerators. Low-energy, linear- or planar-beam accelerators are more often (successfully) found in curing of thin layers on flexible substrates (ESI, RPC). It is therefore interesting to have a look at the activities of *EBC on building materials in Japan*.[47]

In 1979, a production line for glossy *cement-based roof-tiles* was already started. Base and top coat with good weatherability of about 65 μm each are applied by spraying and are EB cured.

In 1984, a producer of *gypsum panels* in cooperation with a paint company demonstrated in their Ginzha shop the possibilities of beautiful enamel-like decoration effects, obtained even with metal-pigmented solvent-free coatings and EBC. Since the gypsum company lacked the know-how of making the panels water- and humidity-proof, they could not be used for wet rooms or outdoors. This market limitation was probably the reason why an industrial production was not becoming a reality.

Learning from the above weaknesses (no water-resistance, poor weight-to-strength ratio, banning of asbestos as a reinforcing fibre), another Japanese company created in 1989 a production line for EBC *decorated melamine–cement panels*. The mechanical strength of these melamine-impregnated panels is three to four times higher than that of equivalent cement panels; they are used as new housing construction and decoration element with good water-resistance.

A similar, apparently more sophisticated product was created and patented at about the same time by a Dutch company.[48] It took years of confidential chemical and process development, it was a great challenge for the second American manufacturer of EB processors (RPC) to install their first large unit for the production of *decorative building panels*.

Based on the patent description and some hearsay, it will be tempting to describe the major elements of this EBC success story. A fibre- and resin(?)-reinforced inorganic panel material is base- and top-roller-coated with weather-resistant formulations. An optional PETP film is laminated on top. The wet 'sandwich' passes under a smooth or structured roll and is immediately EB cured to freeze the imprinted structure. Depending on the film finish and the lamination roll surface, the product will become flat or structured, glossy or with a silk finish. Up to now, e.g. in the Falco process, it was thought that a rough outdoor structure could only be realised by (very difficult) airless spraying. But this is not the only advantage of the PETP film: it also reduces inert-gas consumption and will protect the decorative panel till it is peeled for installation. If weather-resistance was not evident in the beginning, it was clear that EBC would result in improved abrasion and chemical resistance. These panels will become strongly competitive to powder-coated metallic building elements, maybe with a slightly higher weight and limited or non-existing shapability.

EBC OF COATINGS ON METALS

Why do we find interest in EBC coatings and installations for metal coils, sheets, tapes and three-dimensional parts? It cannot be the advantage of heat-sensitivity of the substrate; we will see below that even some heat would be of advantage. The major reasons for EB processing are

1. getting away from solvents and their costly recycling systems;
2. saving space needed for huge drying tunnels, particularly for coil coating; and
3. the creation of glossy decorated surfaces with improved resistance to stain, chemicals and scratching or abrasion.

In the early 1980s it has been attempted to coat galvanised steel (Laminoires de Strasbourg), to prime-coat aluminium coil (VAW-BASF), to varnish aluminium foil, and to decorate different types of steel sheets. The major problem encountered is the lack of adhesion, worse, the impossibility of shaping such coated sheets. The basic advantages of radiation curing—no solvent, no heat and almost no time—become the real handicaps when rigid, pore-free, uncleaned or unprimered metallic surfaces are to be coated. EBC coatings have in general poor adhesion because of reduced wetting and because they shrink by about 10% during instant polymerisation.

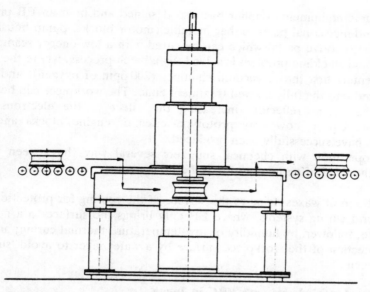

Fig. 4.15. Four-station carousel with EB scanner for the curing of metallic coatings on car wheel rims (Polymer-Physik/Dürr/Volkswagen).

Researchers have recognised that the metallic surfaces must be at least as clean as for thermal curing, that the reactivity of the coatings must be reduced, because lower reactivity means less shrinkage and higher dose, which means higher temperature and more time for the curing process. In the case of can coatings (UV cured) it was found that a thermal post-cure—to dry the interior varnish—resulted in improved adhesion of the outer coating and inks.

Since deformation of such coatings is an added difficulty, an important equipment supplier to the automotive industry guided a large feasibility project at Volkswagen to paint and EBC already formed *car wheel rims*:[35] EBC of three-dimensional objects brings three other difficulties: curing of surfaces parallel to the beam, inerting and shielding of a larger process zone. After spray painting with a metallic coating, the rims were lowered into a cassette of a four-station carousel: S1 = load/unload, S2 = evacuate, S3 = inert and S4 = cure (Fig. 4.15).

With a 200 kV, 40 mA scanned beam it was possible to cure 6 rims/min. A pilot run during several months (on a few square metres floor space) allowed the production of nicely decorated wheel rims for the 'Golf', which had good abrasion, stone impact and salt-spray resistance. These good results can be explained by a perfected thermally cured base coating and by a slow and warming dose of ~300 kGy!

The project was (temporarily?) interrupted due to an exaggerated cost price for nitrogen, due to unacceptable (odour) paint components and due to a lack of maintenance of the handling equipment—not easy in a harsh environment of vacuum, irradiation, etc. Anyway, it is not an easy task to get approval for a new process from a customer in the automotive industry.

The same equipment supplier has also designed and built an EB processor for three-dimensional parts, either metallic (motor blocks, pump housings) or plastic (automotive parts) which can be cured with a low-energy scanner. The inerting and shielding problem is solved, by a batch process, where the painted object, enters first into a vacuum chamber (200 ppm of oxygen), and is then introduced into the fully inerted treatment zone. The workpiece can be rotated under the beam, reflectors increasing the effect of the electrons on its side-walls. Up to now only prototype-series of engine blocks and pump housings have successfully been produced.

In cooperation with chemical suppliers several ways have been tried to reduce the need of an inerted atmosphere:

- addition of waxes rising to the surface of the coating for protection;
- hybrid curing systems, where EB only brings the surface to a non-tacky state, followed by humidity or low-temperature thermal curing; and
- protection of the workpiece surface by a water layer to avoid 'supply' of oxygen.

Finishing of steel sheets with EBC in Japan

Several steel companies have succeeded after long development efforts to create new products and production lines with EBC. Sheets were selected, rather than coil or three-dimensional objects, despite a certain experience from a production line for three-dimensional motorcycle parts which was in operation between 1973 and 1980. Improvements over existing technology were looked for, therefore the traditional primering was maintained.

The first line started in 1982;[49] it produced the so-called *Ellio-sheet* with a top-coat cross-linked by EB to obtain higher hardness and stain resistance. A special version is the *'white-boards'* with excellent stain resistance used for erasable marking with felt-pens for classrooms and seminar rooms.

In cases where the sheet had to be bent or shaped, the coating was replaced by an EBC lamination of white or coloured PVC film. The adhesive and PVC film allowed a limited deformation without delamination, but the surface hardness/scratch resistance was evidently not very high.

Another plant started in 1988 for the *decoration of pre-coated steel sheets*.[50,51] On a galvanised, chromated and traditionally primed steel sheet, EBC top coats (25 μm) are applied by a curtain coater. This (white) layer is printed with a decor and covered with a 50 μm protective coating. These two to three layers are cured together at an average dose of 50 kGy.

In an alternative method a traditionally coated and printed PETP film is laminated and EB cured on in-line, chemically primered steel sheets or coil at a residual temperature of about 100°C from the primering operation.[52]

Panels for tunnel walls[53]

The producer of Ellio-sheets found a diversified application which could be manufactured on the same line. To improve illumination and make maintenance easier, Japanese highway tunnel side-walls are clad with a row of protective panels.

Up to now these panels were asbestos-based and painted. They did not offer much resistance to mechanical impact and after a few cleaning cycles the paint became dull and lost reflectance. Metallic sheets of the 'white board' type described above gave too much dazzling. After looking at different shapes of glass fillers to optimise optical and maintenance/life aspects, small glass beads were selected as an additive to an EBC white coating on traditionally primered stainless-steel panels. The glass-bead size and quantity had also to fill two other requirements: no reduction of pot life, as happens with aluminium oxide additions, and no significant increase of the viscosity to allow curtain coating. With this solution long-lasting products offering additional safety and tunnel illumination were realised.

The idea of using glass beads in combination with *traffic (signs)* and EBC is not new, but it has been kept quite secret by 3M, who produce such surfaces for weather- and UV-resistant highway signs in the US and Europe. This development was based on a large study about the thermal (and UV) resistance of a large number of monomers, which led to a prize-winning paper at the Radcure Conference in Basel (1985).[54] Short-time tests at elevated temperature showed which monomers 'evaporate' (10–30%) and should be avoided for such critical applications.

Varnishing of aluminium foil

The aluminium converting industry shows a strong interest to replace todays varnishing technology based on polluting chemicals and an energy-wasting double curing process by radiation curable, space- and energy-saving methods. It has rapidly become clear that the aluminium foil must be free of lamination oil traces and even optimally chemically treated (e.g. chromated). Only very few of the available clear varnishes give reasonable adhesion. This may however change with the introduction of cationic curable varnishes.

The few varnishes with acceptable adhesion were submitted to a half-way deep drawing test: because tons of these varnished foils are transformed into deep drawn containers for butter, marmalade, quick meals, etc. If the varnish supported needed to be exceptionally stretched, none of them could be compressed in the rim area. For this reason the samples did not even go to the next test: steam or hot water sterilisation at 130°C for 30 min.

This is another example which shows that many more potential applications would exist if the right chemical, physical and physiological properties could be found.

Wire enamelling

As shown with the example of aluminium foils, it has always been difficult to obtain good adhesion of EBC coatings on metallic surfaces because of poor wetting and instant shrinkage during curing. On copper wire this is even more difficult, otherwise EBC would certainly have been used years ago to replace the multilayer thermal curing in large towers of very polluting phenol-based coatings on enamelled wire for electrical motors. Should adhesion be found, the potential users might complain about the price of the coatings or the fact that they are not as heat-resistant as the types used today, which are at least in a class of 150°C.

Equipment for handling fine wires is known; it is used for the cross-linking with EB of a polymer insulation layer of hook-up wires (for a description, see chapters on EB equipment and cross-linking). UV curing as it is used successfully for coatings on optical fibres with special geometry UV lamps, maybe combined with a cationic curing mechanism is another way. A small EB pilot unit is used in Germany for the development of such enamel coatings, but no details can be obtained at this time. A breakthrough in this and many other metal coating fields would be very welcome, especially to those who have to find less-polluting technologies as soon as possible.

Other applications of EBC

Replacement of metal for prototypes or a small series by radiation cured resins will become a growing field of application. Stereolithography prototypes are built up from a monomer-resin bath by programmed curing, layer by layer, of a three-dimensional object with a laser beam. More details will be given in the chapter on laser applications (see Chapter 00).

A Japanese company suggests replacing an electrocast and plated *stamper for the embossing* of PVC film holograms by EBC resins.[55] Polyarylate films are used as a new base because of their temperature resistance (T_g 215°C) and their surface tension which gives the best adhesion for the EBC layer. In the monomer selection a compromise between two-functional (good peeling) and three-functional monomers (better heat resistance during the embossing process at 80–120°C) is the best choice.

For at least the last 15 years the field of *abrasive products* and their production is a field of R&D, patents and some limited applications.[56] Early patents and limited production dealt with the *impregnation* of cloth for abrasive tapes and discs. To assure through cure, scanning accelerators with up to 550 kV were used. It is still very desirable to replace huge curing ovens, release of polluting solvents and chemicals by the introduction of EBC. There may be a limited number of EB applications in the US, but they are not publicly acknowledged.

The following problems have been encountered in the development of EBC abrasive products.

- Low-energy accelerators do not have enough penetration to cure the binder under medium to large sized grains (high thickness and density). Curing from the reverse has been suggested; in this case the selected paper or cloth thickness should not exceed 200 g/m².
- Traditional base-papers show a tendency to embrittle and discolour under

EBC. EB resins may have poor adhesion on normally prepared papers. Several parameters will have to be changed to succeed.
- EB resins are more expensive and not as heat- or water-resistant compared to the melamine resins used up to now.

Probably the first industrial application of linear cathode accelerators is the curing of the binder of electrostatically sprayed *flock on plastic substrates* used in the shoe industry. In many ways it is similar to the abrasive applications; several advantages were claimed for this never-duplicated technique:

- no solvents, therefore no explosion risks;
- short treatment zone, no heat damage to the substrate; and
- better anchorage of the flock in a thinner layer of binder, combined with improved solvent resistance.

This pioneering application found a lot of interest in the automotive suppliers industry: many surfaces in a car are flocked for decorative, sound-dampening and tightness effects (ceilings, glove boxes, and in a very large volume: EPDM window profiles). During the development steps three major drawbacks of EBC after flocking were observed.

1. Polyamide, the usual flock material and its colourants often changed their colour by EBC—not acceptable in the very demanding car industry.
2. The flock layer carries an air cushion which may lead to inhibition and insufficient cure of the binder.
3. The binders known some years ago, did not give sufficient adhesion on vulcanised or oxidised EPDM profiles. They did not pass the severe abrasion tests.

Producers of these profiles could not yet realise their logical dream of combining the vulcanisation of the EPDM with the curing of the binder by EBC in one operation.

CONCLUSIONS

These descriptions of EBC applications and attempts are certainly not exhaustive, because many developments and in-house applications are still kept secret. Despite the reluctance and difficulties encountered when changing from traditional/thermal curing to room temperature curing processes, their number regularly increases and their description follows with a certain delay.

When taking two industrial breakthroughs of EB applications as an example, we can feel encouraged that this will happen in other fields. We are thinking of the cross-linking of polyolefines to induce shrink memory and improved properties to packaging materials (200–300 accelerators used world-wide by W. R. Grace, DuPont and others) and the major application of EBC in the 1980s: safe and high-speed curing of litho inks and varnishes on packaging cardboard (30–40 compact low-energy accelerators used by Tetrapak and others). These successful markets will encourage other groups to go from a single to multiple or market applications.

Simplified and less expensive accelerators, increased ecological pressure, the huge potential in the adhesive fields and converting of packaging materials will bring additional and new EBC applications at a growing rate.

REFERENCES

1. Läuppi, U. V. *Radiat. Phys. Chem.*, **35** (1990) 30.
2. French Patent 2.347.987, 1976.
3. French Patent 2.447.232, 1980.
4. O'Neill, J. R. *Proc. RadTech Int. Conf.,* New Orleans, (1988) p. 245.
5. Pfahl, K. A. *et al.*, Euro. Patent 0 130 659, 1984.
6. Beer, E. *et al. Hoechst High Chem.*, **4** (1987) 42.
7. Globic, F. Cl. & Peter, D. R. St. *Radiat. Phys. Chem.* **31** (1986).
8. Sack, W., German Patent 33 00 025, 1983.
9. Gray, L. S., Euro. Patent 0 106 695, 1983.
10. Anon. RadTech Int. Report, Sept./Oct., 1990, 8.
11. Patent. Application for filter paper (MD-Paperworks).
12. Tripp, E. In *Proc. RadTech Int. Conf.,* New Orleans, (1988) p. 37.
13. Bickford, R. H. In *Proc. RadTech Int. Conf.,* Chicago, **2** (1990) p. 109.
14. McIntyre, F. In *Proc. RadTech Int. Conf.,* Chicago, **2** (1990) p. 73.
15. Eckberg, R. P. & Riding, K. D. *Proc. RadTech Int. Conf.,* Chicago, **1** (1990) p. 357.
16. Eckberg, R. P. & Riding, K. D. *Proc. RadTech Europe Conf.* (1989) p. 311.
17. Anon. *Adhesive Age,* Dec. (1985) 20.
18. Anon. *PFF Converter,* Sept. (1987).
19. Grosso, J. S. & Wilheim, M. J. *Printed Circuit Fabr. Mag.*, Feb. (1985).
20. Anon., *Packaging Digest,* **18** (1981).
21. Rodrigues, A. M. & Newcomb, W. T. **25** (1985) 617.
22. O'Brian, T. *Radiat. Phys. Chem.*, **25** (1985) 609.
23. Bly, J. H., Weiss, H. K. & Stichling, R. B. RDI-Euro 102.
24. Barlow, A., Biggs, J. & Maringer, M. *Radiat. Phys. Chem.*, **9** (1977) 685.
25. Barlow, A., Meeks, L. & Hill, L. *Radiat. Phys. Chem.*, **14** (1979) 783.
26. Kruger, Euro. Patent 0 279 722, 1988.
27. Keough, A. H. *Conf. Proc. Radcure '84,* Atlanta, 1984, p. 1038.
28. Keough, A. H., US Patent 5,002,795, 1991.
29. Delaney, W. H. & Keough, A. H. *Proc. RadTech Int. Conf.,* Chicago, **2** (1990) p. 105.
30. Ohta, T. *et al.*, Euro. Patent 0147 061, 1984.
31. Ohta, T. *Radiat. Phys. Chem.*, **22** (1983) 795.
32. Läuppi, U. V. *Proc. RadTech Asia,* Osaka, **1** (1991) p. 23.
33. Cermak, M. L. & Rand, W. M. Preprint Radcure Conf. Basel (1985).
34. Tabata, Y. *Proc. RadTech Conf.,* New Orleans, **1**, (1988) p. 43.
35. Dieck, E. L., Holl, P. *et al. Conf. Proc. Radcure '84,* Atlanta, (1984), p. 1212.
36. Kopchonov, Grozdov, Krajzman *et al. Proc. RadTech Int. Conf.,* **2** (1990), p. 10.
37. Häring, E. Eurisotop Series No 105 (1976).
38. Quintal, B. S. & French, D. *Radcure Conf.*, SME Tech. Paper FC83-277 (1983).
39. McBosch, J., LeFors, J. D. & Hetzel, T. *Beta-Gamma,* Feb. (1988) 4.
40. Anon., RadTech Int. Report (1990) Sept./Oct., 8.

41. Holl, P. *Coating,* March, (1988) 85.
42. Jeanmonod, Ch., Swiss Patent 664 707, 1988.
43. Holl, P. (1991) described at Radcure Seminar at TA Esslingen (D).
44. Jun, X. M. *et al. Proc. Conf. Rad. Curing Asia,* (1988) p. 197.
45. Seman, A. Sh. M. *Proc. Conf. Rad. Curing Asia,* (1988) p. 202.
46. Pikaev, A. K. *Proc. RadTech Asia Conf.* (1991) p. 35.
47. Maruyama, T. & Sugimoto, K. *Proc. Rad. Curing Conf. Asia,* (1988) p. 275.
48. Van der Hoeven, J. C. W., Euro. Patent Applic. 0 166 153, 1985.
49. Ellio-sheet '86? *(86 or 88) Proc. Conf. Rad. Curing Asia.*
50. Masuhara, K., Koshiishi, K. & Tomosue, T. *Proc. Conf. Rad. Curing Asia,* (1988) p. 231.
51. Masuhara, K., Koshiishi, K. & Tomosue, T. *Proc. RadTech Int. Conf.* (US) (1990) p. 281.
52. Yamada, H. & Sumita, M. *Proc. Conf. Rad. Curing Asia,* (1986) p. 281.
53. Ueno, N. & Oka, J. *Conf. Proc. RadTech Asia,* (1991) p. 109.
54. Thalacker, V. Radcure Conf., Basel, SME Tech. Paper, (1985) p. 85.
55. Inaba, Y. & Okano, S. *Proc. RadTech Conf.,* Florence, (1989), p. 137.
56. Hesse, W., US Patent 4,047,903, 1977.

Chapter 5

UV and Electron Beam Curable Pre-Polymers and Diluent Monomers: Classification, Preparation and Properties

NORMAN S. ALLEN & MICHELE EDGE

Centre for Archival Polymeric Materials, Chemistry Department, Manchester Metropolitan University, Jon Dalton Building, Chester Street, Manchester, M1 5GD, UK

Introduction 226
Epoxy Acrylates 226
    Bisphenol A epoxy acrylates 227
    Applications and examples 228
Acrylated Oils 228
Urethane Acrylates 228
    Polyether urethane acrylates 229
    Applications 231
Polyester Acrylates 231
    Applications and examples 232
    Polyether acrylates 232
    Preparation 232
Water-Miscible Pre-polymers 233
Vinyl/Acrylic Pre-polymers 234
Unsaturated Polyesters 235
    Applications 235
Polyene/Thiol/Silicone Systems 236
Diluents/Monomers 237
    Reactive monomers 237
    First-generation reactive monomers 237
        Monofunctional monomers 237
        Difunctional monomers 238
        Trifunctional monomers 238
    Second-generation monomers 238
        Monofunctional monomers 238
        Difunctional monomers 238
            Tripropyleneglycol diacrylate (TPGDA) 238
            Dianol diacrylate (CDDA) 238
        Trifunctional monomers 239
    Other monomers 239

Vinyl functional monomers	240
Non-reactive monomers	240
Amine Synergists	240
EB-Curing Technology	241
EB accelerators	242
Scanned-beam accelerators	242
Electron-curtain accelerators	242
Operational conditions	243
Dose	243
Electron energy and penetration	244
Energy losses	245
Backscatter effects	245
Mechanism of EB curing	247
Post-cure Stability of EB-Cured Coatings	248
Thermal stability	249
Photostability	252
Photostabilisation	259
References	261

INTRODUCTION

Several types of polymer systems suitable for curing by means of either high-intensity UV or visible light or electron beam (EB) curing have been, and continue to be, developed. Insofar as the polymer system is concerned both curing processes are essentially the same. To date several types of radiation-curable pre-polymers and monomers have been developed.[1-4] These are epoxy acrylates, acrylated oils, urethane acrylates, unsaturated polyesters, polyether acrylates, vinyl acrylates and polyene/thiol/silicone systems. The main objective in the preparation of these materials is to obtain a molecule which contains unsaturation capable of reacting with other unsaturated molecules when subjected to the irradiation conditions defined above to give a solid coherent film. Obviously, the molecules containing unsaturation must remain stable until the cross-linking reaction is required. This is normally achieved by the addition of 500–1000 ppm of an oxygen inhibitor or free-radical scavenger.

In this chapter, we review the types of pre-polymer systems and diluent monomers, their preparation, formulation, properties and applications. Particular emphasis will be placed on EB-curing technology with reference to the thermal and photostability of the cured coatings.

EPOXY ACRYLATES

The reaction of an epoxy group with other acrylic or methacrylic monomers will give rise to an epoxy acrylate (or methacrylate). There are a wide range of epoxy acrylates available including acrylates of bisphenol A glycidyl ether, acrylates of epoxidised oils such as soya or linseed and acrylates of epoxy novolaks.

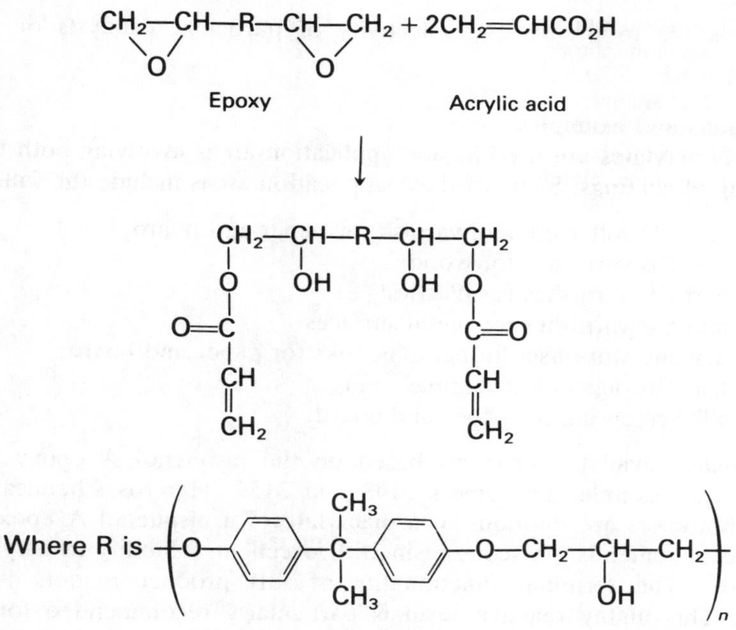

$$CH_2\!-\!CH\!-\!R\!-\!CH\!-\!CH_2 + 2CH_2\!=\!CHCO_2H$$
$$\underset{O}{\diagdown\diagup} \qquad \underset{O}{\diagdown\diagup}$$

Epoxy　　　　　　　Acrylic acid

↓

```
      CH₂—CH—R—CH—CH₂
      |   |      |   |
      O   OH     OH  O
      |              |
O=C             C=O
      |              |
      CH             CH
      ||             ||
      CH₂            CH₂
```

Where R is $\left(\!-O\!-\!\!\!\bigcirc\!\!\!-\underset{\underset{CH_3}{|}}{\overset{\overset{CH_3}{|}}{C}}\!-\!\!\!\bigcirc\!\!\!-O\!-\!CH_2\!-\!\underset{\underset{OH}{|}}{CH}\!-\!CH_2\!-\right)_n$

Scheme 5.1

Bisphenol A epoxy acrylates

The preparation of a simple epoxy acrylate is illustrated in Scheme 5.1 where R is a bisphenol A glycidyl ether.

Epoxy resins which can be acrylated, range from liquid epoxy resins such as the pure bisphenol A diglycidyl ether, to the solid high melting point ones. This range of epoxy acrylates forms the bulk of the epoxy acrylates currently in use. Their major restriction is processing viscosity, and for the higher-molecular-weight resins a diluent monomer must be present.

Catalysts can be used to promote the acid–epoxy reaction and include tertiary amines,[5-9] potassium hydroxide,[8] N,N-benzyl methylamine,[8] N,N-dimethylaniline[8] and benzyl tri-methyl ammonium hydroxide.[8] These catalysts would be used at about 1% by weight of the reactants although in general catalyst levels are best determined by an empirical approach and are normally used at 1–2%. Mixtures of catalysts can also be used and are often more effective than single ones and some function synergistically.

The reaction temperature is an important factor to be considered here since the catalyst efficiency will vary with temperature. The types of catalysts used are inorganic alkaline salts such as sodium carbonate, organo-metalllic salts such as lithium octanoate and basic organic compounds such as tertiary amines and phosphines. Generally, alkaline substances catalyse the addition reaction (i.e. acid and epoxy), whilst organo-metal compounds promote both addition and condensation (i.e. acid and alcohol) esterification. However, in general, it is not possible to predict the selectivity of particular catalysts or catalyst mixtures.

Applications and examples

The epoxy acrylates are used in side application areas involving both UV and EB curing of coatings. Some of these application areas include the following:

- UV and EB roller ink and varnish for paper and board;
- UV and EB varnishes for wood;
- UV and EB varnishes for plastics;
- UV and EB varnishes for metal surfaces;
- UV dry and wet offset lithographic inks for paper and board;
- UV and EB pigmented coatings; and
- UV silk-screen ink for paper and board.

Commercially available products based on the bisphenol A epoxy acrylate include, for example, Photomers 3148 and 3159 (Harcros Chemicals Ltd). These photomers are dilutions of a diacrylate of a bisphenol A epoxy resin. The base oligomer is a viscous resin with excellent solubility in acrylic ester monomers. The terminal functionality of this product renders it highly reactive. This highly reactive resin is particularly recommended for use in wood-finishing applications and varnishes for paper and board. The base resin has been found to exhibit excellent sanding properties when formulated into base coats for wood applications. Photomer 3049 (Harcros Chemicals Ltd) is a fatty acid modified aromatic epoxy resin. The fatty acid modification improves the hydrophobic characteristics of the resin and hence has significantly improved its lithographic performance. Photomer 3049 should be added at a level of 10% to improve the water resistance of a varnish or ink.

ACRYLATED OILS

Oils containing epoxy groups, and particularly epoxidised soya bean oil, can be readily acrylated under conditions, which are similar to those used for epoxy acrylate manufacture. Generally, the viscosity of acrylated oils is sufficiently low (200–300 P (20–30 Pa . s) to allow a diluent-free preparation.

Epoxidised acrylated oils have several important properties such as slow cure, give soft films and good adhesion, have very low skin irritancy and provide excellent pigment wetting.

Other specialised epoxy acrylates are epoxy novolak oligomers which are used in UV solder-resist formulations for the electronics industry.

URETHANE ACRYLATES

The reaction of an isocyanate group with the hydroxyl group of an acrylic or methacrylic monomer will give the corresponding urethane acrylates or methacrylates. The diisocyanates which may be used for this purpose have

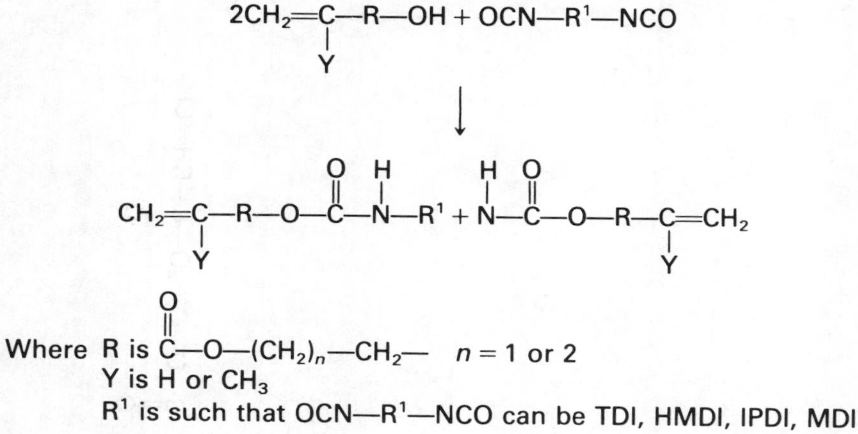

$$2CH_2=\underset{Y}{\underset{|}{C}}-R-OH + OCN-R^1-NCO$$

$$\downarrow$$

$$CH_2=\underset{Y}{\underset{|}{C}}-R-O-\overset{O}{\overset{\|}{C}}-\overset{H}{\overset{|}{N}}-R^1 + \overset{H}{\overset{|}{N}}-\overset{O}{\overset{\|}{C}}-O-R-\underset{Y}{\underset{|}{C}}=CH_2$$

Where R is $\overset{O}{\overset{\|}{C}}-O-(CH_2)_n-CH_2-$ $n = 1$ or 2
Y is H or CH_3
R^1 is such that $OCN-R^1-NCO$ can be TDI, HMDI, IPDI, MDI

Scheme 5.2

included tolylene diisocyanate (TDI), hexamethylene diisocyanate (HMDI), isophorone diisocyanate (IPDI) and diphenylmethane diisocyanate (MDI) while the monomers include hydroxyethyl acrylate, propyl acrylate and ethyl methacrylate. If other hydroxy compounds (e.g. polyethers, polyesters, polyols) are also present, which contain more than one hydroxyl group per molecule, then chain lengthening occurs giving a wide range of pre-polymers which vary not only in functionality and molecular weight, but also in their final film properties.[1]

The simplest urethane pre-polymers are formed by reacting a diisocyanate with a hydroxy monomer in stoichiometric proportions according to the general mechanism in Scheme 5.2. A catalyst is normally used in this process by most workers such as dibutyl tin dilaurate[10] although for TDI a catalyst, is not essential.[8] Amine catalysts can be used, but they are generally far less effective than the organo-tin ones.[11] Inhibition of the acrylic addition polymerisation reaction can be achieved by using only air,[10] because reaction temperatures are generally significantly lower than those for epoxy acrylate preparations. However, in practice an inhibitor is normally incorporated in the formulation.

Flexible urethane acrylates can also be made by the reaction of a diisocyanate with a long-chain glycol. The half adduct of a diisocyanate acrylate may also be used. Aliphatic diol polyethers such as polyethylene glycols, polypropylene glycols and caprolactone polyols are widely used for this purpose as are polyesters with an excess of hydroxyl groups. As the flexibility increases the hardness, speed of cure and solvent resistance of the films decrease. Reaction of an aliphatic polyester as HO^OH with TDI and hydroxyethyl acrylates is shown in Scheme 5.3.

Polyether urethane acrylates

In this case an isocyanate functional adduct is produced by capping a polyether with a diisocyanate which is then reacted with an unsaturated hydroxy

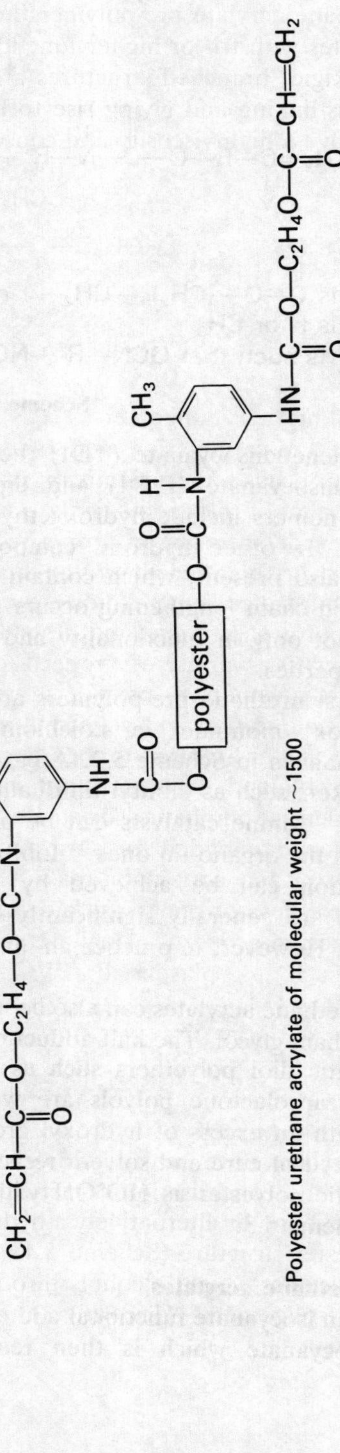

HO—[polyester]—OH + 2
$\begin{array}{c}\text{CH}_3\\ \text{NCO}\\ \\ \text{NCO}\end{array}$
 + 2 CH$_2$=CH—C—O—C$_2$H$_4$OH
                            ‖
                            O

→

CH$_2$=CH—C—O—C$_2$H$_4$—O—C—NH—[Ar(CH$_3$)]—NH—C—O—[polyester]—O—C—NH—[Ar(CH$_3$)]—NH—C—O—C$_2$H$_4$—O—C—CH=CH$_2$

Polyester urethane acrylate of molecular weight ~1500

Scheme 5.3

compound such as 2-hydroxyethyl acrylate. Similar processes to the polyester urethane acrylate can be used.

Hard urethane acrylate pre-polymers are also possible through the reaction of diisocyanates with tri- or higher functionality polyols such as **I** and hydroxy monomers. Rigid, branched structures are produced that are capable of high levels of cross-linking and giving rise to hard, durable and brittle films. Such systems do have a high viscosity and consequently a diluent monomer must be added.

$$\text{HOH}_2\text{C}-\underset{\underset{\text{CH}_2\text{OH}}{|}}{\overset{\overset{\text{CH}_2\text{OH}}{|}}{\text{C}}}-\text{CH}_2\text{OH}$$

I

Applications

In the electronic area oligomers such as a trifunctional aliphatic urethane acrylate is used in formulations for solder-resistant printed circuit boards and screen inks. Because the chemistry of urethane acrylates is much more versatile than that of the epoxy acrylates it is possible to produce pre-polymers which offer a wide spectrum of properties. As a class of compounds, urethane acrylates offer very high reactivity, toughness, chemical resistance and adhesion to 'difficult' substrates. Furthermore, by varying the 'backbone' composition, i.e. the level of unsaturation, and other functionality parameters coatings can be produced with a variety of properties. Sometimes formulations are mixed with other pre-polymers to provide more versatility and quite often diluents are used. Discolouration and yellowing are major problems (see later) and if aromatic isocyanates are used to incorporate the urethane group, colour-stable films are not obtained. In fact, colour stability is achieved through the use of the more expensive aliphatic isocyanates, such as IPDI or HMDI. Urethane acrylates are recommended for abrasion-resistant formulations for poly(vinyl chloride) (PVC) and cork floor tiles. They are also used in wood coatings, overprint varnishes, printing ink and in adhesive applications. Due to their excellent adhesion properties they are suitable for a variety of flexible plastic substrates, such as plasticised PVC, polyester film and polyurethane 'leathercloth'.

POLYESTER ACRYLATES

Polyester acrylates cover a very wide range of products and formulations and are made by the condensation of acrylic acid with the hydroxyl groups on a polyester backbone or alternatively a hydroxyacrylate with residual acid groups on the polyester structure (Scheme 5.4). Products range from those produced with low-molecular-weight alcohols through to di- and multifunctional polyols to polyester oligomers. Viscosities can be a high as 1000 P (100 Pa . s)

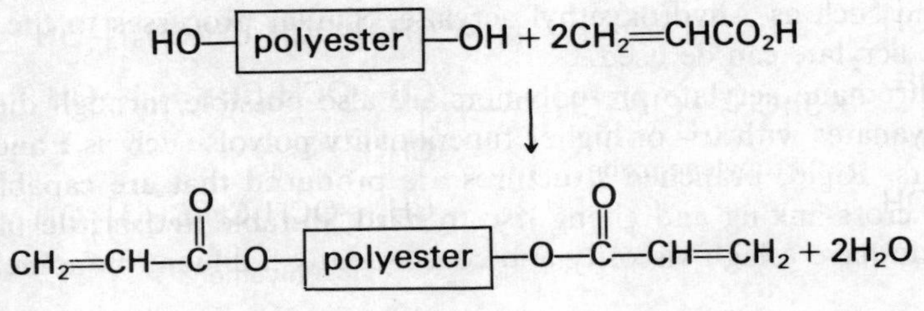

$$\text{HO}-\boxed{\text{polyester}}-\text{OH} + 2\text{CH}_2{=}\text{CHCO}_2\text{H}$$

$$\downarrow$$

$$\text{CH}_2{=}\text{CH}-\overset{\overset{\text{O}}{\|}}{\text{C}}-\text{O}-\boxed{\text{polyester}}-\text{O}-\overset{\overset{\text{O}}{\|}}{\text{C}}-\text{CH}{=}\text{CH}_2 + 2\text{H}_2\text{O}$$

Scheme 5.4

depending upon the formulation and in some cases an organic solvent may have to be used in order to azeotropically aid the removal of water from the reaction followed by vacuum distillation of the solvent.

The polyester acrylates used as radiation pre-polymers, are characterised by their low viscosities and relative economy, compared with other acrylated pre-polymers. All of the polyesters may be considered for reaction with isocyanates and can be used for reaction with acrylic acid because they are all rich in hydroxyl groups. For reaction with an unsaturated hydroxy monomer an acid-rich polyester is required.

Condensation telomerisation may also be used to obtain polyesters with carefully controlled molecular weights[12] through the use of a monofunctional acid which acts as a chain stopper. This may be achieved, using acrylic or methacrylic acid. By varying the molecular weight of the polyester, it is possible to obtain acrylates from low viscosity, to hard solids, at ambient temperatures. Compatibility with other pre-polymers is good, so that they can be used in many formulations.

Applications and examples
Polyester acrylates are mainly used in UV roller coat varnishes for paper and board and UV wood coatings. Examples of these are the tetrafunctional Photomers 5018 and 5029 (Harcros Chemicals Ltd).

Polyether acrylates
Polyether acrylates are lower viscosity resins compared with the polyester types and are relatively inexpensive to prepare. In the acrylation of polyethers however, a transesterification technique is used to prevent the polyether links from degrading. Polyethers may also be reacted with isocyanate groups and this is often an easier method of introducing polyether linkages into an acrylated system since the problems of removing the by-products of trans-esterification such as ethanol are eliminated.

Preparation
The reaction of ethylene or propylene oxide with a polyol in the presence of a basic or acidic catalyst such as boron fluoride or sodium hydroxide[1] will give a

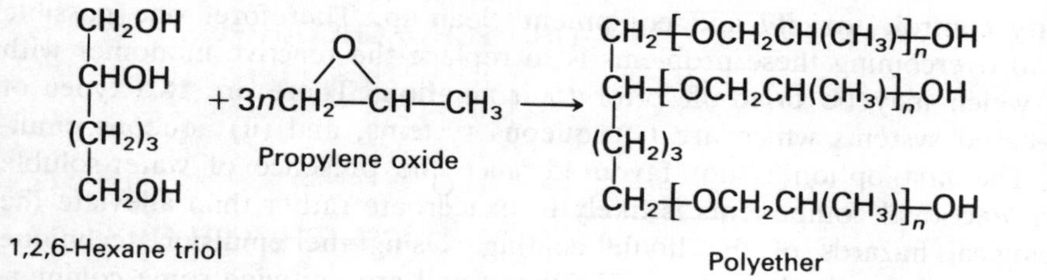

Scheme 5.5

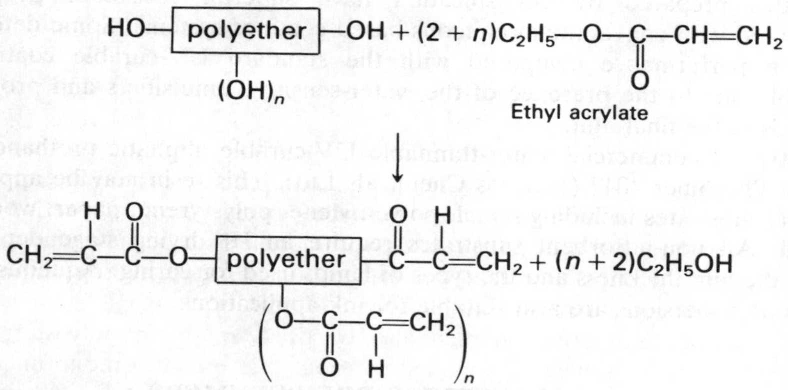

$$\begin{array}{c}CH_2OH\\|\\CHOH\\|\\(CH_2)_3\\|\\CH_2OH\end{array} + 3nCH_2\overset{O}{-\!\!\!\overline{}\!\!\!-}CH-CH_3 \longrightarrow \begin{array}{c}CH_2\!\!-\!\![OCH_2CH(CH_3)]_n\!\!-\!\!OH\\|\\CH\!\!-\!\![OCH_2CH(CH_3)]_n\!\!-\!\!OH\\|\\(CH_2)_3\\|\\CH_2\!\!-\!\![OCH_2CH(CH_3)]_n\!\!-\!\!OH\end{array}$$

1,2,6-Hexane triol   Propylene oxide   Polyether

Scheme 5.6

HO—[polyether]—OH + (2 + n)C_2H_5—O—C(=O)—CH=CH_2
         |
    (OH)$_n$ Ethyl acrylate

↓

CH_2=C(H)—C(=O)—O—[polyether]—C(=O)—C(H)=CH_2 + (n + 2)C_2H_5OH
                  (O—C(=O)—C(H)=CH_2)$_n$

polyether. Where a secondary hydroxyl is available incomplete etherification may occur. The total degree of etherification will depend upon the ratio of propylene oxide and polyol. A typical reaction is illustrated by Scheme 5.5 for 1,2,6-hexane triol with propylene oxide. Acrylation is then typified by Scheme 5.6 using ethyl acrylate as an example.

Many other common polyethers have been acrylated. Amongst them are ethoxylates and propoxylates of trimethylolpropane, pentaerythritol and polyethers of 1,4-butane diol.[1,8] These form the basis of the new generation of low viscosity, low toxicity monomers which are rapidly gaining in importance (see later). In the preparation of polyether acrylates ethyl acrylate is widely used since it forms an azeotrope with the ethanol formed in the transesterification reaction. An efficient column to separate ethyl acrylate and ethanol is essential, and a packed column is normally used to minimise loss of ethyl acrylate with the azeotropic mixture.

WATER-MISCIBLE PRE-POLYMERS

One weak link in the modification of many pre-polymer systems is the use of a reactive diluent. Quite often this can give difficulties in the areas of odour, viscosity control, toxicity and equipment clean-up. Therefore, one possible route to overcoming these problems is to replace the reactive monomer with water which may be dried-off prior to irradiation. There are two types of water-based systems which are (i) aqueous systems, and (ii) aqueous emulsions. The first option is not favoured since the presence of water-soluble acrylic functional components is likely to exacerbate rather than alleviate the toxicological hazards of the liquid coating. Using the emulsion technique polyesters and acrylics have been developed and are enjoying some commercial success particularly in the field of wood coatings. The epoxy and urethane acrylates prepared by emulsification have superior resistance properties compared with conventional aqueous-based coatings systems. Some deterioration in performance compared with the standard UV-curable coatings is possible due to the presence of the water-sensitive emulsifiers and protective colloids in the final film.

A typical commercial water-thinnable UV-curable aliphatic urethane acrylate is Photomer 4047 (Harcros Chemicals Ltd). This resin may be applied to several substrates including metal, polyethylene, polystyrene, paper, wood and board. All non-adsorbent substrates require an IR drying stage depending upon the film thickness and the types of lamps used for curing. Aqueous-based pigment dispersions are also suitable for ink applications.

VINYL/ACRYLIC PRE-POLYMERS

These pre-polymers contain residual double bonds capable of further cross-linking. They can be prepared from vinyl or acrylic polymers by copolymerisation with monomers containing pendant acid, anhydride, hydroxy or glycidyl groups. Reaction of these groups with an unsaturated monomer is then possible. Vinyl double bonds may be introduced using a wide selection of compounds some of which are maleic anhydride, fumaric anhydride, itaconic anhydride, allyl alcohol, cyclopentadiene, acrylic acid, methacrylic acid, acrylamide, hydroxyethyl acrylate, glycidyl acrylate or glycidyl methacrylate. It is well known that the vinyl groups present in such compounds as maleic anhydride are less responsive to UV than those based on acrylic acid. The following order of reactivity has been established:

$$\text{vinyl} < \text{allyl} < \text{methacrylic} < \text{acrylic}$$

The exothermic nature of the free-radical reaction of acrylics makes bulk polymerisation impractical and the base acrylic polymer is normally prepared in solution which must be distilled off at the end of the reaction. Residual peroxide is removed, by holding the product at 80–130°C for up to 5 h, before cooling, and the addition of the unsaturated component. The incorporation of the unsaturated components would follow one of the techniques already discussed in previous sections on polyester acrylates and urethane acrylates.

Pre-polymers of this type offer good colour, durability and chemical resistance. Some of these resins may be dissolved in reactive diluents and then used in adhesive applications. Whilst the thermoplastic component may not take part fully in the curing reaction it has been shown that some grafting may occur; the extent of which depends upon the curing conditions and the nature of the polymer system.

UNSATURATED POLYESTERS

Unsaturated polyesters have been known for some time and were one of the earliest systems used in radiation curing. Unsaturated polyesters can be broadly defined as condensation products of organic diacids and glycols. Unsaturation can be incorporated in a variety of ways to produce terminal, pendent and internal double bonds. Of these types, internal unsaturation provided by maleic anhydride is the most common.

Among the alternatives which have been mentioned are stilbenes, cinnamates, allyl, acrylamide, norbornenyl, and acrylic. Whilst some of these types provide a reactivity advantage it tends to be reduced by slow copolymerisation with styrene. Replacement of styrene is generally economically unattractive, and this usually leaves the styrene/maleic systems in the most favourable cost–performance position.

Maleic derived polyesters span a broad performance range. Among the critical variables are maleic content, ratio of aromatic to aliphatic 'saturated' diacid, molecular weight, monomer content, glycol, inhibitor and ratio of maleic to fumaric esters.

Film properties can be changed and improved through the selection and ratio of the acids and glycols and this is achieved only at an increased cost, and normally without an improvement in the rate of cure.

Unsaturated polyesters, suffer from air inhibition whether peroxide cured, UV cured or EB cured. One method of overcoming this is to incorporate a small amount of paraffin wax (0·2–2%) into the polyester solution, and as the curing process proceeds, the wax comes out of solution and rises to the surface, to form a barrier. In this case curing is carried out in two stages. Firstly, low-pressure mercury vapour lamps cause partial gelation and dissolution of the wax; secondly, medium-pressure mercury vapour lamps are used to complete the cure. Another method, which is widely used, is to incorporate allyl ether groups into the polyester backbone which produces a polyester, which is not inhibited by air. Such 'air drying' types are much more expensive than the conventional unsaturated polyesters.

Applications

By far the biggest coating market for unsaturated polyesters is in wood finishing, e.g. fillers, sealers and top coats. Here, due to the problems of handling large sections of wood or chipboard, production-line speeds are relatively slow and this is compatible with the longer cure times associated with polyester/styrene formulations. The finishes are very hard, tough, and solvent and heat resistant, but they suffer from a lack of flexibility. Chemical resistance is also poor, due to the ester groups in the backbone of the polyester. Whilst styrene is easily the most common monomer used with polyesters, vinyl esters, acrylates and N-vinylpyrrolidone are common alternatives.

POLYENE/THIOL/SILICONE SYSTEMS

Free-radical addition of mercaptans to olefins has been known for many years. As presently understood, mercaptan olefin polymerisation occurs according to Scheme 5.7. The company W. R. Grace carried out much of the original work in this area and they are consequently very heavily patented.[13-15] This chemistry depends upon the rate of hydrogen transfer from mercaptan being competitive with the rate of olefin polymerisation. Olefin polymerisation would, of course, produce a vinyl polymer and an unreacted mercaptan.

In principle, this type of polymerisation affords some attractive alternatives as compared with acrylate polymerisation. The thiol-ene polymer can be regarded as a polysulphide, but, more importantly, if R Scheme 5.7 is a urethane then one has an essentially thermoplastic urethane. Thermoset characteristics can be inserted by the addition of a trifunctional monomer or a trimercaptan. In contrast to the polymer derived from thiol-ene, an acrylic polymer results from vinyl polymerisation and the backbone remains fundamentally an acrylic polymer. In actual practice urethane acrylate oligomers do provide highly elastomeric polymers, but in some respects they probably are inferior to conventional urethane elastomers. In addition, to make a highly elastomeric urethane acrylate a relatively high-molecular-weight oligomer with an attendant high viscosity would be required. It would seem therefore that if the thiol-ene chain length is sufficiently long, then polymers comparable to conventional urethane elastomers could be produced.

There are two advantages to the thiol/polyene systems:

1. they are non-air-inhibited; and
2. flexible cured films can be obtained from relatively low viscosity mixture without the need to incorporate a diluent.

$$\phi_2CO + RSH \xrightarrow{h\nu} \phi_2\dot{C}OH + RS^\cdot$$

$$RS^\cdot + CH_2{=}CHX \longrightarrow RSCH_2\dot{C}HX$$

$$RSCH_2\dot{C}HX + RSH \longrightarrow RSCH_2CH_2X + RS^\cdot$$

$$RSCH_2\dot{C}HX + MCH_2{=}CHX \longrightarrow RSCH_2{-}\underset{X}{CH}{-}\left[CH_2\underset{X}{CH}\right]_{M-1}{-}CH_2\dot{C}HX$$

Scheme 5.7

Polyfunctional thiol compounds have been developed by W. R. Grace and these give very tough abrasive-resistant coatings which are ideal for applications such as flooring compounds. It is also possible to modify acrylic UV-curable formulations with polythiol polymers to improve their properties.

Radiation-curable silicone resins and monomers are now widely established with numerous applications requiring resistance to hydrolysis and heat, and possessing good sealant and mechanical properties. Coatings and optical fibres are one of the major applications.

A commercial example of a difunctional silicone acrylate for use in UV- and EB-curable systems is Photomer 7020 (Harcros Chemicals Ltd) which shows improved compatibility due to its acrylic functionality. This product is particularly recommended for formulations where good slip is required in the cured film, e.g. record sleeves.

DILUENTS/MONOMERS

There are two types of diluents or monomers employed in radiation-curable systems, namely, reactive and non-reactive and these will be discussed separately.

Reactive monomers

Early monomers developed for radiation-curable coatings in the late 1960s suffered from the fact that they were highly skin irritant. The types of monomers employed were monofunctional (containing one double bond per molecule) and multifunctional (containing two or more double bonds per molecule).

The reactive monomer or diluent plays several important roles in the UV and EB curing of coatings. These include the following:

- viscosity reducer for high molecular weight propolymers;
- the determination of the cure rate in conjunction with the photoinitiator system;
- higher-molecular-weight pre-polymer molecules are linked by the monomers which also, depending on their functionality, contribute towards decreasing the time required for full cure;
- the determination of the film properties which are to some extent dependent on the type of monomer employed; and
- the control of adhesion and flexibility of the cured film.

Examples of the early monomers used include the following mono- and difunctional products.

First-generation reactive monomers

Monofunctional monomers
These include 2-ethylhexyl acrylate, phenoxyethyl acrylate and hydroxyethyl acrylate all of which exhibit strong characteristic odours.

Difunctional monomers
These include hexanediol diacrylate, butanediol diacrylate, neopentylglycol diacrylate, diethyleneglycol diacrylate and tetraethyleneglycol diacrylate. Again, these monomers have a strong odour, are skin irritants and some are carcinogenic.

Methacrylic analogues are also available but they suffer from the disadvantage of having slow reactivity, as a result of oxygen inhibition and also exhibit strong odour. The only major area where the methacrylates are used is in the UV-adhesive industry.

Trifunctional monomers
These include trimethylolpropane triacrylate and pentaerythritol triacrylate both of which are little used and the latter is a suspect carcinogen.

Second-generation monomers
Major raw-material suppliers have, since the 1960s, developed new monomers which have low Draize values and are widely used in radiation-curable coatings. These areas include varnishes for paper and board, inks for paper and board, wood coatings, metal decoration, cork and PVC tiles, electronics, e.g. solder and etch resists, adhesives and plastics coatings. Such low Draize monomers used include the following products.

Monofunctional monomers
These photomers are monofunctional aromatic acrylates having low viscosity and odour. They are used in radiation-cured coatings where good adhesion is essential. Commercial types include examples such as Photomers 4039 and 7031 made by Harcros Chemicals Ltd, UK.

Difunctional monomers
These difunctional monomers are prepared by reacting one mole of a diol, e.g. tripropylene glycol, with two moles of acrylic acid. The low Draize difunctional monomers which are widely used in the radiation-cured coatings industry include the following.

Tripropyleneglycol diacrylate (TPGDA). This reactive low viscosity monomer has good solvency properties with most acrylated pre-polymers as well as a low Draize rating and high reactivity.

$$CH_2=CH-CO-O-(CH_2CH_2CH_2O)_3-CO-O-CH=CH_2$$
II

Dianol diacrylate (CDDA). This is a low Draize product and is used in a number of radiation-cured coatings areas.

$$(CH_2=CHCO-O-CH_2CH(OH)CH_2-O-p-C_6H_4-p-)_2C(CH_3)_2$$
III

Trifunctional monomers

The two most likely types of trifunctional monomers which are likely to be used in various UV and EB applications areas are a triacrylate of propoxylated glycerol (GPTA) (**IV**) and ethoxylated trimethylolpropane triacrylate (**V**).

$$CH_2=CHCO-(O-CH_2-CH(CH_3))_n-O-$$
$$CH(CH_2-O-(CH(CH_3)CH_2-O-)_n-CO-CH=CH_2)_2$$
$$\mathbf{IV}$$

$$CH_2=CH-CO-O-(OCH_2CH_2)_n-O-CH_2-CH_2-CH-O-(CH_2CH_2O)_n-CO-CH=CH_2$$
$$CH_3CH_2\overset{|}{C}H-CH_2-O-(CH_2CH_2O)_n-CO-CH=CH_2$$
$$\mathbf{V}$$

Other monomers

More sophisticated monomers have been developed to obtain products which exhibit, low toxicity, low volatility, low viscosity, high reactivity, good mechanical properties and good weatherability. This has been done recently by incorporating in the monomer backbone ethoxy or propoxy groups. The introduction of polar groups in the molecule of this monomer leads to improvement in the reactivity and the end-use properties. The polar groups introduced in the monomer backbone include carbamate or carbonate groups.

Recently, the influence of the polar groups in three different acrylic monomers containing either carbamate (CL 960, CL 986) or carbonate (CL 1000) functional groups has been determined.[13] The formulae and characteristics of these monomers are illustrated in Table 5.1, with comparison to other usual monoacrylates, and 2-ethylhexylacrylate (EHA). The very large differences of viscosity and boiling point for the different monomers is clearly apparent in the table. The position of the carbamate function also appears to be very important for the physical properties of the pure material. Comparison with

Table 5.1
Monomers containing polar groups

$$CH_2=CH-CO-O-X-R$$

Monomers	X	R	Mol. wt	Boiling point (°C)	Viscosity (cP)^a	d (at 25°C)	n (at 25°C)
CL 960	$(CH_2)_2-NH-CO-O$	$CH(CH_3)_2$	201	115°C	70	1·11	1·456
CL 986	$(CH_2)_2-O-CO-NH$	$CH(CH_3)_2$	201	50°C	Solid		
CL 1 000	$(CH_2)_2-O-CO-O-$	$CH(CH_3)_2$	202	98°C	12	1·08	1·432
EHA		$CH_2CC_4H_9$ $\|$ C_2H_5	114	75°C	4	0·89	1·433

^{a}IP = 0·1 Pa . s.

other usual mono-acrylate monomers showed excellent performance of carbamate acrylic monomers with regard to reactivity, volatility, usefulness and weatherability.

Such formulations were also found to exhibit a similar reactivity to the conventional diacrylic diluent monomers but in addition, they allowed one to obtain much lower residual unsaturation in the cured coatings.

Vinyl functional monomers

Vinyl toluene and styrene are slow curing. These are usually incorporated with unsaturated polyesters for wood coatings' application areas. Both are used because of their low cost but suffer from the disadvantage of giving odour on the production line. 2N-Vinylpyrrolidone (2NVP) is far more important because of its higher reactivity and good viscosity-reducing powers when used with other acrylate monomers. 2NVP can also be used as a water-soluble monomer with water-based oligomers as well as flexible coatings because of its low shrinkage value. However, it suffers from the disadvantage of being costly and having high odour.

Non-reactive monomers

This class of diluents is divided into solvent and plasticiser types. The solvent dissolves the pre-polymer and is then flashed-off prior to curing after application to the substrate. The pre-polymers are soluble in the uncured state. The role of the solvent is to act as a carrier for the pre-polymer as well as a viscosity reducer for application purposes.

The concentration of solvent employed in a UV coating can range from 5 to 50% depending on the type of application. Low levels of solvent employed in the coating can be absorbed into the substrate or can be flashed off by the IR present in the lamps used. High levels of solvent in the coating, especially when applied as a thick coating, will lead to the necessity of flashing-off the solvent before passing under the UV lamps. High levels of solvent can also influence the properties as well as the gloss level of the film. The plasticiser's role is to modify tack, viscosity, flexibility and hardness of the cured film. The formulator must also consider a number of properties of the monomers which meet the final properties of the coating. These include, toxicity, shrinkage and surface tension, viscosity and solubility, odour and stability, and hydrophobic and hydrophilic properties.

AMINE SYNERGISTS

Amine synergists are used in conjunction with photoinitiators to accelerate the rate of cure when incorporated into UV formulations for the following:

- roller ink varnishes;
- paper and board;

- UV dry offset inks;
- UV silk-screen inks;
- UV coatings for plastics; and
- metal surface.

These are prepared by the Michael addition of a secondary alkyl, aryl, aliphatic or acyclic amine to an unsaturated acrylate group. A typical synthetic procedure is described below.[14]

The acrylate compound is charged into a 250 cm^3 four-necked flask equipped with a condenser, stirrer, thermometer and dropping funnel. The required amount of amine is added slowly from the dropping funnel with stirring. An immediate exotherm is observed. Periodically, a sample of the reaction mixture is removed for titrimetric analysis as outlined below. When the reaction is complete and the correct amine value obtained the amine acrylate is steam stripped to remove unreacted secondary amine. Water (5% w/w) is added, and the reaction vessel fitted with an air bleed and a distillation head leading to a trap and a high-vacuum system. The temperature of the reaction mixture is then raised to 40°C and the water striped-off under a reduced pressure of 13 mmHg (1·73 kPa). Remaining traces of water are removed at 75°C. Stripping is considered complete when less than 0·2% w/w water is found to be present by Karl–Fischer titration.

There are a number of such products available commercially. These include Photomers 4116 and 4182 (Harcros Chemicals Ltd). A typical structure of an amine synergist is as follows:

$$CH_2=CH-CO-O-CH_2-\underset{\underset{CH_2-O-CO-CH=CH_2}{|}}{\overset{\overset{CH_2-O-CO-CH_2-CH_2-N(C_2H_5)_2}{|}}{C}}-CH_2-CH_3$$

VI

These monomers are usually mixed with oligomers in formulations between 8 and 20% by weight.

EB-CURING TECHNOLOGY

EB-curing technology is closely related to UV-curing technology with respect to the raw materials which can be used to provide the cured coating.[15,16] The development of EB curing, however, has lagged behind that of UV curing due to the lack of suitable electron accelerators which need to be compact and inexpensive.[17,18] The capital cost of EB-curing equipment is far greater than that for UV curing although this cost has reduced in recent years with the introduction of new accelerators.[1,16,19–21] Consequently, EB-curing has become a viable process in a number of applications which have been dominated by

UV curing or conventional coatings techniques. There is now a wide range of applications in which EB-curing techniques are used.[22,23]

The basic concept of EB curing is the use of high-energy electrons to induce free-radical polymerisation.[24] EB-curing systems are designed to allow the unpolymerised coating to pass beneath an electron source at varying rates in order to facilitate polymerisation. This technology will be elaborated on further in the following sections.

EB accelerators

An EB accelerator can be regarded as a vacuum triode in which low-velocity electrons are produced at the surface of the cathode or a filament by applying a current and accelerated to the anode by application of a high sensitive voltage.[19–21,25] An electric or magnetic focusing device is used to ensure maximum concentration of electrons at the intended surface. The electrons emitted from the cathode individually carry the full charge of the accelerating potential in eV as they strike the anode. The electron density is controlled by the amperage passing through the cathode. This process of producing accelerated electrons is carried out in a vacuum, which is maintained while allowing for the efficient transmission of electrons by a thin metallic window usually made of titanium or aluminium foil. The EB passes through this window and strikes the substrate to be cured, which is blanketed under nitrogen. At the present time there are two basic types of EB-curing accelerators commercially available: the scanned-beam and electron-curtain types.[16,18,21,25]

Scanned-beam accelerators

The basic design of this type of accelerator is illustrated in Fig. 5.1. The electrons are produced at a pointed cathode and are accelerated towards the anode. This produces a narrow beam of electrons which are scanned across the direction of the substrate by an electromagnetic scanning device. These scanned-beam accelerators can have a single or multistage electron accelerator. The single-stage acceleration is used for accelerating voltages of less than 250 kV. The multistage accelerator is used for accelerating voltages in excess of 250 kV.

Electron-curtain accelerators

The basic design of this type of accelerator is illustrated in Fig. 5.2. This accelerator uses a linear filament type of cathode which is in the centre of a cylindrical vacuum chamber. The electrons emitted from the cathode are accelerated over a single stage to pass through the metallic window producing a continuous stream (curtain) of electrons along the length of the filament. This type of accelerator is used for accelerating voltages between 150 and 300 kV and is generally cheaper than the scanning-beam accelerators.

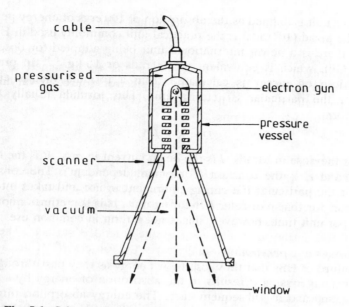

Fig. 5.1. Schematic diagram of a scanned-beam EB accelerator.

Operational conditions

With EB curing there are a number of parameters which need to be considered and defined with regard to the efficiency of the curing process and operational conditions such as dose, electron energy and penetration, electron energy losses and backscatter effects.

Dose

The radiation dose is the amount of energy absorbed per unit mass of material. The unit used to measure an absorbed dose of electron energy was formerly

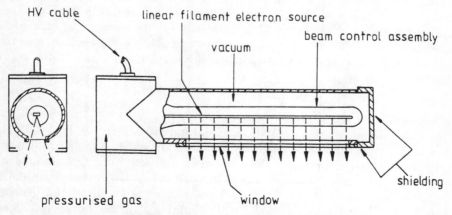

Fig. 5.2. Schematic diagram of an electron-curtain EB accelerator.

the rad. This unit is defined as the absorption of 100 ergs of energy per gram of sample. The Mrad (10^6 rads) is the practical unit commonly used in EB curing. However, there is a newer international unit being adopted for dose; namely the gray (Gy), which is equivalent to 100 rads or $1 J kg^{-1}$.[18] In practice the absorbed dose of energy is calculated from a formula that is empirically derived for the particular EB-curing unit. This formula usually takes the following form:

$$D = K \times I/V$$

where D is the dose in Mrads, I is the beam current in mA, V is the line speed in ft/min, and K is the constant (equipment dependent). The constant K is derived for the particular EB-curing equipment in use and takes into account energy losses for that particular EB-curing unit. It is sometimes quoted as the dose rate per unit time, however, this is not a term in common use.

Electron energy and penetration

It is the nature of EBs that they lose their energy as they pass through matter. This is advantageous as it results in the absorption of energy by a radiation-curable coating and its subsequent cure. The energy absorption throughout a coating (its depth in dose) is not, however, uniform because of this continual loss of energy. The depth to which a material can be penetrated by an EB is directly proportional to the accelerating voltage and inversely proportional to the specific gravity of the material being irradiated. These parameters are generally related by a depth–dose curve which is used to evaluate the minimum and maximum doses required throughout the coating, usually working through a required minimum dose. An example of a depth–dose curve is shown in Fig. 5.3. The 300 kV curve indicates that this EB has lost all

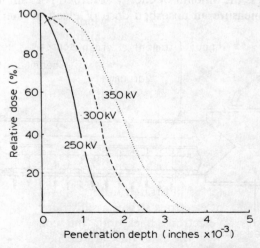

Fig. 5.3. Depth versus dose curves for an electron-curtain EB-curing equipment.

of its energy in passing through a unit density material 0·025 in thick (0·0625 g/cm^2). However, as can be seen from the graph the surface of the coating is likely to be fully cured whereas the bottom of the coating is likely to be appreciably undercured due to the relatively small dose received by this area of the coating. It is necessary, therefore, to consider dose uniformity, which involves selecting a maximum and minimum dose level acceptable for curing and then selecting an accelerating voltage which will deliver the required dose range. The ideal situation is to obtain the smallest percentage variation of dose going from the top to the bottom of the coating.

Energy losses

Since an EB loses its energy on passing through matter, the most likely source of energy loss in EB-curing equipment is at the titanium or aluminium window. Although the windows are very thin (0·05 mm) they are made of relatively high-density material. The consequence of this energy loss is the production of X-rays by interaction of the accelerated electrons with the electron shells of the elements in the window. There are basically two types of possible interactions producing either a bremsstrahlung or a characteristic X-ray. The bremsstrahlung occurs when accelerated electrons are within the coulomb field of an atom core. This causes a deflection from their original direction with a consequent loss of kinetic energy. This loss in kinetic energy is then emitted in the form of photons. Since there are a number of separate coulombic interactions possible for each accelerated electron there are a number of photon energies emitted. The bremsstrahlung is therefore a continuous spectrum. The characteristic X-ray is produced when an electron is ejected out of an inner shell of the metal atom. The vacancy produced is then filled by electrons from higher shells and the energy differences which then occur are quantised and are emitted as X-rays with characteristic energies producing a linear spectrum. The graph shown in Fig. 5.4 illustrates the relationship between beam energy loss and operating voltage at a typical titanium-foil window. It can be seen that low operating voltages beam energy losses are high whereas at high operating voltages the beam energy losses are low.

Energy can also be lost in the air gap between the foil window and the substrate. As with window losses, these energy losses are more pronounced at lower accelerating voltages. However, these losses are much smaller than those occurring at the window. If a unit was operating at 125 kV then the 125 kV curing beam would experience an approximately 40% energy transfer loss with 30% occurring at the window and 10% in the air gap.

Backscatter effects

The concept of 'backscatter' is illustrated in Fig. 5.5.[26] The effect is produced when an energetic beam of electrons strikes a substrate with significant density to deflect some of the electrons back into the coating. These deflected electrons have lost energy by the time they are deflected so their penetration potential is less than the original beam. However, they do account for an

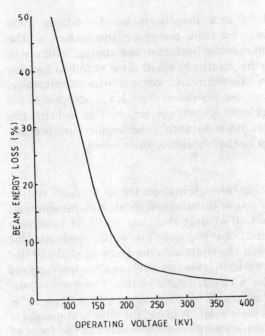

Fig. 5.4. Percentage energy loss at the titanium window versus operating voltage for an electron-curtain EB-curing unit.

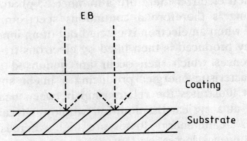

Fig. 5.5. EB backscatter.

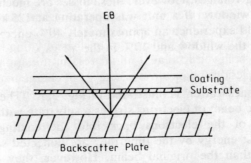

Fig. 5.6. Use of backscatter plate in EB curing.

additional dose delivered within the bottom portion of the coating. The backscatter effect will depend on the substrate material, its thickness and its density. If the effect is pronounced it could result in cure being achieved at lower accelerating voltages, thus saving energy. This effect is readily adaptable to thin coatings and thin substrates when the equipment utilises a backscatter plate (see Fig. 5.6).

In this application the EB passes through the coating and the substrate to strike a backscatter plate and passes back through the substrate and coating, thus delivering a second, albeit reduced, dose of electrons.

Mechanism of EB curing

The mechanism of EB curing is fundamentally different from that of UV curing. The latter requires the use of photoinitiators which are activated by UV or visible light with energies in the range 3–6 eV whereas EB curing is a consequence of the direct interaction between radiation and monomer in the energy range 150–500 000 eV.[1,24,27] Since chemical bond energies lie in the range 1·5–8·5 eV it is obvious that UV initiation results in specific bond rupture whereas EB initiation results in an unpredictable pattern of bond scission.

Thus, on exposure of an organic coating to an accelerated EB an exchange of energy takes place in which the kinetic energy of the electron is transferred to the absorbing medium. This encounter sets up a coulombic interaction which results in several pathways of producing highly reactive radicals capable of initiating polymerisation. The exact mechanism of energy transfer and production of energised molecular species is unknown and likely to be highly complex. Gaseous-phase mass spectroscopy has been used as a model to elucidate the processes that take place in the condensed phase as in a curable coating. There are two basic primary processes which can occur, namely, ionisation or excitation. Electron impact can give rise to ionisation with molecular ionic and molecular radical formation through fragmentation (Scheme 5.8).

Molecular excitation may also occur thus:

$$AB + e^- \longrightarrow AB^+$$

These primary species can now undergo a number of secondary processes such as recombination or further ionisation resulting in slower electrons. The secondary processes of Scheme 5.9 can occur.

The radicals produced are believed to be the initiating species, whereas the ionic species are believed to play a minor role in initiation and undergo rapid recombination reactions by electron capture and charge neutralisation. Radical formation can be enhanced by including specific atoms in the medium. For

$$AB + e^- \longrightarrow AB^+ + e^-$$
$$\searrow A^+ + B^\cdot$$

Scheme 5.8

$$AB + e^- \longrightarrow A^{\cdot} + B^{\cdot}$$

$$AB^+ + e^- \longrightarrow AB^*$$
(electron capture)

$$AB^+ + CD^- \longrightarrow AB^* + CD^*$$
(neutralisation)

$$AB^+ + CD \longrightarrow ABC^+ + D^{\cdot}$$

$$AB^+ + AB \longrightarrow AB_2^+ + A^{\cdot}$$
(intermolecular energy transfer)

$$AB^* \longrightarrow A^{\cdot} + B^{\cdot}$$
(molecular dissociation)

Scheme 5.9

example, as represented above is favoured for molecules containing halogen atoms or cyano groups. However, aromatic molecules act as electron traps resulting in low radiation yields. With a high-intensity EB a very high local concentration of free radicals is formed in the coating. This results in an inefficient conversion of radiation energy to chemical energy because a large percentage of the free radicals recombine instead of initiating the curing process. This problem can be overcome by making use of several passes or by pulsing the EB. Some workers have studied the effect of EB dose on the rate of double bond disappearance in an unsaturated polyester/styrene solution (65/35%).[28] They found a linear increase in the rate of double bond disappearance with an increase in the dose rate up to abut 20 Mrads/min. At high rates this decrease in unsaturation levelled out and then decreased slightly up to 85 Mrads/min.

POST-CURE STABILITY OF EB-CURED COATINGS

One area regarding the post-cure performance and properties of radiation-cured coatings which is currently of particular interest is the post-cure stability, with the main areas being thermal stability and photostability. Applications requiring thermal stability include solder masks, abrasives, wire and cable coatings, solar collectors or reflectors, adhesives and optical-fibre coatings whereas good photostability includes applications such as no-wax flooring, automotive finishes, solar collectors or reflectors and various varnishing applications.

The thermal and photochemical degradation and oxidation of polymeric systems are subjects of considerable academic and industrial interest. Although some facets of these subjects are reasonably well understood there are some that have received little attention, are complex and/or complicated by various parameters.[29] The latter may include the nature of the industrial manufacturing operation, processing conditions and additives. Studies on the stability of radiation-cured coating systems are of more recent interest in view of the industrial development in this field over the last 10 years.[30] In this section we review the post-cure thermal and photostability of radiation-cured coatings with emphasis on EB-cured systems although occasional reference will be made to UV-cured systems for comparative purposes. In order to help with the discussion and the naming of complex chemical structures the following abbreviations will be used:

BZ	Benzophenone
Di-PETA mono/DPMHP	Dipentaerythritol monohydroxypentaacrylate
GPTA	Glycerolpropoxylate triacrylate
HALS	Hindered amine light stabiliser
HDDA/HDODA/1,6-HDODA	1,6-Hexanediol diacrylate
IOTG	Isooctyl thioglycolate
ITX	2-Isopropylthioxanthone
MDEA/MDEOA	Methyldiethanolamine
MDI	4,4'-Diphenylmethylene-4,4'-diisocyanate
PETA/PETA-tetra	Pentaerythritol tetra-acrylate
PETA-tri/hydroxy-PENTA	Pentaerythritol triacrylate
TEGDA	Triethyleneglycol diacrylate
TMPEOTA	Trimethylolpropane ethylene oxide triacrylate
TMPPOTA/TMPPTA	Trimethylolpropane propoxylate triacrylate
TMPTA	Trimethylolpropane triacrylate
TPGDA	Tripropyleneglycol diacrylate

Thermal stability

As seen from the above section the majority of radiation-curable resins/oligomers are acrylate based and often complex in structure. It is envisaged therefore that because of this, and the fact that three-dimensional networks are produced on curing, the thermal degradation and/or oxidation processes will be more complex in radiation-cured coatings than those of simpler polyacrylates.

Thermal methods have been used to study the thermal stability of both UV- and EB-cured coatings.[31] Here the effects of multifunctional monomer, diluent, oligomer and curing conditions were investigated in air and nitrogen. The effect of monomer structure was found to follow the trend of decreasing thermal stability:

TMPTA = PETA > HDODA ≫ TMPPOTA
$\qquad\qquad\qquad$ = TMPEOTA > TPGDA = TEGDA

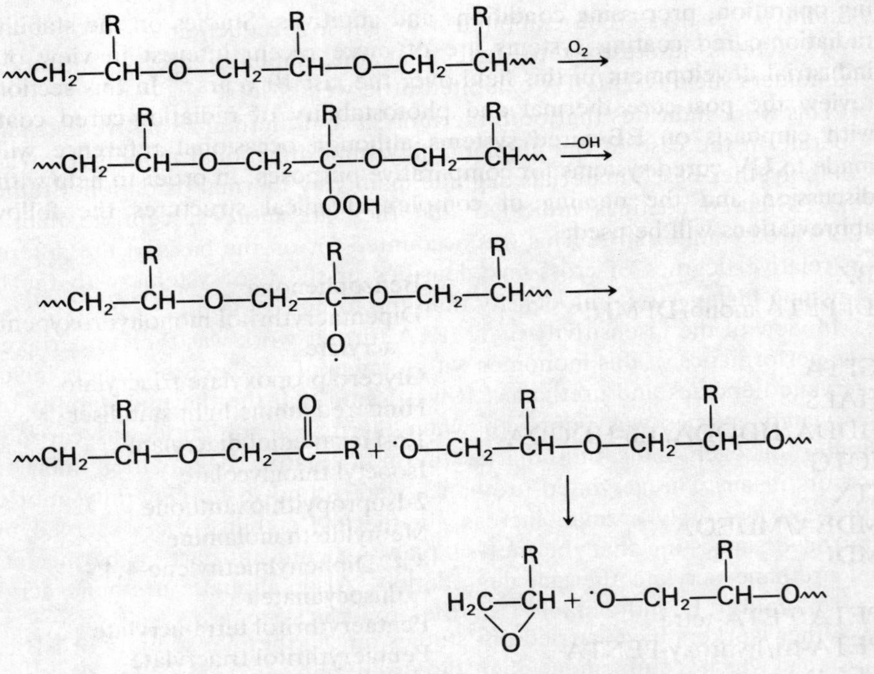

Scheme 5.10

From this work two factors appeared to govern the order in thermal stability. The first relates to monomer functionality and cross-link density. For a given series such as alkyl or ether type, the higher functionality monomers were found to be more stable than the difunctional monomers. On this basis therefore, the cross-link density is likely to increase with increased functionality leading to lower oxygen permeability and improved thermal stability. The second factor is the backbone structure of the monomer. Thus, those multifunctional monomers that contain a polyether backbone are more prone to thermal degradation than their alkyl counterparts. For example, TMPTA is more stable than TMPPOTA. It is well known that the CH bonds α to ether linkages in polymer backbones are highly susceptible to thermal oxidative attack by Scheme 5.10. Photoinitiator concentration was also found to have an effect on the thermo-oxidative stability of the cured resin. Thus, as the concentration of the photoinitiator was reduced from 5 to 2% w/w so the stability of TMPTA was reduced. This effect is associated with the lower cross-link density, in the latter sample, due to under-cure. The fact that the multifunctional monomers containing a polyether backbone are more susceptible to thermal oxidation than their alkyl counterparts was shown by thermal experiments carried out under nitrogen. Under these conditions the polyether-based multifunctional monomers exhibited a dramatic increase in thermal stability. Thus, TEGDA exhibited a weight loss reduction from 5 to 7·4% at 350°C. It is interesting to note that the aliphatic alkyl multifunctional monomers such as TMPTA exhibited little weight loss.

This work indicates that both thermolysis and thermal oxidative mechanisms operate in the case of the aliphatic alkyl multifunctional monomers, whereas the polyether-based materials degrade mainly by thermal oxidation. Studies on the EB-cured coatings indicated that they are more susceptible than their UV-cured counterparts. This was accounted for on the basis of the differences in relative degrees of cross-link densities of the two systems with the latter having a higher cross-link density than the former system.

In view of the insensitivity of TMPTA further work was then undertaken on the performance of this monomer with a range of acrylated oligomers including acrylated epoxies and urethanes. It was found that the thermal stability of the acrylated epoxies was improved when thermal degradation was carried out under nitrogen indicating that a thermo-oxidative degradation mechanism occurs in air. The acrylated urethanes were found to be generally inferior to the epoxies. Only a small increase in thermal stability was observed under nitrogen suggesting that thermolysis plays a significant part in the mechanism of urethane acrylate thermal degradation. Also, aliphatic urethane acrylates were found to be more stable than the aromatic systems.

Other workers have carried out studies on the thermal stability of EB-cured systems which complement that discussed above.[32] Thus, the difunctional acrylate monomers such as 1,6-HDODA, TPGDA and TEGDA had recorded weight losses at 350°C of 22, 37 and 30%, respectively, compared with the more stable multifunctional acrylates such as TMPTA, PETA-tri and PETA-tetra which had recorded weight losses of 23, 3·5 and 2%, respectively. The polyether containing multifunctional acrylates such as TMPPTA were found to be more susceptible to thermal oxidation than their corresponding aliphatic multifunctional analogues such as TMPTA. It was postulated that high functionality monomers produce three-dimensional polymer networks of higher cross-link densities than those produced by difunctional monomers. The effect of EB dose on the thermal stability of 1,6-HDODA showed that with increasing dose for curing the thermal stability of the coating is improved due to increased cross-link density. Thus, at 350°C recorded weight losses were found to be 22, 7 and 5% for dose rates of 3, 5 and 10 Mrads, respectively. Evidence for the increased cross-link density was provided by solid-state ^{13}C-NMR and oxygen permeability measurements.

Low-temperature degradation studies have also been carried out on both UV- and EB-cured clear and pigmented systems.[33] In this work the effect of heat ageing at 100°C for 30 min was examined on the colour and gloss of UV- and EB-cured TPGDA on an aluminium base coated with an acrylic/isocyanate base coat. As one might expect, the EB-cured coatings were superior to those cured with UV radiation with regard to colour stability. There was also a trend between the gloss changes on heating and dose rate. Also, the gloss difference of the UV-cured coatings before and after heat ageing was found to be generally worse than that of EB-cured coatings before and after ageing. It was suggested that the poor performance of the UV-cured coatings, in relation to the gloss, may be due to the inhibition of photopolymerisation by oxygen at the coating surface. This would produce different extents of polymerisation at the surface compared with the bulk of the coating. This effect is significantly reduced in the case of the EB-cured coatings due to a more effective through-cure. EB-cured pigmented coatings gave similar results to those of the corresponding clear lacquers.

Relatively little work has been published on the stabilisation of radiation-cured coatings against thermal degradation. In one study,[31] two approaches were adopted to achieve stabilisation. The first involved the use of ethylthioethyl methacrylate which copolymerises with the multifunctional monomers and/or acrylate oligomers used. The second approach was to include a chain transfer agent, isooctyl thioglycolate which is converted to a thioether after chain transfer. The latter are believed to act as thermal stabilisers by scavenging oxygen, thus oxidising the thioether to a sulphoxide and a sulphone. It was further found that these compounds stabilised multifunctional polyether acrylate monomers against thermal oxidation, which further indicates that these multifunctional monomers degrade by an oxidative route. These stabilisers exhibited some improvement with epoxy acrylate oligomers and a marked improvement with urethane acrylate oligomers. This suggests that, since the urethane is degraded predominantly by a thermolysis process, these stabilisers must also be operating as free-radical scavengers.

Several other more conventional stabilisers have been examined on the thermal stability of UV-cured epoxy acrylate based formulations using differential scanning calorimetry and oxygen absorption.[34] Three types of stabiliser system were investigated, namely, hindered phenolic compounds, hindered amine stabilisers and thio/phosphate-based peroxide decomposers. Some of the stabilisers proved to be incompatible while others altered the photoinitiator conversion rate. To summarise it was found that the HALS, bis(2,2,6,6-tetramethyl-4-piperidinyl) sebacate and the hindered phosphonate namely, bis(n-octadecyl-3,5-ditertbutyl-4-hydroxybenzyl) phosphonate markedly improve the thermal stability of the cured coatings.

Photostability

The photostability of radiation-cured coatings particularly with regard to outdoor weatherability is of major importance. A number of applications require the UV- or EB-cured coating, either clear or pigmented, to provide stable colour and gloss performance under ambient or artificial lighting conditions. Other applications require the physical and mechanical properties of the films to be maintained under these conditions. For UV-cured coatings the nature of the photoinitiator is particularly important and this has been discussed elsewhere in the series. Photoyellowing and photooxidative stability are two main areas of interest.

Each component of a formulation is likely to produce its own contribution to

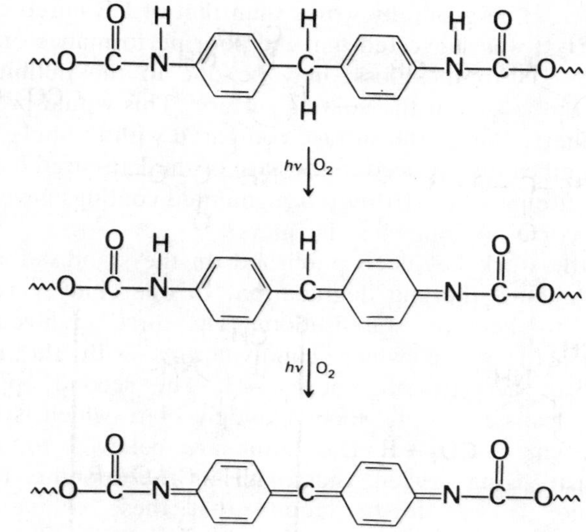

Scheme 5.11

the photoyellowing and photooxidative stability/instability of the system. In one study several commercially available acrylate monomers were studied for their resistance to UV-induced yellowing.[35] Coatings containing aromatic groups or nitrogen in their functionality were found to be prone to yellowing, whereas binary mixtures of TEGDA and TMPTA were the most resistant to yellowing. In another study on coating structure,[33] epoxy acrylate oligomers were found to be inferior to urethane acrylates based on the bisphenol-A structure. On the other hand, oligomers based on aromatic urethane and aromatic epoxy were found to be more prone to photoyellowing than their aliphatic analogues.[36] The aliphatic urethane oligomer was found to be the most resistant to photoyellowing. The photoyellowing of aromatic-based oligomers is believed to be due to the formation of quinoid structures. Aromatic urethanes based on MDI have been shown to oxidise to a highly coloured quinone-imide product by reaction Scheme 5.11.[37]

Oligomers based on toluene diisocyanate can undergo α cleavage reactions which may be followed by a photo-Fries type rearrangement with the production of highly coloured quinoid structures as shown by Scheme 5.12.[38]

Other workers have studied the photochemistry or bisphenolic polymers and model compounds.[39] It was found that the aromatic chromophores induced photocleavage and photooxidation reactions. The production of quinoid products was concluded to be responsible for the photoyellowing. Far-UV irradiation of UV-cured epoxy acrylate coatings has been shown to result in the destruction of the phenyl rings of the bisphenol-A[40] and cause strong photoyellowing which bleached on subsequent exposure to long-wavelength UV irradiation. Detailed intercomparative studies have been carried out on

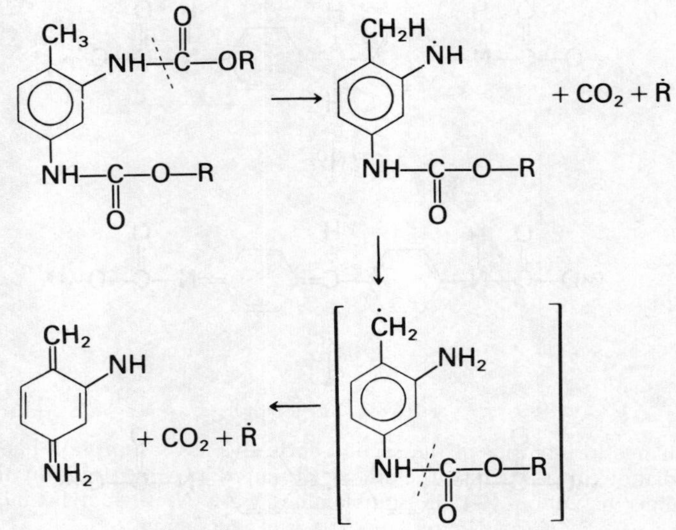

Scheme 5.12

the photoyellowing of both UV- and EB-cured coatings composed of multi-functional monomers and oligomers.[41–46] During preliminary work on both types of radiation-cured systems severe photoyellowing was observed with amine-terminated diacrylate monomers such as that shown in **VI**. This growth in yellowing was associated with the growth of an absorption band at 275–280 nm and associated absorptions in the IR absorption spectrum at 1605 and 1680 cm^{-1}. The former was found to be more sensitive and specific if measured using second-derivative absorption spectroscopy. Irradiation under nitrogen and oxygen conditions showed that the absorptions were due to an oxidative process and only associated with the amine functionality for the series studied. Figures 5.7–5.9 show the relative rates and levels of photo-yellowing for a wide range of EB-cured amine diacrylates based on TMPTA using the second-order derivative absorption technique. The absorption band is seen to increase initially over the first 200 h of irradiation and then gradually decrease (photobleach). UV-cured systems showed a different pattern of behaviour in that the absorption increased rapidly during the first 20 h and then photobleached rapidly depending on the type of residual photoinitiator present. An example is shown in Fig. 5.10 for comparative purposes for three different amine-terminated TMPTA coatings using 5% w/w of BZ as the photoinitiator. Confirmation that the photoyellowing is due solely to that of the amine functionality is shown by the results in Fig. 5.11 where it is seen that the changes in the absorption band at 275 nm are unaffected by the nature of the multifunctional monomer structure. Studies on model structures for the amine-terminated acrylates such as lauryl-3-(diethylamine) propionate showed

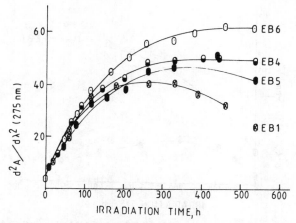

Fig. 5.7. Change in intensity of the second-derivative UV absorption band at 275 nm during irradiation in a Microscal Unit of EB-cured (4 Mrad) (12 μm thick) amine diacrylate films based on (◓) di-n-butylamine, (⊗) N-ethylbutylamine, (●) di-n-propylamine, and (○) di-n-pentylamine.

that similar degradation products were produced both thermally and photochemically and were identified as conjugated unsaturated carbonyl structures.[41,42,44] From the data shown in Figs 5.7–5.9 the rates of photoyellowing were found to follow the order:

di-n-pentylamine > di-n-butylamine > di-n-propylamine > N-ethylbutylamine

> diethanolamine > diethylamine > diisopropylamine

> N-ethylethanolamine > N-methylethanolamine > dicyclohexylamine

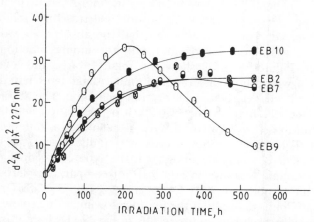

Fig. 5.8. Change in intensity of the second-derivative UV absorption band at 275 nm during irradiation in a Microscal Unit of EB-cured (4 Mrad) (12 μm thick) amine diacrylate films based on (⊗) diisopropanolamine, (◓) N-ethylethanolamine, (○) diethanolamine and (●) diethylamine.

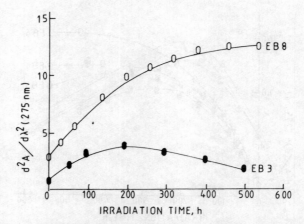

Fig. 5.9. Change in intensity of the second-derivative UV absorption band at 275 nm during irradiation in a Microscal Unit of EB-cured (4 Mrad) (12 μm thick) amine diacrylate films based on (●) dicyclohexylamine, and (○) N-methylethanolamine.

It is noted from this order that all the simple alkylamines exhibit the greatest degree of photoyellowing whilst hydroxyl-containing amines are generally lower. In the former case the extent of photoyellowing follows the order:

di-*n*-pentyl > di-*n*-butyl > di-*n*-propyl > diethyl

indicating that the longer alkyl chain length is giving rise to increased conjugation. Amines with bulky substituents also exhibit a lower degree of photoyellowing probably due to their resistance towards oxidation. The key initiators in the photoyellowing phenomenon were assigned to hydroperoxides

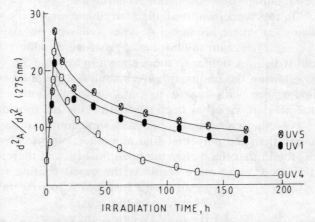

Fig. 5.10. Change in intensity of the second derivative UV absorption band at 275 nm during irradiation in a Microscal Unit of UV-cured (5% w/w BZ) (12 μm thick) amine diacrylate films based on (○) di-*n*-butylamine, (●) N-ethylbutylamine, and (⊗) di-*n*-propylamine.

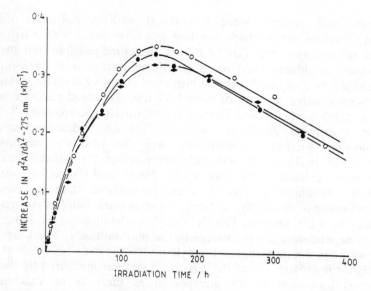

Fig. 5.11. Change in intensity of the second-derivative UV absorption band at 275 nm during irradiation in a Microscal Unit of EB-cured (4 Mrad) (4 μm thick) amine diacrylate films based on (○) GPTA, (◆) TMPETA, and (●) TMPTA after reaction with diethylamine.

formed in the oxidation of the terminal amine functionality. This was confirmed by the observation of inhibition of photoyellowing after pre-treating the coatings with sulphur dioxide gas which destroys the hydroperoxides to inactive products.[42]

Coatings containing the bisphenol-A structure were also found to photoyellow.[43] In this work, whilst aliphatic urethane oligomers were found to be non-yellowing, aromatic bisphenol-A types yellowed on extended ageing. The latter was associated with the bisphenol-A structure undergoing oxidation to give quinoid structures similar to those shown in Scheme 5.8.

Like photoyellowing the photooxidative stability of radation-cured coatings is also of importance with regard to outdoor applications. The effect of multifunctional acrylate type has been studied on the photooxidative stability of coatings.[38] Polyether-containing monomers were found to be more susceptible to oxidation than the polyether-free monomers. In the absence of ether linkages it was found that the higher the functionality, and therefore cross-link density, of the coating the more stable is the cured coating to weathering. Polyether-containing monomers are more susceptible to oxidation due to the ease of oxidation of the α CH bonds to the ether linkage in the structure. During the photoageing of EB- and UV-cured epoxy and urethane acrylates the former were surprisingly found to be less stable as indicated by film failure. However, the aliphatic urethane systems were found to be less stable than the bisphenol-A based epoxy acrylate systems. During this work it is noted that the epoxy-based systems were formulated with a polyether diacrylate (TPGDA) whereas the urethane acrylate was formulated with a trifunctional polyether diluent monomer (GPTA). It is therefore possible that the higher concentration of diluent used in the urethane system is the reason for its greater instability compared with the bisphenol-A epoxy acrylate system.

The photooxidative stability of both UV- and EB-cured coatings has been studied in depth using IR and Fourier transform IR spectroscopy.[41–46] Apart from the absorption bands at 1605 and 1680 cm^{-1} discussed above for photoyellowing, increased absorptions are also observed at 3540, and 1720 cm^{-1} due to hydroxyl and carbonyl product formation. Three main features were evident from this work. First, multifunctional monomers terminated with amine groups are more photostable than the non-amine-terminated systems. Secondly, EB-cured systems are more photostable than those cured by a UV process. Thirdly, the photostabilities of the latter systems are highly dependent upon the nature of the photoinitiator. The latter is more evident with non-amine-terminated acrylates.

The higher photostability of the EB-cured systems measured by the growth in hydroxyl formation is not unexpected as there is no residual active photoinitiator present. In amine-terminated acrylates the greater stability is associated with the ability of the tertiary amine group to effectively scavenge oxygen and therefore concentrate oxidation on the amine site thus protecting the remaining acrylate structure. This was confirmed by a detailed hydroperoxide study.[43] In contrast to the effect on photoyellowing sulphur dioxide treatment was found to influence the photooxidative stability of coatings made from multifunctional acrylates. It was concluded that species other than hydroperoxides must be the key photoinitiators of oxidation. Such species could be impurity carbonyl groups. The addition of aromatic epoxy and aliphatic urethane oligomers to diluent di- and triacrylate monomers were found to increase the stability of the subsequent UV- and EB-cured coatings. This was associated with either an antioxidant effect of the two oligomers or more likely the higher cross-link density of the films.

More recent work on the photooxidative stability of a series of EB-cured multifunctional monomers has shown that stability is controlled by both the EB dose and monomer structure.[46] Thus, for GPTA increasing the EB dose markedly increased the post-cure stability of the coating as determined by the growth in the hydroxyl index at 3450 cm^{-1}. This effect on stability is associated with two effects on the coating, namely, increased cross-link density and a reduction in residual unsaturation. The structure of the multifunctional monomer is also important and again using a hydroxyl index the following order in stability has been observed (Fig. 5.12):

$$\text{TMPTA} > \text{GPTA} > \text{TMPETA}$$

This order in stability is linked directly to the level of methylene protons α to the ether oxygen in each system. Thus, with TMPTA there is no ether functionality whereas GPTA contains three α methylene hydrogen atoms

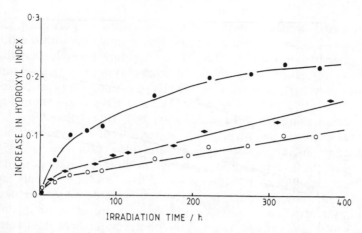

Fig. 5.12. Increase in hydroxyl index during irradiation in a Microscal Unit of EB-cured (4 Mrad) (4 μm thick), (◆) GPTA, (●) TMPETA, and (○) TMPTA after reaction with diethylamine.

adjacent to an ether oxygen while the TMPETA contains four α methylene hydrogens adjacent to an ether oxygen. These methylene hydrogen atoms are relatively labile and can be easily abstracted by alkoxy or hydroxy radicals during oxidation of the coating system. The nature of the amine functionality in amine-terminated diacrylates is also an important factor in controlling coating stability and this is illustrated by the data in Fig. 5.13 on the hydroxyl index for GPTA reacted with various amine structures. Here, the general order in stability is

di-n-butylamine = diethylamine < dimethylamine < morpholine

< dicyclohexylamine < ethylethanolamine < methylethanolamine

It is interesting to note that of the four most stable amine acrylates three of them have amine functionality which contains an oxygen atom either as ether or as a hydroxyl group. Here, the adjacent methylene hydrogens may be acting as sacrificial oxidation sites thus protecting the remaining polymer structure. The dicyclohexylamine may be inducing increased stability due to the reduced number of abstractable hydrogen atoms α to the nitrogen in the amine. The three least stable alkylamino structures all have the same number of abstractable hydrogen atoms α to the nitrogen atom.

Photostabilisation

Apart from investigating the coating parameters such as the nature of the monomer/oligomer structure or curing process stability may also be achieved through conventional methods of stabilisation involving the use of additives provided they do not, of course, interfere with the rate of cure or the coating properties.

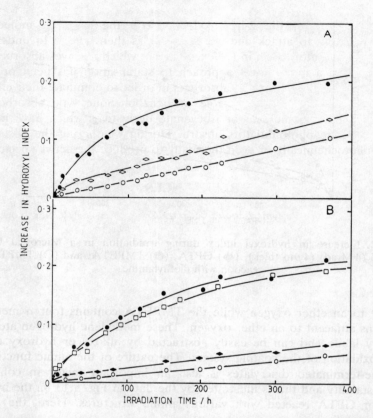

Fig. 5.13. Increase in hydroxyl index during irradiation in a Microscal Unit of EB-cured (4 Mrad) (12 μm thick) GPTA (A) after reaction with (●) diethylamine, (◇) morpholine, and (○) dicyclohexylamine; and (B) after reaction with (●) di-n-butylamine, (◆) ethylethanolamine, (○) methylethanolamine, and (□) dimethylamine.

A number of workers[47,48] have examined the influence of various photo-stabilisers on the rate of cure of various monomers and found that whilst absorbers such as the orthohydroxyaromatics will impair UV curing the hindered amine type stabilisers have little or no influence. The latter have also been reported to impair the photoyellowing of urethane acrylates. However, some reports are conflicting with other studies showing that absorbers have little influence on the cure rate of epoxy acrylates.[33] Many stabilisers have been eliminated on the basis of compatibility whilst others such as cyanodiphenylacrylate and dimethyl-p-methoxybenzylidenemalonate significantly reduce photoyellowing.[49] HALS now appear to be the most practical systems in commercial use for long-term weatherability and performance.[38,50] With hydrophenylbenzotriazoles they exhibit synergism due to their protective effect on the HALS and vice versa. In the latter case the HALS will react with active free radicals in the oxidation of the coating thus protecting the absorber from radical attack. Here, the 2-hydroxyl group in the absorber molecule is particularly prone to attack and the stabiliser is then unable to undergo its facile keto-enol tautomerism in the excited state which removes absorbed light energy. Finally, a more novel approach to stabilisation has been proposed based on the use of blocked UV absorbers in order to eliminate their effect on the cure rate.[51] Thus, benzotriazole or benzophenone type absorbers are converted into sulphonic acid or isocyanate derivatives which have minimal absorption where photoinitiators absorb. During weathering the absorber is released photochemically or even thermally to produce the active species.

REFERENCES

1. Roffey, C. G. *Photopolymerisation of Surface Coatings*, J. Wiley, New York, USA, 1982.
2. 'UV2', 1978, OCCA Conference, 61.
3. Kirland, J. *J. Polym. Sci. Polym. Chem. Edn*, **18** (1980) 1139.
4. Rybny, C. B., De Fazio, C. A., Shadidi, J. K., Trehelis, J. C. & Vona, J. A. *J. Paint Technol.*, **46** (1974) 60.
5. UK Patent 1,006,587.
6. UK Patent 1,030,760.
7. Caldwell, R. G. & Ihring, J. C. *J. Amer. Chem. Soc.*, **84** (1952) 2878.
8. US Patent 3,297,745.
9. US Patent 4,070,262.
10. US Patent 3,673,140.
11. US Patent 3,567,494.
12. Talemichi, H. & Ogaswara, T. *Chem. Econ. Eng. Rev.*, **10** (1978) 37.
13. Chevallier, F., Chevallier, S., Decker, C. & Moussa, K. (1987) SME Technical Paper, **9,** 1987, FC 87.
14. Allen, N. S., Lo, C. K. & Salim, M. S., Radtech Europe '89, Florence, 1989.
15. O'Hara, K. *Polym. Paint Colour J.*, **171** (1981) 11.
16. Skelhorne, G., Smith, P. & Harrison, F. *The Brit. Ink. Maker*, Feb–Mar. (1981) 60.
17. McCarty, W. H. *J. Radiat. Curing*, **4** (1977) 7.
18. DuPlessis, T. A. & DeHollain, G. *J. Oil. Colour Chem. Assoc.* **62** (1979) 239.
19. Morganstern, K. H. In *Non-Polluting Coatings and Coating Processes*, ed. J. L. Gordon & J. W. Prane. Plenum Press, New York, USA, 1972.
20. Teesdale, D. H. *Coat. Curing Conv.*, **2** (1985) 2.
21. Nablo, S. V. Soc. Man. Eng., Tech. Pap. FC-82, 1982.
22. Nablo, S. V. *J. Paint Technol.*, **46** (1974) 51.
23. Tripp, E. P. & Weisman, J. *J. Mod. Paint Coatings*, Mar. (1982) 51.
24. Heine, H. C., Rosenkranz, H. J. & Rudolph, H. *Angew. Chem.*, **84** (1972) 1032.
25. Nablo, S. V. & Tripp, E. P. *Radiat. Phys. Chem.*, **9** (1977) 325.
26. Schmidt, C. K. *J. Radiat. Curing*, **6** (1979) 8.
27. Dowbenko, R., Friedlander, C., Gruber, G., Prucnal, P. & Wismer, M. *Prog. Org. Coatings,* **11** (1983) 71.
28. Burlant, W. & Hinsch, J. *J. Polym. Sci.*, **2** (1964) 2135.
29. Grassie, N. & Scott, G. *Degradation and Stabilisation of Polymers*. CUP, Cambridge, UK, 1985.
30. White, N. J., Allen, N. S., Robinson, P. J. & Clancy, R. In *Photopolymerisation and Photoimaging Science and Technology,* ed. N. S. Allen. Elsevier, London, UK, 1989.

31. Ambruster, D. C. & Molina, J. F. *J. Radiat. Curing,* **8** (1981) 4.
32. Thalacker, V. P. & Boettcher, T. E. Proc. Radcure Europe, SME Tech. Pap. FC85-450, 1985.
33. O'Hara, K. J., Proc. Radcure Europe, SME Tech. Pap., FC85-421, 1985.
34. Heyward, I. P., Chan, M. G. & Ludwick, A. G. *Polym. Preprints,* **25** (1984) 44.
35. Gismondi, T. E. *J. Radiat. Curing,* **11** (1984) 14.
36. Schmid, S. T. *J. Radiat. Curing,* **11** (1984) 19.
37. Allen, N. S. & McKellar, J. F. *J. Appl. Polym. Sci.,* **20** (1976) 1411.
38. Gatechair, L. R. In *UV Curing Science and Technology* (Vol. II), ed. S. P. Pappas. Technology Marketing Corp., Stanford, CT, USA, 1985, Chpt. 7.
39. Pappas, S. P. & Winslow, F. H. (eds) *Photodegradation and Stabilisation of Coatings* (ACS Symp. Ser. 151). American Chemical Society, Washington DC, USA, 1986.
40. Decker, C. & Bendaika, T. *J. Radiat. Curing* **11** (1984) 6.
41. Allen, N. S., Robinson, P. J., White, N. J. & Skelhorne, G. G. *Euro. Polym. J.,* **20** (1984) 13.
42. Allen, N. S., Robinson, P. J., White, N. J. & Skelhorne, G. G. *Euro. Polym. J.,* **21** (1985) 107.
43. Allen, N. S., White, N. J. & Swale, D. W. Proc. Radcure Europe, SME Techn. Pap. FC85-412, 1985.
44. Allen, N. S., Robinson, P. J., White, N. J., Clancy, R. & Gardette, J. L. *J. Photochem. Photobiol. A,* **47** (1989) 223.
45. Allen, N. S., Lo, D., Salim, M. S. & Jennings, P. *Polym. Degrad. Stabil.,* **28** (1990) 105.
46. Allen, N. S., Robinson, P. J., White, N. J. & Clancy, R. *Euro. Polym. J.,* **25** (1989) 145.
47. Hult, A. & Ranby, B. *Polym. Degrad. Stabil.,* **9** (1984) 1.
48. Puglisi, J. & Vigeant, F. *J. Radiat. Curing,* **7** (1980) 31.
49. Tu, R. S. Proc. Radcure V, SME Techn. Pap., FC80-575, 1980.
50. Gatechair, L. R. & Wostratzky, D. *J. Radiat. Curing,* **10** (1983) 4.
51. Olson, D. R. & Schroeter, S. H. *Polym. Sci.,* **22** (1978) 165.

Chapter 6

Role of Grafting in UV- and EB-Curing Reactions

PAUL A. DWORJANYN[a] & JOHN L. GARNETT[b]

[a]School of Chemistry, [b]School of Chemical Engineering and Industrial Chemistry, University of New South Wales, PO Box 1, Kensington, 2033 NSW, Australia

Introduction 264
Grafting Reactions 265
    EB grafting of monomers 265
    UV grafting of monomers 266
    UV/EB grafting of oligomers 267
    Post-irradiation grafting with UV/EB 268
    Homopolymer during UV/EB grafting 268
Typical Examples of Grafting Systems 268
    Grafting with mutual technique using ionising radiation 269
    Grafting with mutual technique using UV 270
    Grafting with pre-irradiation technique 271
Physical Parameters Affecting Grafting Yield 272
    Dose and dose rate with ionising radiation 272
    Intensity of light with UV 274
    Air, temperature 274
Effect of Additives on Grafting Yields 276
    Solvent 277
    Polyfunctional monomers—ionising radiation 280
    Polyfunctional monomers—UV 281
    Mineral acids and inorganic salts—ionising radiation 283
    Mineral acids and inorganic salts—UV 288
    Synergistic effect of acids and salts with polyfunctional monomers—ionising radiation 290
    Synergistic effect of acids and salts with polyfunctional monomers—UV 291
Effect of Additives on Homopolymerisation Yields 291
    Homopolymerisation in the presence of acid and polyfunctional monomers—ionising radiation 293
    Homopolymerisation in the presence of acid and polyfunctional monomers—UV 296
    Homopolymerisation suppression using co-monomer techniques 299

Effect of Structure of Monomer and Backbone Polymer in Grafting with Additives	301
Mechanisms Involving Additive Effects in Grafting	305
Effect of acids and salts—partitioning phenomena	305
Effect of polyfunctional monomers	308
Effect of cationic salts in suppressing homopolymer—partitioning phenomena	308
Relevance of Grafting in EB and UV Curing	310
Grafting of monomers used in curing—free-radical systems	311
Grafting of monomers used in curing—cationic systems	315
Grafting of oligomers used in curing—free-radical and cationic initiators	317
Effect of commercial additives used in curing on grafting	319
Concurrent Grafting During Curing—Future Developments	320
Mechanistic studies	320
Role of ions	320
Partitioning effects—additives	323
Commercial applications	324
Acknowledgements	325
References	325

INTRODUCTION

Grafting and curing initiated by either ionising (γ or electron beam (EB)) or non-ionising (UV) radiation are mechanistically similar processes.[1,2] Grafting is essentially the copolymerisation of a monomer/oligomer (M/OL) to a backbone polymer (eqn (1)), whereas curing is the rapid polymerisation of an oligomer/monomer mixture onto the surface of a substrate (eqn (2)). There is no time-scale theoretically associated with grafting reactions, which can take minutes, hours or even days, whereas curing reactions are usually very rapid, occurring in a fraction of a second.

$$\zeta \xrightarrow{M/OL} \zeta\text{-}\!\!\sim\!\!\sim\!\! M/OL \qquad (1)$$

$$\sim\!\!\sim\!\! \xrightarrow{M/OL} \sim\!\!\sim\!\!\underset{\sim\!\!\sim}{M/OL} \qquad (2)$$

In radiation curing, the film, as it is formed, does not usually follow the contours of the surface as in thermal solvent-based cured processes but effectively fills in the valley of the substrate, especially those with uneven surfaces such as cellulose. One of the differences between grafting and curing is the nature of bonding occurring in each process. In grafting, covalent C—C bonds are formed, whereas in curing, bonding usually involves weaker van der Waals or London dispersion forces. In some curing systems the ability to achieve concurrent grafting with cure is important, since adhesion between the substrate and the cured film is improved as a consequence of the associated covalent bonding. For other applications concurrent grafting with curing can be an impediment to the final properties of the finished product. It is thus necessary to examine the parameters which influence the grafting yield during these curing processes. This aspect of grafting is the subject of this paper. The basic principles of radiation grafting using both EB and UV systems will be treated initially, then these concepts will be applied to the special case of grafting which occurs simultaneously with curing.

GRAFTING REACTIONS

UV- and EB-grafting processes are similar except that with UV, in order to achieve satisfactory rates of reaction in reasonable periods of time, photoinitiators and/or photosensitisers are required. The presence of these additives can influence the nature of the reaction, particularly the properties of the finished product. There are a variety of grafting processes which are capable of being initiated by UV and EB; however, for the effect of grafting on curing, the in-situ simultaneous grafting technique is of most relevance and will be emphasised here.

EB grafting of monomers

Grafting initiated by ionising radiation of which EB is a typical source, is depicted in eqns (3)–(8) for free-radical processes (X and Y are usually H).

$$PX \rightsquigarrow P^\cdot + X^\cdot \qquad (3)$$

$$P^\cdot + M \longrightarrow PM^\cdot \longrightarrow \text{graft copolymer} \qquad (4)$$

$$P^\cdot + O_2 \longrightarrow PO_2^\cdot \longrightarrow PO_2H \longrightarrow PO^\cdot + OH^\cdot \qquad (5)$$

$$PO^\cdot + M \longrightarrow POM^\cdot \longrightarrow \text{graft copolymer} \qquad (6)$$

$$SY \rightsquigarrow S^\cdot + Y^\cdot \qquad (7)$$

$$S^\cdot + PX \longrightarrow P^\cdot + SX \qquad (8)$$

These free-radical processes have been extensively reviewed.[1-6] There are essentially three techniques for grafting by EB, namely, (i) the mutual, direct or simultaneous technique, (ii) the pre-irradiation procedure, and (iii) the peroxide process. In the mutual or simultaneous method (eqns (3) and (4)) the backbone polymer (PX) is irradiated whilst in direct contact with the monomer (M) which may be present as vapour, liquid or in solution. Irradiation can occur in air or an inert atmosphere and leads to the formation of active free radicals in the backbone polymer, resulting in graft copolymerisation.

In the pre-irradiation technique, PX is irradiated *in vacuo* or in the presence of an inert gas prior to exposure to the monomer, which may be present as either vapour or liquid. On subsequent heating of the monomer now in contact with the irradiated polymer, radicals formed during irradiation are mobilised and react with monomer to yield graft copolymer. This reaction can also occur at room temperature if sufficient time is allowed for the monomer to diffuse to the active centres of the backbone polymer.

In the peroxide technique (eqns (5) and (6)), the polymer is irradiated in the presence of air to produce mainly hydroperoxides which are often quite stable but can be decomposed in the presence of monomer to produce graft copolymer. It is obvious that the peroxide method is also a pre-irradiation process, the irradiation being performed in the presence of air. In both of these processes grafting sites can be formed by direct bond rupture as a result of the primary effects of the radiation. Active sites can also be formed by abstraction reactions involving radicals resulting from the effects of EB on other components (SY) in the grafting system, particularly the solvent (predominantly) and monomer (eqns (7) and (8)).

In the ionising radiation system, ions as well as free radicals can participate in these processes.[7-11] Typical of these ionic systems are those shown in eqns (9)–(11) for cationic initiation using $CHR=CH_2$ as typical monomer. A similar process can also be described for anionic initiation.

$$CHR=CH_2 \rightsquigarrow [CHR=CH_2]^+ + e^- \qquad (9)$$

$$CHR=CH_2 + [CHR=CH_2]^+ \longrightarrow {}^{\cdot}CHR-CH_2-CH_2-\overset{+}{C}HR \qquad (10)$$

$$P^{\cdot} + {}^{\cdot}CHR-CH_2-CH_2-\overset{+}{C}HR \longrightarrow P-CHR-CH_2-CH_2-\overset{+}{C}HR \qquad (11)$$

UV grafting of monomers

UV can initiate analogous grafting reactions to those previously discussed for EB if sources of the appropriate UV wavelength are used,[2,5,12-17] and, like EB, the most frequently utilised method and the one of most relevance to curing work, is the mutual or simultaneous technique. Such UV-initiated reactions are also essentially free radical in nature, radicals, where grafting may occur, being formed directly in the backbone polymer as a consequence of the incident UV, especially if such materials contain appropriate functional groups (or impurities) which can efficiently absorb the incident UV (eqn (12)). Initiation via ionic species has also recently been reported under UV conditions.[18]

$$PX \xrightarrow{h\nu} P^{\cdot} + X^{\cdot} \qquad (12)$$

Because of limitations with the penetration of UV through a backbone polymer, photoinitiators (RI) are usually incorporated in the system to improve the efficiency of the grafting process. Such reagents yield high concentrations of radicals under appropriate UV conditions. The resultant radicals can then diffuse into the backbone polymer, abstract hydrogen atoms and create grafting sites for monomer reaction (eqns (13) and (14), where $X = H$).

$$RI \xrightarrow{h\nu} R^{\cdot} + I^{\cdot} \qquad (13)$$

$$R^{\cdot} + PX \longrightarrow P^{\cdot} + RX \qquad (14)$$

In some instances, photosensitisers, which themselves do not yield radicals but excited states with UV, are incorporated with the photoinitiator to improve the efficiency of the grafting process by energy transfer from excited photosensitiser to other components in the system, particularly the photoinitiator. Many monomers, particularly acrylates, which are used in grafting can also act as photosensitisers or photoinitiators in these grafting systems. Charge transfer complexes involving donor–acceptor systems can also be used to initiate photografting. Typical of the reagents in this category are materials like benzophenone and the amines, particularly the tertiary derivatives, such as triethylamine which sensitise UV-grafting reactions by forming exciplexes with the triplet state of the aromatic ketone.[19,20]

In a recent development, cationic photoinitiators have been used to graft certain types of monomers to backbone polymers, like cellulose.[18] These photoinitiators decompose under appropriate UV conditions to give ions and radicals, both species being capable of initiating grafting reactions. By judicious choice of scavengers, either ionic or free-radical reactions can be made to predominate in these processes. Typical of the cationic catalysts used are triaryl sulphonium salts, particularly the hexafluorophosphate, which decomposes under UV to yield radicals and ions according to eqns (15) and (16) (where RH represents a hydrogen donor, usually a specific monomer).

$$Ar_3S^+X \xrightarrow{h\nu} [Ar_3S^+X^-]^* \qquad (15)$$

$$[Ar_3S^+X^-]^* + RH \longrightarrow Ar_2S + Ar^\cdot + R^\cdot + H^+X^- \qquad (16)$$

Grafting by this cationic process is particularly useful for monomers like the vinyl ethers such as triethyleneglycol divinylether (TEGDVE) and cycloaliphatic diepoxides (CADE).

UV/EB grafting of oligomers

The previous discussion has been confined to the grafting of monomers to substrates. Of increasing significance is the extension of this monomer work to the grafting of oligomers. This aspect is of particular relevance to curing since the components of coating mixtures used to form cured films usually contain a high percentage of oligomers and the possibility of concurrent grafting of these oligomers is important. Theoretically, both pre-irradiation and mutual techniques are suitable for grafting oligomers; however, preliminary published work in this field has emphasised the latter technique. Both free radical and ionic processes are relevant to these oligomer grafting systems. Not only are conventional oligomer acrylates available which are capable of grafting by free-radical or ionic processes using UV or EB, but also vinyl ether oligomers are now being produced commercially which are capable of being grafted, predominantly via cationic processes, with UV. Initial work in this field has involved essentially free-radical initiation (eqns (12)–(14) and (17)) using UV irradiation.

$$P^\cdot + OL \longrightarrow POL^\cdot \longrightarrow \text{graft copolymer} \qquad (17)$$

Post-irradiation grafting with UV/EB

An important property of grafting reactions is the possible occurrence of post-irradiation effects, i.e. at the completion of the irradiation procedure, when the source is removed and the grafting system is allowed to stand, grafting continues as a post-irradiation phenomenon. The extent of such post-irradiation grafting usually depends upon the time of standing after irradiation and also the temperature of the system. The process can occur with both ionising radiation and UV. Analogous phenomena in curing are also found; the well-known post-cure effect can either be an advantage or disadvantage to the properties of the finished product. Concurrent grafting can also be a component in such post-irradiation curing. Post-irradiation effects usually accompany grafting when the mutual technique is used. Trapped radicals are formed in the backbone polymer by irradiation and are not all utilised in the primary grafting step, particularly if these sites are in more difficultly accessible positions in the bulk polymer. With the additional time available after the completion of the irradiation monomer molecules can diffuse to these sites and graft.

Homopolymer during UV/EB grafting

A competing detrimental reaction occurring with grafting is homopolymerisation (eqns (18) and (19)). Homopolymer is generally an unwanted by-product in most grafting reactions, necessitating removal in the purification step after grafting. Homopolymerisation can be minimised by the use of copper and iron salts,[8,21-23] however, grafting yields are also reduced in the presence of these metal ions, but usually to a lesser extent than homopolymer. The inhibitory effect of these ions has been attributed to the transfer of electrons from the propagating chain into the empty d shells of the ion.

$$M \rightsquigarrow M^{\cdot} \qquad (18)$$

$$M^{\cdot} + M \longrightarrow M_n^{\cdot} \longrightarrow \text{homopolymer} \qquad (19)$$

TYPICAL EXAMPLES OF GRAFTING SYSTEMS

In this section, data from representative grafting systems will be discussed to indicate the type of reactivity and characteristics of grafting profiles to be expected. Because of the size of the grafting literature, most of the data shown and discussed here will be taken from the authors' own work; however, where relevant, related published studies will be referenced. For most of the grafting data obtained with ionising radiation, [60]cobalt and spent fuel element facilities have been the predominant sources of radiation used. However, extrapolation from these sources to EB generally involves an increase in dose rate and from dose rate effects with such sources, the type of reactivity to be expected with EB can generally be predicted. Cellulose and the polyolefins (polypropylene, polyethylene) will be used predominantly as representative backbone polymers in this discussion since these two materials are typical of naturally occurring and synthetic substrates found in practice. In addition, paper and the polyolefins are amongst the most important substrates used in UV- and EB-curing processes and ease of grafting to these two backbone polymer systems is relevant to the possible observation of concurrent grafting during curing. Styrene will be the representative monomer for most of the reactions reported here since it is relatively inert chemically and contains no functional groups which may influence the grafting pattern. The scope of this work will subsequently be extended to include the role of functionality of monomer in grafting, particularly that of acrylates which are of importance in curing.

Grafting with mutual technique using ionising radiation

The results in Fig. 6.1 show the data obtained when styrene is grafted to cellulose in a variety of alcohols in the presence of ^{60}cobalt. The grafting profile for styrene in methanol to cellulose in Fig. 6.1 is typical for many systems, the peak observed in this particular instance at 30–40% being the Trommsdorff effect.[24-26] Variables which affect the position and intensity of the peak include monomer concentration, type of solvent, type of monomer, backbone polymer and irradiation source conditions. In a preparative context the Trommsdorff peak is important, since not only do the grafting yields attain a maximum at this point, the grafted chains are longer.

When cellulose is replaced by a synthetic backbone polymer such as polypropylene, a similar grafting profile is observed for the styrene in methanol system under the irradiation conditions used in Table 6.1. The data in the table illustrate the magnitude of the grafting yields to be expected under the radiation dose and dose rate conditions reported.

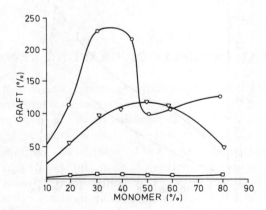

Fig. 6.1. γ-Radiation grafting of styrene in alcohols to cellulose; dose rate $8 \cdot 3 \times 10$ Gy/h, dose $2 \cdot 0 \times 10^3$ Gy. (○) Methanol; (▽) n-propanol; and (□) n-butanol.

Table 6.1
Grafting of styrene in methanol to polypropylene using ionising radiation and UV[a]

Styrene (% v/v)	Graft (%)	
	Ionising radiation[b]	UV[c]
20	33	12
40	89	47
60	47	22
80	31	17

[a] Polypropylene film (0·12 mm, ex-Shell).
[b] Irradiated at $5·0 \times 10^2$ Gy/h to $5·0 \times 10^3$ Gy (^{60}cobalt).
[c] Irradiated 6 h at 24 cm from 90 W medium-pressure mercury vapour lamp using benzoin ethyl ether as photoinitiator (1% w/v).
UV intensity approximately $3·3 \times 10^4$ mJ cm^{-2}.

Grafting with mutual technique using UV

The essential difference between UV and ionising radiation grafting in a practical context is the need to use photoinitiators/photosensitisers in the UV system in order to achieve fast rates of reaction in reasonable periods of time. This situation can be directly attributed to the lower energy of the UV, leading to poorer penetration of the substrate and the subsequent difficulty in rapidly creating large numbers of active sites. By contrast, with ionising radiation, complete penetration of any backbone polymer is possible leading to the direct formation of grafting sites within the substrate.

When UV is used to promote the grafting of styrene in methanol to cellulose, using uranyl nitrate as photosensitiser (Fig. 6.2), analogous results to

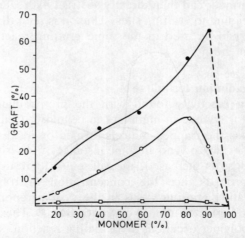

Fig. 6.2. Photosensitised grafting of styrene in alcohols to cellulose; Uranyl nitrate (1% w/v) as photosensitiser and irradiated 24 h at 24 cm from 90 W lamp. (●) Methanol, (○) n-propanol, and (□) n-butanol.

the ionising radiation system in Fig. 6.1 are obtained. Under the UV conditions reported, the Trommsdorff peak shifts to higher monomer concentration, presumably reflecting the role of diffusion processes related to the ability of the photosensitiser to diffuse into the backbone polymer and create sites for grafting. With polypropylene as backbone polymer, similar results to the ionising radiation system are found for the UV grafting of styrene in methanol (Table 6.1). In this instance, benzoin ethyl ether (B) is the photoinitiator for the reaction, contrasting with the previous cellulose example where uranyl nitrate was used as photosensitiser (Fig. 6.2). The mechanisms whereby photosensitiser and photoinitiator operate are different. Photosensitisers generally act via excited states whereas photoinitiators directly form free radicals. In the cellulose system (CeOH) studied in Fig. 6.2, uranyl nitrate is the photosensitiser and acts via processes depicted in eqns (20)–(22), when methanol is used as solvent.

$$UO_2^{2+} \xrightarrow{h\nu} [UO_2^{2+}]^* \tag{20}$$

$$[UO_2^{2+}]^* + CH_3OH \longrightarrow CH_3O^{\cdot} + H^+ + UO_2^+ \tag{21}$$

$$[UO_2^{2+}]^* + CeOH \longrightarrow UO_2^+ + H^+ + CeO^{\cdot} \tag{22}$$

By contrast, photoinitiators such as B act via eqn (23) through the direct formation of free radicals.

$$\text{Ph-CO-CH(OEt)-Ph} \xrightarrow{h\nu} \text{Ph-CO}^{\cdot} + {}^{\cdot}\text{CH(OEt)-Ph} \tag{23}$$

The radicals, once formed, can then directly abstract hydrogen atoms from the backbone polymer leading to grafting sites. Thus, it is seen that photoinitiators and photosensitisers can be used in the same grafting reaction in a manner similar to that used in curing.

Grafting with pre-irradiation technique

The type of grafting reactivity found with the pre-irradiation technique is shown in Table 6.2 for styrene in methanol to cellulose. Cellulose is ideal for this type of work since the trapped radicals formed by irradiation can possess relatively long lifetimes, up to years in some instances.[26] As is to be expected, if the pre-irradiation dose is increased, the subsequent pre-irradiation grafting yields are also improved since the concentration of trapped radicals is increased. Pre-irradiation grafting yields in the system reported in Table 6.2 are also higher at the higher monomer concentrations.[27] These results indicate that, because homopolymer occurs as a competing reaction, styrene monomer is lost from the grafting system at low monomer concentrations by precipitation of low-molecular-weight oligomers in the methanol solvent. Inclusion of homopolymer inhibitors such as copper or iron salts in the monomer solution

Table 6.2
Grafting of styrene in methanol to cellulose using the pre-irradiation technique with ionising radiation[a]

Styrene (% v/v)	Graft (%) at a dose (Gy) of	
	5.0×10^3	1.0×10^4
10	3	4
15	5	6
20	6	8
25	6	11
30	6	18
35	10	22
40	51	84
45	46	76

[a]Cellulose Whatman 41 paper irradiated in ^{60}cobalt source at dose rate of 3.3×10^2 Gy/h followed by reaction for 24 h in dark at 20°C.

helps to overcome this problem in a manner similar to that discussed for the mutual technique (see the section entitled 'UV grafting of monomers'—p. 266).

PHYSICAL PARAMETERS AFFECTING GRAFTING YIELD

The essential physical parameters which affect the grafting yields in the mutual method, which is of most relevance to curing work, are radiation conditions such as dose and dose rate (for ionising radiation), intensity of light absorbed (for UV) and the effect of air and temperature. For the ionising radiation work and its relevance to EB conditions, dose rate is particularly important since most data in grafting by this technique have utilised ^{60}cobalt and similar sources. These results can be readily extrapolated to the effects expected from irradiation at the much higher dose rate conditions delivered by EB machines. The same situation applies to UV, where low-intensity sources are usually used to initiate grafting, compared with the higher powered (up to 600 W/in (~240 W/cm)) lamps currently available for curing.

Dose and dose rate with ionising radiation

The effect of dose and dose rate for the grafting of styrene in methanol to cellulose is shown in Table 6.3. At the lowest dose rate (1·0 kGy/h) and total dose (10·0 kGy), a Trommsdorff peak in grafting occurs at 40% monomer concentration. As the total dose is increased to 50 kGy at constant dose rate, the shape of the grafting profile changes, with the Trommsdorff peak shifting to monomer concentrations above 80%. The grafting yield is also increased with increasing dose at the higher monomer concentrations. As the dose rate is increased by an order of magnitude, from 1·0 to 10 kGy/h, at a constant dose

Table 6.3
Effect of dose and dose rate on grafting of styrene in methanol to cellulose using ionising radiation[a]

Dose rate (Gy/h)	Dose (Gy)	Vacuum (V) or air (A)	Graft (%) at monomer concentration (% v/v)			
			20	40	60	80
1.0×10^3	1.0×10^4	V	52	105	102	94
		A	51	123	91	98
	5.0×10^4	V	69	109	173	220
		A	76	113	184	250
	1.0×10^5	V	61	121	166	247
		A	66	115	179	231
2.5×10^3	1.0×10^4	A	20	49	61	68
	5.0×10^4	A	52	85	115	155
	1.0×10^5	A	55	90	143	193
5.0×10^3	1.0×10^4	A	9	21	40	55
	5.0×10^4	A	24	51	68	100
	1.0×10^5	A	44	76	159	201
7.5×10^3	1.0×10^4	A	6	16	29	50
	5.0×10^4	A	17	42	55	76
	1.0×10^5	A	35	73	81	107
1.0×10^4	1.0×10^4	V	4	13	29	38
		A	3	10	16	32
	5.0×10^4	V	12	30	47	63
		A	15	34	51	68
	1.0×10^5	V	31	70	80	105
		A	32	57	85	101

[a]Irradiation in spent fuel element facility at the Australian Atomic Energy Commission.

of 10 kGy, a dramatic effect on grafting yield is observed. Yields are significantly reduced at all monomer concentrations studied with the Trommsdorff peak shifting to greater than 80% monomer concentration. If the dose rate is maintained at 10 kGy/h and the total dose increased from 10 to 100 kGy the expected uniform increase in grafting yield at all monomer concentrations studied is observed.

Similar dose rate effects are found when styrene in methanol is grafted to synthetic backbone polymers such as polyethylene (Table 6.4). At constant total dose, yields are highest at the lowest dose rate (100 Gy/h) with a large Trommsdorff peak occurring at 50% monomer concentration. This peak shifts to greater than 70% monomer concentration at the highest dose rate (750 Gy/h). Overall, when these ^{60}cobalt data are extrapolated to higher dose rate EB conditions, grafting yields/unit dose are significantly reduced. However, to compensate for this reduction, because the EB dose rate is so much higher than with ^{60}cobalt, the radiation dose can be easily increased to bring the yield up to the required level in very short processing time. Further, at the

Table 6.4
Effect of dose rate on grafting of styrene in methanol to polyethylene using ionising radiation[a]

Styrene (% v/v)	Graft (%) at dose rate (Gy/h) of		
	1.0×10^2	2.1×10^2	7.5×10^2
20	24	24	9
30	61	48	18
40	51	92	27
50	409	216	39
60	—	196	43
70	223	159	53
80	—	130	51

[a]Total dose of 2.3×10^4 Gy; film thickness 0·12 mm (Union Carbide); irradiated with ^{60}cobalt.

highest dose rate reported in Table 6.4 with ^{60}cobalt (7.5×10^2 rad/h), the grafting yield increases with increasing monomer concentration reaching a maximum in the highest styrene solutions studied (70–80%). With EB, and its high dose rates, reasonable grafting yields can be achieved using essentially neat monomer or monomer containing small percentages of solvent (5–10%) and a monomer impregnation technique (Garnett, J. L., et al., unpublished). The similarity in application between such a grafting process with systems currently in use in curing is obvious, the ultimate being achieved when monomer and/or oligomer, which is being cured in a solvent-free coating, is all grafted. This concept of optimising concurrent grafting during curing is critical for the commercial viability of a number of industrial EB curing applications, i.e. loss of significant amount (20–30%) of monomer or oligomer from the coating after curing and subsequent treatments or use, cannot be tolerated since the final properties of the finished product can be impaired (Garnett, J. L., et al., unpublished).

Intensity of light with UV
Analogous to the ionising radiation work, grafting with UV is proportional to the intensity of the incident radiation, as the data in Table 6.5 show for the photografting of styrene in methanol to cellulose. The principle is applicable to systems that utilise either photoinitiators (e.g. benzoin ethyl ether, B) or photosensitisers (e.g. uranyl nitrate, UN). The experimental system used for the reported systems was a rotating rack holding the tubes containing the styrene in methanol and cellulose. A 90 watt medium-pressure mercury vapour lamp was positioned at the centre of the rack and the intensity of the light was varied by locating the samples at different distances from the lamp.

Air, temperature
Under certain experimental conditions, the presence of air has an inhibitory effect on grafting in both ionising radiation and UV systems.[2,4,26,28] The results

Table 6.5
Effect of intensity of UV on photografting of styrene in methanol to cellulose[a]

Photoiniator/sensitiser[b]	Graft (%) at distance (cm) from UV lamp	
	24 cm	12 cm
Nil	0·0	0·0
Uranyl nitrate	3·4	10·9
ADS[c]	1·8	3·7
Biacetyl	3·8	11·0[d]
Benzoin ethyl ether	1·4	5·6

[a] Solutions of styrene (30% v/v) in methanol in Pyrex tubes with cellulose (Whatman 41), irradiated 3 h from 90 W UV lamp.
[b] Concentration (1% w/v).
[c] ADS—anthraquinone-2,6-disulphonic acid, disodium salt.
[d] Graft (0·0%) with UV excluded.

of grafting styrene to cellulose using ^{60}cobalt (Table 6.3) indicate that air can affect the intensity of the Trommsdorff peak and also the monomer concentration at which the peak is observed (Fig. 6.3). This is particularly evident at low dose rates. Similar observations can be made for the analogous photografting data (Table 6.6) where the strong peak in grafting, in tubes sealed under evacuated conditions at 20% monomer concentration, is strongly diminished in the presence of air.

If the temperature is increased in a typical mutual radiation grafting system, the grafting yield generally increases.[4,26] For the grafting of styrene in methanol to cellulose using ^{60}cobalt, the apparent overall activation energy for grafting at monomer concentrations of 80, 60, 40 and 20% styrene are 22·6, 23·8, 41·8 and 52·3 kJ/mol, respectively[26] (Fig. 6.4). A similar increase in grafting yield with temperature is found when polyethylene is used instead of cellulose as backbone polymer for the same monomer–solvent system.[29]

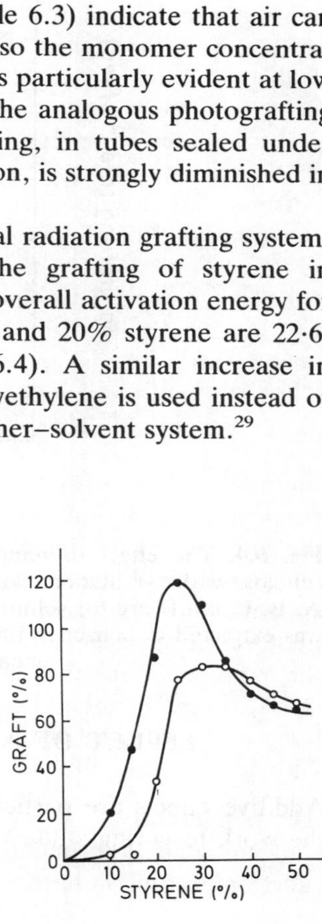

Fig. 6.3. Effect of oxygen on Trommsdorff peak in grafting styrene to cellulose in methanol using gamma radiation; dose rate $6·8 \times 10$ Gy/h, dose $1·12 \times 10^3$ Gy. (●) Vacuum, and (○) air.

Table 6.6
Effect of air on photografting styrene in methanol to cellulose

Styrene (% v/v)	Graft (%)	
	Air[a]	Vacuum[b]
20	13	30
40	28	19
60	34	33
80	53	42
90	64	66

[a] Solutions contained UN (1% w/v) in stoppered tubes; irradiated for 24 h at 24 cm from 90 W medium-pressure UV lamp.
[b] Same conditions as in footnote a except tubes sealed at 10^{-2} torr (1·33 Pa).

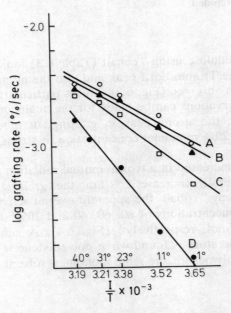

Fig. 6.4. The effect of temperature on grafting yields for styrene in methanol to cellulose with γ-radiation; dose rate $8·3 \times 10$ Gy/h, dose $2·0 \times 10^3$ Gy. Arrhenius plots A, B, C and D are for solutions of 80, 60, 40 and 20% styrene, respectively. Cellulose was extracted in benzene (100 h) then chloroform (48 h) to constant weight. Data are mean values of five determinations.

EFFECT OF ADDITIVES ON GRAFTING YIELDS

Additive effects are particularly important in grafting when the relevance of the work to curing studies is considered. In curing, a variety of additives are utilised to improve a range of properties in the finished product, such as gloss, flow, slip and adhesion. The effect of these additives on concurrent grafting during curing is significant. In this section, the role of additives including solvents in grafting by ionising radiation and UV will be discussed. Subsequently, the implication of the data to curing will be considered. Although curing systems are essentially solvent-free, in some instances it is beneficial to include small percentages of solvent to improve the actual application of the coating. Where high pigment or filler loadings are required, such as EB magnetic-tape work, coatings may contain high levels of solvent to permit uniform application of the material, the solvent being removed prior to EB curing. Even in those curing systems which are solvent-free the present solvent work will be of significance since the functionality of the solvent indicates the type of functional group, which if present in the oligomer/monomer mixture, may enhance grafting during cure.

Solvent

Solvents not only influence the yield in grafting reactions, they can also affect the properties of the finished products, i.e. whether the material is grafted predominantly on the surface or within the bulk of the backbone polymer. Cellulose is a typical backbone polymer whose behaviour illustrates the role of solvents in these reactions. The data in Fig. 6.1 for the effect of alcohol structure on styrene grafting to cellulose using ionising radiation is typical. The dramatic effect of chain length of these solvents on grafting yield is obvious. There is a progressive decrease in grafting yield with increasing chain length and degree of branching of the alcohol, grafting being most effective in methanol and being severely lowered with butanol. The size of the alcohol molecule is thus important in these processes, a small molecule such as methanol not only being capable of swelling the trunk polymer but also possessing the additional advantage of being miscible in all proportions with styrene monomer, thus simplifying access to grafting sites in the cellulose trunk polymer. By contrast, butanol is a relatively poor swelling solvent for cellulose and grafting is correspondingly low. However, grafting in butanol is essentially a surface phenomenon only, thus this solvent can be useful for preparing such surface-grafted copolymers, longer irradiation times than with methanol being required to obtain significant grafting yields. The longer-chain alcohols above butanol are progressively poorer as solvents for this grafting reaction;[26] however, if such alcohols are mixed with a lower-molecular-weight active solvent such as methanol, enhancement in grafting over the whole monomer concentration range is observed, exemplified by the methanol–octanol system for radiation grafting styrene to cellulose (Fig. 6.5).

When UV is used as radiation source for the same grafting system involving styrene and the alcohols to cellulose, analogous results to those obtained with ionising radiation are observed. Thus the low-molecular-weight alcohols like methanol are most effective (Fig. 6.2) and mixing of methanol with a poor grafting solvent such as n-octanol (1:1) leads to enhancement in grafting yield

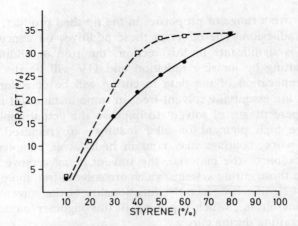

Fig. 6.5. γ-Grafting of styrene to cellulose using (●) Methanol and (□) methanol/octanol (1:1) as solvents; dose rate 8.0×10^2 Gy/h, dose 2.0×10^3 Gy.

at all monomer concentrations studied (Fig. 6.6). When the ratio of methanol to n-octanol is varied in the system a Trommsdorff peak in grafting is observed at 66% n-octanol in methanol, this peak being virtually independent over the monomer range (40–80% v/v) studied (Fig. 6.7).

If cellulose is replaced with a synthetic backbone polymer like polypropylene, the principle of wetting and swelling by the solvent still applies. However, in this instance, the low-molecular-weight alcohols do not swell the backbone polymer, the swelling is caused by the styrene monomer and this leads to facile grafting. In contrast, the higher-molecular-weight alcohols such as n-octanol are more compatible with polystyrene and grafting yields with

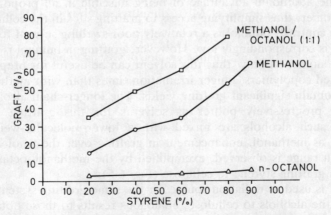

Fig. 6.6. Photosensitised grafting of styrene in methanol/octanol mixtures to cellulose. UN (1% w/v) as photosensitiser and irradiated 24 h at 24 cm from 90 W medium-pressure UV lamp.

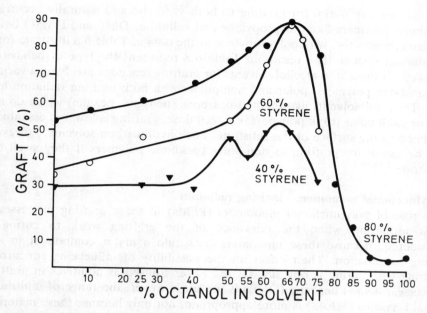

Fig. 6.7. Photosensitised grafting of styrene to cellulose in various octanol/methanol mixtures. Conditions as in Fig. 6.6.

polypropylene are enhanced with these solvents (Table 6.7). In these data, it is also significant that the Trommsdorff peak occurs at 20% monomer concentration with n-butanol and 30% with n-octanol, the yield in both of these solvents being very much higher than with the lower-molecular-weight analogues.

When non-alcohol solvents are used for the above grafting processes, the principles developed for the alcohols are still relevant. Thus, polar solvents such as dimethyl formamide (DMF), dimethyl sulphoxide (DMSO), acetone

Table 6.7
Effect of alcohol on radiation grafting of styrene to polypropylene[a]

Styrene (% v/v)	Graft (%)			
	Methanol	Ethanol	n-Butanol	n-Octanol
20	29	44	123	49
30	94	89	74	107
40	50	65	34	68
50	37	47	40	42
60	36	36	33	32
70	35	32	29	29
80	29	30	28	26
100	22	22	22	22

[a] Film (0·12 mm, ex-Shell); total dose $3·0 \times 10^3$ Gy, dose rate $4·0 \times 10^2$ Gy/h.

and dioxane are useful for grafting to both synthetic and naturally occurring backbone polymers like polypropylene and cellulose, DMF and DMSO being particularly effective with polypropylene as the data in Table 6.8 illustrate for a γ-radiation system. The results in Table 6.8 represent the type of behaviour expected of these non-alcohol solvents for grafting reactions involving a variety of backbone polymers, polar and non-polar, with both ionising radiation and UV. Thus, for solvents like the hydrocarbons (benzene, hexane) which do not wet or swell polar trunk polymers, e.g. cellulose, grafting is low and essentially confined to the surface of the substrate. Such hydrocarbon solvents, however, can be useful for grafting to non-polar backbone polymers if they swell the substrate.[2,28,30-35]

Polyfunctional monomers—ionising radiation

The role of polyfunctional monomers (PFMs) in these grafting processes is most important when the relevance of the grafting work to curing is considered, because these monomers constitute a major component in any curing formulation. Their effect on the possibility of influencing concurrent grafting with curing is obvious. Of the PFMs utilised as additives in grafting processes, divinyl benzene (DVB) is useful; however, the range of multifunctional acrylates (MFAs) is more appropriate, not only because these materials are extensively used in curing but because of the variety of monomers of different structure available in this group.[31,36] The PFMs, especially the MFAs, are particularly valuable in these radiation grafting reactions, since when used in additive amounts (1% v/v), they can increase grafting yields significantly. Using trimethylolpropane triacrylate (TMPTA) as reference MFA, the enhancement effect of this monomer for the grafting of styrene in methanol to cellulose and polypropylene using ionising radiation is shown in Table 6.9. With cellulose, enhancement at all monomer concentrations studied is observed, the Trommsdorff peak occurring at 40% monomer concentration. Data

Table 6.8
Radiation grafting of styrene to polypropylene using solvents other than alcohols[a]

Styrene (% v/v)	Graft (%)			
	DMF	DMSO	Acetone	1,4-Dioxane
20	24	11	13	6
30	40	29	20	12
40	43	66	24	15
50	44	61	25	17
60	40	56	22	19
70	39	42	24	21
80	33	31	25	23
100	22	22	22	22

[a]Conditions as in Table 6.7.

Table 6.9
Effect of TMPTA as additive for enhancing radiation grafting of styrene in methanol to cellulose and polypropylene

Styrene (% v/v)	Graft (%)					
	Cellulose				Polypropylene	
	No additive[a]	TMPTA[a]	No additive[b]	TMPTA[b]	No additive[c]	TMPTA[c]
10	20	23	—	—	—	—
20	60	88	30	78	17	51
40	65	205	45	145	62	179
60	55	145	38	96	35	108
80	60	164	31	72	24	68

[a] Whatman 41 paper; dose rate $5 \cdot 0 \times 10^3$ Gy/h to $5 \cdot 0 \times 10^3$ Gy. TMPTA (1% v/v).
[b] Whatman 41 paper; dose rate $5 \cdot 0 \times 10^2$ Gy/h to $2 \cdot 0 \times 10^3$ Gy.
[c] Film (0·12 mm); dose rate $5 \cdot 0 \times 10^2$ Gy/h to $2 \cdot 0 \times 10^3$ Gy.

for two different dose rates and total doses are described in order to indicate the magnitude of the enhancement effect, which is consistent in both cellulose radiation systems. Similar results are obtained with polypropylene, the position of the Trommsdorff peak occurring at the same monomer concentration as with cellulose (Table 6.9).

Polyfunctional monomers—UV
For the corresponding MFA enhancement in UV processes, tripropyleneglycol diacrylate (TPGDA) is used as representative monomer as this is the preferred MFA for many curing formulations. TMPTA is more reactive in curing; however, the product from such materials is generally brittle whereas with TPGDA a more flexible film is obtained. The results from TPGDA for the photografting of styrene in methanol to both cellulose and polypropylene are shown in Table 6.10 and are consistent with the ionising radiation data previously reported for TMPTA (Refs 18 and 37, and Dworjanyn, P. A., *et al.*, unpublished). Enhancement in photografting is observed at all monomer concentrations studied for both cellulose and polypropylene, the Trommsdorff peak again occurring at 40% monomer.

A wide range of MFAs is available for curing reactions, each MFA possessing certain unique properties to be imparted to the cured film. Each of the MFAs exhibit different rates of cure and their corresponding reactivities in grafting processes are thus related. Using grafting with UV as representative system, because UV curing is currently a bigger field than EB curing, the effect of MFA structure on photografting styrene in methanol to polypropylene is shown in Table 6.11. Of the MFAs reported, 1,6-hexanediol diacrylate (HDDA) is interesting since this monomer has excellent properties in curing formulations except for a tendency to possess a skin irritancy problem which

Table 6.10
Effect of TPGDA as additive for enhancing photografting of styrene in methanol to cellulose and polypropylene

Styrene (% v/v)	Graft (%)				
	Cellulose		Polypropylene		
	B[a]	B + TP[b]	B	B + TP[b]	B + TP[c]
20	16	80	12	76	79
40	30	111	47	289	113
60	28	72	22	101	52
80	29	33	17	63	39

[a]B (1% w/v).
[b]TP—TPGDA (1% v/v); cellulose Whatman 41 paper, polypropylene 0·12 mm; irradiated 6 h at 24 cm from 90 W lamp.
[c]TP—TPGDA (0·2% v/v); other conditions as in footnotes a and b.

limits its use. TMPTA is the most active of the MFAs in this grafting system followed by HDDA and propoxylated glyceryl triacrylate (PGTA). For comparison purposes, a multifunctional methacrylate (MFMA) is included in the data, the monomer chosen, namely trimethylol propane trimethacrylate (TMPTMA) being used extensively in radiation cross-linking work. For analogous curing studies, the methacrylates are usually slower to cure than the corresponding acrylates, thus MFAs are generally preferred in curing applications. The data in Table 6.11 indicate that TMPTA is the most effective of all MFAs studied in the table for enhancing the photografting of styrene in methanol to polypropylene, particularly at the Trommsdorff peak. However, TMPTMA is a marginally better additive than TMPTA for enhancing yields in this same grafting reaction.

Table 6.11
Effect of structure of multifunctional acrylate and methacrylate as additives in photografting styrene in methanol to polypropylene[a]

Styrene (% v/v)	Graft (%)					
	B	B + T	B + TM	B + P	B + TP	B + H
20	5	100	95	50	17	53
30	24	297	282	113	99	81
40	45	496	513	289	336	370
50	31	283	285	131	167	315
60	22	281	180	98	118	113

[a]Irradiated 8 h at 24 cm from 90 W lamp at 20°C; additives (1% v/v); B (1% w/v); T—TMPTA; TM—TMPTMA; P—PGTA; TP—TPGDA; H—HDDA.

Table 6.12
Effect of acid and salt as additives in the simultaneous radiation grafting of styrene in methanol to cellulose[a]

Styrene (% v/v)	Graft (%)		
	No additive	Sulphuric acid (0·1 M)	Lithium perchlorate (0·2 M)
15	32	34	54
20	66	120	155
25	106	153	159
30	112	95	96
35	110	60	80
40	82	52	72

[a]Irradiated at $3·3 \times 10^2$ Gy/h to total dose of $5·0 \times 10^3$ Gy at 24°C.

Mineral acids and inorganic salts—ionising radiation

With certain concentrations of particular mineral acids, grafting enhancement can be observed for a wide variety of monomer/polymer systems in the presence of ionising radiation.[2] Similarly, the inclusion of selected inorganic salts has beneficial effects on grafting yields. Of particular relevance to curing are the acid and salt effects found in the mutual, pre-irradiation and post-irradiation grafting techniques. In the mutual method, typical of the acid effects observed are those found for the radiation grafting of styrene in methanol to cellulose using [60]cobalt initiation (Table 6.12). In this system, predominant grafting enhancement is observed at low monomer concentrations with a Trommsdorff peak shifting from 30% monomer (no acid) to 25% monomer (0·1 M sulphuric acid). Similar results are obtained with polyethylene as trunk polymer (Table 6.13), the respective Trommsdorff peaks occurring at

Table 6.13
Effect of acid and salt as additives in the simultaneous radiation grafting of styrene in methanol to polyethylene[a]

Styrene (% v/v)	Graft (%)		
	No additive	Sulphuric acid (0·1 M)	Lithium perchlorate (0·2 M)
15	31	32	44
20	64	70	81
25	103	148	192
30	187	240	196
35	193	212	140
40	150	157	114

[a]Irradiated at $3·3 \times 10^2$ Gy/h to total dose of $2·0 \times 10^3$ Gy at 24°C.

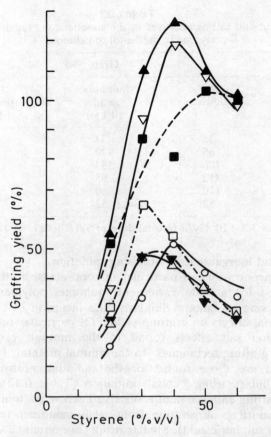

Fig. 6.8. Effect of acid additives on γ-grafting of styrene in methanol to polyethylene; dose rate $8 \cdot 0 \times 10^2$ Gy/h, dose $2 \cdot 0 \times 10^3$ Gy; all acid concentrations 0·1 M. (○) Without acid, (△) acetic acid, (▼) formic acid, (□) hydrochloric acid, (■) perchloric acid, (▽) sulphuric acid, and (▲) nitric acid.

35% monomer concentration (no acid) and 30% monomer (0·1 M sulphuric acid). Thus it can be seen that acid, not only increases the grafting yield, it also leads to an enhancement in the intensity of the peak under the conditions corresponding to the Trommsdorff effect. Strong acids are particularly effective in this enhancement process, as the data in Fig. 6.8 show for the comparison of the enhancement effects of different acids on the grafting yield of styrene in methanol to polyethylene. The results indicate that ionic strength in the solution can be an important property for explaining the role of acid.

When mineral acids are replaced with inorganic salts, such as lithium perchlorate, increases in radiation grafting yield are also observed.[32] The data in Tables 6.12 and 6.13 show that acid and salt effects are similar, although for cellulose the salt is marginally more effective than acid in the enhancement

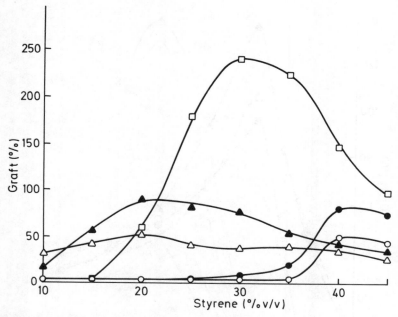

Fig. 6.9. Pre-irradiation γ-grafting of styrene in methanol to cellulose with acid and salt additives. Dose rate 3.3×10^2 Gy/h, dose 5.0×10^5 Gy: (○) no additive, (△) sulphuric acid (0.1 M), (□) lithium perchlorate (0.2 M). Dose rate 3.3×10^2 Gy/h, dose 1.0×10^6 Gy: (●) no additive, (▲) sulphuric acid (0.1 M).

process, whereas the reverse occurs when polyethylene is used as the trunk polymer under the same radiation grafting conditions. Inclusion of salts such as lithium perchlorate also increases the grafting yield at the Trommsdorff peak. In earlier preliminary studies, lithium perchlorate was shown to be the best of the range of inorganic salts studied for this work; however, other salts can be as effective in grafting enhancement under the appropriate radiation conditions.

With pre-irradiation and post-irradiation grafting techniques analogous acid and salt effects are also found.[34,35] For the pre-irradiation grafting of styrene in methanol to cellulose, lithium perchlorate is markedly more effective than acid at monomer concentrations greater than 20% (Fig. 6.9). Inclusion of salt leads to a large induced Trommsdorff peak at 30% styrene concentration, a result of value in a preparative context. In post-irradiation grafting of styrene in methanol to polypropylene, significant enhancement in grafting yield is observed at certain monomer concentrations as a result of standing at room temperature after the completion of the irradiation (Fig. 6.10). Inclusion of acid enhances post-irradiation grafting at monomer concentrations up to 50%, the additive being particularly effective for the styrene solution corresponding to the Trommsdorff peak.

Acid effects in radiation grafting are also influenced by the type of solvent

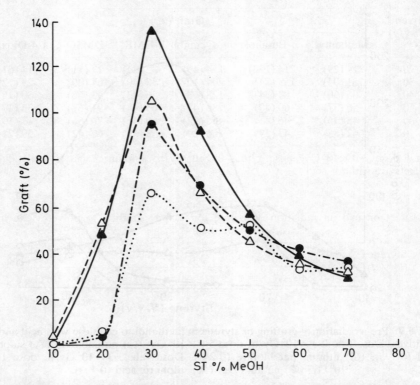

Fig. 6.10. Post-irradiation γ-grafting of styrene in methanol to polypropylene in presence of acid additive; dose rate 3.9×10^2 Gy/h, dose 2.0×10^3 Gy. (○) No additive, (△) sulphuric acid (0.1 M): 0 h after irradiation. (●) No additive, (▲) sulphuric acid (0.1 M): 48 h after irradiation.

used and the physical parameters such as dose rate, air and temperature. Solvent is particularly important as shown by the data in Table 6.14 for the radiation grafting of styrene to polypropylene. Of the solvents reported, the alcohols are most effective with acid for enhancing the grafting yield.

As the dose rate is increased, at constant total dose, the grafting yield is reduced for a particular monomer concentration. Inclusion of mineral acid accentuates this trend, as the data show for the grafting of styrene in methanol to polyethylene (Table 6.15).

The presence of air retards grafting under certain experimental conditions (see the section entitled 'Air temperature'—p. 274), this effect being accentuated in acid solution as the data in Table 6.16 show for the grafting of styrene in methanol to polypropylene, particularly at the lowest monomer concentration studied.

Addition of acid to the grafting solution also has a marked effect as the temperature of grafting is increased (Table 6.17). Using both polyethylene and cellulose as representative backbone polymers for the grafting of styrene in

Table 6.14
Acid enhancement for radiation grafting styrene in various solvents to polypropylene[a]

Styrene (% v/v)	Graft (%)					
	Methanol	n-Butanol	n-Octanol	DMF	DMSO	1,4-Dioxane
20	214 (29)	214 (123)	154 (49)	32 (24)	17 (11)	14 (6)
30	151 (94)	159 (74)	204 (107)	58 (40)	64 (29)	23 (12)
40	86 (50)	83 (34)	128 (68)	56 (43)	93 (66)	30 (15)
50	56 (37)	60 (40)	86 (42)	63 (44)	91 (61)	33 (17)
60	48 (36)	46 (33)	66 (32)	60 (40)	70 (56)	17 (19)
70	42 (35)	43 (29)	66 (29)	53 (39)	62 (42)	39 (21)

[a]Total dose 3.0×10^3 Gy, dose rate 4.0×10^2 Gy/h; sulphuric acid (0.2 M); data in brackets are without acid.

Table 6.15
Effect of acid on radiation dose rate for styrene grafting to polyethylene in methanol[a]

Styrene (% v/v)	Graft (%)					
	1.0×10^2 Gy/h		4.1×10^2 Gy/h		11.2×10^2 Gy/h	
	N	H$^+$	N	H$^+$	N	H$^+$
20	24	32	14	19	7	8
30	61	82	37	51	14	17
40	51	344	76	81	23	27
50	409	543	109	134	25	35
60	—	—	89	119	28	36
70	223	211	89	73	35	37
80	—	—	68	62	35	37

[a]Total dose 2.3×10^4 Gy, film thickness 0.12 mm (ex-Union Carbide); N—neutral; H$^+$—sulphuric acid (0.2 M).

Table 6.16
Effect of air on acid enhancement in radiation grafting to styrene to polypropylene in methanol[a]

Styrene (% v/v)	Graft (%)			
	Air		Vacuum[b]	
	N	H$^+$	N	H$^+$
20	8	15	11	66
40	54	68	59	81
50	39	46	41	52
60	30	31	36	50

[a]Total dose 2.0×10^3 Gy, dose rate 4.0×10^2 Gy/h; N—neutral; H$^+$—sulphuric acid (0.1 M).
[b]Sample tubes evacuated to 10^{-3} torr (0.133 Pa) after three freeze thaw cycles and sealed prior to irradiation.

Table 6.17
Effect of temperature on acid enhancement in radiation grafting of styrene to polyethylene[a] and cellulose[b] in methanol

Substrate	Temperature (°C)	Graft (%)			
		20% Styrene (v/v)		30% styrene (v/v)	
		N	H⁺	N	H⁺
Polyethylene	14	42	44	135	172
	20	64	70	187	240
	29	84	118	244	267
Cellulose	0	17	25	33	56
	15	40	62	85	94
	20	51	70	101	112
	25	60	121	—	—

[a]Total dose $2 \cdot 0 \times 10^3$ Gy, dose rate $3 \cdot 3 \times 10^2$ Gy/h; low-density polyethylene (0·12 mm); N—neutral; H⁺—sulphuric acid (0·1 M).
[b]Total dose $5 \cdot 0 \times 10^3$ Gy, dose rate $3 \cdot 3 \times 10^2$ Gy/h; Whatman 41 paper.

methanol, acid enhancement is observed at all monomer concentrations studied and at all temperatures reported in the table. With cellulose, in particular, at the lowest monomer concentration, the intensity of the acid enhancement increases with increasing temperature.

Mineral acids and inorganic salts—UV

In like manner to ionising-radiation systems, photografting processes are also enhanced under certain experimental conditions using acid and inorganic salts as additives. The difference with the UV systems is the necessity to include photoinitiators which are used to accelerate the grafting reaction as the data in Table 6.18 for styrene in methanol to polypropylene show. Without inclusion of the photoinitiator, B, yields are very low; however, addition of acid to such a system leads to finite graft under the UV conditions shown. When B is included in the monomer solution, photografting is appreciable, the yields being further increased at certain monomer concentrations by the addition of mineral acid (Table 6.18). The presence of B induces a Trommsdorff peak at 35% monomer which is strongly enhanced in the presence of acid and shifts to 30% styrene with this additive.

When inorganic salts such as lithium perchlorate replace mineral acid, similar photografting enhancement is achieved. For the photografting of styrene in methanol to polyethylene (Table 6.19), data show increases in yield for all monomer concentrations studied, except for the 40% solution, the largest enhancement occurring at the gel peak. With cellulose as backbone polymer, photografting yields for styrene in methanol are also increased for all monomer concentrations reported (Table 6.19). When the inorganic salt data are compared with acid, both additives enhance photografting to polyethylene in

Table 6.18
Effect of photoinitiator and acid on photografting styrene to polypropylene in methanol[a]

Styrene (% v/v)	Graft (%)			
	Without B		With B (1% w/v)	
	No additive	H$^+$	B	B + H$^+$
20	<2	4·0	7·2	19
25	—	—	—	48
30	<2	5·4	29	65
35	—	—	41	31
40	<2	6·1	39	19
50	<2	7·4	26	18
60	<2	4·0	—	—

[a]Irradiated for 18 h at 30 cm from 90 W lamp at 20°C; B (1% w/v); H$^+$—sulphuric acid (0·2 M).

monomer solutions up to 35% concentration, however, with cellulose, lithium perchlorate is superior with grafting yields increasing in all solutions up to 60%. This observation may reflect the role that acid has in the presence of cellulose. Cellulose is, under certain acid conditions, degraded and therefore the inorganic salt may be the preferred additive for this enhancement work if the objective is to use the technique to prepare cellulose graft copolymers in high yields.

Table 6.19
Effect of acid and lithium perchlorate as additives in photografting styrene in methanol to polyethylene and cellulose

Styrene (% v/v)	Graft (%)					
	Polyethylene[a]			Cellulose[b]		
	B	B + H$^+$	B + L	B	B + H$^+$	B + L
10	—	—	—	5	8	11
15	6	5	9	—	—	—
20	8	14	16	11	23	38
25	10	16	28	—	—	—
30	17	25	31	25	20	41
35	20	35	24	—	—	—
40	19	24	17	32	18	38
60	—	—	—	38	15	40

[a]Irradiated for 10 h at 24 cm from 90 W lamp at 20°C; B (1% w/v); H$^+$—sulphuric acid (0·05 M); L—LiClO$_4$ (0·1 M).
[b]Irradiated 15 h; H$^+$—sulphuric acid (0·1 M); L—LiClO$_4$ (0·2 M).

Synergistic effect of acids and salts with polyfunctional monomers—ionising radiation

Grafting profiles for the enhancement in radiation grafting of a typical system such as styrene in methanol to polyethylene are different for both acid and MFA additives, thus inclusion of the two additives in the same monomer solution leads to synergistic effects in grafting by ionising radiation (Fig. 6.11). Because of the relevance of this work to curing, TMPTA is used as representative MFA in Fig. 6.11. The synergistic effect is particularly significant at monomer concentrations corresponding to the Trommsdorff peak.[29]

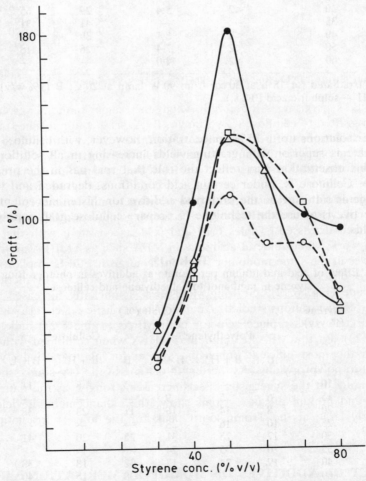

Fig. 6.11. Effect of TMPTA and acid additives on radiation grafting of styrene in methanol to polypropylene; dose rate 4.1×10^2 Gy/h, dose 2.4×10^3 Gy. (○) No additives; (△) sulphuric acid (0.2 M); (□) TMPTA (1% v/v); (●) TMPTA (1% v/v) + sulphuric acid (0.2 M).

Table 6.20
Synergistic effect of TMPTA with lithium nitrate in radiation grafting of styrene to cellulose[a] in methanol

Styrene (% v/v)	Graft (%)							
	Dose 2.0×10^3 Gy				Dose 1.0×10^4 Gy			
	No additive	L	T	L+T	No additive	L	T	L+T
15	8	8	9	10	32	30	135	168
20	11	12	19	26	44	38	141	181
25	14	15	24	22	58	46	228	142
30	19	16	20	26	82	72	351	430
35	28	39	28	42	130	96	370	578
40	21	24	41	67	112	161	392	350
60	27	22	43	52	123	98	415	390

[a]L—lithium nitrate $LiNO_3$ (0.2 M); T—TMPTA (1% v/v); dose rate 5.0×10^2 Gy/h.

Analogous synergism is observed when the same monomer solution is radiation grafted to cellulose (Dworjanyn, P. A., *et al.*, unpublished) using an inorganic salt instead of acid with TMPTA (Table 6.20). For this latter process lithium nitrate is effective in enhancing radiation grafting yields at certain monomer concentrations; however, it is less efficient than lithium perchlorate.

Synergistic effect of acids and salts with polyfunctional monomers—UV
The role of MFAs and MFMAs in accelerating photografting processes has already been discussed (Tables 6.10 and 6.11). Consistent with the ionising radiation work, when mineral acid and an MFA such as TMPTA are included as additives in the same monomer solution, synergistic effects in photografting yields are observed for a typical system such as styrene in methanol to polyethylene (Fig. 6.12). Grafting yields are significantly increased at most monomer concentrations studied. Analogous synergistic effects are also found when inorganic salts replace acid for both MFAs (Table 6.21) and MFMAs (Table 6.22). Of the MFAs examined, TMPTA when combined with lithium nitrate is the most efficient enhancing system for photografting styrene in methanol to polypropylene. As the length of time in the UV source increases, the magnitude of the synergistic effect increases strongly, as the data for the TMPTA and lithium nitrate systems show, the enhancement in yield being particularly large at the Trommsdorff peak for the longest irradiation time (Fig. 6.13).

EFFECT OF ADDITIVES ON HOMOPOLYMERISATION YIELDS

As previously discussed in the section entitled 'Homopolymer during UV/EB grafting'—see p. 268, a problem with radiation grafting generally, and the

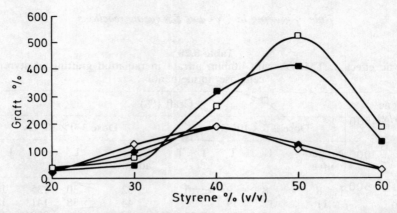

Fig. 6.12. Synergistic effects of TMPTA and acid as additives in photografting of styrene in methanol to polyethylene. All samples contain B (1% w/v); irradiated for 24 h at 30 cm from 90 W lamp. (♦) No additives; (■) TMPTA (1% v/v); (◇) sulphuric acid (0·2 M); (□) TMPTA (1% v/v) + sulphuric acid (0·2 M).

Table 6.21
Synergistic effect of multifunctional acrylates with lithium nitrate in photografting styrene in methanol to polypropylene[a]

Styrene (% v/v)	Graft (%)							
	T	T+L	P	P+L	TP	TP+L	H	H+L
20	100	126	50	55	17	94	53	115
30	297	529	113	147	99	217	81	239
40	496	337	289	146	336	187	370	180
50	283	221	131	79	167	98	315	91
60	281	81	98	46	118	57	113	67

[a]Irradiated for 8 h at 24 cm from 90 W lamp at 24°C; B in all solutions (1% w/v); all monomers (1% v/v); T—TMPTA; L—lithium nitrate $LiClO_4$ (0·2 M); P—PGTA; TP—TPGDA; H—HDDA.

Table 6.22
Synergistic effect of multifunctional methacrylates with lithium nitrate in photografting styrene in methanol to polypropylene[a]

Styrene (% v/v)	Graft (%)							
	TM	TM+L	EM	EM+L	TEM	TEM+L	PM	PM+L
20	95	111	30	95	25	105	30	77
30	282	376	66	225	61	296	70	178
40	513	274	214	192	217	141	169	104
50	285	144	211	131	132	101	185	94
60	180	93	157	76	104	64	94	53

[a]Conditions as in footnote a of Table 6.21; TM—TMPTMA; EM—EGDMA (ethylene glycol dimethacrylate); TEM—TEGDMA (triethylene glycol dimethacrylate); PM—PEGDMA (polyethylene glycol dimethacrylate).

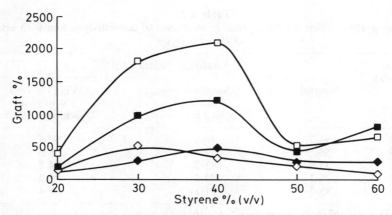

Fig. 6.13. Role of UV exposure in synergistic effects of acrylate monomer and inorganic salt in photografting of styrene in methanol to polypropylene; all samples contain B (1% w/v); other conditions as in Fig. 6.12. After 8 h UV exposure: (◆) TMPTA (1% v/v), (◇) TMPTA (1% v/v) + lithium nitrate (1% w/v). After 16 h UV exposure: (■) TMPTA (1% v/v), (□) TMPTA (1% v/v) + lithium nitrate (1% w/v). For control data with photoinitiator only, grafting yields at 40% styrene concentration were 45% (8 h) and 116% (16 h).

mutual technique specifically, is that homopolymer is a potential detrimental competing reaction. The same situation pertains to photografting, where photoinduced homopolymer consumes appreciable amounts of the monomer required for grafting. The role of additives in grafting reactions with respect to their effect on the homopolymer is thus most important, particularly if they can reduce homopolymer yields and subsequently improve grafting efficiency.

Homopolymerisation in the presence of acid and polyfunctional monomers—ionising radiation

Acid and PFM additives already discussed increase both grafting yields and homopolymer formation; however, grafting efficiency, defined as the ratio of graft to graft-plus-homopolymer, is also improved in the presence of both acid and PFM and thus grafting is preferentially favoured under these conditions at certain monomer concentrations. The data in Table 6.23 for the grafting of styrene in methanol to polyethylene in the presence of acid and DVB as additives confirm these conclusions. In the presence of acid alone the grafting efficiency, especially at 40% styrene, is significantly improved, whilst with DVB, in the presence of sulphuric acid, the grafting efficiency is increased even further at 50% styrene concentration. Although the grafting efficiency is improved in the presence of these additives at particular concentrations, i.e. grafting is preferentially enhanced, homopolymer yields still remain high and need to be reduced. Previous authors have used salts of copper and iron, also Mohr's salt ($FeSO_4(NH_4)_2SO_4 \cdot 6H_2O$), to suppress homopolymer in the simultaneous technique;[8,21] however, such salts usually reduce grafting and homopolymer concurrently, the latter more efficiently. If acid is added to a grafting

Table 6.23
Radiation grafting efficiency for styrene in methanol to polyethylene film with acid and DVB additives[a]

Styrene (% v/v)	Grafting efficiency (%)[b]		
	Neutral	Sulphuric acid (0·2 M)	DVB (1% v/v) + sulphuric acid (0·2 M)
20	51·7	49·6	54·5
30	56·2	59·4	58·8
40	73·9	83·0	74·2
50	75·1	79·4	85·2
70	45·7	41·2	65·4

[a] Dose rate $1·0 \times 10^2$ Gy/h, total dose $2·4 \times 10^3$ Gy.
[b] Ratio of graft/(graft + homopolymer) × 100.

system of styrene in methanol to polypropylene with copper sulphate already present, relatively high grafting yields are obtained under conditions where homopolymer is strongly suppressed (Table 6.24). More importantly, at certain monomer concentrations studied, there is even a synergistic effect in grafting efficiency with both cationic salt and acid, the predominant effect being observed at the 30% monomer concentration. This synergistic property is independent of the cationic salt structure in terms of the three salts studied (i.e. cupric sulphate, ferrous sulphate and Mohr's salt) although the copper salt appears to be marginally best in enhancement (Table 6.25). Optimum grafting yields for specific monomer concentrations occur at cationic salt concentrations of between 5×10^{-4} M and 1×10^{-2} M whilst acid concentrations of approxim-

Table 6.24
Effect of copper sulphate and sulphuric acid on radiation grafting of styrene to polypropylene[a]

Styrene (% v/v)	Graft (%)			
	Control	Cu^{2+}	H^+	$Cu^{2+} + H^+$
20	10·6	4·6	14	11·5
30	42	25	40	64
40	67	41	108	94
50	64	32	73	41
60	51	27	40	45
70	49	42	44	29

[a] Radiation dose of $2·5 \times 10^3$ Gy at $5·2 \times 10^2$ Gy/h; Cu^{2+}—copper sulphate (5×10^{-3} M); H^+—sulphuric acid (0·2 M); methanol as solvent; polystyrene precipitation studies[22] also visual observation of the monomer solutions confirms the suppression of homopolymer.

Table 6.25
Comparison of copper sulphate with ferrous sulphate and Mohr's salt in radiation grafting styrene to polypropylene in presence of acid[a]

Styrene (%v/v)	Graft (%)			
	Control	$Cu^{2+} + H^+$	$Fe^{2+} + H^+$	$M + H^+$
20	10.6	11.5	11.9	3.5
30	42	64	56	35
40	67	94	84	91
50	64	41	60	35
60	51	45	28	15
70	49	29	32	39

[a]Radiation grafting conditions as in footnote a of Table 6.24; cationic salts (5×10^{-3} M); Cu^{2+}—copper sulphate; Fe^{2+}—ferrous sulphate; M—Mohr's salt.

ately 10^{-1} M appear to be close to the optimum to maximise grafting yields under these conditions.[22]

When TMPTA replaces acid in the monomer solution, the expected MFA enhancement is observed for the radiation grafting of styrene in methanol to polypropylene (Table 6.26). Addition of copper ions to this system significantly reduces homopolymer yield, at the same time maintaining appreciable grafting yields. The effect of copper ions on the synergistic effect of acid with TMPTA is even more significant. Not only is the homopolymer yield reduced in the presence of copper ions, the grafting yield is actually enhanced at the 30% monomer concentration (Table 6.26). This is an interesting observation since it is one of the few results in the presence of both acid and MFA additives where grafting is not reduced when copper ions are added to the grafting solution to reduce homopolymer formation.

Table 6.26
Effect of multifunctional acrylate, acid and copper sulphate as additives in radiation grafting styrene in methanol to polypropylene[a]

Styrene (% v/v)	Grafting (%)							
	Blank	Cu^{2+}	H^+	$H^+ + Cu^{2+}$	T	$T + H^+$	$T + Cu^{2+}$	$T + H^+ + Cu^{2+}$
20	23	16	38	39	32	50	28	49
30	53	92	111	112	114	198	86	212
40	142	149	233	167	272	269	129	205
50	145	122	97	104	303	256	214	145
60	116	89	78	60	248	160	170	128

[a]Dose rate 5.0×10^2 Gy/h to 2.5×10^3 Gy; T—TMPTA (1% v/v); H^+—sulphuric acid (0.2 M); Cu^{2+}—copper sulphate (5×10^{-3} M).

Homopolymerisation in the presence of acid and polyfunctional monomers—UV

In UV systems, inclusion of acid in the monomer solution, not only enhances grafting yields but also increases homopolymer formation in a manner similar to the ionising radiation work (Table 6.27). Using the photografting of styrene in methanol to cellulose as representative UV system, the data in Table 6.27 show that, in the absence of acid, the grafting efficiency increases with increasing monomer concentration to 90% styrene, this trend being strongly accentuated when nitric acid is included in the monomer solution.

The role of photoinitiator is important in speeding-up UV grafting and the effect of copper and iron ions as well as Mohr's salt on reducing homopolymerisation in the presence of the photoinitiator is relevant, particularly when additives such as acid and MFAs are included in the monomer solution.[22] For a typical UV system, such as the photografting of styrene in methanol to polypropylene, the data in Table 6.28 show that photografting without photoinitiator is low under the UV radiation conditions used. Inclusion of mineral acid in the grafting solution in the absence of photoinitiator leads to an increase in photografting at almost all monomer concentrations studied, particularly with the monomer solution corresponding to the Trommsdorff peak (50%). Addition of copper sulphate to these acidified monomer solutions in the absence of photoinitiator lowers the grafting yield marginally at the lower monomer concentrations up to 50%, except for the monomer solution corresponding to the Trommsdorff peak (35%) where a significant enhancement is observed. In the presence of acid the Trommsdorff peak shifts from 50 to 30% monomer. When photoinitiator is added to the styrene in methanol solutions, an extraordinary effect in photografting with polypropylene is observed (Table 6.29). In the presence of ferrous sulphate and without other additives the graft is virtually eliminated, especially at low monomer concentrations. However, addition of acid to these solutions, leads to a large enhancement in grafting at all monomer concentrations studied. More particularly, an extremely large increase in grafting yield, by almost two orders of

Table 6.27
Acid effect on homopolymer in photografting styrene in methanol to cellulose[a]

Styrene (% v/v)	Graft (%)		Homopolymer (%)	
	Neutral	Nitric acid (1·0 M)	Neutral	Nitric acid (1·0 M)
20	13	20	10	18
40	28	23	18	19
60	34	37	24	24
80	53	77	24	24
90	64	107	34	32

[a] Irradiated 24 h at 24 cm from 90 W lamp at 20°C; using UN (1% w/v).

Table 6.28
Effect of photoinitiator, acid and cationic salt in photografting styrene in methanol to polypropylene[a]

Styrene (% v/v)	Graft (%)			
	Without B			With B
	No additive	H^+	$Cu^{2+} + H^+$	
20	<2	4.0	3.1	7.2
30	<2	5.4	9.0	29
40	<2	6.1	5.0	39
50	<2	7.4	5.1	26
60	<2	4.0	4.3	25
70	<2	1.8	4.0	23

[a] Irradiated for 18 h at 30 cm from 90 W lamp; B (1% w/v); H^+—sulphuric acid (0.2 M); Cu^{2+}—copper sulphate (5×10^{-3} M).

magnitude, is observed at the Trommsdorff peak (30%). A similar result is obtained when copper sulphate or Mohr's salt is used in the monomer solution instead of ferrous sulphate. Not only is the grafting yield improved under these conditions but also the grafting efficiency (Table 6.30), which demonstrates that grafting is preferentially favoured over homopolymerisation in these grafting systems, i.e. in the presence of both cationic salts and acids. This trend is particularly evident at the 40 and 50% styrene monomer concentrations where the homopolymer yields are relatively low especially with the 40% solution. This result is of preparative significance, since the UV conditions are defined in Table 6.30 for obtaining relatively high grafting yields with minimum loss of monomer from homopolymer.

The final feature involving the use of these cationic salts to preferentially

Table 6.29
Comparison of cationic salts in photografting of styrene in methanol to polypropylene in presence of acid and photoinitiator[a]

Styrene (% v/v)	Graft (%)					
	B	$B + H^+$	$B + Fe^{2+}$	$B + H^+ + Cu^{2+}$	$B + H^+ + Fe^{2+}$	$B + H^+ + M$
20	7.2	19	1.0	25	22	14
25	22	48	—	49	83	72
30	29	65	1.6	117	165	102
35	41	31	—	72	30	39
40	39	19	1.6	64	35	36
50	26	18	2.8	31	25	41

[a] Conditions as in footnote a of Table 6.28; M—Mohr's salt (5×10^{-3} M); Fe^{2+}—iron sulphate (5×10^{-3} M).

Table 6.30
Effect of copper sulphate on homopolymer yields for the photografting of styrene in methanol to polypropylene in presence of acid[a]

Styrene (% v/v)	B		$B + H^+ + Cu^{2+}$	
	Graft (%)	Homo-polymer (%)	Graft (%)	Homo-polymer (%)
20	7·2	12·8	25	31
30	29	17·5	117	42
40	39	35	64	12·2
50	26	14·4	31	7·7

[a]Conditions as in footnote *a* of Table 6.28.

suppress homopolymerisation is shown by the data in Table 6.31 where inclusion of both acid and TMPTA in the same monomer solution containing photoinitiator leads to synergistic effects in photografting enhancement for styrene in methanol to polypropylene similar to that already observed in the previous section with ionising radiation. Addition of copper sulphate to the grafting solution containing both TMPTA and acid results in increased grafting yields with a shift in the Trommsdorff peak to the lower monomer concentration. Under these conditions homopolymerisation remains suppressed. In the absence of acid, grafting yields are strongly reduced even with TMPTA at all monomer concentrations up to 50%. The remarkable result in Table 6.31 is the data for the 30% styrene solution where in the presence of all additives, including copper, the highest grafting yield of all the UV conditions used in the table is found. Thus, in the presence of copper ions, there is a synergistic effect with all additives leading to a very significant increase in grafting yield.

Table 6.31
Synergistic effect of acid, TMPTA and cationic salt for enhancing yields in photografting of styrene in methanol to polypropylene[a]

Styrene (% v/v)	Graft (%)			
	B	$B + Cu^{2+} + T$[b]	$B + T + H^+$	$B + T + H^+ + Cu^{2+}$
20	7	2	25	30
30	29	2	70	380
40	39	9	306	121
50	26	11	200	121
60	25	10	150	98
70	23	51	75	80
80	15	121	70	59

[a]B (1% w/v); H^+—sulphuric acid (0·2 M); T—TMPTA (1% v/v); Cu^{2+}—copper sulphate ($5·0 \times 10^{-3}$ M); irradiated 18 h at 30 cm from 90 W lamp.
[b]Level of graft for $B + Cu^{2+}$ low, similar to data for $B + Fe^{2+}$ (Table 6.29).

Homopolymerisation suppression using co-monomer techniques

An additional method for reducing homopolymerisation in radiation grafting work is the co-monomer technique,[38] which is particularly effective for reducing homopolymer yields with radiation sensitive monomers such as acrylic acid, methacrylic acid, the polyfunctional acrylates and their esters. With the low-molecular-weight acrylate esters, such as ethyl acrylate, the homopolymerisation problem is evidenced not so much by high yields but by erratic and irreproducible grafting.[38] The co-monomer technique is particularly valuable for these types of monomers and, if styrene is one of the monomers used in the mixture, copolymerisation of the second reactive monomer can be achieved by the mutual irradiation method with a minimum of homopolymer formation. From the data for the radiation grafting of ethyl acrylate to wool in the absence of styrene (Table 6.32) it is observed that when ethyl acrylate is grafted at 2×10^3 Gy, copolymerisation is very inefficient due to homopolymer formation. Inclusion of styrene, not only leads to satisfactory grafting of ethyl acrylate with little homopolymer formation, but at high ethyl acrylate to styrene ratios (5:1), the graft consists predominantly of ethyl acrylate (74%) and the properties exhibited by the copolymer are essentially those of an ethyl acrylate graft with little contribution from styrene. Analogous results are also obtained with the styrene co-monomer technique when used for the radiation grafting of methyl methacrylate and acrylonitrile to wool (Fig. 6.14), virtually no homopolymer being found in the copolymers.

Consistent with the previous acid additive work, if mineral acid is added to the monomer solution, enhancement in the styrene co-monomer technique is observed (Table 6.33). In this table, the radiation induced grafting of ethyl acrylate to Belmerino wool cloth is reported, using the styrene co-monomer technique in the presence of mineral acid. When the data in Tables 6.32 and 6.33 are compared, it is obvious that there is grafting enhancement with the inclusion of mineral acid, e.g. the data for the monomer solution containing 37·5% styrene and 37·5% ethyl acrylate at 2.0×10^3 Gy where the yield is increased from 2·7 to 27·9% by the inclusion of mineral acid. More

Table 6.32
Grafting of ethyl acrylate and styrene to Belmerino wool cloth using ionising radiation[a]

Grafting solution				Styrene/ethyl acrylate in grafted copolymer	
Ethyl acrylate	Styrene	Dose (Gy)	Graft (%)	Found	Calculated
25·0	50·0	1.0×10^3	1·2	0·91	2·24
37·5	37·5	2.0×10^3	2·7	0·40	1·44
37·5	37·5	4.0×10^3	9·5	0·87	1·44
62·5	12·5	1.0×10^3	2·1	0·26	0·56

[a] Dose rate 1.0×10^3 Gy/h; air not excluded; liquid/wool ratio (25:1).

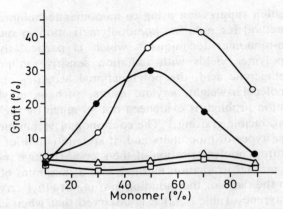

Fig. 6.14. Radiation grafting of methyl methacrylate (MMA) and acrylonitrile (AC) in methanol to wool using styrene (ST) co-monomer method; dose rate 2.5×10^2 Gy/h, dose 2.0×10^3 Gy. (○) ST/AC (1:1), (●) ST/MMA (1:1), (△) MMA and (□) AC.

importantly, in the presence of acid, the ratio of ethyl acrylate in the copolymer grafted is very much higher. In run number six (Table 6.33) the inclusion of styrene not only leads to satisfactory graft and very low homopolymer yields, but at high ethyl acrylate to styrene ratios (5:1) the percentage of ethyl acrylate in the graft copolymer is very high (87%), even higher than in the run without acid. This result is of significance in a preparative context especially for the synthesis of ethyl acrylate copolymers reproducibly and in high yields. The styrene co-monomer technique has also recently been used to study the radiation induced graft copolymerisation of

Table 6.33
Effect of acid on grafting of ethyl acrylate and styrene to Belmerino wool cloth using ionising radiation[a]

Grafting solution		Dose (Gy)	Graft (%)	Styrene/ethyl acrylate in grafted copolymer	
Ethyl acrylate	Styrene			Found	Calculated
37.5	37.5	Nil	7.0[b]	—	—
37.5	37.5	5×10^2	8.1	0.11	1.44
37.5	37.5	1.0×10^3	13.1	0.17	1.44
37.5	37.5	2.0×10^3	27.9	0.14	1.44
37.5	37.5	4.0×10^3	80.1	0.87	1.44
62.5	12.5	2.0×10^3	17.5	0.13	0.56

[a]Dose rate 1.0×10^3 Gy/h; air not excluded; liquid/wool ratio (25:1); all grafting solutions contain sulphuric acid (0.2 M).
[b]Acid-catalysed copolymerisation 4 h at 43°C (same time as highest dose radiation sample).
See Table 6.32 for corresponding data without acid.

ethyl acrylate onto sisal fibres. The process was also extended to include the role of acid which was shown to be beneficial for the technology.[39] This styrene co-monomer thus appears to be of general applicability in the radiation grafting of a wide range of monomers and backbone polymers.

EFFECT OF STRUCTURE OF MONOMER AND BACKBONE POLYMER IN GRAFTING WITH ADDITIVES

The technique of grafting using ionising radiation or UV is theoretically applicable to any monomer or backbone polymer,[3,4] a wide range of monomers and backbone polymers already having been investigated in this field. The data in Table 6.34 indicate the type of reactivity to be expected for a variety of backbone polymers. Grafting yields with PVC are amongst the highest under the irradiation conditions used. The data in the table reflect the relative ease of breaking of carbon–halogen and carbon–hydrogen bonds of the backbone polymer to create sites where grafting can occur. Relative bond dissociation energies are C–Cl, 3·49 eV; C–H, 4·2 eV. Cellulose esters, as represented by the acetate, appear to be as reactive as cellulose itself. Acid enhancement is observed for all the backbone polymers reported, the result indicating that the effect could be a general phenomenon for all trunk polymers independent of their backbone structure. Solvents have previously

Table 6.34
Comparison of substrates for radiation grafting of styrene in methanol with and without acid

Styrene (% v/v)	Graft (%)					
	Cellulose[a]	Wool[b]	PE[c]	PVC[d]	Leather[e]	Cellulose acetate[f]
20	8 (24)[g]	8 (38)	29 (128)	0 (10)	5·7	21
30	17 (33)	—	94 (150)	13 (34)	4·6, 22[h] (27)[h]	25
40	22 (31)	27 (38)	50 (85)	66 (70)	4·0	25
50	21 (27)	—	37 (55)	6 (25)	—	—
60	28 (25)	28 (73)	36 (45)	3 (10)	7·5	19

[a]Cellulose: total dose $2·0 \times 10^3$ Gy, dose rate $4·0 \times 10^2$ Gy/h, $4·0 \times 10^{-1}$ M sulphuric acid.
[b]Wool: total dose $2·0 \times 10^3$ Gy, dose rate $2·5 \times 10^2$ Gy/h, $2·0 \times 10^{-1}$ M sulphuric acid.
[c]Polyethylene: total dose $3·0 \times 10^3$ Gy, dose rate $4·0 \times 10^2$ Gy/h, $2·0 \times 10^{-1}$ M sulphuric acid.
[d]PVC: total dose $1·0 \times 10^3$ Gy, dose rate $1·0 \times 10^3$ Gy/h, $7·5 \times 10^{-2}$ M sulphuric acid.
[e]Leather: total dose $1·0 \times 10^3$ Gy, dose rate $4·0 \times 10^2$ Gy/h, $1·0 \times 10^{-1}$ M sulphuric acid.
[f]Cellulose acetate: total dose $2·0 \times 10^3$ Gy, dose rate $1·84 \times 10^3$ Gy/h.
[g]Figures in brackets denote graft with acid.
[h]Leather: total dose $3·0 \times 10^3$ Gy, dose rate $4·0 \times 10^2$ Gy/h, $1·0 \times 10^{-1}$ M sulphuric acid.

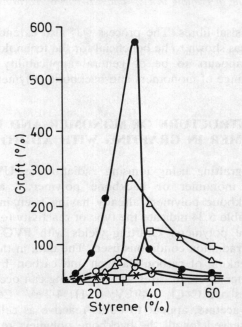

Fig. 6.15. Effect of solvent on radiation grafting of styrene to PVC; dose rate 1.0×10^3 Gy/h, dose 5.0×10^3 Gy. (●) Methanol, (△) n-butanol, (○) t-butanol, (□) octanol, and (×) DMF.

been shown to be important additives in grafting work, using cellulose and the polyolefins as typical backbone polymers for this current paper. In Fig. 6.15 the solvent effect for the grafting of styrene to another backbone polymer, PVC, is shown and again, consistent with the studies on cellulose and the polyolefins, significant enhancement in grafting with certain solvents, particularly the low-molecular-weight alcohols, is observed. The polarity of the backbone polymer and the solvent is thus important in determining the grafting profiles for monomer to backbone polymer.

Typical of the variety of backbone polymers studied in this current grafting work are the wide range of ethylene propylene terpolymers (EPDMs).[23] Representative data for the grafting of hydroxyethyl methacrylate (HEMA) onto a variety of commercially available rubber vulcanisates is shown in Table 6.35. These results were obtained with ionising radiation in the presence of copper nitrate to suppress homopolymerisation. The solvent used was water and, in the absence of copper nitrate, grafting yields were very low because of competing homopolymerisation. As the concentration of HEMA in the aqueous solution is increased, grafting yields are correspondingly increased for all backbone polymers studied, the Vistalon 6505 being the most effective. When multifunctional acrylates are added to the grafting solution (Table 6.36), very large increases in grafting yield are obtained at certain monomer concentrations with particular EPDMs. Of the MFAs used in the table,

Table 6.35
Effects of monomer concentration on radiation grafting of HEMA to EPDM rubber[a]

EPDM type	Graft yield (mg/cm^2)			
	20% HEMA[b] (w/w)	40% HEMA[b] (w/w)	60% HEMA[b] (w/w)	HEMA[c]:H$_2$O (20:80)
Polysar (826X)	4·0	6·0	31·0	0·5
Vistalon (3708)	4·0	10·0	50·0	0·6
Vistalon (7000)	3·5	6·0	25·0	0·5
Vistalon (6505)	4·5	10·0	75·0	2·0
Polysar (585)	2·5	2·6	13·0	0·3

[a] Dose rate 5·0 × 10^2 Gy/h to 7·5 × 10^3 Gy.
[b] 1·0 M copper nitrate aqueous solution.
[c] No copper nitrate.

PEGDMA is generally better than TMPTA, particularly with Vistalon 3708 as backbone polymer. However, of the three MFAs used, an alkoxylated triacrylate commercially available monomer was even better. The data show that the principles of grafting developed for the simple styrene and cellulose–polyethylene systems can be extended to more complicated, industrially orientated systems.

A wide range of monomers has also been utilised in these grafting studies with ionising radiation and UV, typical of the type of monomers used are those listed in Table 6.37. Each of the monomers was grafted in 40% v/v methanol to cellulose in the presence of ionising radiation. Under these conditions,

Table 6.36
Effect of multifunctional acrylate on the grafting of HEMA to EPDM using ionising radiation[a]

EPDM	HEMA (% w/w)	Graft yield (mg/cm^2)		
		TMPTA	PEGDMA	ALTA[b]
Polysar (826X)	20	5·2	4·5	3·0
	40	8·0	11·0	11·5
Vistalon (3708)	20	6·0	6·0	4·0
	40	28·0	82·0	78·0
Vistalon (7000)	20	6·0	4·5	4·0
	40	18·0	26·0	26·5
Vistalon (6505)	20	5·0	5·0	3·5
	40	34·0	20·0	114·0
Polysar (585)	20	3·0	2·5	0·5
	40	3·0	5·3	3·5

[a] Dose rate 5·0 × 10^2 Gy/h, dose 7·5 × 10^3 Gy; copper nitrate (1·0 M).
[b] ALTA (alkoxylated triacrylate; trade name—Sartomer 9035).

Table 6.37
Radiation grafting of miscellaneous monomer[a] to cellulose[b]

Monomer	Graft (%)
Styrene	40
o-Methylstyrene	110
p-Methylstyrene	6
2,5-Dimethylstyrene	19
o-Bromostyrene	27
o-Chlorostyrene	74
Vinylpyridine	3
Methyl methacrylate	18[c]
Vinyl acetate	11[d]

[a] 40% v/v in methanol.
[b] Dose rate 5.4×10^3 Gy/h to 4.5×10^2 Gy.
[c] Dose rate 1.0×10^4 Gy/h to 4.5×10^2 Gy.
[d] Dose rate 1.0×10^4 Gy/h to 3.9×10^3 Gy.

relative to styrene, 2-vinylpyridine is strongly deactivated whereas o-methyl styrene is strongly activated. The effect of group functionality on grafting is shown by the reactivity of the substituted styrenes reported in Table 6.37, o-methylstyrene being strongly activating as is o-chlorostyrene. These substituent properties have been related to the Hammett sigma function.[26] Changing the solvent, introducing additives and varying a wide range of processing parameters can markedly improve the grafting efficiency of these monomers, particularly those which are deactivated such as 2-vinylpyridine. In

Table 6.38
Photografting of acrylic acid, acrylonitrile, 4-vinylpyridine and vinyl acetate in methanol to cellulose[a]

Run	Monomer	Sensitiser	Irradiation time (h)	Graft (%)
1	Acrylic acid	UN	3	15.3
2	Acrylic acid	B	3	10.6
3	Acrylonitrile	UN	3	15.8
4	Acrylonitrile	B	3	10.3
5	Acrylonitrile	Benzophenone	24	7.5
6	4-Vinylpyridine	UN	3	10.1
7	4-Vinylpyridine	B	3	11.1
8	Vinyl acetate	UN	3	2.0[b]

[a] Solution of monomer (30% v/v) in methanol containing sensitiser/photoinitiator and cellulose irradiated at 24 cm from 90 W lamp; except run 5 where 90% v/v monomer was used.
[b] Vinyl acetate was studied at all concentrations in the range 1–100%, but graft was always less than 2% obtained with 30% monomer concentration.

like manner to the ionising radiation system, a wide range of monomers can also be used in photografting. A representative number of these monomers are shown in Table 6.38 for photografting to cellulose using methanol as solvent. These include acrylic acid, acrylonitrile, 4-vinylpyridine and vinyl acetate in the presence of UN and B as photosensitiser and photoinitiator, respectively.

MECHANISMS INVOLVING ADDITIVE EFFECTS IN GRAFTING

Additive effects are not only significant in grafting, *per se*, they are also relevant to curing processes, since many additives which accelerate grafting are also used extensively in curing formulations, e.g. the MFAs. The mode of action of the additives reported in the current work is important since, for practical grafting applications, each of the additives needs to be optimised in performance. Since the reactivity of the additives discussed in this paper is similar for both ionising radiation and UV grafting systems, a common treatment for the phenomena can be applied to the two radiation processes. For the purposes of this discussion, additives can be classified into three groups, (i) acids and salts such as lithium perchlorate and nitrate, (ii) PFMs including predominantly MFAs and MFMAs, and (iii) cationic salt suppressors. Although the mode of operation of each group of additives appears to be different, there are parameters common to all three which influence their reactivities. In addition, when all three types are present in the one grafting solution, the overall mechanism of the enhancement process can be complicated by possible secondary reactions between, not only the additives themselves, but also with the other components in the grafting solution. In this respect, the backbone polymer is particularly important since many such polymers can be sensitive to certain of the additives, as is the case with the detrimental affect of acid on cellulose and wool over the relatively long periods of time experienced in grafting solutions during irradiation.

Effect of acids and salts—partitioning phenomena

The mechanism of acid and salt effects in radiation and UV grafting is complicated by the variety of components present in any one grafting system.[35,40] The inclusion of solvents, monomers, backbone polymers and other additives needs to be considered in any thorough treatment. Solvents are particularly relevant since materials that wet and swell the backbone polymer generally enhance grafting. The mechanism of these additive effects is also influenced by the level of monomer already grafted; thus, during the grafting of styrene to cellulose, initially the styrene is grafted to cellulose alone whereas subsequent styrene molecules will be grafted to a progressively enriched styrenated cellulose. In earlier work,[2] the acid enhancement effect was attributed to two predominant factors namely, the radiolytic yield of hydrogen atoms and the extent to which grafted monomer (polystyrene) was solubilised in the bulk solution. Solvent properties were also important, since those with high radiolytic yields of hydrogen atoms, such as methanol and related low-molecular-weight alcohols, were considered to be effective in grafting because of their relatively high yield G(H) values (eqns (24) and (25)).

$$CH_3OH + H^+ \longrightarrow CH_3OH_2^+ \qquad (24)$$

$$CH_3OH_2^+ + e^- \longrightarrow CH_3OH + H^\cdot \qquad (25)$$

The hydrogen atoms produced can abstract hydrogen atoms from the trunk polymer yielding grafting sites where monomers can be attached. This radiation chemistry mechanism for grafting enhancement fails to explain a number of further observations that have already been reported.[1,2,10] These include the occurrence of graft enhancement only at certain acid and monomer concentrations, the efficacy of sulphuric, nitric and perchloric over other acids and the presence of grafting enhancement in the pre-irradiation and post-irradiation techniques where the concept of radiolytically produced hydrogen atoms is not relevant.

From current results and more extensive evidence recently reported,[41,42] a new model has been proposed to explain the effect of acids and salts in enhancing yields in these grafting reactions. The basis of the new model originated from observations of the effect of acids and salts on the swelling properties of polyethylene in the presence of methanolic solutions of styrene.[32] In this system, partitioning of styrene into polyethylene is significantly improved by the inclusion of mineral acid or lithium salt in the grafting solution (Table 6.39). Styrene labelled with tritium was used for these experiments which indicate that most swelling occurs within the first few minutes of exposure of backbone polymer to the solution. At 25°C at least 80% of the swelling is achieved during the first hour, the system asymptotically approaching equilibrium after 10 h. In each series of data the lithium salt is more efficient than sulphuric acid in the partitioning of monomer into the polyethylene. As representative monomer solution for these swelling experiments, 30% styrene in methanol was chosen since, for most radiation grafting

Table 6.39
Variations in styrene absorption by polyethylene with time: initial swelling behaviour[a]

Time of swelling (min)	Styrene absorption (mg styrene/g polyethylene)		
	No additive	Sulphuric acid	Lithium perchlorate
0·0	0	0	0
0·5	11·9	12·4	12·9
2·0	24·9	27·2	31·3
4·0	38·0	41·2	43·1
equilibrium[b]	45·4	54·0	54·4

[a]Styrene in methanol (30% v/v); sulphuric acid (0·1 M); lithium perchlorate (0·2 M).
[b]Value determined after 13 h swelling; temperature 25°C.

conditions, peak grafting and enhancement occur around this region. The swelling behaviour exhibited in Table 6.39 indicates that the presence of electrolytes significantly affects the concentration of the monomer within the graft region. In effect, these results demonstrate that partitioning of non-polar monomer into non-polar media may be significantly improved by the presence of dissolved electrolyte. This partitioning behaviour may be interpreted as an example of the 'salting out' technique employed in solvent extraction. In the present grafting systems, the driving force for the increased partitioning of monomer into substrate is the reduced solubility of styrene in the bulk solution due to the presence of dissolved electrolyte. The net result of this driving force is higher rates of monomer diffusion into, and equilibrium monomer concentration within, the substrate.

The new partitioning model described above is consistent with previous grafting models discussed in the literature, all of which include monomer concentration dependence factors.[1] The experimental findings of the current work are consistent with these earlier grafting models, in that increased monomer supply has been shown to result in enhanced grafting rates. The higher concentration of monomer at the grafting site favours propagation of growing chains. The enhanced grafting yield is usually associated with a corresponding increase in molecular weight and there is no evidence for an increase in the number of chains initiated.[42] Thus, overall, according to this model,[42] it is proposed that increased partitioning of monomer occurs in the graft region of the backbone polymer when acids or salts are dissolved in the bulk grafting solution. This permits higher concentrations of monomer to be available for grafting at a particular backbone site in the presence of these additives. The extent of improvement in this monomer partitioning depends on the polarities of monomer, grafted polymer, backbone polymer and solvent, as well as the type of acid or salt used and its concentration. The polarity of the backbone polymer is particularly relevant since it leads to a further extension of the partitioning concept.

In addition to influencing the partitioning of monomer between backbone polymer and bulk grafting solution, electrolytes, like the acids and salts studied in the current work, can, themselves, be partitioned between the two phases. This results in changes in concentration of electrolyte, particularly in the bulk grafting solution when the electrolyte concentration would be lowered. Such a reduction in concentration may materially affect the efficiency of the electrolyte in partitioning monomer into the backbone polymer. In the specific example of a polar backbone polymer, electrolyte levels are correspondingly increased and the back reaction occurs where monomer may even be repelled out of the grafting environment until equilibrium is attained. Such a situation is favoured as the polarity of the backbone polymer is increased, leading to a greater tendency to absorb electrolyte from the monomer solution. Thus, for highly polar backbone polymers, the experimental conditions for achieving significant electrolyte enhancement in grafting may be more subtle than with non-polar substrates and it is the relative concentration of electrolyte in both phases which will markedly influence the level of grafting enhancement attained. Overall, it is the effect of these ionic species in partitioning which is essentially responsible for the observed increase in grafting yields initiated by ionising radiation and UV in the presence of acid and salt additives. Radiolytically generated free radicals can also be expected to make some contribution to this enhancement effect in a system in which initiation occurs by ionising radiation; however, this radiation chemistry explanation does not appear to be the predominant pathway to enhancement. Further, in photografting studies, radiolytically produced free radicals are not generated and the latter point is not relevant to UV systems. Thus, the concept is not valid for any general mechanism proposed to cover both ionising radiation and UV grafting systems.

Effect of polyfunctional monomers

The presence of PFMs, particularly MFAs such as TMPTA, has been shown to significantly enhance grafting in the present systems especially at the Trommsdorff peak. The mechanistic role of these additives is complicated since they not only enhance copolymerisation, they can also cross-link the grafted polystyrene chains. In the grafting experiments, branching of the growing grafted polystyrene occurs when one end of the PFM (using a difunctional monomer as an example), immobilised during grafting, is bonded to the growing chain. The other end is unsaturated and free to initiate new chain growth via a scavenging reaction. The new branched polystyrene chain may eventually terminate by cross-linking with another polystyrene chain or by reaction with an immobilised PFM radical. Grafting is thus enhanced mainly through branching of the growing polystyrene chains.[33] In addition, in photografting work alone, MFAs such as TMPTA can enhance grafting by the absorption of UV in primary processes, i.e. they act as photoinitiators and photosensitisers in such processes. The other feature of the PFM enhancement effect is the possibility that such reagents may also be partitioned between the bulk grafting solution and the solution absorbed within the backbone polymer in a manner similar to that already discussed for acids and salts. The parameters which influence monomer partitioning in the presence of electrolytes will thus also be relevant to the partitioning of MFAs. However, because of the relatively low concentration of MFA required to achieve enhancement and also the size of the molecule, the partitioning effect with MFAs may not be so significant. This important mechanistic point is currently the subject of further extensive work (Dworjanyn, P. A., et al., unpublished).

Effect of cationic salts in suppressing homopolymer—partitioning phenomena

The role of cationic salts in preferentially suppressing homopolymer formation in grafting systems can also be attributed to partitioning effects similar to those found for acid and salt additives discussed in the section entitled 'Effects of acids and salts—partitioning phenomena'—see p. 305. The effect of cationic salt partitioning can be observed visually, especially its relationship to the polarity of the backbone polymer. In recent work,[43] the relative intensities of colour in the bulk grafting solution and the backbone polymer due to the distribution of cationic salt between the two phases could be clearly seen (the blue copper salts are particularly effective in this respect). Thus, using a polar backbone polymer like cellulose and comparing it to a non-polar substrate, like polypropylene, in the presence of copper salts, the cationic salt is absorbed strongly into the polar cellulose. The reverse occurs with polypropylene, the cationic salt tending to remain predominantly at the surface of the less polar polyolefin. This difference in distribution of cationic salt in the backbone polymer can markedly influence the relative reactivities of cellulose and polypropylene with additives in grafting.[43]

The following mechanism has been proposed to explain the observed effects of cationic salts in these radiation grafting systems in the presence of acid and salt additive.[18,22] The mechanism is an extension of the previous partitioning model developed to explain electrolyte enhancement effects in the current radiation grafting reactions. It explains the effect of cationic salts in preferentially suppressing homopolymer formation and under certain experimental conditions, actually increasing grafting yields. Using the grafting of styrene in methanol to non-polar polypropylene as model grafting system and, in the absence of electrolytes such as mineral acids and lithium salts, the effect of cationic salts in suppressing homopolymer formation in such a system has been attributed to scavenging of monomer radicals by the cation.[44,45] The behaviour of such cationic salts in grafting reactions can be explained in the following manner.

1. In the absence of cationic salts more polystyrene homopolymer is formed in the grafting solution and this competes with the backbone polymer for the styrene monomer. When the cationic salt is present homopolymer yields are reduced, the competing reaction is minimised leaving more styrene available for grafting.
2. Due to partitioning of the cationic salt between the bulk monomer solution and the solution absorbed by the backbone polymer, lower concentrations of cationic salts are found in the less polar backbone polymer, thus more monomer/oligomer radicals remain unscavenged by the cations in the backbone polymer and these are available for enhancement in grafting. Because of steric considerations and mobility problems, such monomer/oligomer radicals would be expected to preferentially graft rather than homopolymerise within the backbone polymer.
3. In the specific case of UV grafting, reduction in homopolymer in the presence of cationic salts leads to lower turbidity in the monomer solution, due to minimising polystyrene precipitation in the methanol as the irradiation proceeds, thus UV transmittance through the grafting solution is improved and grafting increases.

When electrolytes such as mineral acids and lithium salts are added to this grafting solution containing cationic salts, the enhancement effect of these electrolyte additives is now superimposed on the effect of the cationic salt. The essential consequence of the role of the cationic salt is to maintain in the graft solution, as the irradiation proceeds, a larger pool of monomer which is available for partitioning in the presence of electrolyte and thus enhanced grafting.

If the non-polar polypropylene is replaced by polar cellulose as backbone polymer, then compensation is made for the polarity of cellulose in the partitioning model. With cellulose, the concentrations of cationic salt suppressors and electrolytes will be markedly increased within the backbone polymer. The polarities of monomer solution and backbone polymer will thus be changed and partitioning of reagents will be correspondingly affected.

When an MFA such as TMPTA is added to the acidified grafting solutions containing cationic salts, the above partitioning processes will occur. Theoretically, TMPTA could also be partitioned between bulk grafting solution and backbone polymer however, the concentration of this additive is relatively low (1%) and, because of molecular size, the partitioning effect of TMPTA would be minimised. The predominant role of TMPTA and related MFAs in these reactions appears to be by branching of the polystyrene chains.[33]

RELEVANCE OF GRAFTING IN EB AND UV CURING

Grafting and curing processes initiated by ionising radiation and UV are mechanistically related since both involve predominantly free-radical reactions with a possible contribution from ionic processes under certain experimental conditions. Radiation curing is the rapid polymerisation of an oligomer/monomer mixture to give a film which can also be cross-linked. The film can be attached (grafted) by covalent carbon–carbon bonds to the substrate or the bonding can involve simple physical forces. The rapid polymerisation process is initiated by EB sources (ionising radiation) with electron energies up to 400 keV and delivering 250 kGy m/s. With UV, fast curing rates are achieved with the aid of photoinitiators, photosensitisers and high-performance (up to 600 W/in (~240 W/cm)) UV lamps. With both EB and UV systems, the actual curing process utilises essentially 100% solid resin systems, is basically solvent-free with minimal environmental pollution problems, possesses low energy consumption with compact equipment and cures at room temperature, even in the presence of pigments and fillers.

Although radiation curing processes are effectively solvent-free, for some applications small, and even reasonable, amounts of solvent are required to achieve good application to the substrate, e.g. magnetic-tape processing. The solvent effect in curing is thus important since it may affect the nature of the bonding between cured film and substrate, i.e. the possibility of concurrent grafting. Solvent may also be trapped in the cured film with subsequent slow migration, after curing, affecting the properties of the finished product. The well-known blocking effect experienced in certain UV/EB cured products can also be attributed to residual solvent under some circumstances. Solvent effects in radiation grafting discussed earlier in this paper are also of importance in curing since they can act as a guide to the functionality needed in curing formulations to achieve particular properties in the finished film. Critical to the industrial performance of these oligomer/monomer mixtures is the role of additives, predominantly proprietary commercial materials, which are used to improve properties such as slip, flow, level of gloss and adhesion. For a discussion of the relevance of grafting in curing, the roles of monomer, oligomer and additives each need to be considered in the overall mechanism of the curing process.

Grafting of monomers used in curing—free-radical systems

Monomers are used in UV- or EB-curing formulations as reactive diluents to adjust the rheological properties of the coating for application and also to speed up cure by cross-linking. The structure of the reactive diluent can thus be critical in determining the properties of the finished film. The reactive diluent can be either monofunctional or multifunctional, usually a mixture of the two, the latter generally in higher proportion because of the possibility of enhancing concurrent cross-linking during cure. The principal parameters involved in choice of diluent are solubility, odour, viscosity and the ability to reduce the viscosity of the medium efficiently, volatility, functionality, surface tension, shrinkage during cure, glass transition temperature (T_g) of the homopolymer, effect on speed of the overall cure and toxicological properties.

MFAs are currently the most useful monomers utilised as reactive diluents in curing formulations. The data in Table 6.40 indicate that these MFAs and their methacrylate analogues (MFMAs) can themselves also be photografted to substrates such as cellulose. Thus, in any curing formulation, the reactive diluent cannot only accelerate rates of polymerisation by cross-linking, it is also capable of participating in concurrent grafting during cure. The data in the table illustrate the relationship between grafting yield and structure of MFA and MFMA. In this respect TMPTA is one of the most reactive monomers in the table and, for this reason, TMPTA has been used as representative MFA in the preceding grafting sections of this paper. Thus, MFAs can not only accelerate grafting of other monomers (see the sections entitled 'Polyfunctional monomers—ionising radiation'—p. 280, and 'Polyfunctional monomers—UV'—p. 281), they can also participate in grafting themselves.

In this respect, in many curing formulations, mixtures of monomers of differing structures, rather than one specific monomer, are frequently used with oligomers to impart specific properties to a film. The relative rates of reactivity in grafting and curing of the two monomers can obviously affect the composition of the cured film, thus its structure and hence its physical properties. As a guide to the type of behaviour to be expected in such

Table 6.40
Photografting of various acrylates in methanol to cellulose[a]

Monomer[b]	Graft (%)	
	1 h irradiation	2 h irradiation
AMA	1·7	10·2
TEGDMA	8·7	34·0
EGDMA	4·4	
EGDA	4·9	[c]
DEGDMA	6·2	14·5
DEGDA	3·6	
HDDA	4·0	
BGDMA	7·6	
CMA	2·2	
PETA	13·1	16·3
PEGDMA	14·3	
TMPTA		22·3

[a] Solution of acrylate (30% v/v) in methanol containing UN (1% w/v) and cellulose irradiated at 24 cm from a 90 W high-pressure UV lamp.
[b] AMA—allyl methacrylate; EGDA—ethyleneglycol diacrylate; DEGDMA—diethyleneglycol dimethacrylate; DEGDA—diethyleneglycol diacrylate; BGDMA—1,3-butyleneglycol dimethacrylate; CMA—cyclohexyl methacrylate; PETA—pentaerythritol triacrylate. (See text for other acronyms.)
[c] Homopolymer formation is severe in all runs reported, however, most of the copolymers can be recovered (after grafting) for experimental purposes, although this was not possible with the EGDA sample.

situations, styrene and hydroxyethyl methacrylate (HEMA) have been chosen as simple examples to illustrate the problems associated with relative reactivities of monomers in radiation grafting and by implication, their effect on concurrent grafting during cure. The two monomers are appropriate since both have previously been used in radiation curing work. Thus, styrene has been an

Table 6.41
Grafting of styrene and HEMA onto polypropylene in the presence of methanol[a]

Monomer (% v/v)	Graft (%)	
	Styrene	HEMA
20	8·1	0
30	18	0
40	50	15
50	43	40
60	29·2	Solid

[a] Dose rate of $5·2 \times 10^2$ Gy/h, total dose $2·0 \times 10^3$ Gy.

Table 6.42
Effect of concentration and feed ratio of styrene and HEMA on grafting to polypropylene using methanol[a]

Feed ratio		Graft (%)		
Volume Styrene/ HEMA	Mole Styrene/ HEMA	20% Monomer (v/v)	40% Monomer (v/v)	60% Monomer (v/v)
0:5	0:1	0	15	—
1:4	0.21:0.79	5.5 (1.2[b]/4.3[c])	12.1 (2.8/9.3)	20.6 (4.5/16.1)
2:3	0.41:0.59	6.2 (2.2/4.4)	12.8 (5.2/7.6)	40.4 (14.5/25.9)
2.5:2.5	0.51:0.49	7 (3/4)	14.3 (7.5/6.8)	38.4 (18.4/20)
3:2	0.61:0.39	7.3 (4/3.3)	19.4 (11.8/7.6)	45.8 (25/20.8)
4:1	0.81:0.19	8.2 (5.7/2.5)	37.5 (27.8/9.7)	53.5 (33.6/19.9)
5:0	1:0	8.1	50	29.2

[a] Dose rate of 5.2×10^2 Gy/h, total dose 2.0×10^3 Gy
[b] Graft percent of styrene in copolymer.
[c] Graft percent of HEMA in copolymer.

essential reactive diluent in UV-curable polyester wood finishes for many years and HEMA has been a component in UV/EB oligomer manufacture.

When styrene and HEMA are independently radiation grafted to polypropylene in methanol (Table 6.41), yields of copolymer with styrene alone at the radiation dose used are reasonable; however, analogous data with HEMA alone are erratic and irreproducible because of competing homopolymer formation consistent with earlier acrylate work discussed in the section

Table 6.43
Effect of polyfunctional monomers on grafting of styrene and HEMA to polypropylene using methanol[a]

Feed ratio		Graft (%)		
Volume Styrene/ HEMA	Mole Styrene/ HEMA	Control	TMPTA[b]	DEGDMA[c]
0:5	0:1	15	7	7.7
1:4	0.21:0.79	12.1 (2.8[d]/9.3[e])	7.7 (2/5.7)	10 (2.5/7.5)
2:3	0.41:0.59	12.8 (5.2/7.6)	12.7 (5.8/6.9)	13 (5.6/7.4)
2.5:2.5	0.51:0.49	14.3 (7.5/6.8)	15.2 (8/7.2)	15 (8/7)
3:2	0.61:0.39	19.4 (11.8/7.6)	23 (13.6/9.4)	19.2 (12.2/7)
4:1	0.81:0.19	37.5 (27.8/9.7)	49 (34.5/14.5)	38.8 (28.4/10.4)
5:0	1:0	50	88.4	72

[a] Dose rate of 5.2×10^2 Gy/h, total dose 2.0×10^3 Gy; total monomer 40% v/v.
[b] TMPTA (1% v/v).
[c] DEGDMA (1% v/v).
[d] Graft percent of styrene in copolymer.
[e] Graft percent of HEMA in copolymer.

Table 6.44
Effect of sulphuric acid on grafting of styrene and HEMA in methanol to polypropylene[a]

Feed ratio		Graft (%)	
Volume Styrene/HEMA	Mole Styrene/HEMA	Control	Sulphuric acid (0·2 M)
0:5	0:1	15	22
1:4	0·21:0·79	12·1 (2·8[b]/9·3[c])	15·4 (3·5/11·9)
2:3	0·41:0·59	12·8 (5·2/7·6)	22 (9·7/12·3)
2·5:2·5	0·51:0·49	14·3 (7·5/6·8)	24·8 (13·2/11·6)
3:2	0·61:0·39	19·4 (11·8/7·6)	30·5 (17·7/12·8)
4:1	0·81:0·19	37·5 (27·8/9·7)	52·8 (31/21·8)
5:0	1:0	50	66

[a]Dose rate of $5·2 \times 10^2$ Gy/h, total dose $2·0 \times 10^3$ Gy; total monomer 40% v/v.
[b]Graft percent of styrene in copolymer.
[c]Graft percent of HEMA in copolymer.

entitled 'Homopolymerisation suppression using co-monomer techniques—see p. 299. When the two monomers are mixed in the same solution, grafting of HEMA readily occurs, the composition of the monomers in the graft varying when compared with the monomer composition in solution (Table 6.42). When TMPTA and DEGDMA are included as additives in the monomer solution, the expected increase in grafting yield is observed (Table 6.43), also the composition of the monomers grafted is altered in the presence of MFA and MFMA. The same situation is observed when mineral acid is used as additive

Table 6.45
Comparison of tetrahydrofuran with methanol as solvents for the radiation grafting of styrene and HEMA to polypropylene[a]

Feed ratio		Graft (%)	
Volume Styrene/HEMA	Mole Styrene/HEMA	Methanol	THF
0:5	0:1	15	—
1:4	0·21:0·79	12·1 (2·8[b]/9·3[c])	224 (3/221)
2:3	0·41:0·59	12·8 (5·2/7·6)	59 (5·7/53·3)
2·5:2·5	0·51:0·49	14·3 (7·5/6·8)	35·5 (6/29·5)
3:2	0·61:0·39	19·4 (11·8/7·6)	20 (4·5/15·5)
4:1	0·81:0·19	37·5 (27·8/9·7)	7·5 (3·2/4·3)
5:0	1:0	50	5·3

[a]Dose rate of $5·2 \times 10^2$ Gy/h, total dose $2·0 \times 10^3$ Gy; total monomer concentration 40% v/v.
[b]Graft percent of styrene in copolymer.
[c]Graft percent of HEMA in copolymer.

Table 6.46
Monomer reactivity ratios for styrene (r_1) and HEMA (r_2) in radiation grafting in methanol to polypropylene[a]

Grafting conditions		r_1	r_2
Total monomer concentration	20%	0·45	0·75
	40%	1·33	0·65
	60%	0·45	0·56
Total monomer concentration 40%	+ TMPTA	0·82	0·4
	+ DGDMA	1·33	0·65
	+ sulphuric acid	0·7	0·39
	+ water (5%)	0·98	0·25
	+ water (10%)	1·42	0·31
	+ THF	0·58	6·0

[a] Conditions as in Tables 6.41–6.45.

in the monomer solutions, the enhancement in grafting yield at certain monomer concentrations being particularly significant (Table 6.44).

An extraordinary effect in composition of the grafted mixture and also the grafting yield is observed when tetrahydrofuran (THF) is used as solvent. Not only is the yield extremely high at the low styrene concentration (20% in the mixture) the composition of the grafted film is 98% HEMA and the properties of the film resemble those of a HEMA copolymer alone (Table 6.45). The composition of a film where two monomers are grafted simultaneously from solution is reflected in their monomer reactivity ratios. A summary of the monomer reactivity ratios of styrene and HEMA when grafted under different conditions is shown in Table 6.46. Such observations are important for extrapolating to curing work since the results show the wide variations in grafted film composition depending on the experimental conditions used. The results with the inclusion of water are particularly interesting since water-based radiation-curable systems are currently being used. The presence of water again markedly changes the composition of the monomers in the grafted film. The results described here are for two simple monofunctional monomers and indicate the complexity of the chemistry involved when these data are extrapolated to curing systems with multifunctional monomers.

Grafting of monomers used in curing—cationic systems

Most of the monomers used in EB and UV curing react via free-radical processes and the preceding discussion of monomer reactivity ratios is based on that assumption. More recent work using cationic photoinitiators indicates that curing can be achieved in UV by ionic mechanisms. Exploratory work with these cationic photoinitiators has also been extended to EB systems; however, most current commercial applications predominantly involve UV.[46,47] Typical of the cationic photoinitiators used are triarylsulphonium salts, such as the hexafluorophosphate, as discussed previously (see the section entitled 'UV grafting of monomers'—p. 266). In the presence of UV and hydrogen donors,

Table 6.47
Grafting and curing of vinyl ether and epoxide monomers to cellulose with cationic and free-radical photoinitiators

System	Passes[a]	Monomer (% v/v)	Graft (%)					
			No additive	I	I+T	C	C+T	D
TEGDVE	1	30	—	0	2·6	0	3	—
		70	—	0	0·9	4	5	—
		100	—	—	—	27	—	—
	3	100	—	<1 (M)	—	100 (M)	—	—
		100	—	<1 (F)	—	123 (F)	—	—
CADE	3	100	—	0 (M)	—	140 (M)	—	—
		100	—	0 (F)	—	150 (F)	—	—
PEA[b]	1	100	80	—	—	125	—	145

[a]Number of passes under Minicure (M) or Fusion (F) lamp systems of 200 W/in (~80 W/cm) at speed of 15 m/min (one pass) and 30 m/min (three passes) using the draw-down bar application technique.
[b]PEA—polyester acrylate; methanol solvent. With 1% w/v B (parentheses, no initiator) at 20% v/v in methanol, solutions irradiated in Pyrex tubes at 24 cm for 6 h from a 90 W lamp, results were polyester acrylate 13(1); urethane acrylate 12(2) and epoxy acrylate 11(2); I—Irgacure 184 (1-hydroxy-1-cyclohexylphenyl ketone, Ciba-Geigy); T—TMPTA (1% v/v); C—cationic FX512 (triaryl sulphonium hexafluorophosphate, 3M); D—Darocur 1173 (2,2-dimethyl, 2-hydroxy acetophenone, Merck); all photoinitiators (1% w/w).

both ions and free radicals are formed from these cationic initiators (eqns (15) and (16)) and subsequent polymerisation can theoretically occur via either ionic or free-radical pathways. However, the essential advantages of the system using the ionic pathways is that non-acrylate chemistry can be used as an alternate process in curing to produce new polymers and thus new products not capable of readily being obtained by other techniques. Of the non-acrylate monomers used[18] vinyl ethers and epoxides as typified by TEGDVE and CADE are currently the most important. The results show that, using the draw-down bar technique for application, neat TEGDVE and CADE when cured in this manner also concurrently undergo grafting to cellulose using triarylsulphonium hexafluorophosphate as cationic initiator (Table 6.47). However, using the same technique when Irgacure 184 free-radical photoinitiator is used, no significant graft is observed (Table 6.47). Dilution of TEGDVE with methanol in the presence of cationic photoinitiator reduces the graft significantly. Addition of TMPTA to these methanolic solutions enhances the graft, particularly at low monomer concentrations. Similar results are obtained when Irgacure 184 replaces the cationic photoinitiator in these TMPTA–methanolic solutions. With respect to the possibility of concurrent grafting occurring during cationic curing of these monomers, it is important to note that the results in Table 6.47 have been obtained from exhaustive solvent extraction of the graft copolymer after irradiation. Further studies are currently being pursued to elucidate the nature of the bonding in these graft copolymers (Dworjanyn, P. A., et al., unpublished).

Grafting of oligomers used in curing—free-radical and cationic initiators

The fact that oligomers usually constitute the major component in any radiation curing formulation raises the interesting possibility of these materials (i) grafting themselves, and (ii) enhancing the concurrent grafting of other components in the formulation during the curing process. The three predominant types of oligomers used in commercial curing are PEAs, urethane acrylates (UAs) and epoxy acrylates (EAs). Data for the grafting of three such commercially available oligomers, using both cationic and free-radical photoinitiators, are shown in Table 6.48. With the PEA under UV-curing conditions, high grafting is achieved with cationic photoinitiators (125%) well above the level of the control (80%). As expected, the yield is even higher (145%) with the free-radical photoinitiator (Darocure 1173). When PEA is dissolved in methanol and UV irradiated using the simultaneous method in stoppered Pyrex tubes at low UV intensities, significant grafting is also obtained with B as photoinitiator. Similar results are found for the aromatic UA tested and the EA, however, with these last two oligomers up to 30% of TPGDA as reactive diluent is present, hence part of the copolymer may be contaminated with this monomer. By contrast, PEA is neat and the graft will contain no monomer in the copolymer. This aspect of grafting involving commercially available oligomers is important since for manipulative purposes in curing formulations, the oligomers, generally being viscous, are pre-diluted with monomer, this step usually being needed with UAs and EAs. In the only other reported photografting of these oligomers,[48] a commercially available epoxy diacrylate

Table 6.48
Grafting and curing of oligomers to cellulose with cationic and free radical photoinitiators[a]

System[b]	Passes[c]	Monomer (% v/v)	Graft (%)		
			No additive	F	M	B
PEA	1	100	80	125	145	—
PEA	—	20[d]	1	—	—	13
UA	—	20[d]	2	—	—	12
EA	—	20[d]	2	—	—	11

[a] Conditions as in Table 6.47.
[b] All acrylate commercial curing oligomers supplied by Polycure Pty Ltd, Sydney, Australia.
[c] Curing conditions as in footnote b of Table 6.47; graft data for column F obtained with Fusion lamp, for column M with Minicure facility.
[d] In methanol using B (1%), solution irradiated in Pyrex tubes at 24 cm for 6 h from 90 W lamp with simultaneous technique.

Table 6.49
Effect of commercial acrylate oligomers as additives in photografting styrene in methanol to polypropylene[a]

Styrene (% v/v)	Graft (%)					
	No additive	EA	UA	PEA	TPGDA[b]	TPGDA[c]
20	12	35	53	53	76	79
40	47	123	131	131	289	113
60	22	59	69	69	101	52
80	17	36	41	37	63	39

[a]Irradiated 6 h at 24 cm from 90 W UV lamp; polypropylene film (0·12 mm, ex-Shell); B (1% w/v) photoinitiator in all runs; EA (1% w/v); UA (1% w/v); PEA (1% w/v); all oligomers from Polycure Pty Ltd, Sydney, Australia.
[b]1% w/v.
[c]0·2% w/v.

pre-polymer based upon the glycidyl ether of bisphenol A was grafted in the presence of 25% w/w of TMPTA as reactive diluent and using 1,2-diphenyl dimethoxy ethanone (Irgacure 651) as free-radical photoinitiator for the process. Again in this latter work the copolymer may be contaminated with reactive diluent TMPTA.

In addition to grafting themselves during curing, oligomers in a curing formulation may influence the ease of grafting of the other components, particularly the other monomers that are reactive diluents such as the MFAs. The ability to influence concurrent grafting of these reactive diluents can be important to the properties of the finished cured product. The data (Dwornjanyn, P. A., et al., unpublished) for the effect of three typical commercial curing oligomers, a PEA, UA and EA, on the photografting of styrene in methanol to polypropylene is shown in Table 6.49. Analogous data are depicted in Table 6.50 for cellulose. All three oligomers enhance styrene

Table 6.50
Effect of commercial acrylate oligomers as additives in photografting styrene in methanol to cellulose[a]

Styrene (% v/v)	Graft (%)					
	No additive	EA	UA	PEA	TPGDA[b]	TPGDA[c]
20	16	24	54	61	80	100
40	30	53	94	119	111	108
60	28	35	74	37	72	99
80	29	47	51	63	33	62

[a]UV conditions as in Table 6.49; cellulose Whatman 41 filter paper.
[b]1% w/v.
[c]0·2% w/v.

grafting yields at almost all monomer concentrations examined, the increase in yield being particularly strong at the Trommsdorff peak. When compared with the enhancement properties of a typical MFA used in curing, namely TPGDA, the three oligomers are not as efficient in increasing grafting yields; however, the enhancement with the oligomers is significant and of mechanistic importance. The lower relative efficiency of the oligomers when compared to an MFA such as TPGDA is not unexpected, since the oligomers may be considered to be polymeric MFAs and, as such, their ability to diffuse into substrates would be more restricted than that of the lower-molecular-weight MFAs.

Effect of commercial additives used in curing on grafting

The role of additives is vital in any EB or UV industrial-curable formulation. Silanes such as Z-6020 (Dow), fluorinated alkyl esters like FC-430 (3M) and organic compounds similar to urea are the most frequently used for a variety of purposes, mainly to adjust gloss, slip, flow and viscosity in curing formulations. Without the inclusion of these additives, a commercially acceptable product is frequently difficult to obtain. The effect of these additives on concurrent grafting during cure is thus important.

When the silane is added to the monomer solution in a typical photografting system, such as styrene in methanol to polypropylene, the presence of the silane leads to a retarding effect on the grafting yield at all monomer concentrations studied (Fig. 6.16). By contrast the fluorinated alkyl ester is an activator and leads to mild enhancement in the same system. In addition, inclusion of organic materials like urea also improves grafting yields at certain styrene concentrations as the data in the same figure illustrate. When the three

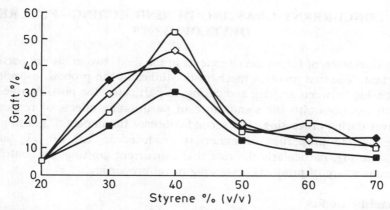

Fig. 6.16. Effect of additives used in commercial curing formulations on photografting of styrene in methanol to polypropylene. Irradiated for 8 h at 30 cm from 90 W lamp at 20°C; all samples contain B (1% w/v). (◆) No additive, (◇) urea (1% w/v), (■) urea + silane (z-6020, Dow, 1% v/v), and (□) urea + fluorinated alkylester (FC-430, 3 M, 1% v/v).

Table 6.51
Synergistic effects of TMPTA with additives used in commercial curing formulations (urea, silane, fluorinated alkyl ester) on photografting styrene in methanol to polypropylene[a]

Styrene (% v/v)	Graft (%)			
	No additive	B	B + U + FE	B + U + Si + FE + T
20	<5	<5	<5	260
30	<5	35	30	588
40	<5	39	46	711
50	<5	17	19	368
60	<5	14	13	283
70	<5	14	11	131

[a]Conditions as in Fig. 6.16; T—TMPTA (1% v/v); U—urea; Si—silane; FE—fluorinated alkyl ester.

additives, namely the silane, fluorinated alkyl ester and urea, are combined with TMPTA in the one monomer grafting solution a dramatic increase in grafting yield by almost two orders of magnitude is observed (Fig. 6.16, Table 6.51). This synergistic TMPTA effect is unique since large enhancement in graft is achieved even in the presence of retarders such as the silane. The results are of mechanistic significance in curing reactions since the data suggest that the effect of multifunctional acrylate will override the presence of other additives, particularly those that are detrimental in determining the adhesive properties of the finished material. Choice of MFA is thus critical in curing work.

CONCURRENT GRAFTING DURING CURING—FUTURE DEVELOPMENTS

In any discussion of future developments in the field, two areas are particularly important. The first involves mechanistic studies of the process, especially the relationship between grafting and curing, in particular, the possible role of ions in both reactions also the significance of partitioning effects of reagents and additives during processing. The second concerns the impact of grafting during curing on the properties of materials produced in EB or UV industrial manufacturing, particularly the role that concurrent grafting can contribute to the manufacture of products possessing novel properties.

Mechanistic studies

Role of ions
Basic studies of EB- or UV-curing systems are complicated by the variety of chemical components present in such formulations. With respect to the

occurrence of concurrent grafting during cure, both grafting and curing processes are considered to proceed by similar mechanisms involving free-radical intermediates. The possible role of ions in both processes has already been discussed (see the section entitled 'EB grafting of monomers'—p. 265, and 'Grafting of monomers used in curing—cationic systems'—p. 315). In EB, the recent development of new low-energy, high-performance machines delivering high electron dose rates has increased the interest in the possibility of ion participation in such curing.[10,11,43] Since polymerisation in free-radical processes is proportional to (dose rate)$^{0.5}$ whereas ionic processes are directly proportional to dose rate, it is logical to assume that as the dose rates from these new machines is increased, ionic processes if present, will be accentuated and contribute to the curing mechanism. Such a proposal, if valid, could lead to novel processing conditions yielding new products. With UV, cationic photoinitiators are now available for curing by ionic mechanisms. This development has broadened the scale of the processing chemistry available in UV curing to include non-acrylate materials such as vinyl ethers and epoxides.

The work involving the possible role of ionic intermediates with EB is particularly promising, since it is based on electron interactions with monomers in the recently developed technique of Fourier transform ion cyclotron resonance mass spectrometry (FTICRMS). The progressive formation of oligomers from monomers by ion–molecule reactions can be observed.[10,11] Typical of the data obtained are spectra in Fig. 6.17 for the FTICRMS of TPGDA and butyl acrylate (BA). With TPGDA, the ion at m/z 113 is intense in both spectra; however, as the time allowed for ion–molecule reactions is increased, ions at higher m/z, namely 502, 413 and 359 are progressively formed, presumably by ion–molecule reactions involving fragments of TPGDA. Typical of these cationic reactions is that shown in eqn (26).

$$R^+ + M \longrightarrow RM^+ \qquad (26)$$

Mechanistically the ion at m/z 413 can be attributed to an ion–molecule reaction involving fragment R which is $-CH_2(CH_3)CHOCOCH=CH_2$ with M for TPGDA. There is thus an increase in higher-molecular-weight species, i.e. oligomerisation with time. The results with BA are similar, the dimer and higher-molecular-weight species are readily formed. The data for TPGDA and BA have been observed in the gas phase and demonstrate that monomers such as TPGDA which are used in EB curing can readily polymerise in the presence of electrons via ionic processes. This model developed in the gas phase has been extended to condensed phases such as are experienced in EB curing using kinetic theory.[10,16,43] The results suggest that polymerisation processes under EB-curing conditions are capable of proceeding via ionic intermediates. Further EB work is required to confirm this proposal in practical EB-curing systems. If proven, and the ionic process can be optimised in EB curing, a number of advantages in processing can be utilised. Thus, by emphasising either the free-radical or ionic process, coatings with different structures and therefore potentially different properties could be made from the same coating

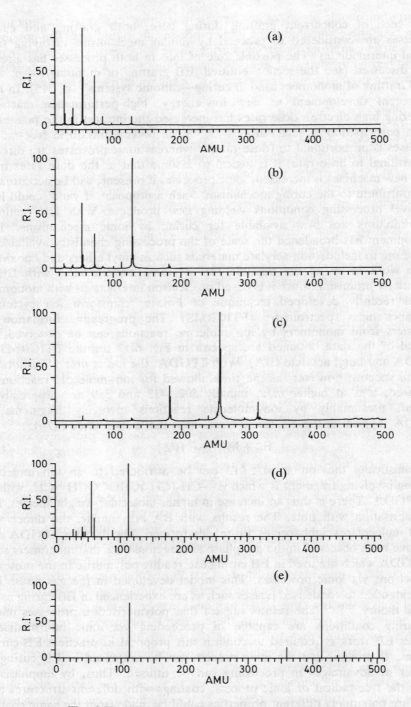

Fig. 6.17. FTICR mass spectra of BA and TPGDA.

mixture. Since ionic processes are oxygen-insensitive, expensive inerting in EB curing could be eliminated if the ionic processes are optimised. The possibility of concurrent grafting occurring during these ionic-curing processes would need to be investigated. The effect of this grafting process on the properties of the finished product cured via ionic mechanisms could then be evaluated.

Partitioning effects—additives
The use of novel additives to increase performance and scope of radiation curing processes continues to be relevant. Currently, additives are used to assist gloss, flow, slip and adhesion in radiation cured systems. In basic grafting work, additives have also been found to enhance grafting yields as previously discussed in this paper, partcularly the use of acids and MFAs.[2,18,29,41,42,49-54] This enhancement has been interpreted in terms of partitioning effects of additives and reagents between grafting solution and trunk polymer.[42] MFAs are particularly effective as additives in accelerating grafting. These monomers are also major components used to speed-up rates of reaction in curing formulations. It is thus logical to suggest that common parameters exist between grafting and curing, and the concepts of partitioning of reagents developed to explain the role of additives in grafting may also be relevant to curing systems. Thus, it is proposed that in curing, partitioning occurs between components in the coating and the substrate.

In curing, rates of reaction are much faster than in most grafting systems on which evidence for the partitioning has been obtained. Because of these fast line processing speeds, the time interval involving the application and cure cycle is so short that the establishment of equilibrium conditions for partitioning of reagents may not be attained and the system may be displaced from equilibrium at the instant of cure, with partitioning of reagents being incomplete. However, with most curing systems, especially those involving UV with organic substrates, the top surface layers of the substrate only are involved in reaction, thus, in the cured film, partitioning of components from the coating will only involve these layers and equilibrium partitioning of reagents *with respect to these layers* may be achieved. The effect of partitioning in these layers leads to coatings where the composition of the film closest to the substrate may differ significantly from that in the remainder of the cured coating. Such coatings may be a continuum of polymerised species differing in composition with depth of film. Alternatively, if chain termination occurs at any level, the coating may consist of layers. From monomer reactivity ratios in grafting, it is observed that in most systems, the monomer composition of the grafted films is not the same as the monomer composition in the bulk solution, i.e. monomers are partitioned into the backbone polymer in a preferred way. Similarly, in curing, the structure of the cured film may vary because of primary partitioning of particular coating components at the surface of the substrate where curing occurs.

In practical curing applications, evidence to support the partitioning model can be found. Partitioning of reagents is a well-known general phenomenon experienced in industrial processing such as in solvent extraction and textile coating, particularly the padding operation. Two typical examples of partitioning in UV curing have come to the authors' attention. Thus, in the UV curing of a coating on one side of paper board, occasionally, after curing, odour from one of the monomers in the cured coating could be detected at the back of the board where no coating had been applied. In these coatings, where mixtures of monomers are used, it is the monomer of lowest functionality which is invariably detected. This is particularly evident when the coating mixture contains monofunctional and trifunctional monomers. Presumably the smaller lower-molecular-weight monofunctional monomer is preferentially partitioned at the surface of the board on application. This monomer then migrates through the board. Because of penetration difficulties with UV, it is obvious that some of the monomer escapes cure and diffuses through the board to atmosphere. Analogous examples have also been found with components of UV inks, e.g. benzophenone. In one instance, with a printed label on a plastic bottle, the benzophenone from the UV-cured ink, after 6 months, had migrated through the label and even through the plastic bottle to contaminate the contents of the bottle. Such diffusion behaviour would be consistent with the partitioning model proposed. However, in this latter example, concurrent grafting with curing of the components would minimise the observed problem. In this respect, the time lag between application of coating and cure would also be important and more work needs to be performed in this area.

Commercial applications

The present work involving the occurrence of concurrent grafting during cure has implications in industrial manufacturing. For many curing processes utilising organic substrates, it is generally accepted that concurrent grafting is an advantage. However, with the rapid expansion of UV and EB technology, the necessity to recycle products processed by this technology is becoming increasingly important. In certain industries, particularly printing and packaging, there is concern with the problems of ease of stripping UV- and EB-cured inks and coatings from substrates to permit ready recycling of the product. For this purpose, concurrent grafting during cure can be an impediment, since grafting leads to essentially covalent bonding between film and substrate and, for ease of recycling in many processes, this bond must be readily broken. This is more easily achieved if bonding between film and substrate involves weak physical or van der Waals forces. There is thus a need to expand the range of coatings available which will possess good adhesion after curing without concurrent grafting. This problem has already surfaced in the label field where difficulty has been experienced in recycling EB-coated labels on bottles. For the success of this application, novel oligomers have been developed for use in these label coatings. Such coatings possess good adhesion without grafting during curing and are soluble in hot alkali so that the coating on the labels can be readily removed during recycling of the bottles. In a related paper application, an even more stringent alkali solubility property was required. Thus, the coating had to be soluble in 0·025% alkali at room temperature but insoluble in water. One method of achieving was to incorporate —COOH or

—SO$_3$H groups into the oligomer. This property enabled the finished product to be recycled in a paper-mill.

In contrast to the printing and packaging industry, the requirements of the plastics field are generally different. The above type of recycling problem is not so critical and covalent bonding is an advantage in many cured products, i.e. the attainment of concurrent grafting during cure is useful. In this respect, there is a significant difference between EB- and UV-processing systems. With EB, electrons can form radical sites in the backbone polymer by direct bond rupture during processing and thus concurrent grafting is facile under these conditions. With UV, grafting sites are created by abstraction and related reactions involving the radicals or ions formed from the primary absorption step involving the photoinitiator. This latter process is energetically more difficult than with the EB system thus concurrent grafting with cure is not as readily achieved with UV.

Overall, in the manufacture of materials with radiation curing techniques the role of concurrent grafting during cure will, in many applications, be determined by the end-use of the product. In the examples cited, if recycling is important then there may be a need to eliminate grafting during curing whereas if strong bonding is required between film and substrate, the presence of concurrent grafting during cure will be an asset. There is a need for further work in this area to differentiate processes where concurrent grafting during cure is useful from those where the necessity to recycle is more important. The development of new monomers and oligomers will be essential for rapid progress in this field.

Finally, the use of novel additives to increase the performance and scope of radiation curing processes continues to be relevant. Currently, additives are used to assist gloss, slip, flow and adhesion in radiation-cured systems. In this context, these additives can also be of use in recycling, since they can influence the strength of bond between film and substrate during curing, e.g. in the present work silanes have been shown to retard grafting by repulsion due to the silicon atom whereas in the presence of MFAs this retardation effect can be overcome. Much work with additives needs to be done not only in developing new materials, but also in understanding the mechanism of operation of those currently used.

ACKNOWLEDGEMENTS

The authors thank the Australian Institute of Nuclear Science and Engineering, the Australian Research Grants Committee and Firenews Pty Ltd for financial assistance.

REFERENCES

1. Stannett, V. T., Silverman, J. & Garnett, J. L. In *Comprehensive Polymer Science*, ed. G. Allen. Pergamon Press, New York, USA, 1989, p. 317.
2. Garnett, J. L. *Radiat. Phys. Chem.*, **14** (1979) 79.

3. Charlesby, A. *Atomic Radiation and Polymers*. Pergamon Press, Oxford, UK, 1960.
4. Chapiro, A. *Radiation Chemistry of Polymeric Systems*. Wiley-Interscience, New York, USA, 1962.
5. Hebeish, A. & Guthrie, H. T. *The Chemistry and Technology of Cellulosic Copolymers*. Springer-Verlag, Berlin, Germany, 1981.
6. Kabanov, V. Y. *Radiat. Phys. Chem.*, **33** (1989) 51.
7. Kabanov, V. Y., Sidorova, L. P. & Spitsyn, V. I. *Euro. Polym. J.*, **10** (1974) 1153.
8. Huglin, M. B. & Johnston, B. L. *J. Polym. Sci. A-1*, **7** (1969) 1379.
9. Lampe, F. W. *J. Phys. Chem.*, **63** (1959) 1986.
10. Garnett, J. L., Dworjanyn, P. A., Nelson, D. J. & Bett, S. J. In *Proceedings of Radtech Asia*, ed. Y. Tabata. Tokyo, 1988, p. 343.
11. Bett, S. J., Dworjanyn, P. A., Greenwood, P. F. & Garnett, J. L. In *Proceedings RadTech '90—North America* (Vol. 1), Radtech International, Northbrook, Illinois, 1990, p. 313.
12. Geacintov, N., Stannett, V., Abrahamson, E. W. & Hermans, J. J. *J. Appl. Polym. Sci.*, **3** (1960) 54.
13. Reine, A. H. & Arthur, J. C. *Text. Res. J.*, **32** (1962) 918.
14. Needles, H. L. & Wasley, W. L. *Text. Res. J.*, **39** (1969) 97.
15. Kubota, H., Murata, T. & Ogiwara, T. *J. Polym. Sci.*, **11** (1973) 485.
16. Davis, N. P., Garnett, J. L. & Urquhart, R. *J. Polym. Sci. Polym. Lett. Edn*, **14** (1976) 537.
17. Hon, D. N.-S. & Chan, H.-C., In *Graft Copolymerization of Ligno-Cellulose Fibres*. (ACS Symp. Ser. 187), ed. D. N. S. Hon, American Chemical Society, USA, 1982, p. 101.
18. Dworjanyn, P. A. & Garnett, J. L. In *Proceedings Grafting Processes onto Polymeric Films and Surfaces: Scientific and Technological Aspects* (Supplement No. 85), ed. I. R. Bellobono. Ricera Scientifica ed Educazione Permanente, University of Milan, Milan, Italy, 1990, p. 63.
19. Garnett, J. L. In *Cellulose: Structure, Modification and Hydrolysis*, ed. R. A. Young & R. M. Rowell. John Wiley, New York, USA, 1986, p. 159.
20. Allen, N. S., Hardy, S. J., Jacoline, A., Glasser, D. M., Catalina, F., Navaratnam, S. & Parsons, B. J. In *Proceedings Grafting Processes onto Polymeric Films and Surfaces: Scientific and Technological Aspects*, (Supplement No. 85), ed. I. R. Bellobono, Ricera Scientifica ed Educazione Permanente, University of Milan, Milan, Italy, 1990, p. 89.
21. Chapiro, A. & Seidler, P. *Euro. Polym. J.* **1** (1965) 189.
22. Nho, Y. C., Garnett, J. L. & Dworjanyn, P. A. *J. Polym. Sci., Part A, Polym. Chem.* **30** (1992) 1219.
23. Katbab, A. A., Burford, R. P. & Garnett, J. L. *Radiat. Phys. Chem.*, **39** (1992) 293.
24. Trommsdorff, E., Kohle, H. & Lagally, P. *Makromol. Chem.*, **1** (1948) 169.
25. Odian, G., Sobel, M., Rossi, A. & Klein, R. *J. Polym. Sci.*, **55** (1961) 663.
26. Dilli, S., Garnett, J. L., Martin, E. C. & Phuoc, D. H. *J. Polym. Sci.*, **C37** (1972) 57.
27. Dilli, S. & Garnett, J. L. *Aust. J. Chem.*, **24** (1971) 981.
28. Dilli, S. & Garnett, J. L. *Aust. J. Chem.*, **23** (1970) 1163.
29. Dworjanyn, P. A. & Garnett, J. L. In *Progress in Polymer Processing* (Vol. 3), ed. J. Silverman & A. Singh. Hanser, New York, 1992, p. 93.
30. Dilli, S. & Garnett, J. L. *J. Appl. Polym. Sci.*, **11** (1967) 859.
31. Ang, C. H., Garnett, J. L., Levot, R. & Long, M. A. In *Initiation of Polymerzation* (ACS Symp. Ser. 212), ed. F. E. Bailey, Jr., American Chemical Society, USA, 1983, p. 209.

32. Garnett, J. L., Jankiewicz, S. V., Long, M. A. & Sangster, D. F. *J. Polym. Sci. Polym. Lett. Edn*, **23** (1985) 563.
33. Dworjanyn, P. A., Fields, B. & Garnett, J. L. In *The Effects of Radiation on High-Technology Polymers* (ACS Symp. Ser. 381), ed. E. Reichmanis & J. H. O'Donnell. American Chemical Society, USA, 1989, p. 112.
34. Ang, C. H., Garnett, J. L., Levot, R., Long, M. A. & Yen, N. T. In *Biomedical Polymers, Polymeric Materials and Pharmaceuticals for Biomedical Use*, ed. E. P. Goldberg & A. Nakajima. Academic Press, New York, USA, 1980, p. 299.
35. Garnett, J. L., Jankiewicz, S. V., Levot, R. & Sangster, D. F. *Radiat. Phys. Chem.*, **25** (1985) 509.
36. Ang, C. H., Garnett, J. L., Long, M. A. & Levot, R. *J. Appl. Polym. Sci.*, **27** (1982) 4893.
37. Dworjanyn, P. A. & Garnett, J. L. *J. Polym. Sci. Polym. Lett. Edn*, **26** (1988) 135.
38. Garnett, J. L. & Kenyon, R. S. *J. Polym. Sci. Polym. Lett. Edn*, **15** (1977) 421.
39. Zaharan, A. H. & Zohdy, M. H. *J. Appl. Polym. Sci.*, **31** (1986) 1925.
40. Chappas, W. J. & Silverman, J. *Radiat. Phys. Chem.*, **14** (1979) 847.
41. Bett, S. J., Dworjanyn, P. A., Fields, B. A., Garnett, J. L., Jankiewicz, S. V. & Sangster, D. F. In *Radiation Curing of Polymeric Materials*. (ACS Symp. Ser., 417), ed. C. E. Hoyle & J. F. Kinstle. American Chemical Society, USA, 1990, p. 128.
42. Garnett, J. L., Jankiewicz, S. V. & Sangster, D. F. *Radiat. Phys. Chem.*, **36** (1990) 571.
43. Kiatkamjornwong, S., Dworjanyn, P. A. & Garnett, J. L. In *Proceedings of Radtech Asia '91*, Osaka, RadTech Japan, 1991, p. 384.
44. Collinson, E., Dainton, F. S., Smith, D. R., Trudel, G. J. & Tazuke, S. *Discuss. Faraday Soc.*, **29** (1960) 188.
45. Hoffman, A. S. & Ratner, B. D. *Radiat. Phys. Chem.*, **14** (1979) 831.
46. Crivello, J. V. *Adv. Polym. Sci.*, **62** (1984) 1.
47. Lapin, S. C. & Snyder, J. R. In *Proceedings RadTech '90—North America* (Vol. 1), ed. RadTech International, Northbrook, IL, USA, 1990, p. 410.
48. Bellobono, I. R., Zeni, M., Selli, E. & Marcandalli, B. *J. Photochem.*, **35** (1986) 367.
49. Mukherjee, A. K. & Gupta, B. D. *J. Appl. Polym. Sci.*, **30** (1985) 2643.
50. Hong Fei, H., Ying Dong, P., Jilan, W., Fuliang, Y. & Xinde, F., *Radiat. Phys. Chem.*, **25** (1985) 501.
51. Gupta, B. D. & Chapiro, A. *Euro. Polym. J.*, **25** (1989) 1137.
52. Stannett, V. T. *Radiat. Phys. Chem.*, **35** (1990) 82.
53. Misra, B. N., Chauhan, G. S. & Rawat, B. R. *J. Appl. Polym. Sci.*, **42** (1991) 3223.
54. El-Assy, B. N., *J. Appl. Polym. Sci.*, **42** (1991) 885.

Chapter 7

Experimental and Analytical Methods for the Investigation of Radiation Curing

JAN F. RABEK

Karolinska Institute, Polymer Research Group, Department of Dental Materials and Technology, Alfred Nobels Alle 8, Box 4064, S-14104 Huddinge, Sweden

Introduction 330
Photocuring 330
UV–vis Lamps and Radiation Devices Used in the Research Laboratories . . . 333
EB-Radiation-Curing Devices 343
Absorption of Light by Photocured Formulations 346
Analysis of Purity of Monomers, Solvents, Diluents and Initiators 350
Kinetics of Photoinitiated Polymerisation 351
   Determination of photoinitiator decomposition 354
   Determination of polymerisation rate 356
   Determination of polymerisation rate (monomer conversion) by spectroscopical methods 361
   Determination of polymerisation rate (monomer conversion) by liquid chromatography 366
   Determination of polymerisation rate (monomer conversion) by DSC measurements 368
   Photocalorimetry 371
   Determination of k_p and k_t rate constants 381
Mechanistic and Kinetic Studies of Photopolymerisation by Holographic Method 382
Microwave Dielectrometry 385
Applications of Piezoelectric Quartz Crystal Microbalances 388
Viscosity Measurements during Curing 389
Determination of Sample Thickness of Cured Thin Polymer Films 391
Shrinkage Measurements 393
Dilatometric Measurements 396
Methods for the Study of Polymerisation Accompanied by Cross-Linking Sol–Gel Analysis 399
Measurements of Swelling of Polymer Networks 402
Glass Transition Temperature (T_g) for Characterisation of the Degree of Cross-Linking 405
Determination of the Photocrosslinking Number 406

Applications of Modern FTIR Spectroscopy 407
ATR IR Spectroscopy 407
RA FTIR Spectroscopy 410
DRIFT Spectroscopy 412
Photoacoustic FTIR Spectroscopy 413
NMR Spectroscopy 415
ESR Spectroscopy 416
ESCA for the Study of Polymer Surfaces 418
   Application of the ESCA for depth profiling 423
   Chemical tagging for functional-group analysis in ESCA 424
Measurements of Physical and Mechanical Properties of Cured Samples 425
Hardness Measurements 426
   Indentation hardness 426
   Scratch hardness 427
   Pencil hardness 427
   Ploughing test 428
   Pendulum hardness 428
   Dynamic methods 431
   Cure depth profiles 431
Contact Angle Measurements 432
   Contact angle hysteresis 436
Adhesion of Water to the Surface of Cured Polymers 437
   Water absorption measurements 438
Measurements of the Lubrication of Surfaces 438
Dynamic Mechanical Spectrometry 439
Distortions of Radiation-Cured Samples 444
References 444

'The reasonable man adapts himself to the conditions that surround him. The unreasonable man adapts surrounding conditions to himself. All progress depends on the unreasonable man . . .'

George Bernard Shaw

INTRODUCTION

Radiation curing is a method of polymerisation initiated by UV–vis or electron radiation i.e. electron beam (EB), during which cross-linked structures are formed. Both these methods differ essentially in the radiation sources used for the initiation of polymerisation, and in the initiation step (Table 7.1). However, all of the experimental techniques used for the study of kinetics and mechanism of polymerisation, chemical and physical (mechanical) testing of cured materials, and other analyses are almost the same.

PHOTOCURING

Photocuring is a photoinitiated polymerisation of formulations (liquid binder systems) containing mainly unsaturated polymerisable monomers, polymers

Table 7.1
Comparison of photocuring and EB curing

	Photocuring	EB curing
Energy source	Mercury lamps	Electron accelerators
Energy consumption	High	Low
Curing energy activity	Low	Low–high
Type of initiation process	Free radicals from photolysis of a photoinitiator	Free electrons
Polymerisation initiators	Photoinitiators	None
Monomers	Mono- and multifunctional acrylic monomers and others	
Inert atmosphere	Not required	Required
Radiation penetration		
Clear	130 μm	500 μm
Pigmented	50 μm	400 μm
Conversion (%)	90	95–100
Capital cost	Low	High

(pre-polymers) or compounds and photoinitiators during very short exposure time upon UV and/or vis radiation. A solid, non-soluble (cross-linked) polymer is obtained. In addition, formulations may consist of the following compounds:

- co-initiator (reducing agent, chain transfer agent, spectral sensitiser);
- light and/or thermal stabiliser; or
- colourants, plasticisers and additives (e.g. pigments, fillers).

The photoinitiated polymerisation which takes place, occurs by a highly complex mechanism (free radical, cation radical, etc.) which includes a photoinitiation step, propagation and termination reactions. The rate of each reaction in which a liquid system is converted to the solid state depends on many factors:

- wavelength and intensity of radiation;
- absorption efficiency of a photoinitiator; and
- quantum yield of the photodecomposition of a photoinitiator into free radicals.

For industrial applications it is not necessary to determine the rate constants for individual polymerisation reaction stages; however, it is necessary to determine a quantity which characterises the overall polymerisation reaction:

- reactivity of individual components (photoinitiator, monomer or polymerising composition);
- efficiency of curing (cross-linking); and
- properties of cured material.

Radiation-curing systems (physical and chemical properties of finished product) are dependent on the following:

- type of formulation for curing;
- concentration of initiator (photoinitiator);
- light source (wavelength, radiation energy);
- time of exposure;
- temperature; and
- presence or absence of oxygen (air).

In practice it is always necessary to provide extensive experiments and tests in order to optimise a system for a given application. The analytical methods employed in radiation curing consist of the following:

- purity test of monomers (usually by chromatographic methods);
- determination of UV–vis absorption spectra of photoinitiators;
- IR spectroscopy for determination of non-converted double bonds;
- thermal post-curing in a dynamic differential calorimeter in order to obtain information on completion of the reaction;
- determination of the glass transition temperature (T_g) for characterisation of the degree of cross-linking;
- cure depth profiles; and
- physical and mechanical test of final cured product, e.g. impact strength, hardness, elasticity, gloss, gel content, solvent or chemical resistance.

At present, all of the commercial instruments available on the market are very sophisticated, which should put the techniques of analysis and measurements

Table 7.2
Time requirements and equipment for evaluation of photoinitiator reactivity[225]

Method	Time consumption	Equipment	Remarks	Cost of equipment
Measurement of gel content	Approx. 6 days	UV-curing system (with conveyor belt)	Very time consuming Useful in investigations of coatings applied on uneven substrates (e.g. glassfibre coatings)	Very low
Measurement of pendulum hardness	Approx. 1·6 days	UV-curing system (with conveyor belt) pendulum apparatus (according to König)	Time-consuming method Useful in investigation of clear and pigmented coatings applied on plane substrates (e.g. glass or sheet)	Reasonable
Photocalorimetric measurement	Approx. 1–2 h	Photo-DSC[a] and PC[b] cryostate microbalance	Rapid method Useful in screening Monochromatic irradiation Measurement in inert atmosphere However, definite film thickness cannot be generated	High

[a] DSC—differential scanning calorimetry.
[b] PC—photocolorimetric.

into the hands of polymer scientists who are not necessarily specialists in a given experimental technique (e.g. Fourier transform IR analysis (FTIR), Fourier transform nuclear magnetic resonance (FTNMR), electron spin resonance (ESR) and other methods). Practical applications of different measuring methods depend on the accuracy of obtained results; however, the time required and the cost of instruments also play an important role. For example, in Table 7.2 are collected some comparative data on different methods and equipments used for the evaluation of initiator reactivity.

UV–VIS LAMPS AND RADIATION DEVICES USED IN THE RESEARCH LABORATORIES

UV–vis lamps used in most of the research laboratories dealing with photocuring, differ significantly from those used in industrial production processes (cf. Chapter 8). The main differences are as follows:[171]

- emission spectra;
- intensity of emitted radiation;
- working temperature; and
- size and construction.

Due to the different emission spectra and intensity of emitted radiation of lamps in research and for industrial application, it is difficult to compare all quantitative results from photocuring kinetic measurements.

The most common lamps used for the research are as follows:[20,236,241,242]

- Low-intensity (125 W), high-pressure mercury vapour lamps, air-cooled (type HPK 125 W, Philips) (Fig. 7.1), which have emission spectra as shown in Fig. 7.2.

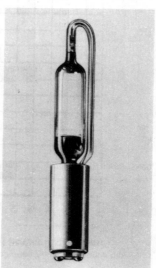

Fig. 7.1. The high-pressure mercury lamp, type HPR 125 W, Philips.[327]

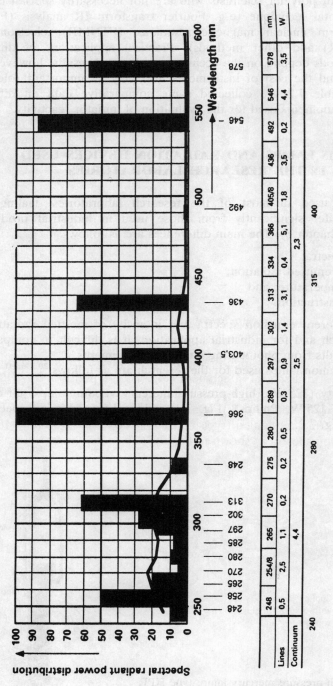

Fig. 7.2. Spectral energy distribution of the UV lamp, type HPR 125 W, Philips.[327]

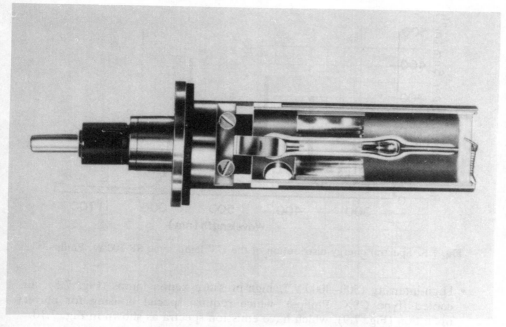

Fig. 7.3. The high-pressure mercury lamp, type SP 500 W, Philips.[326]

- High-intensity (500 W), high-pressure mercury vapour lamps (Fig. 7.3), water-cooled in a special attachment (Fig. 7.4) (type SP 500 W, Philips), which have emission spectra as shown in Fig. 7.5.
- High-intensity (1000–4000 W), high-pressure mercury-doped metal-halide lamps (Fig. 7.6), air-cooled (type HPM 17, 1000–4000 W, Philips), which have emission spectra as shown in Fig. 7.7.

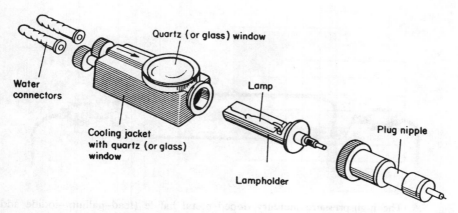

Fig. 7.4. Water-cooling accessories for the lamp type, SP 500 W, Philips.[326]

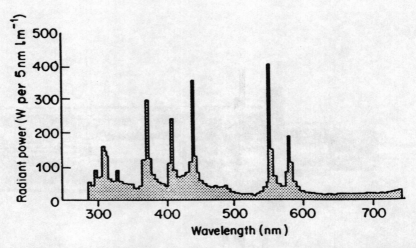

Fig. 7.5. Spectral energy distribution of the UV lamp, type SP 500 W, Philips.[326]

- High-intensity (500–4000 W), high-pressure xenon lamps (Fig. 7.8), air-cooled (type CSX, Philips), which require special housing for correct operation (Fig. 7.9), which have emission spectra as shown in Fig. 7.10.

A required spectral (UV and/or vis) range can be selected from the emitted radiation from a lamp, with the help of transmittance (band pass edge) filters (Fig. 7.11) (solid or liquid), whereas a monochromatic wavelength can only be selected with the help of interference solid filters, which are designed separately for the UV region (Fig. 7.12) and the visible region (Fig. 7.13), or monochromators.[241]

'Mini-conveyor UV-curing units' are commonly used in laboratory research (Fig. 7.14). These devices are very useful in the evaluation of inks and coatings. The radiation emitted by tube lamps (which can have the same

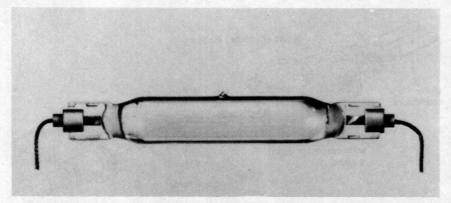

Fig. 7.6. The high-pressure mercury doped metal halide (lead–gallium–iodide additives) UV lamp, type HPM 17, Philips.[326]

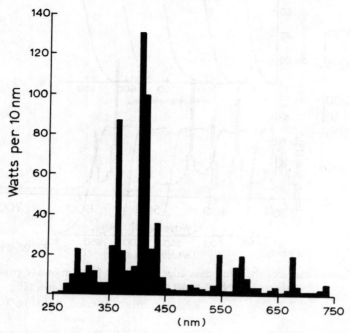

Fig. 7.7. Spectral energy distribution of the lamp, type HPM 17, Philips.[326]

spectral characteristics as used in industrial production devices (see Chapter 8)) is focused on the cured surface by an elliptical reflector. The conveyor system is constructed of enamelled aluminium, sometimes with a coated fibreglass belt. A solid-state variable-speed control enables the operator to accurately adjust the cure rate to the particular formulation of ink or coating.

Intensity of UV radiation is measured in radiometric units (Tables 7.2 and 7.3), whereas intensity of visible light is measured in photometric units (Table 7.2) by radiometers.[241]

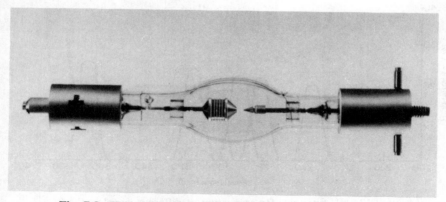

Fig. 7.8. The high-pressure xenon lamp, type CSX, Philips.[326]

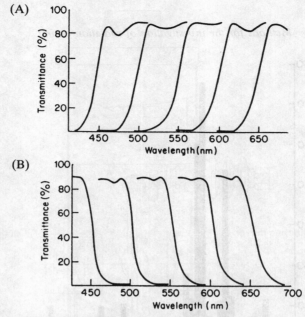

Fig. 7.11. Transmittance of the band pass edge filter sets: (A) long-wave pass, and (B) short-wave pass.[329]

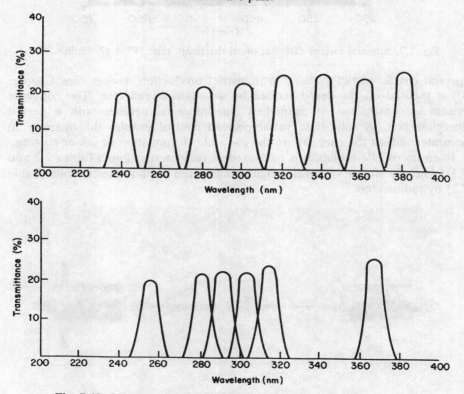

Fig. 7.12. UV filter sets for separation of different wavelengths.[329]

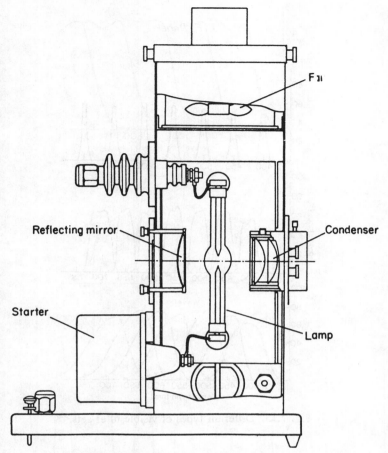

Fig. 7.9. Housing for a high-pressure xenon lamp.[328]

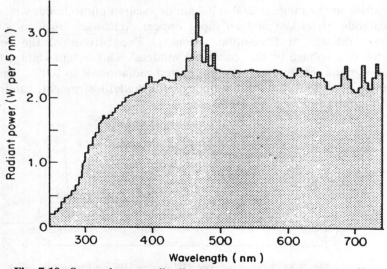

Fig. 7.10. Spectral energy distribution of the lamp CSX, Philips.[326]

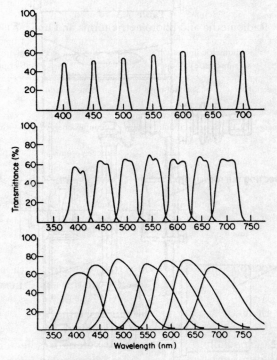

Fig. 7.13. Different types of visible filter sets.[329]

Radiometers consist of a filter, a sensor (radiation silicon detectors, which are sensitive in the range of 200–1100 nm or vacuum photodiodes with specific photocathode materials) and an input optical attachment such as a cosine-correction diffuser or fibre-optic assembly. The current of the sensor is converted to a voltage by the 'convertor module'. This output voltage (0–5 V) can monitor the relative intensity of emitted radiation of 0–100%. This value can simply be displayed using an analogue or digital instrument, calibrated in any radiometric and/or photometric units.

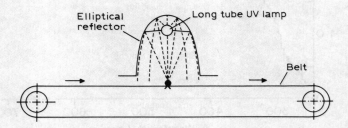

Fig. 7.14. Mini-conveyor UV-curing laboratory unit.

Table 7.3
Radiometric and photometric terms and units[241]

		Radiometric units (subscript e)		Photometric units (subscript v)	
		Non-spectral units	Spectral units	Non-spectral units	Spectral units
Energy (Q)	Radiant energy (Q_e)		Spectral radiant energy ($Q_e(\lambda)$)	Luminous energy (Q_v)	Spectral luminous energy ($Q_v(\lambda)$)
	Joules		Joules per unit wavelength	Lumen-seconds	Lumen-seconds per unit wavelength
Flux (Φ)	Radiant flux (Φ_e)		Spectral radiant flux ($\Phi_e(\lambda)$)	Luminous flux (Φ_v)	Spectral luminous flux ($\Phi_v(\lambda)$)
	Watts		Watts per unit wavelength	Lumens	Lumens per unit wavelength

The cosine-correction diffuser provides the detector with a cosine spatial response which means it will 'see' irradiance in the same manner as the flat surface of the substrate when it is exposed to UV-curing irradiation.

Measurements of UV lamps are very important in industrial processing installations (see Chapter 8). However, in many research laboratories, the intensity of light, especially for a given wavelength selected by filters, is determined by the number of photons emitted by a given lamp, using chemical

Table 7.4
Spectral radiometric units[241]

Radiometric units	Symbol	Unit	Spectral radiometric units	Related to a difference of the wavelength ($d\lambda$)	Unit
Radiant energy	Q_e	J	Spectral radiant energy	$Q_e(\lambda) = \dfrac{dQ_e}{d\lambda}$	J/nm
Radiant flux	Φ_e	W	Spectral radiant flux	$\Phi_e(\lambda) = \dfrac{d\Phi_e}{d\lambda}$	W/nm
Radiant emittance	M_e	W/m^2	Spectral radiant emittance	$M_e(\lambda) = \dfrac{dM_e}{d\lambda}$	W/m^2 nm
Radiant intensity	I_e	W/sr	Spectral radiant intensity	$I_e(\lambda) = \dfrac{dI_e}{d\lambda}$	W/sr nm
Radiance	L_e	W/m^2 sr	Spectral radiance	$L_e(\lambda) = \dfrac{dL_e}{d\lambda}$	W/m^2 sr nm
Irradiance	E_e	W/m^2	Spectral irradiance	$E_e(\lambda) = \dfrac{dE_e}{d\lambda}$	W/m^2 nm
Irradiation	H_e	J/m^2	Spectral irradiation	$H_e(\lambda) = \dfrac{dH_e}{d\lambda}$	J/m^2 nm

actinometers.[241] The knowledge of the number of photons produced by a lamp for a given wavelength is important for the determination of the quantum yield of a photochemical process.

Chemical actinometry has the advantage over physical methods in that the actinometer can be irradiated under conditions similar to those of the photoreaction to be studied. This eliminates the need to make corrections due to reflectance and non-uniformity of the incident light beam. The liquid-phase chemical actinometer traditionally used is the potassium ferrioxalate system, but its use requires the careful preparation of standard solutions and of a calibration graph before measurements can be made.[127,241]

Aberochromics Ltd and the School of Chemistry and Applied Chemistry, University of Wales College of Cardiff (Cardiff, UK), have developed the thermally stable fatigue-resistant photochromic reagent, 'Aberochrome 540', which is well suited for chemical actinometry within the ranges 310–370 and 435–545 nm, and is easy to use.

The reversibility of the photochemical process, eqn (1),[128] allows the actinometer to be used repeatedly, thus obviating the need for a fresh sample of solution for each measurement.

$$\text{(structure)} \rightleftharpoons \text{(structure)} \tag{1}$$

After photocoloration, the solution can be bleached by exposure to white light and reused.

A practical procedure of using this actinometer is as follows: an approximately 5×10^{-3} M solution is prepared by dissolving Aberochrome 540 (25 mg) in pure toluene (20 ml). A known volume (V) of this solution is pipetted into a cuvette and the absorbance (if any) at 494 nm is noted. The stirred solution is then irradiated with a given monochromatic light (e.g. 366 nm) for a known period of time. A magnetic stirrer is recommended and care should be taken that this does not enter the light beam. It is essential that the solution absorbs all the incident light. After irradiation, the absorbance at 491 nm is measured. The quantum yield for colouring (ϕ_c) is temperature independent. The increase in absorbance (A) at 494 nm enables the photon flux (I_0) to be calculated from eqn (2):

$$I_0 = \frac{AVN_A}{\phi_c \varepsilon t} \quad \text{(photons/s)} \tag{2}$$

where V is the volume of solution irradiated (litres), N_A is Avogadro's number ($6 \cdot 023 \times 10^{23}$ mol^{-1}), ϕ_c is 0·20 (from 310–370 nm), ε is 8200 (dm^3 mol^{-1} cm^{-1} at 494 nm) (i.e. molar absorptivity or the molar extinction coefficient), and t is time (s).

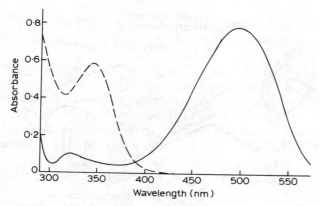

Fig. 7.15. UV–vis absorption spectra of Aberochrome 540: (– – –) before irradiation, and (——) after UV irradiation.[128]

In Fig. 7.15 are shown the UV and visible spectra of a 1×10^{-3} M solution of Aberochrome 540 before and after quantitative conversion into the coloured form.

EB-RADIATION-CURING DEVICES

EB curing offers a solvent-free and initiator-free process for coating and film applications. EB techniques usually involve the use of reactive formulations composed of an unsaturated monomer or oligomer, either by itself, or by compounding with other reactive components. Relatively low-energy EB radiation generated from a linear filament (electrocurtain) can easily polymerise materials containing reactive double bonds by electronic and ionic excitation.

Typical EB apparatus (Fig. 7.16) is equipped with a linear tungsten filament which is used to irradiate the samples. It can operate at a voltage between 100 and 500 kV giving an EB of 10–50 mA. Penetration depths at these voltages is between 5 and 20 μm. A conveyor speed of 10 m/min can be controlled (increased or decreased).[158,286]

A dosage (D) (Mrads/pass or MGy/pass) can be calculated from eqn (3):

$$D = KIS^{-1} \qquad (3)$$

where K is the value determined by the calibration of the dosage, received by the sample (Mrad (or MGy)/mA min), I is the EB current (mA), and S is the speed of conveyor (m/min).

The chemical efficiency measured in the yield of free radicals, or of any other product formed by high-energy radiation, is generally determined in the

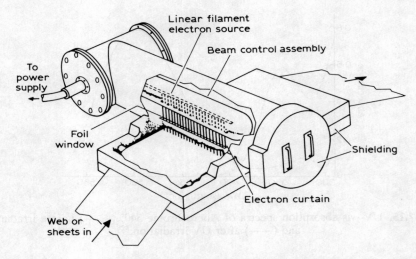

Fig. 7.16. General schematic of an electron curtain device used in radiation curing.[158]

form of the G value:[231,270,309]

$$G = \frac{\text{Number of molecules (free radicals) formed (or destroyed)}}{\text{Number of electron volts (eV) of energy absorbed}} \quad (4)$$

The G value is generally expressed as a number of free radicals formed (or consumed) by absorption of 100 eV of high-energy radiation.

The number of molecules reacting (formed or destroyed) can be determined by any conventional method (e.g. IR and/or Raman spectroscopy), and the number of electron volts of energy absorbed can be obtained by integration of the Gaussian function for the EB shape (Fig. 7.17).

In a particular EB processing unit, a polymer film coated with a given formulation is moving under an EB (Fig. 7.18). A sample moving under a Gaussian beam does not experience a uniform dose rate. A sample moving from the left to right in the EB curtain at first receives a rather small dose (Fig. 7.17, point X_p) when it is in the right wing of the Gaussian function. The dose rises until the sample reaches the peak of the Gaussian function (X_{max}) and then decreases as the sample passes away from the beam ($-X_p$). The total absorbed dose (D) which the sample receives as a result of passing under the beam is obtained by integration of the Gaussian function for the beam shape:[231]

$$D = \frac{P_0 l \sqrt{\pi}}{2\rho h L X_p v} \quad (5)$$

where P_0 is the incident power density at the centre of the EB (cf. Fig. 7.17), l is the half-width of the Gaussian beam (cf. Fig. 7.17), ρ is the sample

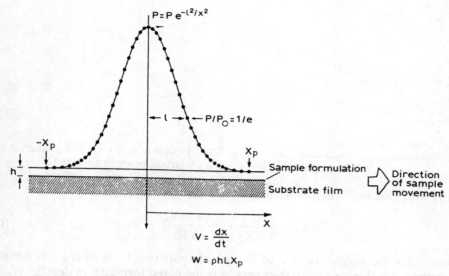

Fig. 7.17. Illustration of a sample moving under a Gaussian-shaped EB.[231]

(formulation) density, L is the width of sample, h is the height of sample, X_p is the width of the X direction (*cf.* Fig. 7.17), and v is the line velocity.

Equation (5) shows that the absorbed dose is directly proportional to the incident power density at the centre of the beam, the width of the beam, and inversely proportional to the velocity of the line.

It is important to obtain the cross-linking G value, since both cross-linking and degradation occurs in cured formulations upon EB irradiation. The relation between solubility (s) and cross-linking G value ($G(X)$) and degradation G value ($G(S)$) is expressed according to Charlesby's theory[45–48] by the

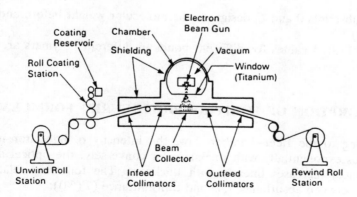

Fig. 7.18. A typical commercial EB processing unit.[231]

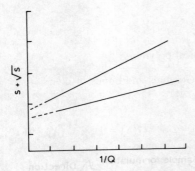

Fig. 7.19. Typical plot of $s + \sqrt{s}$ as a function of the reciprocal of electron dose $(1/Q)$.

following relation, when molecular weight distribution is random:

$$s + \sqrt{s} = \frac{G(S)}{2G(X)} + \frac{100 l d N_A e}{2G(X) \Delta E \bar{M}_n} \times \frac{1}{Q} \quad (6)$$

where s is the weight fraction of the soluble part ($s = 1 - g$ where g is the gel fraction), l is the cured film thickness, d is the cured formulation density, N_A is the Avogadro's number, e is the electron charge, $\bar{M}_n$ is the number average molecular weight, ΔE is the electron energy loss, when an electron passes through a cured polymer,[18] and Q is the exposure dose.

The value of $s + \sqrt{s}$ is, generally, plotted as a function $1/Q$ (Fig. 7.19). The slope of each experimental line gives the $G(X)$ value and the intercept gives the $G(S)$ value.

The $G(S)$ and $G(X)$ values can also be calculated from the change of $\bar{M}_n$ and $\bar{M}_w$ with dose by the following relationships:[82,156]

$$(\bar{M}_n^{-1})_Q = (\bar{M}_n^{-1})_0 + (G(S) - G(X)) \frac{Q}{100 N_A} \quad (7)$$

$$(\bar{M}_w^{-1})_Q = (\bar{M}_w^{-1})_0 + (G(S) - 4G(X)) \frac{Q}{100 N_A} \quad (8)$$

where subscripts 0 and Q designate the molecular weight before and after the dose (Q).

Typical $G(S)$ values for different bonds in different polymers are shown in Table 7.5.

ABSORPTION OF LIGHT BY PHOTOCURED FORMULATIONS

According to the Beer–Lambert law, the intensity of light transmitted (T) decreases exponentially with thickness, or conversely, the optical density (or absorbance) increases linearly with thickness. The following relationship is defined between absorbance (A) and transmittance $(T(\%))$:

$$A = \log_{10}(I_0/I) = -\log_{10} T = \log_{10}(100/T) \quad (9)$$

Table 7.5
Bond breaking (G) values in typical polymers[257]

Polymer	Formula unit	G values and bonds broken		
Polypropylene	$-CH_2\overline{m}\underset{\underset{CH_3}{\mid}}{\overset{\overset{H}{\mid s}}{C}}-$	$G(H_2) = 2.8$, s	$G(S) = 0.2$, m	$G(X) = 0.16$ s
Poly(vinyl chloride)	$-\underset{}{\overset{\overset{Cl}{\mid s1}}{CH}}-\underset{}{\overset{\overset{H}{\mid s2}}{CH}}-$	$G(HCl) = 13$, s1, 2	$G(X) = 0.1$ s2	
Polyethylene	$-\overset{\overset{H}{\mid s}}{CH}\overline{m}\overset{\overset{H}{\mid s}}{CH}-$	$G(H_2) = 3.7$, s	$G(S) = 0.2$, m	$G(X) = 1.0$ s
Polyoxymethylene	$-O\overline{m}\overset{\overset{H}{\mid s}}{CH}-$	$G(H_2) = 1.7$, s	$G(S) = 11$, m	$G(X) = 6$ s, m
Poly(methyl methacrylate)	MeAcO $-CH_2\overline{m}\underset{\underset{CH_3}{\mid}}{\overset{\overset{\mid s}{}}{CH}}-$	$G(\text{ester}) = 2.5$, s	$G(S) = 3.5$ m	
Polystyrene	$-\underset{\underset{H}{\mid s}}{\overset{\overset{H}{\mid s}}{C}}\overline{m}\underset{\underset{Ph}{\mid}}{\overset{\overset{\mid s}{}}{C}}-$	$G(H_2) = 0.03$, s	$G(S) = 0.01$, m	$G(X) = 0.03$ s
Polytetrafluoroethylene	$-CF_2\overline{m}CF_2-$	$G(S) = 0.3$ m		
PDMS	$H\overset{s2}{-}CH_2$ $\underset{\mid s1}{}$ $-O-Si-$ $\underset{\mid s1}{}$ $H\underset{s2}{-}CH_2$	$G(H_2, CH_4) = 3.0$, s1, 2	$G(X) = 3.0$ s1, 2	
Polyisobutylene	$-CH_2-\underset{\underset{CH_3}{\mid s}}{\overset{\overset{CH_3}{\mid s}}{C}}-$	$G(H_2, CH_4) = 2.1$, s	$G(S) = 4.0$ m	
Polybutadiene	$-CH_2-CH$ s $=CH-CH_2-$ m	$G(C{=}C) = 12$, m2	$G(X) = 2.0$ m2	

In spectrophotometric practice, quantitative analysis (i.e. study of photoinitiator photolysis kinetics) is based on the application of the Beer–Lambert law, which is given by eqn (10):

$$A = \log_{10}(I_0/I) = \varepsilon c l \quad \text{(dimensionless)} \tag{10}$$

where I_0 is the intensity of incident light, I is the intensity of the light transmitted through the sample film or sample solution, l is the thickness of the sample (path length), ε is the molar absorptivity (also called the molar extinction coefficient), and c is the sample concentration.

ε is constant for a particular compound at a given wavelength. Where values for ε are very large, it is convenient to express it as its logarithm ($\log_{10} \varepsilon$). The absorbance (extinction) is a cumulative property for a mixture of two or more components (e.g. for hybrid photosensitisers):

$$A = l(\varepsilon_1 c_1 + \varepsilon_2 c_2 + \cdots + \varepsilon_n c_n) \tag{11}$$

The concentration of two components may be determined when four values of ε are known, and when the measurements are made for two wavelengths:

$$A' = l(\varepsilon'_1 c_1 + \varepsilon'_2 c_2) \quad \text{for } \lambda' \tag{12}$$

$$A'' = l(\varepsilon''_1 c_1 + \varepsilon''_2 c_2) \quad \text{for } \lambda'' \tag{13}$$

The wavelengths λ' and λ'' are chosen such that one component strongly absorbs these wavelengths whereas the other component absorbs them much less (Fig. 7.20).

The validity of the Beer–Lambert law should always be tested before using it for quantitative analysis. To test eqn (10) it is necessary to measure absorbance (A) as a function of concentration (c) at a fixed wavelength (λ) and

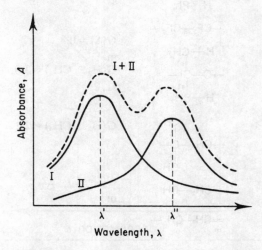

Fig. 7.20. Overlapping spectra of two components I and II and the spectrum of a mixture of two components I + II.

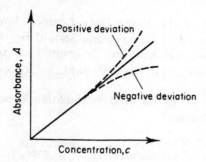

Fig. 7.21. Testing the validity of the Beer–Lambert law, i.e. calibration curves for quantitative analysis.

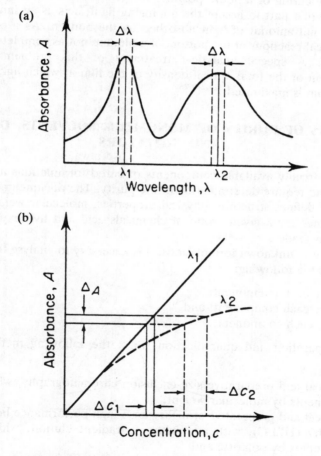

Fig. 7.22. Test of the Beer–Lambert law: (a) for the absorption spectrum; (b) at wavelengths λ_1 and λ_2.

cell (or film) path (Fig. 7.21). If the Beer–Lambert law is obeyed over the concentration range tested, a straight line should be obtained through the origin. Deviations from the law are designated as positive or negative. Sometimes the Beer–Lambert law is obeyed at one wavelength (λ_1) but not at another (λ_2) (Fig. 7.22).

Absorption of light by photocured formulation depends on many parameters, such as the following:

- UV source emission spectrum;
- absorption spectra of photoinitiators, monomer, polymer, pigments and other additives;
- substrate reflectance; and
- thickness of a photocured sample.

During photocuring of a liquid polymer formulation film, the top of the film, which is the first part to accept the impinging light, acts as an inner filter and reduces the net amount of light absorbed by the bottom. As a consequence, photochemical reactions at the bottom of a film are not as complete as they are at the top. A special analytical treatment of this problem allows the determination of the total optical density of the film at which light absorption by the bottom is maximised.[124,249,285]

ANALYSIS OF PURITY OF MONOMERS, SOLVENTS, DILUENTS AND INITIATORS

Most commercially available components of cured formulations are industrial chemicals and require determination of the purity. The documentation of these components defines structure, physical properties, molecular weight, viscosity and sometimes for a given group of chemicals acid and hydroxyl values and esterification grades.

In the case of unknown formulations, it is necessary to analyse them in order to carry out the following:

- separate each component;
- quantify each component; and
- identify each component.

For the separation and quantification steps, the following methods can be employed:

- analytical and preparative size-exclusion chromatography, which separate components by molecular weight;
- analytical and preparative straight-phase high-performance liquid chromatography (HPLC), with and without gradient elution, which separates components by polarity; and
- analytical capillary gas chromatography, which separates components by volatility.

For the identification of the components the following techniques can be used;

- UV–vis, FTIR and/or Raman spectroscopy;
- proton NMR; and
- electron impact and chemical ionisation mass spectrometry.

The details of these methods and techniques can be found in a number of books.[35,75,92,208,240]

KINETICS OF PHOTOINITIATED POLYMERISATION

A kinetic analysis of reactions which occur under UV–vis radiation is generally based on the following definitions.

- *Photoinitiated polymerisation* is a polymerisation process initiated by free radicals formed from photolysis of a photoinitiator.
- *Photopolymerisation* is a polymerisation process in which every chain-propagating step involves a photochemical reaction.

From these statements, it follows that photopolymerisation should be considered a multistep reaction, in contrast to photoinitiated polymerisation, which is a chain process.

Photopolymerisation can, in principle, be divided into two classes:

1. photopolymerisation, in which propagation occurs by the reaction of a molecule in an excited state with another molecule in the ground-state; and
2. photopolymerisation, in which propagation occurs by the reaction of a reactive ground-state species, formed earlier by a photochemical reaction (e.g. radical recombination).

Only photoinitiated polymerisation is responsible for the curing process in which are formed cross-linked structures. There are many experimental methods used for kinetic investigation of photoinitiated polymerisation (Table 7.6). Some of them are discussed in the following sections.

The photoinitiated free-radical polymerisation, can be considered in terms of four elementary steps with assumption that chain transfer reaction is ignored.[72,76,160,180,226,242]

1. Formation of free radicals (R·) from photoinitiator (I):

$$I + h\nu \xrightarrow{k_d} R^{\cdot} + R^{\cdot} \qquad (14)$$

The rate of the photoinitiator decomposition (R_d) is given by eqn (15):

$$R_d = k_d[I] \qquad (15)$$

where k_d is the radical formation rate constant (litres/mol s), and [I] is the concentration of photoinitiator (mol).

Table 7.6
Performance analysis of various methods used for kinetic investigation of light-induced polymerisations[69]

Characteristics of analytical methods	Real-time analysis	Response time (s)	High-intensity operation	Rate (R_p) evaluation	Unsaturation measurements	Post-polymerisation monitoring	Properties evaluation	Ref.
IR spectroscopy	No	>3	Yes	Yes	Yes	No	Yes	57, 67
^{13}C NMR spectroscopy	No	>10	Yes	No	Yes	No	No	232
Photoacoustic spectr.	No	>1	No	Yes	No	No	Yes	254, 266
Calorimetry (DSC)	Yes	>1	No	Yes	No	No	No	62, 181, 294
Dilatometry	Yes	>1	No	Yes	No	No	Yes	64, 235
IR radiometry	Yes	0.5	Yes	No	No	No	Yes	282
Laser nephelometry	Yes	0.1	Yes	No	No	Yes	No	68, 99
Laser interferometry	Yes	0.001	Yes	No	No	Yes	No	26, 66
Real-time IR spectr.	Yes	0.1	Yes	Yes	Yes	Yes	Yes	69, 72, 73

2. Addition of the first monomer molecule (M) to the initiator radical (R·):

$$R^{\cdot} + M \xrightarrow{k_i} RM^{\cdot} \qquad (16)$$

The rate of chain initiation (R_i) is given by eqn (17).

$$R_i = k_i[R^{\cdot}][M] \qquad (17)$$

where k_i is the chain initiation rate constant (litres/mol s). Because the number of moles of free radicals (R·) formed per second by decomposition of photoinitiator (I) (eqn (14)) is twice as large as the number of moles of photoinitiator disappearing per second, the effective rate of the chain initiation (R_i) is given by eqn (18):

$$R_i = 2k_i[I] \qquad (18)$$

Equation (16) has no influence on the kinetics of initiation of free-radical polymerisation, because it occurs much more rapidly than eqn (14). For the simplest case in which all radicals formed by eqn (14) can react further by eqn (16) with monomer molecules:

$$R_i = R_d \qquad (19)$$

3. Propagation reaction in the next monomer molecules are added to the growing $RM_n^{\cdot}$ radical:

$$RM^{\cdot} + nM \xrightarrow{k_p} RM_{n+1}^{\cdot} \text{ (or P}^{\cdot}) \qquad (20)$$

The rate of propagation (R_p) is given by eqn (21):

$$R_p = -\frac{d[M]}{dt} = k_p[RM^{\cdot}][nM] = k_p[P^{\cdot}][M] \qquad (21)$$

where k_p is the propagation rate constant (litres/mol s), [M] is the monomer concentration (mol), and [P·] is the sum of concentrations of all RM· radicals (including radicals R·)(mol).

4. Bimolecular termination reaction:

$$M_n^{\cdot} + RM_m^{\cdot} \xrightarrow{k_t} \text{unreactive species} \qquad (22)$$

The rate of termination (via a combination and/or a disproportionation reaction) is given by eqn (23):

$$R_t = -\frac{d[M]}{dt} = 2k_t[P^{\cdot}]^2 \qquad (23)$$

where k_t is the termination rate constant (litres/mol s). Bimolecular termination consists of three definable steps.

(i) Two radicals migrate together via translational diffusion.

(ii) The radical centres reorientate by segmental diffusion.
(iii) They overcome chemical activation barrier and react.

The activation energy for chemically controlled free-radical reactions is often very low. The termination reaction is, therefore, diffusion controlled for most of the monomer conversion range.[252,319,322]

Determination of photoinitiator decomposition

Photoinitiators (I) used in photocuring can be divided into three groups.

1. Photoinitiators which intramolecularly cleave into radicals ($R_1^{\cdot}, R_2^{\cdot}$), which initiate free-radical polymerisation:

$$I \xrightarrow{h\nu} I^* \tag{24}$$

$$I^* \longrightarrow R_1^{\cdot} + R_2^{\cdot} \tag{25}$$

where I^* is the excited singlet (S_1) or triplet (T_1) state of an initiator.

2. Photoinitiators which abstract hydrogen intermolecularly from hydrogen-donating molecules (RH):

$$I^* + RH \longrightarrow {}^{\cdot}IH + R^{\cdot} \tag{26}$$

3. Photoinitiators which form, with a co-initiator (AH), a charge transfer (CT) complex and further dissociate into radicals, which initiate free-radical polymerisation:

$$I^* + AH \longrightarrow [I \cdots AH]^* \longrightarrow {}^{\cdot}IH + A^{\cdot} \tag{27}$$

Note: The terminology employed to describe a photoinitiator or photosensitiser is unfortunately wrongly used by several authors.

- The term *photoinitiator* should be used to describe a chemical compound or a chemical system which absorbs light and dissociates (photolyses) into free radicals.
- The term *co-initiator* (*activator*) should be used to describe a chemical compound or a chemical system which does not absorb light, but is nevertheless indirectly involved in the production of initiator radicals in a photochemical reaction.
- The term *photosensitiser* should be reserved to describe a chemical compound or a chemical system which sensitises (photosensitises) photoreaction by an energy transfer mechanism from an acceptor to the donor molecule.

Each photoinitiator is characterised by the following:

- absorption spectrum;
- energy and lifetimes of S and/or T state(s);
- efficiency of radiation and/or radiationless physical processes; and
- quantum yield of photodecomposition.

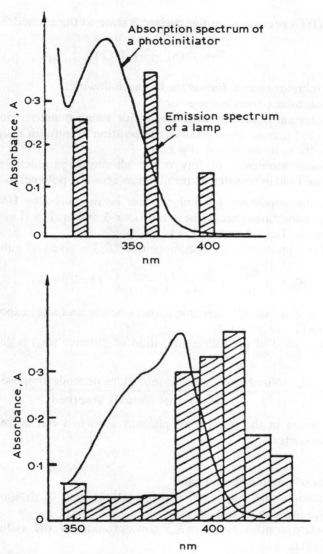

Fig. 7.23. Adaptation of absorption spectra of different photoinitiators and emission spectrum of a given lamp used for irradiation.

Photoinitiator can only be photodecomposed if its absorption spectrum covers (even partially) the emission spectrum of a lamp used for irradiation (Fig. 7.23), i.e. photoinitiator must absorb effectively the incident radiation.

Photoinitiator (decomposition) efficiency depends on several factors.

- Efficient population of the reactive excited S and/or T state(s) requires desirable absorptivity characteristics, and a high efficiency of intersystem crossing (ISC) process from the excited S state to the excited T state:

$$I \xrightarrow{h\nu} I^*(S) \xrightarrow{ISC} I^*(T) \qquad (28)$$

- Efficient initiator radical formation by the following:
  - intramolecular photocleavage; or
  - intermolecular hydrogen abstraction with a high quantum yield.
- Radicals (R·) formed from photodecomposition of initiator must be highly reactive with monomers and oligomers.
- The initiator molecule, or any of its photolysis products, should not function as a chain transfer or termination agent in polymerisation.

Photoinitiator decomposition (photolysis) can be measured by UV–vis spectroscopy under conditions that it obeys the Beer–Lambert law (i.e. there is no deviation from the Beer–Lambert law).

The rate of the photoinitiator decomposition (R_d) is given by eqn (29):

$$R_d = -\frac{d[I]}{dt} = \left(\frac{A_1 - A_2}{A_0}\right)\left(\frac{[I]}{t_1 - t_2}\right) \quad \text{(mol/litre s)} \qquad (29)$$

where A_0, A_1 and A_2 are UV–vis absorbances before and after exposure to the light time t_1 and t_2, respectively.

The quantum yield of photodecomposition of initiator (ϕ_d) is given by eqn (30):

$$\phi_d = \frac{R_d}{I_a} = \frac{\text{Number of initiator molecules photodecomposed}}{\text{Number of photons absorbed}} \qquad (30)$$

Note: I_a expressed in the number of photons absorbed can be measured by chemical actinometers.

Determination of polymerisation rate

The polymerisation rate (i.e. rate of propagation—rate of disappearance of monomer) (R_p) is given by eqn (21).

At steady-state conditions ($R_i = R_t$) concentration of the radicals [P·] is given by eqn (31):

$$\frac{d[P^\cdot]}{dt} = R_d - R_t = 0 \qquad (31)$$

where R_i is the rate of chain initiation (for the simplest case in which all radicals formed by photodecomposition of initiator react with monomer molecules, $R_i = R_d$), R_t is the rate of termination (eqn (23)), and R_d is the rate of the photoinitiator decomposition:

$$R_d = \phi_d I_a \qquad (32)$$

where ϕ_d is the quantum yield of photodecomposition of initiator (i.e.

quantum yield for production of radicals; see the section entitled 'Determination of Photoinitiator Decomposition'—p. 354), and I_a is the absorbed intensity.

By rearranging eqn (31), the following can be obtained:

$$R_d = R_t = \phi_d I_a = 2k_t[P^\cdot]^2 \tag{33}$$

and

$$[P^\cdot] = \left(\frac{\phi_d I_a}{2k_t}\right)^{0.5} \tag{34}$$

The rate of the photoinitiator decomposition (i.e. the rate of radical production) is given by eqn (32),[263,294] where I_a is the absorbed intensity:

$$I_a = I_0 A \tag{35}$$

where I_0 is the intensity of incident light, and A is the absorbance:

$$A = \varepsilon[I]l \tag{36}$$

where ε is the molar extinction coefficient of the photoinitiator, $[I]$ is the molar photoinitiator concentration before irradiation, and l is the thickness of the sample (path length).

Assuming the validity of the Beer–Lambert law:

$$I_a = I_0(1 - \exp(-\varepsilon[I]l)) = I_0(1 - \exp(-2 \cdot 303\varepsilon[I]l)) \tag{37}$$

However, when a laser is used as a source of radiation:[65]

$$I_a = I_0(1 - \exp(-2 \cdot 303\varepsilon[I]l)) \frac{10^7 p}{l} \tag{38}$$

where I_0 is the intensity of incident light (or photon flux) of the laser beam (einstein/s cm) (note: 1 einstein = 6×10^{23} photons), and p is the ratio of the measured power output of the laser to the maximum power available.

Finally, the polymerisation rate (R_p) is expressed as eqn (39):

$$R_p = -\frac{d[M]}{dt} = k_p\left(\frac{\phi_d I_a}{2k_t}\right)^{0.5}[M] \tag{39}$$

i.e. the polymerisation rate depends on the square root of the absorbed light intensity (I_a) and on the first power of monomer concentration $([M])$.[32,145,152,300] Because, according to eqn (40) I_a and hence the polymerisation rate also depends on the absorbance (A) (or molar extinction coefficient (ε)), photoinitiator concentration $([I])$ and sample thickness (l).

$$I_a = I_0 A = I_0 \varepsilon l[I] \tag{40}$$

The ratio $R_p^2/(I_0(1 - \exp(-2 \cdot 303\varepsilon l[I])[M]^2)$ is directly related to the initiation quantum yield (ϕ_d):[67]

$$\frac{R_p^2}{I_0(1 - \exp(-2 \cdot 3\varepsilon l[I])[M]^2} = \phi_d \frac{k_p^2}{2k_t} \tag{41}$$

Table 7.7
Light absorption characteristics of various photoinitiators[152]

Initiator	λ_{max} (nm)	ε (mol/cm)	Concentration (mM)
Azobisisobutyronitrile	360	11·9	18·03
2,2'-Azobis(4-cyanovaleric acid)	345	20·2	10·62
Benzoin methyl ether	341	215	0·993
Benzoin isopropyl ether	325	280	0·766
2,2'-Diethoxyacetophenone	335	72	2·98

can be used for the comparison of initiation effectiveness of different photoinitiators used for photocuring of a given formulation.

An increase of the photoinitiator concentration causes the following.

- A decrease in the induction period, due to more radicals produced, which can initiate polymerisation, and react with inhibiting oxygen and other inhibitor molecules.
- A decrease of the amount of residual uncured monomer molecules. For example, increasing the initiator concentration from 1 to 5%, can decrease amount of unreacted monomer molecules from 22 to 8%.[72]
- According to eqn (39), as the photoinitiator concentration is increased, the initial rate of polymerisation will first increase and then asymptotically approach a constant value. After a certain threshold photoinitiator concentration is exceeded, the rate of polymerisation will be essentially independent of initiator concentration. This threshold photoinitiator concentration (Table 7.7) will depend on the values of ε and l (see eqn (40)). For low initiator concentrations exists the square-root dependence of initial rate on the initial photoinitiator concentration.[152]
- For photoinitiators with high molar absorptivity, the initial rate of polymerisation will not exhibit half-order dependence on initial initiator concentration. The exact relationship involving the exponential function should be used for interpreting the rate data.[152]

The polymerisation rate (R_p) can be presented in a number of plots:

- polymerisation rate (R_p) (mol/litre s) versus square root of intensity (I_0) (W/m^2)$^{0.5}$ of incident UV radiation (at constant concentration of photoinitiator (mol/litre));
- polymerisation rate (R_p) versus square root of photoinitiator molar concentration ([I]) (mol/litre)$^{0.5}$;
- polymerization rate (R_p) versus molar concentration of pure monomer ([M]) (mol/litre);
- plot of R_p/[M] (per s) versus [M] (under constant UV irradiation (W/m^2)) at constant concentration of a photoinitiator (mol/litre);

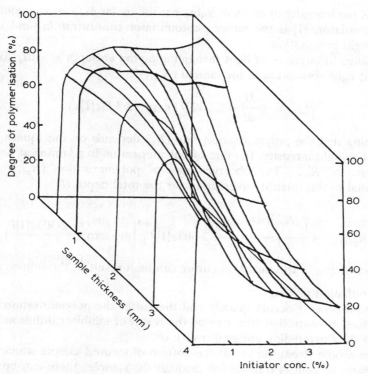

Fig. 7.24. Degree of polymerisation (%) as a function of sample thickness (mm) and initiator concentration (wt %).[123]

- plot of $R_p/[M][I]^{0.5}$ (per s)(mol/litre)$^{0.5}$ versus [M] (under constant UV irradiation (W/m^2) at constant [I]/[M] ratio).

It is very useful to plot a diagram of the degree of polymerisation as a function of photoinitiator concentration and sample thickness (Fig. 7.24). From this diagram it is easy to see that as the concentration of photoinitiator increases the degree of polymerisation reaches a maximum and then decreases. This effect is very marked for film thicknesses greater than 1 mm. The optimum level of initiator corresponding to a maximum cure decreases as the sample depth increases. Photoinitiators with a high extinction coefficient at the principal wavelength of the UV source give rise to optimum concentrations at lower initiator levels. It should, therefore, be possible to predict the appropriate initiator concentration for a particular film thickness.

The rate of polymerisation depends on light intensity, so the thickness of the sample will obviously affect the observed rate. The intensity of light reaching a depth (x) is given by eqn (42):[294]

$$I_x = I_0 \exp(-2\cdot 303\varepsilon[I]x) \tag{42}$$

where I_0 is the intensity of incident light, ε is the molar extinction coefficient of the photoinitiator, [I] is the molar photoinitiator concentration, and x is the depth of light penetration.

The change in intensity of light through a lamina at depth x, $-dI_x/dx$, gives the rate of light absorption in the lamina ($R_{I(x)}$):

$$R_{I(x)} = -\frac{dI_x}{dt} = I_0 2\cdot 303 \varepsilon I \exp(-2\cdot 303 \varepsilon [I] x) \tag{43}$$

By assuming that the polymerisation rate (R_p) depends on the square root of the absorbed light intensity, the rate of polymerisation in a lamina at depth x is proportional to $R_{I(x)}^{0.5}$. The observed rate of polymerisation ($R_{p(x)}$) is then proportional to this quantity averaged over the total depth (l):

$$(R_{p(x)}^{0.5})_{av} = \frac{\int_0^l R_{I(x)}^{0.5} dx}{l} = \left(\frac{4I_a}{l^2} 2\cdot 303 \varepsilon [I]\right)^{0.5} \left[1 - \exp\left(\frac{-2\cdot 303 \varepsilon [I] l}{2}\right)\right] \tag{44}$$

A low conversion in the radiation curing can be a result of the following:

- low radiation power;
- gelation process occurs quickly and thereafter the polymerisation process is diffusion controlled (this can be the result of inhibitor diffusion into gel from the surrounding uncured resin); or
- beam profile (from the top to the bottom of a cured sample where change intensity of light is transmitted through the sample, there can be a point where cure drops off).

In a number of photoinitiated polymerisations the polymerisation rate (R_p) is very low, i.e. monomer conversion is low. The reason for such a limited conversion can be the dead-end polymerisation. The phenomenon of dead-end polymerisation is encountered in a batch, free-radical polymerisation when the initiator is depleted before all of the monomer is converted to polymer. Under such conditions the polymerisation does not proceed to completion and the conversion of monomer approaches a limiting value asymptotically. A number of papers present an interpretation of experimental results to determine rate constants for this type of polymerisation.[13,49,81,111,131,153,261,267,290,291,295]

Most studies of the kinetics of free-radical polymerisations are limited to low conversion. However, several theories have been proposed to explain the features of high conversion polymerisations.[31]

Free-radical polymerisation to high conversions in bulk can be divided into three stages:[105]

1. a low conversion region in which the steady-state radical concentration is very low (10^{-7}–10^{-8} M) and k_p and k_t are independent of conversion to a first approximation;
2. a region above the gel point where the polymer chains become entangled and where the increased viscosity of the polymerisation mixture results in

a marked change in the diffusion characteristics of the growing polymer radicals—these two effects cause a rapid decrease in the magnitude of k_t and an increase in the radical concentration, consequently the polymerisation rate begins to rise rapidly (the Trommsdorff effect); and

3. a region at high conversion where the polymerising mixture becomes a glass and diffusion of monomer becomes restricted, thus resulting in a marked decrease in the polymerisation rate.

Determination of polymerisation rate (monomer conversion) by spectroscopical methods

The polymerisation rate (R_p), i.e. the rate of disappearance of monomer (rate of monomer conversion), can be determined by measuring the decrease by IR and/or Raman absorption for the double bond (existing in an unsaturated monomer) using eqn (45):[65]

$$R_p = -\frac{d[M]}{dt} = \left(\frac{A_1 - A_2}{A_0}\right)\left(\frac{[M]_0}{t_1 - t_2}\right) \text{ (mol/litre s)} \quad (45)$$

where A_0, A_1 and A_2 are IR absorbances before and after exposure to the light during time t_1 and t_2, respectively; and $[M]_0$ is the molar initial concentration of monomer before irradiation.

The quantum yield of polymerisation (ϕ_p) is given by eqn (46):

$$\phi_p = \frac{R_p}{I_a} = \frac{\text{Number of monomer molecules polymerised}}{\text{Number of photons absorbed}} \quad (46)$$

After combination of the equations for R_p (eqn (45)) and for I_a (eqn (37)) the quantum yield ϕ_p is equal to[72]

$$\phi_p = \frac{R_p}{I_a} = \frac{(A_1 - A_2)[M]_0}{(t_2 - t_1)A_0 I_0 (1 - \exp(-2 \cdot 303 \varepsilon [I] l))} \quad (47)$$

This spectroscopical method can be applied under the following conditions:

- when the chosen peak width and position of a given C=C do not change upon polymerisation;
- when the baseline of this isolated band can be determined accurately; and
- when the intensity of the absorption is comparable to the C=C absorbance band of the monomer.

In order to overcome difficulties in obtaining accurate values for the thickness of a thin film, samples require the use of an internal standard, e.g. a particular reference band. The intensity of these bands should not change (increase or decrease) with conversion of monomer double bonds. However, this disadvantage can be corrected mathematically.

The degree of monomer conversion (α) can be calculated from eqn (48):

$$\alpha = \frac{[M]_0 - [M]_t}{[M]_0} \times 100 = \frac{(A_{810})_0 - (A_{810})_t}{(A_{810})_0} \times 100 \quad \text{(mol \%)} \quad (48)$$

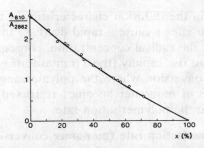

Fig. 7.25. A typical kinetic curve of ratio of IR absorbance of a given monomer at 810 cm^{-1} and 2862 cm^{-1} versus extent of double-bond conversion (α) measured by DSC.[165]

where $[M]_0$ and $[M]_t$ are monomer concentrations before and after time (t) irradiation, respectively; and $(A_{810})_0$ and $(A_{810})_t$ are the absorbances at 810 cm^{-1} of the sample before and after time (t) irradiation, respectively.

In Fig. 7.25 is presented an example of the determination of the extent of conversion (ratio of IR absorbances at 810 cm^{-1} (C=C) and 2862 cm^{-1} (CH$_2$ reference band) versus the extent of double bond conversion (α) measured with DSC). In Table 7.8 are collected some examples of a practical application of this method.

In a quantitative approach, the decrease of the IR absorption band characteristic of the double bond (i.e. 810 cm^{-1}) as a function of the exposure time (Fig. 7.26) allows precise evaluation of how many monomer functions (double bonds) have been polymerised as a function of light dose and then the deduction of the degree of conversion of the polymer formed.

In the real-time IR (RTIR) technique, the sample (1–100 μm) is simultaneously exposed to the polymerising UV-radiation beam and to the analysing IR beam which monitors the resulting drop of the absorbance at 812 cm^{-1} of the reactive double (C=C) bond (Fig. 7.27).[67,69,70,72,73]

The cure profiles (Fig. 7.28) can be directly recorded as a function of the exposure time by using the IR spectrophotometer recorder operated at

Table 7.8
Determination of monomer conversion (C=C disappearance) by IR and Raman spectroscopy

Measured decrease of C=C bond	Method	Internal standard (reference band)	Ref.
1641 cm^{-1}	Raman	C=O (1720–1730 cm^{-1})	52, 322
810 cm^{-1}	IR	CH$_2$ (2862 cm^{-1})	65, 165
1640 cm^{-1}	Raman	Aromatic C=C stretching (1615 cm^{-1})	186
1639 cm^{-1}	IR	Aromatic (1583 cm^{-1})	286
1640 cm^{-1}	IR	?	44, 61

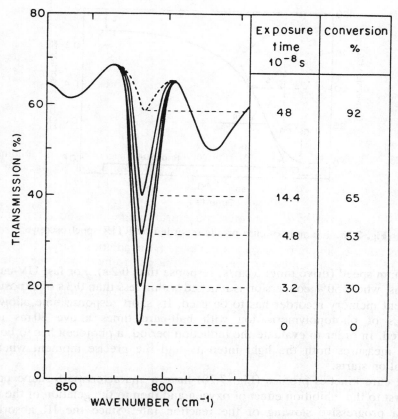

Fig. 7.26. A typical change of IR absorption spectra at 810 cm^{-1} of a given photoformulation exposed for various times (to nitrogen-laser pulses).[65]

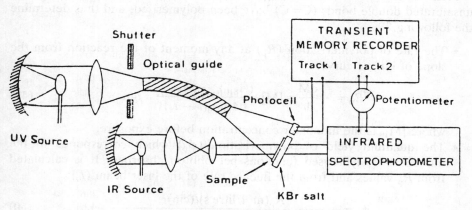

Fig. 7.27. Experimental set-up of the RTIR spectroscopy.[73]

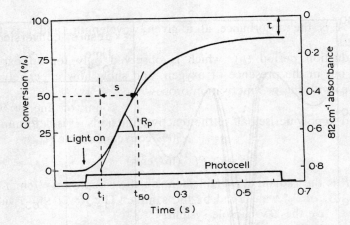

Fig. 7.28. Example of cure profile recorded by RTIR spectroscopy.[73]

maximum speed (drive time: 5 cm/s, response time 0·2 s). For fast UV-curable systems, where 50% conversion is reached within less than 0·3 s of exposure, a transient memory recorder has to be used. Its short response time allows the kinetics of photopolymerisation with half-cure times above 30 ms to be followed. In order to evaluate the induction period, a photocell has to be used which measures both the light intensity and the precise moment when the irradiation starts.

The cure kinetics profiles (Fig. 7.28) have a characteristic S-shape profile, due first to the inhibition effect of oxygen and then to the gelation of the resin, with a progressive slowing of the reaction rate. Since the IR absorbance increment is proportional to the number of reactive double bonds which have polymerised, the recorded curve directly translates into the dependence of percent conversion on irradiation time.

From kinetic curve profiles (Fig. 7.28), it is easy to evaluate how many unsaturated double bonds (C=C) have been polymerised, and thus determine the following.[70]

- The rate of polymerisation (R_p) at any moment of the reaction from the slope of kinetic curve:

$$R_p = -\frac{dM}{dt} = [M]_0 \frac{(A_{810})_{t_1} - (A_{810})_{t_2}}{(A_{810})_0 (t_2 - t_1)} = tg\alpha \tag{49}$$

where $[M]_0$ is the monomer concentration before exposure.

- The quantum yield of polymerisation (ϕ_p) which corresponds to the number of polymerised functions per photon absorbed. It is calculated from R_p values and from the fluence rate of the laser beam (I_0):

$$\phi_p = \frac{R_p(\text{mol/litre s})l(\text{cm})}{10^3(1 - \exp(-2\cdot303 A)I_0(\text{einstein/cm}^2 \text{ s})} \tag{50}$$

where A is the absorbance at a given wavelength, and l is the film thickness.
- The induction period (t_i), which is observed only for an experiment carried out in the presence of oxygen, and show how effectively oxygen interferes with the polymerisation process.
- The photosensitivity (S), defined as the amount of energy which is required to polymerise half of the reactive functions. It is determined from the RTIR curves by measuring the 50% conversion time ($t_{0.5}$):

$$S = t_{0.5}Pf \quad (\text{mJ/cm}^2) \tag{51}$$

where P is the radiant power incident on the sample (mW/cm^2) and f is the fraction of light absorbed by the sample. The lower the S value, the more sensitive the UV-curable system.
- The residual unsaturation content (τ) can be calculated from eqn (52):

$$\tau = \frac{(A_{810})_t}{(A_{810})_0}[M]_0 \tag{52}$$

Its value depends very much on the monomer functionality and the chemical structure of monomer.

The RTIR spectroscopy can also be applied for some ultrafast curing reactions induced by laser radiation.[70–72]

The RTIR spectroscopy has a number of advantages over other existing methods used today to study photopolymerisation (cf. Table 7.6). The most important advantages are as follows:[73]

- Real-time monitoring, providing a precise analysis of the quasi-instant liquid–solid phase change in a fraction of second.
- The great sensitivity of IR spectroscopy allows very small changes in the monomer concentration to be detected, for example, less than 1% conversion in a 1-μm-thick coating (10 μg). By using an IR beam condenser, tiny areas, down to 1 mm^2, can be analysed in order to evaluate the cure regularity throughout the sample or to study small size samples.
- The kinetics of cure reactions can be studied over a very broad range of light intensity, i.e. starting from laboratory exposure lamps to those lamps used in industrial UV-curing devices. In order to reproduce the actual profiles of UV exposure and temperature that are found on a production line, the sample can easily be irradiated with a scanning beam in the spectrophotometer chamber. Repeating the process will simulate, very exactly, the successive exposures of the sample when more than one lamp is used for the curing. After each pass, the cure extent can be determined accurately until a tack-free coating is obtained.
- Quantitative results about the important kinetic parameters such as, induction period, polymerisation rate, photosensitivity, and residual unsaturation of tack-free and scratch-free coatings.

- The cured sample can be examined at any moment of reaction for evaluation of some of its physical characteristics, like hardness, abrasion resistance, flexibility, solvent resistance, gloss, heat resistance, etc.
- The measuring technique is not only restricted to UV-radiation polymerisation, but it can be extended to other types of radiation like lasers, microwaves, and EBs.

In spite of a number of advantages the RTIR spectroscopy has also several limitations, such as those listed below.

- The sample thickness is limited to the range 1–100 μm.
- The method is limited to samples containing black pigments, and a high concentration of other coloured pigments, especially when they have an absorption band in the region of 800 cm^{-1}.
- The coating support must be transparent to IR radiation, i.e. must be sodium chloride or potassium bromide crystal plates.
- The vertical position of examined samples in a number of IR spectrophotometers makes the examined fluid sample flow down. In order to solve this problem a thin monomer layer can be deposited between two polyethylene sheets.

Other very convenient methods which permit the real-time kinetics of the laser-induced polymerisations to be followed are based on the following:

- the variations of the refractive index (interferometric method) that occur upon polymerisations as a result of the shrinkage process;[26,66] and
- the turbidity of the irradiated system (nephelometric method) which grows during the laser-induced polymerisation process.[68]

Determination of polymerisation rate (monomer conversion) by liquid chromatography

High performance liquid chromatography (HPLC) is a very useful method for the determination of the polymerisation rate (R_p), i.e. the rate of disappearance of the monomer (rate of monomer conversion) during polymerisation.

After the polymerisation procedure, the polymer samples must be extracted with a solvent for a given time, e.g. 24–48 h, and the extracts analysed by HPLC for the amount of unreacted monomer. This method allows to operate with the 1 mg samples extracted in 400 μl solvent.[164,168] In Fig. 7.29 are shown, for example, chromatograms of extracts from polymerised sample under different times of photocuring.[168]

This method is particularly useful if the photocuring is studied by a photocalorimetric method (e.g. DSC) (see the section entitled 'Photocalorimetry'—p. 371). The DSC samples after the polymerisation procedure can be further extracted and measured by HPLC for the content of the unreacted monomer. Contribution of these two methods allows for direct comparison of the double bond conversion measured with DSC and amount of converted monomer from HPLC (Fig. 7.30).[162]

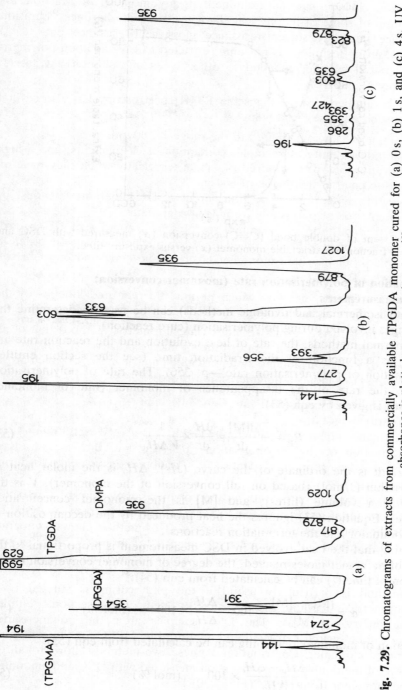

Fig. 7.29. Chromatograms of extracts from commercially available TPGDA monomer cured for (a) 0 s, (b) 1 s, and (c) 4 s. UV absorbance is plotted versus retention time.[168]

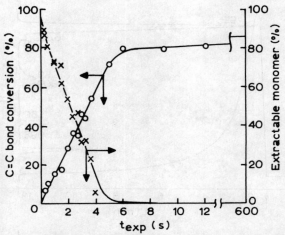

Fig. 7.30. Extent of double bond (C=C) conversion (α) measured with DSC and fraction of extractable monomer (x) versus exposure time.[162]

Determination of polymerisation rate (monomer conversion) by DSC measurements

DSC (both isothermal and dynamic methods) can be used to determine the heat changes involved during polymerisation (cure reaction).[1,2,34,258,296,317]

In isothermal methods, the rate of heat evolution and the reaction rate are recorded as a function of the irradiation time (see the section entitled 'Determination of polymerisation rate'—p. 356). The rate of polymerisation (R_p), i.e. the rate of the disappearance of monomer (rate of monomer conversion) is given by eqn (53):

$$R_p = -\frac{d[M]}{dt} = \frac{dH}{dt} \times \frac{1}{V \Delta H_0} \tag{53}$$

where dH/dt is the ordinate of the curve (J/s), ΔH_0 is the molar heat of polymerisation (J/mol) (based on full conversion of the monomer), V is the total reaction volume (litres), and [M] is the monomer concentration (mol/litre). Equation (53) ignores the heat produced by the decomposition of the photoinitiator and the termination reactions.

Assuming that the heat evolved in DSC measurement is proportional to the number of monomer moles reacted, the degree of monomer conversion (α) as a function of time (t) can be calculated from eqn (54):[34]

$$\alpha = \frac{[M]_0 - [M]_t}{[M]_0} \times 100 = \frac{\Delta H_t}{\Delta H_0} \times 100 \quad (\text{mol \%}) \tag{54}$$

The fraction of monomer remaining can be calculated from eqn (55):

$$\frac{\Delta H_0 - \Delta H_t}{\Delta H_0} \times 100 \quad (\text{mol \%}) \tag{55}$$

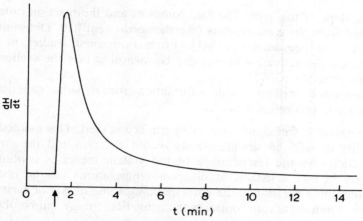

Fig. 7.31. Typical DSC curve measured during the photopolymerisation. The arrow indicates that the UV radiation is switched on.[83]

where $[M]_0$ and $[M]_t$ are monomer concentrations before and after time (t) irradiation, respectively; H_0 is the total heat of polymerisation for full conversion of the monomer (J/mol, J/g); and H_t is the apparent heat of polymerisation with the incomplete monomer conversion derived from the area under the total DSC curve up to time (t) (i.e. the heat evolved to the point of interest).

A DSC curve, such as is shown in Fig. 7.31, can be transformed into a degree of monomer conversion (α) versus time of irradiation (t) plot (Fig. 7.32). The initial polymerisation rate $(R_p)_0$ (mol/litre s) can be obtained from

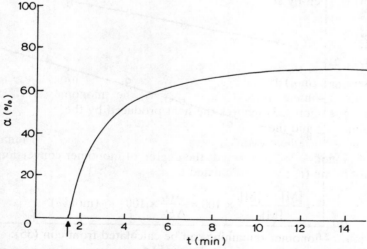

Fig. 7.32. Conversion of monomer (α) versus time plot of the photopolymerisation derived from the DSC curve (Fig. 7.31). The arrow indicates that the UV radiation is switched on.[83]

the initial slope of this plot. The rate constants and the reaction order can be determined from these parameters (see the section entitled 'Determination of k_p and k_t rate constants'—p. 381). From isothermal studies at different temperatures, the activation energy can be measured (see the section entitled 'Photocalorimetry'—p. 371).

The methods to extract kinetic information from dynamic experiments can be divided into two categories.

1. A method in which only one exotherm is analysed. This method uses the ability of DSC to simultaneously record the rate and the enthalpy of reaction. As the temperature of the system increases continuously, a series of rate constants at different temperatures can be obtained by determining dH/dt and the corresponding value of ΔH at each temperature. From these rate constants the activation energy can be obtained.

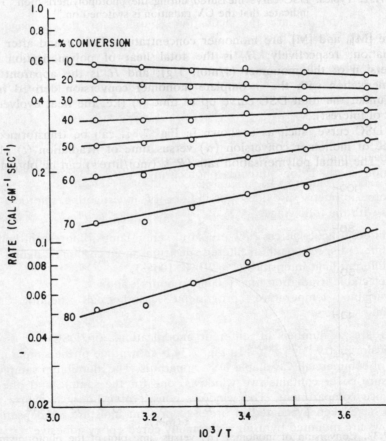

Fig. 7.33. A typical Arrhenius plot obtained from the DSC measurements for a given monomer (lauryl acrylate).[294]

2. This method is based on the fact that the peak exotherm temperature (T_p) varies in a predictable manner with the heating rate. A simple but accurate relationship between activation energy (E), heating rate and peak exotherm temperature can be obtained. This method permits calculation of the activation energy without previous knowledge of the reaction order and is also valuable as a precursor for the isothermal studies.[229,230]

To analyse the temperature dependence of the polymerisation rate of the monomers, an Arrhenius-type relation has to be assumed. The slope of the plot of a log-rate against reciprocal temperature is proportional to the apparent activation energy (E_{app}) for the polymerisation.

For a mechanism which assumes a bimolecular termination step and zero activation energy for photoinitiation, E_{app} can be calculated:

$$E_{app} = E_p - (E_t/2) \tag{56}$$

where E_p and E_t are the activation energies for the propagation and termination steps, respectively.

In Fig. 7.33 are shown Arrhenius plots of the data obtained by DSC for lauryl acrylate photopolymerisation. A negative apparent activation energy implies that the activation energy for termination is greater than the activation energy for propagation.[294]

Photocalorimetry

Photocalorimetry is based on the isothermal DSC method and is the most common method for studying photocuring.[6,7,33,60,83,85,90,96,97,134–140,145,157,161–164,167–170,179,181,210–214,225,256,284,294,305–307]

Photocalorimetry provides possibilities of investigating photocuring processes with the following:

- different light sources (low-pressure, xenon lamps, fluorescent lamps);
- different spectra (full or filtered spectrum, monochromatic light);
- different light intensities (1%, 10%, 100%);
- selectable irradiation time (50 ms to hours); and
- variable temperature programme (isothermal, heating, cooling, multistep).

There are a number of different modifications of DSC for isothermal photocalorimetry.[145,164,294,306] In Fig. 7.34 is shown one of the simple modifications of commercially available DSC apparatus. The aluminium sample holder enclosure cover contains two windows, one for the sample and one for the reference compartment. The windows consist of cylindrical quartz cuvettes which have been evacuated in order to prevent moisture condensation. The windows are mounted by using a thermally cured epoxy adhesive.

The polyurethane-foam-filled outer draught shield contains one large evacuated cuvette. On top of the draught shield an electrically driven shutter is

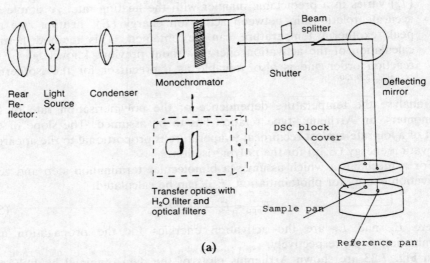

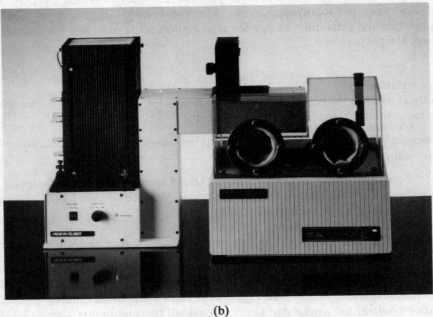

Fig. 7.34. (a) Schematic diagram of the optical set-up for the study of photopolymerisation in the isothermal conditions. (b) A commercially produced irradiation set-up designed to work with a Perkin-Elmer DSC 7 calorimeter, for the study of photopolymerisation in isothermal conditions.

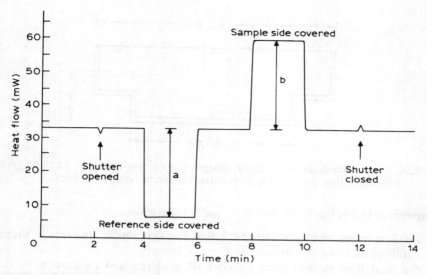

Fig. 7.35. Typical DSC curve obtained during the testing of equal intensity radiation incoming from both the reference-pan side and the sample-pan side.

mounted. The shutter also contains a filter holder to accommodate neutral density filters. On top of the shutter a lamp is mounted.

Application of a low-intensity fluorescent lamp is highly recommended, since the heat production of the fluorescent lamp is very small. It is important that the intensity of radiation incoming from both sample and reference pans is the same. For this purpose, empty graphite pans are inserted as sample, and the sample and reference sides are covered one at a time during measurements (Fig. 7.35). By this measurement, the optics are aligned in a way so that the sample and reference cells are irradiated with the same intensity (in the curve on Fig. 7.35 a = b).

The DSC sample weights are usually about 1 mg, corresponding to about 60 μm, when aluminium lids of standard sample pans are used. In order to maintain a uniform sample thickness with the sample sizes used (generally about 1–10 mg) it is sometimes necessary to use modified aluminium sample pans. Standard aluminium sample cups can be pressed in special forms into the shape shown in cross-section in Fig. 7.36. With an appropriate sample size, meniscus formation is minimised because of the step configuration. Changing the depth (d) of the sample cups it is possible to hold sample volumes of 5, 10, 15 or 20 μl. Actual sample size can be determined by weighing to ±0·05 mg.

Figure 7.37 shows a typical thermogram for the photopolymerisation reaction. The shape of the curve is similar in all kinds of samples. Only the scale factors on the abscissa and ordinate, as well as displacement of the baseline, may differ strongly, depending on the sample. The thermogram shows the flow of heat during the polymerisation as a function of time. The

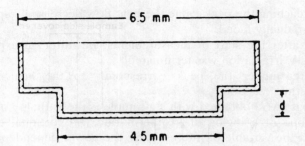

Fig. 7.36. Cross-section of a specially designed aluminium sample pan for photo-polymerisation studies, which maintains uniform sample thickness.[294]

thermogram yields the following data and information:

- the first minute was used to check that the DSC apparatus was in thermal equilibrium before switching on the UV lamp;
- the heat flow started after a switch-off and reached its maximum value after only 1·8 min;

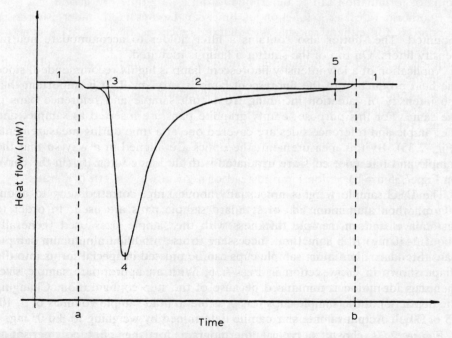

Fig. 7.37. Typical DSC curve showing schematic process of a photocuring reaction:[97] 1—baseline without any incoming light, 2—baseline with irradiation (line from where to start calculation of the peak area), 3—onset (intersection of the peak slope's tangent with 2), 4—peak maximum, 5—baseline displacement, a—time when shutter was opened, and b—time when shutter was closed.

- after reaching the peak value the heat flow decreased, rapidly at first, and then gradually faded;
- 8 min after the start of measurements, i.e. after an exposure time of 7 min, UV irradiation was terminated;
- the area under the peak corresponds to the heat released in the exothermic reaction;
- the heat (ΔH) obtained with 1 g sample is the enthalpy of the reaction;
- the slope of a loglog of exotherm rate (proportional to reaction rate) against the variable of interest gives the power dependence of the rate on that variable;
- rate of polymerisation (R_p) is usually obtained by graphical differentiation of the heat produced during the polymerisation versus time of irradiation;
- degree of monomer conversion and a fraction of monomer remaining can be calculated from the total heat (ΔH_0) (per gram or mol) (for full conversion, i.e. all double bonds were to react) and observed heat (ΔH_t) (per gram or mol) (apparent heat, i.e. heat evolved to the point of interest).

The total heat evolved from the photopolymerisation of a given monomer-initiator formulation can be determined from a long-time irradiation.

Exotherm rates as a function of time can be observed under isothermal conditions for either continuous radiation reactions or for a dark reaction followed a period of illumination. Data can be acquired either as a trace on a strip-chart recorder or in digitised form using a data microprocessor. The data from the strip-chart traces can be analysed by copying, cutting, and weighting to determine areas (heats). After polymerisation the DSC samples can be extracted with solvent and the extracts can be analysed for unreacted monomer using HPLC.

Parameters which can influence the results are as follows.[97]

- Type of lamp used (mercury, xenon or luminescence). Using the whole spectrum of a xenon lamp instead of a given wavelength (separated by a monochromator) for sample irradiation has a tremendous effect on the photopolymerisation rate (R_p) (Fig. 7.38).
- Wavelength of the light (Fig. 7.39).
- Intensity of the light (Fig. 7.40). The more light falling onto the sample, the faster the start of the reaction, and the higher the degree of conversion. However, the increase of light intensity can damage (photo-degrade) the cured material.
- Reaction time (Fig. 7.41).
- Time constant of the instrument.
- Purge gas (e.g. nitrogen or air).[294,305] The kinetics of photopolymerisation of some monomers are very much dependent on the presence of oxygen (air). In the DSC measurements air should be removed by purging by nitrogen for at least 5 min. The longer the purging with nitrogen, the faster is the reaction (Fig. 7.42).

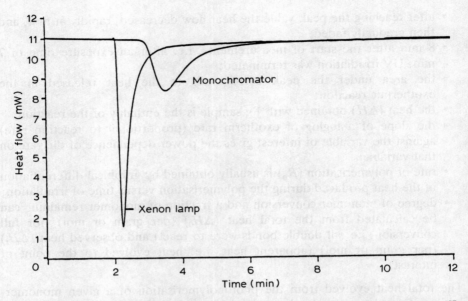

Fig. 7.38. Influence of the type of radiation source on the photocuring reaction.[97]

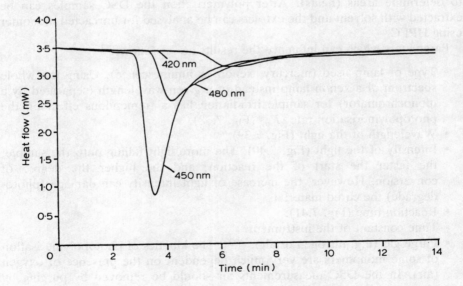

Fig. 7.39. Influence of wavelength on the photocuring reaction.[97]

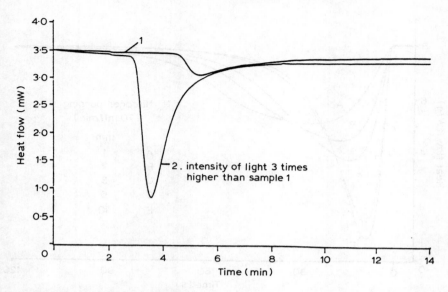

Fig. 7.40. Influence of light energy on the photocuring reaction.[97]

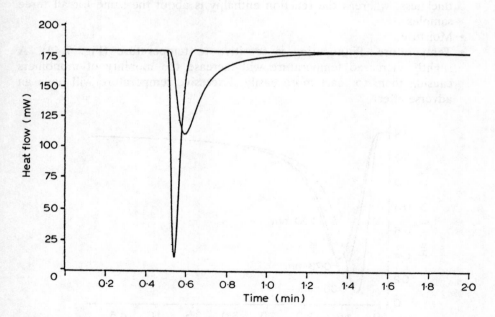

Fig. 7.41. The DSC curve showing photopolymerising systems with different reaction times.[97]

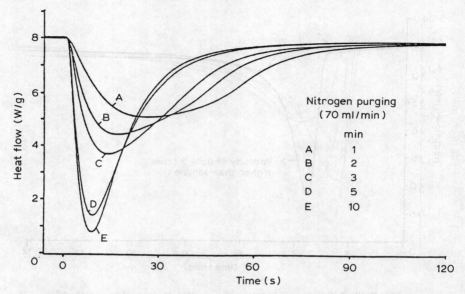

Fig. 7.42. Influence of ambient oxygen on the photocuring reaction.[97]

- Thickness of sample. The sample thickness has a considerable influence on the results (Fig. 7.43). The reaction time decreases with increasing sample thickness, whereas the reaction enthalpy is about the same for all three samples.
- Moisture.
- Temperature. Photocuring is sensitive to temperature (Fig. 7.44). A slightly increased temperature will increase the mobility of monomers causing them to react more easily. Excessive temperature will have an adverse effect.

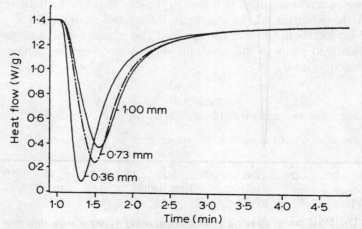

Fig. 7.43. Influence of the sample thickness on the photocuring reaction.[96]

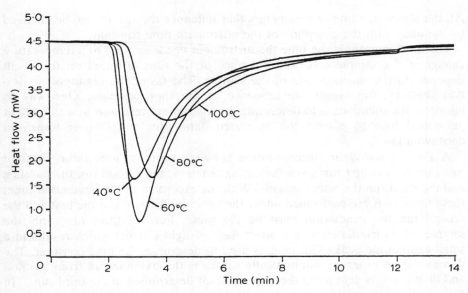

Fig. 7.44. Influence of temperature on the photocuring reaction.[97]

- Pigments. The pigmented samples give different results. As shown in Table 7.9, the cure extent for the filled resin is much lower than for the pure resin.[96] The sample filled with 10% titanium dioxide gives about 30% of the reaction enthalpy of the pure resin, the sample filled with 30% of the same pigment gives about 20% of the reaction enthalpy of the pure resin. The influence of brown pigment is in the same direction but not as large as the influence of the white titanium pigment. The reaction enthalpies given in Table 7.9 refer to the resin content of the sample. This difference in reaction enthalpy is due to the fact that UV light cannot pass through the filled resin. The pigments scatter the incoming light so that the filled systems only cure at the surface. The reaction time is the same as for the unfilled resin so the pigments do not have any influence on the curing reaction itself.

Table 7.9
Influence of the pigment in the sample on the reaction enthalpy[96]

Pigment	Content (%)	Reaction enthalpy (J/g)	Percentage reacted
White	30	−57	19
White	10	−92	31
Brown	30	−67	23
Brown	10	−131	44

All the above-mentioned parameters that influence the results can be changed by the user, with the exception of the instrument time constant.

The time constant is the time the instrument needs to react to a temperature change in the sample cell. The influence of the time constant on the result depends on the reaction rate of the sample. The faster the photopolymerisation reaction, the bigger the influence of the time constant. One way to minimise the influence is to determine the instrument response to a short light pulse and then to correct the measured data using the Fourier transform deconvolution.[96]

A Fourier transform deconvolution software program uses three runs for one sample; the first run gives the curing reaction, the second run the baseline and the third run the time constant. With the exception of the irradiation time, these three runs are performed under the same conditions. For the first and the second run the irradiation must be the same. For the third run, only the sample cell is irradiated with a short flash of light (50 ms) which results in a small exothermic peak. This peak is used to determine the time constant. The deconvolution program automatically subtracts the baseline run from the first and then corrects data using the time constant determined in the third run.[96] In Fig. 7.45 is shown the sample run with and without Fourier transform deconvolution.

The DSC photocalorimetry has the following advantages:

- the rate of reaction can be measured directly, assuming that the heat produced by the polymerisation is proportional to the number of monomer units reacted;

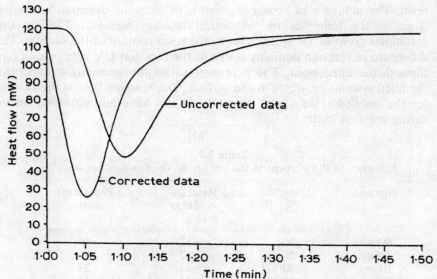

Fig. 7.45. Effect of Fourier transform deconvolution on a sample run.[96]

- because only a small sample is required, good temperature control can be maintained. This simplifies interpretation of the kinetic data; and
- the output of the DSC can be digitised so data can be easily reduced by computer to yield reaction rate as a function of monomer conversion.

A disadvantage of photocalorimetry is that the system to be investigated cannot be applied in a coating of defined thickness.

Determinations of k_p and k_t rate constants

Rearrangement of the expression for the polymerisation rate (R_p), i.e. the rate of disappearance of monomer (rate of monomer conversion) formulated by eqn (39) gives eqn (57):

$$\frac{k_p}{(2k_t)^{0.5}} = \frac{-\frac{d[M]}{dt}}{(\phi_d I_a)^{0.5}[M]} \tag{57}$$

This ratio can be determined empirically from DSC data (see the section entitled 'Determination of polymerisation rate (monomer conversion) by DSC measurements'—p. 368), since $-d[M]/dt$ is related to the exotherm rate, $[M]$ is determined from the monomer conversion, and the quantity $\phi_d I_a$ is determined from measurements of light intensity. These quantities are determined under the conditions of constant radiation yield $k_p/(2k_t)^{0.5}$ under steady-state conditions.[294]

By observing the exotherm rate as a function of time after turning the light off, the ratio $2k_t/k_p$ can be determined for various degrees of conversion (α). When the incident light intensity is zero, the expression for the rate of change of chain radical concentration is given by eqn (58):

$$\frac{d[P^\cdot]}{dt} = -2k_t[P^\cdot]^2 \tag{58}$$

With the boundary conditions that at time $t = 0$ when the light is turned off, the chain radical concentration $[P^\cdot]_0$ is given by eqn (59):

$$[P^\cdot]_0 = \frac{\left(-\frac{d[M]}{dt}\right)_0}{k_p[M]_0} \tag{59}$$

where the subscript zero indicates value at $t = 0$. Equation (58) may be solved to yield eqn (60):

$$\frac{1}{[P^\cdot]} = 2k_t t + \frac{1}{[P^\cdot]_0} \tag{60}$$

Solving eqn (60) for $[P^\cdot]$ using eqn (59) and rearranging leads to eqn (61):

$$\frac{M}{-\frac{d[M]}{dt}} = \frac{2k_t}{k_p} t + \frac{[M]_0}{\left(-\frac{d[M]}{dt}\right)_0} \tag{61}$$

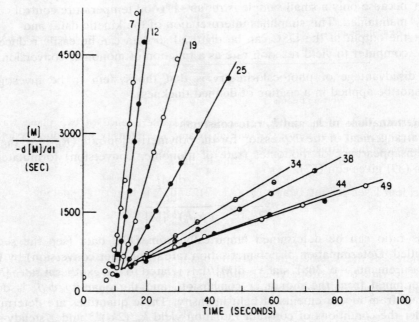

Fig. 7.46. A plot of $[M]/(-d[M]/dt)$ versus time for a given monomer (1,6-hexanedioldiacrylate). Numbers indicate monomer concentration (%).[294]

Since [M] and $d[M]/dt$ as functions of time can be obtained from DSC measurements, the ratio $2k_t/k_p$ is determined from the slope of a plot of $[M]/(-d[M]/dt)$ as a function of time. Having evaluated the ratio $k_p/(2k_t)^{0.5}$ from the steady-state data and $2k_t/k_p$ from the non-steady-rate versus time, it is now possible to calculate values for the individual rate constants k_p and k_t.[294]

In Fig. 7.46 is shown a plot of $[M]/(-d[M]/dt)$ versus time for a series of DSC runs on 1,6-hexanedioldiacrylate. The slopes of these lines were determined by a least-squares calculation to yield values of the ratio $2k_t/k_p$ for various concentrations.

MECHANISTIC AND KINETIC STUDIES OF PHOTOPOLYMERISATION BY HOLOGRAPHIC METHOD

In a holographic method, a photopolymerisation reaction takes place only in the bright areas of the holographic pattern created by the interference of two plane waves of writing (laser) beams (Fig. 7.47).

The result of this photopolymerisation is a spatial modulation of the refractive index that images the incident intensity pattern:[42,188]

$$I(x) = 2I\left(1 + \cos\left(\frac{2\pi}{\Lambda}x\right)\right) \tag{62}$$

where $I(x)$ is the refractive index at point x, I is the intensity of the two

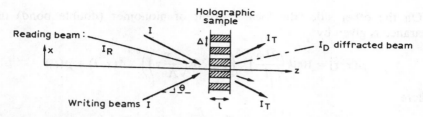

Fig. 7.47. Holographic pattern created by the interference of two plane waves of writing beams.[42] I—intensity of writing beam, I_R—intensity of reading beam, I_D—intensity of the diffracted beam at the instant t, and I_t—intensity of transmitted beam at the instant t.

incident laser writing beams (Fig. 7.47), and Λ is the fringe spacing:

$$\Lambda = \frac{\lambda}{2 \sin \theta} \quad (63)$$

where λ is the wavelength of writing beam, and θ is the angle (Fig. 7.47).

The rate of photopolymerisation (R_p) under holographic conditions (spatially modulated irradiation) is calculated using eqn (64):

$$R_p(x, t) = k[\bar{M}](\phi I_a(x, t) l^{-1})^{0.5} \quad (64)$$

where $R_p(x, t)$ is the local rate of photopolymerisation at the point x and the instant t time under holographic exposure, k is the constant that characterises the matrix, $[\bar{M}]$ is the average monomer concentration, ϕ is the overall initiation yield of the photopolymerisation, i.e. the number of starting chains per photon absorbed, l is the sample thickness (path length), and $I_a(x, t)$ is the spatiotemporal distribution of the absorbed dose:

$$I_a(x, t) = 2I\left(1 + \cos\left(\frac{2\pi}{\Lambda}x\right)\right)(1 - 10^{-A(x,t)}) \quad (65)$$

where $A(x, t)$ is the absorbance (optical density) at time t at the point of abscissa x.

The local percent of monomer (double bond) disappearance $\rho(x, t)$ is given by

$$\rho(x, t) = \frac{100}{[\bar{M}]} \int_0^t R_p \, dt + \rho(t_0) \quad (66)$$

where t_0 is the induction period. The percent of double bonds at any given time t can be determined, e.g. by FTIR (see the section entitled 'Applications of modern FTIR spectroscopy'—p. 407).

$$\rho = 100 \frac{\log T - \log T_0}{\log T_0} = \frac{100}{[\bar{M}]} \int_0^t R_p \, dt \quad (67)$$

where T_0 and T are IR transmission of the sample before and after irradiation, respectively.

On the other side, the local percent of monomer (double bond) disappearance is given by

$$\rho(x, t) = 100k\left(\frac{2I\phi}{l}\right)^{0.5}\left(1 + \cos\left(\frac{2\pi}{\Lambda}x\right)\right)^{0.5} A(x, t) + \rho(t_0) \qquad (68)$$

where

$$A(x, t) = \int_0^t (1 - 10^{-A(x,t)})^{0.5} \, dt \qquad (69)$$

The refractive index of the photopolymerising system (n) increases linearly with the decrease of double bonds as the polymerisation proceeds, and is given under holographic conditions (spatially modulated irradiation) in eqn (70):

$$\begin{aligned} n(x, t) &= n_0 + 2^{0.5} NA(x, t)\left(1 + \cos\left(\frac{2\pi}{\Lambda}x\right)\right)^{0.5} \\ &= n_0 + 2^{0.5} NA(x, t)\left(\cos\left(\frac{\pi}{\Lambda}x\right)\right) \end{aligned} \qquad (70)$$

where N is the constant depending on the photopolymerisable system, and n_0 is the refractive index of the initial polymerisable formulation.

The experimental holographic set-up is shown in Fig. 7.48. A hologram (Fig. 7.47) is created with an He–Ne laser (633 nm). The writing light is split into two beams of equal intensity (I). A system of adjustable mirrors allows the beams to interfere with equal path lengths at a full angle 2θ of about 25°. With such a geometric arrangement, the fringe spacing (Λ) is 1·42 µm.

The hologram is read with an argon laser (488 nm) beam (I_r). The intensity of the diffracted light is measured by a radiation power meter connected to a chart recorder, and the diffraction efficiency is defined as the ratio of the intensity of the diffracted beam (I_D) to that of the reading beam (I_R).

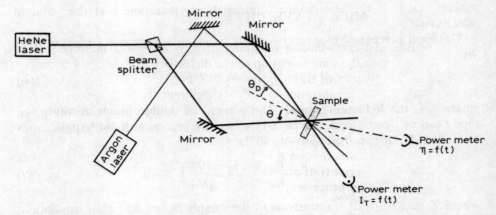

Fig. 7.48. Experimental set-up used to record the growth of the holographic curves.[42]

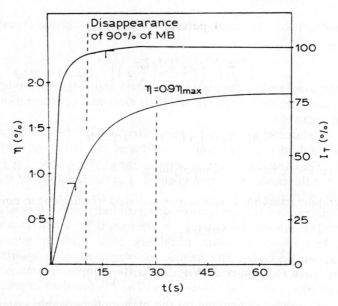

Fig. 7.49. Transmitted intensity (I_T) of each writing beam under holographic exposure and diffraction efficiency (η) as a function of irradiation time.[42] MB—methylene blue.

With a second detector in the path of the red beam it is also possible to record the transmittance of the sample, i.e. the average concentration of the sensitiser, as a function of the irradiation time.

In Fig. 7.49 is shown an example of a study of bleaching of methylene blue (MB) photosensitiser in the presence of diethylamine (DEA) determined by the changing of the transmitted intensity (I_T) of the writing beam and the diffraction efficiency (η).

MICROWAVE DIELECTROMETRY

Polymerisation and cross-linking reactions strongly change a material's dielectric properties, so the time evolution of the dielectric constant can be used to characterise the kinetics of these processes.[36,38–40,133] Usually monomers show characteristic dielectric relaxation times very different from the corresponding polymers. In a monomer–polymer mixture the dielectric constant, measured at frequencies in between the dispersion region of the polymer (low frequencies <100 MHz) and the dispersion of the monomer (high frequencies >10 GHz), can give detailed information on the relative concentration of the reagents.

Dielectric measurements of microwaves have been used for monitoring fast radical polymerisation processes,[36,37,39] residual monomer concentration,[40] and curing processes.[19,40,43,50]

The loss factor (ε''_s) (at frequencies of about 9–10 GHz) (related to the energy absorption) of the polymerising mixture is calculated using eqn (71):[36,39]

$$\varepsilon_s'' = \varepsilon_M''[M]_t[M]_0^{-1} \tag{71}$$

where ε_M'' is the loss factor of the pure monomer, and $[M]_t[M]_0^{-1}$ indicates the residual monomer fraction being $[M]_t$ and $[M]_0$ monomer concentration at time t and $t = 0$, respectively.

Dielectric measurements at microwave frequencies for monitoring the instantaneous monomer concentration are based on a cavity perturbation method.[195–197] The loss factor of a sample inserted in the cavity can be determined from the power reflection coefficient at the input of the microwave resonator at fixed frequency.

The block diagram of the microwave dielectrometry apparatus is shown in Fig. 7.50.[38–40] The microwave circuitry is composed by a microwave source frequency locked to the resonant frequency of a right cylindrical cavity working at about 9·5 GHz. The sample is contained in a quartz cuvette inserted in the cavity as shown in Fig. 7.50. The sample volume is typically 10 µl, with a thickness of about 0·2–0·3 mm. The UV irradiation passes to the

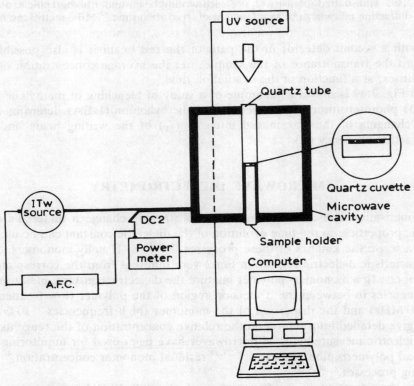

Fig. 7.50. Block diagram of the microwave dielectrometry apparatus.[38]

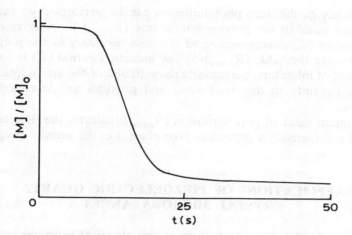

Fig. 7.51. A plot of [M]/[M]$_0$ versus time of polymerisation.[38]

cavity through a quartz tube. The irradiation time is controlled by an electromechanical shutter. A computer is used to control the experiment, and performs the data acquisition and processing. The acquisition of a single measurement takes about 25 ms and the repetitition rate is about 100 ms.

The evolution of a polymerisation process, monitored by the residual monomer fraction $[M]_t[M]_0^{-1}$ versus time is as follows (Fig. 7.51):

$$[M]_t[M]_0^{-1} = 1 - C \tag{72}$$

where C is the conversion factor at time t.

The polymerisation rate (R_p) under stationary conditions is given by eqn (39) (Fig. 7.52).[76]

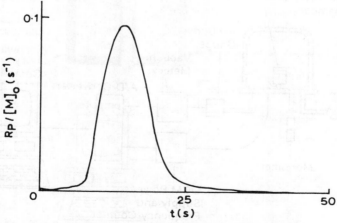

Fig. 7.52. A plot of $R_p/[M]_0$ versus time of polymerisation.[38]

The efficiency of different photoinitiators can be performed by evaluating the maximum value of the polymerisation rate ($R_{p,max}$) in every case as the induction period (T_0) corresponding to the time necessary to the polymerisation rate to reach the value ($R_{p,max}$)/5. The induction period (T_0) is connected to the amount of inhibitors, particularly oxygen, and of the active free radicals formed consequently to the irradiation, and provides an evaluation of the initiation rate.

The maximum value of polymerisation ($R_{p,max}$) indicates the time necessary for the real polymerisation process to take place, i.e. the average propagation rate.

APPLICATIONS OF PIEZOELECTRIC QUARTZ CRYSTAL MICROBALANCES

The progress of many important physical and chemical processes, such as[187] those listed below, can be followed by observing the associated mass changes:

- thin film growth during polymerisation, i.e. kinetics of photopolymerisation;
- condensation, evaporation, adsorption, desorption from the cured surfaces;
- oxidation, decomposition, degradation of cured samples; and
- solubility, surface area, surface tension, surface pressure.

Several highly sophisticated, automatic, microprocessor-controlled piezoelectric quartz crystal microbalances are available commercially. An experimental set-up for the study of kinetics of thin layer deposit photoinitiated polymerisation is shown in Fig. 7.53.[142] This extremely sensitive method allows

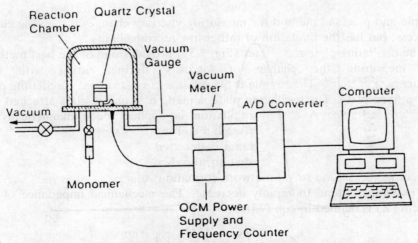

Fig. 7.53. Experimental set-up for polymerisation deposition studies.[142]

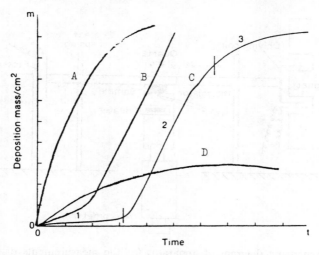

Fig. 7.54. Different forms of the growth curves for depositions of different monomers. The curve (C) can be divided into three regions: (1) induction, (2) rapid growth, and (3) plateau.[142]

the growth curve to be followed as a function of time and allows calculation of the kinetics of polymerisation, determination of induction time, etc. (Fig. 7.54).

VISCOSITY MEASUREMENTS DURING CURING

The viscosity changes considerably over a very short period of time during radiation curing. Application of typical glass (or quartz) viscosimeters is very limited in the study of photocuring. The rolling-ball method provides a very simple and practical method for measuring viscosity changes during the curing process, but has the limitation of rather low reproducibility.[118,218,239,283]

The oscillating-plate rheometer (Fig. 7.55) provides one of the best methods for measuring the change of viscosity during curing with high accuracy.[219,220,227,301] The sample is placed in a gap between an oscillating plate and a fixed plate. The oscillating plate is made of a quartz glass attached to a Duralumin frame. A forced oscillation is applied electromagnetically by passing an alternating current through a coil in a constant magnetic field. The resultant displacement of the frame is detected as a change in capacity of a small condenser. Upon UV irradiation, the dynamic viscosity of a photocured system increases due to the network formation, this causes the movement of the oscillating frame to rapidly decrease. The mechanical impedance of the system (Z) is defined in eqn (73):

$$Z = \frac{F}{S} = \left(\frac{G}{C} - \omega^2 M + K\right) + \frac{\omega \eta}{C} \tag{73}$$

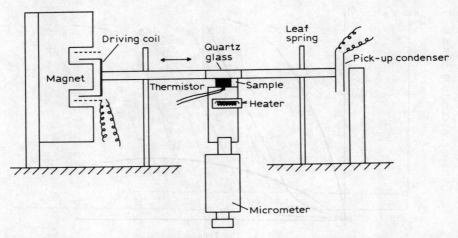

Fig. 7.55. Schematic diagram of apparatus for the measurements of the viscosity of high solid coatings.[219]

where F is the applied sinusoidal force, S is the resultant displacement, G is the dynamic modulus, ω is the angular frequency, M is the mass of the moving components, K is the spring constant, η is the dynamic viscosity, and C is the shape factor given by eqn (74):

$$C = \frac{h}{A} \qquad (74)$$

where h is the thickness of the sample, and A is the area of the sample.

The time dependence of dynamic viscosity $\eta(t)$ can be obtained by eqn (75):

$$\eta(t) = \frac{m(t) h \sin \delta(t)}{A \omega} \qquad (75)$$

where $m(t)$ is the amplitude ratio of force-to-displacement, h is the sample thickness, $\delta(t)$ is the phase angle between them, ω is the angular frequency, and A is the sample area.

The dynamic viscosity is generally plotted against the exposure time and/or light energy.

Displacement (S) can be measured experimentally, and dynamic viscosity at time (t) can be calculated from eqn (76):

$$\frac{\eta_t}{\eta_0} = \sqrt{\frac{(S/S_t)^2 - 1}{(S/S_0)^2 - 1}} \qquad (76)$$

where S, S_0 and S_t are displacements measured in the absence of sample, at time zero, and after a given time t, respectively; and η_0 can be measured separately e.g. by a cone–plate viscometer.

The dynamic viscosity of the pre-polymer (before UV irradiation) is independent of sample thickness. However, if the sample is thick, the curing

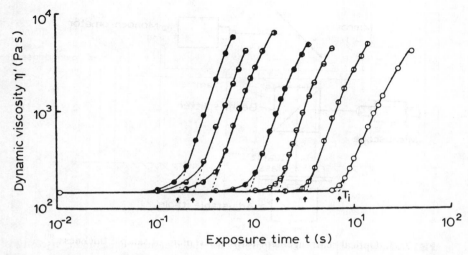

Fig. 7.56. A plot of dynamic viscosity versus time of irradiation for a given monomer.[301]

behaviour under irradiation may vary in the direction perpendicular to the shearing surfaces. To avoid this inconsistency in the photochemical process, the sample has to be kept as thin as possible.

In Fig. 7.56 is shown dependence of dynamic viscosity on the exposure time for a urethane acrylate pre-polymer. The dynamic viscosity rapidly increases with time after a certain period. The induction period (T_i) can be evaluated by the time of the intersection between two lines.

The viscoelastic measurements during UV curing have the following advantages.[301]

- The required amount of sample is extremely small. The measurements can be made even with 0·01 mg of sample.
- The measurements are carried out with liquid polymer, a cell for the sample is not necessary since its surface tension is available.
- Sample thickness is about 10 μm and temperature can be easily controlled.
- By measuring the thickness before and after UV irradiation, the volume contraction of sample can be determined.
- The apparatus may be used for measurement of reaction rates and analysis of reaction mechanisms of photoreactive polymers and thermosetting resins.

DETERMINATION OF SAMPLE THICKNESS OF CURED THIN POLYMER FILMS

In the dynamic viscosity measurements (see the previous section) it is necessary to determine sample thickness with a high accuracy. In Fig. 7.57 is

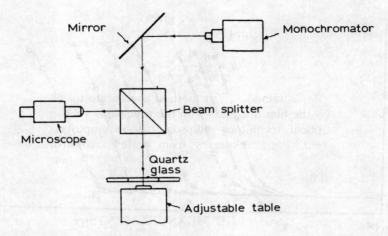

Fig. 7.57. Optical rearrangement for measurement of sample thickness.[301]

shown an optical arrangement for measurement of the sample thickness.[301] A light beam from a monochromator is reflected by a mirror and led to the narrow gap between the quartz glass and sample table through a beam splitter. The principle of operation is shown in Fig. 7.58. Because the quartz glass and sample table are not exactly parallel, but incline very slightly towards each other, interference fringes are formed and detected by microscope. The interference fringes appear at positions where the light reflected by the quartz glass has the reverse phase of that by the sample table:

$$2h = n\lambda_0 \qquad (77)$$

where n is the integer number, and λ_0 is the wavelength. When the wavelength is gradually decreased, the fringes slowly shift. If the wavelength is so decreased that the reference fringe shifts to the position where the neighbouring fringe was located at the wavelength of λ_0, the resultant wavelength λ_1 can be described by eqn (73):

$$2h = (n+1)\lambda_1 \qquad (78)$$

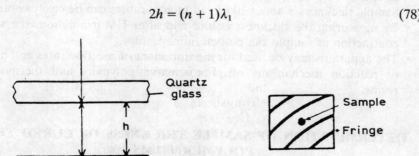

Fig. 7.58. Principle of operation for measurement of sample thickness.[301]

Similarly, at the wavelength λ_m the shift gives the mth fringe and then the sample thickness is given by eqn (79):

$$h = \frac{m}{2} \frac{\lambda_0 \lambda_m}{\lambda_0 - \lambda_m} \qquad (79)$$

The sample thickness obtained by this method is accurate to $\pm 0.1\ \mu m$.

In order to know the film thickness with the precision of a few Ångströms, a non-destructive optical technique—ellipsometry—is employed.[268] With this method, transparent film thicknesses from a few Ångströms to several micrometres, and absorbing film thicknesses of 500–1000 Å, may be measured with a precision approaching ± 1 Å.

SHRINKAGE MEASUREMENTS

Polymerisation of vinyl monomers are accompanied by a relatively large amount of shrinkage (i.e. volume decreasing), for example, in the case of methyl methacrylate it can reach even 20%. A number of polymerising systems (e.g. divinyl compounds) show delayed shrinkage.[160,170]

Shrinkage is not a measure of conversion; however, there is the mutual dependence of shrinkage and conversion. The rate of polymerisation (R_p) (rate of monomer conversion) is expressed as a function of the volume shrinkage (ΔV):[203]

$$R_p = \frac{\Delta V}{V(1/\rho_m - 1/\rho_p) M \Delta t} \qquad (80)$$

where V is the volume of the system; ρ_m and ρ_p are monomer and polymer densities, respectively; M is the molecular weight of the monomer; and Δt is the elapsed time for the volume change (ΔV).

Shrinkage can be measured by the decrease of thickness of a sandwich composed of two glass plates with a thin layer of monomer (50 μm) sample between.[164,170] A sample, consisting of monomer with dissolved photoinitiator will polymerise through one of the plates. Polymerisation shrinkage then causes a decrease of the thickness of the sample layer and therefore of the sandwich. Provided that the polymerising sample adheres well to the glass plates most of the volume shrinkage will occur as a decrease of thickness, with only a minor lateral contribution. The construction of the sandwich is shown in cross-section in Fig. 7.59 and is disassembled in Fig. 7.60.

The lower glass plate consists of a quartz block with dimensions of $2 \times 8 \times 18$ mm. On its top face is a rectangular recess with dimensions of 8×10 mm and a depth of 50 μm. This recess is filled with monomer containing dissolved photoinitiator and covered with the upper glass plate (microscope cover glass with dimensions 8×14 mm and a thickness of 0.14 mm). The cover glass is coated with an aluminium reflective layer. In this way, UV radiation entering the lower plate will pass the sample layer twice.

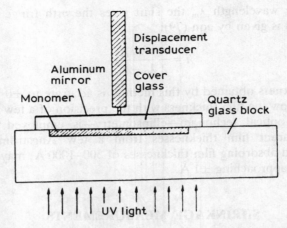

Fig. 7.59. Sample sandwich (in cross-section) for measurement of shrinkage.[164]

During polymerisation, the polymerisation shrinkage will tend to cause a decrease of the sample thickness. Since the thin upper glass is supported at two opposite edges and since the adhesion of the polymerising system to the glass plate is strong enough, the thinner of the two will be curved by the shrinkage process. The displacement of the centre part of the cover glass is continuously monitored by a displacement transducer.

The sample sandwich can be placed in the thermostat flushed with nitrogen (Fig. 7.61).

Rate of shrinkage (R_s) (Fig. 7.62) is usually obtained by graphical differentiation of the displacement versus time curve.[164,170]

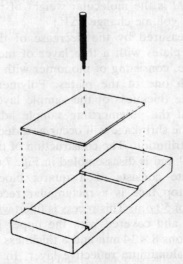

Fig. 7.60. Disassembled sample sandwich for measurement of shrinkage.[164]

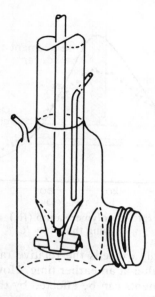

Fig. 7.61. Sample sandwich placed in the thermostat flushed with nitrogen.[164]

In order to allow a proper comparison of rate of shrinkage (R_s) with a rate of polymerisation (R_p) (obtained from DSC curves, see the section entitled 'Determination of polymerisation rate (monomer conversion) by DSC measurements'—p. 368), it is desirable to plot the rate of shrinkage also as a percentage of total shrinkage per second. This requires an estimate of the theoretical amount of total shrinkage at 100% C=C conversion. Due to the occurrence of vitrification this situation cannot, however, be reached.

The rate of shrinkage (R_s) can be the same as a rate of polymerisation (R_p) (Fig. 7.62) or can be faster (Fig. 7.63). As is shown in Fig. 7.62, the relative

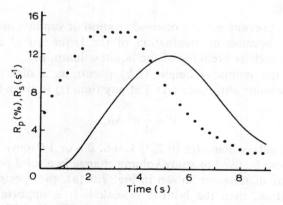

Fig. 7.62. (——) Rate of polymerisation (R_p), and (······) rate of shrinkage (R_s) versus time of irradiation for a given polymer.[164]

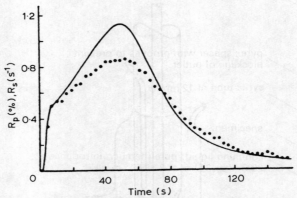

Fig. 7.63. (———) A plot of rate of polymerisation (R_p), and (· · ·) rate of shrinkage (R_s) showing that $R_p \geqslant R_s$.[164]

rate of shrinkage not only exceeds the relative rate of polymerisation, it also reaches the maximum value at an earlier time. However, these results obtained from the DSC measurements can be affected by the inertia of DSC.

DILATOMETRIC MEASUREMENTS

Dilatometry is the standard technique for studying a volume change (e.g. shrinkage) during polymerisation and for the study of a volume change in solid polymer samples.[14,64,201,202–204,235,259]

In spite of the great number of different dilatometer designs, two types are the most common in use:

1. dilatometers in a fused-on capillary (Fig. 7.64(a)), the monomer is inserted through it, with the help of a long syringe needle;
2. dilatometers with a movable-joint capillary (Fig. 7.64(b)); this type is easier to fill.

It is important to prevent wetting or condensation of vapours on the inner wall of the capillary, because contamination of the latter would certainly cause trouble later on, such as breaking of the liquid column, etc.

The drop in the meniscus height (Δh) occurs as a direct result of the shrinkage. The volume shrinkage (ΔV) at any time (t) is given by eqn (81):

$$\Delta V = \frac{\bar{\pi}}{4} d^2 \Delta h_t \qquad (81)$$

where d is the capillary diameter (0·2, 0·4, 0·6, 0·8 or 1·0 mm).

Dilatometers can be used to study volume change in a solid polymer sample. In the case of the dilatometer shown in Fig. 7.64(a), the specimen is inserted through the bottom, then the bottom is sealed. It is important to de-gas a sample during 24 h at 10^{-4}–10^{-5} mmHg (13·3–1·33 mPa). Next, the dilatometer is filled with pure mercury. To calculate specific volumes from the height

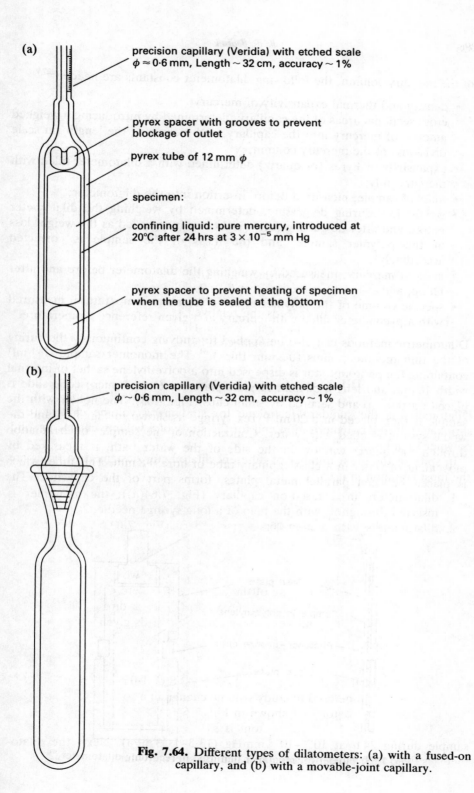

Fig. 7.64. Different types of dilatometers: (a) with a fused-on capillary, and (b) with a movable-joint capillary.

of the mercury column, the following dilatometer constants are needed:[274]

- density and thermal expansivity of mercury;
- cross-sectional areas of the capillary, determined by introducing a weighed amount of mercury into the capillary, and measuring the length (in scale divisions) of the mercury column;
- expansivity of Pyrex (or quartz) determined with a dilatometer filled with mercury only;
- mass of sample, measured before insertion into the dilatometer;
- weight loss during de-gassing, determined by weighing the dilatometer before and after de-gassing (the weight loss is regarded as the weight loss of the polymer sample, and the mass of the sample is corrected accordingly);
- mass of mercury, measured by weighing the dilatometer before and after filling; and
- specific volume of the polymer at some reference temperature, measured with a pycnometer filled with mercury at a given reference temperature.

Dilatometric methods can also be applied to studying continuously the curing of the thin monomer films (0·3 mm thick).[61] The monomer sample (0·1 ml) containing the photoinitiator is dispensed into a polyethylene sachet of internal width 10 mm and length 50 mm. A glass plate cut from a microscope slide is placed on the top and secured by small clips (Fig. 7.65). The holder with the sample is then placed in a 20 ml Pyrex syringe as shown in Fig. 7.66 and the whole system is filled with water. Contraction of the sample, on irradiation through the quartz window in the side of the water bath, is measured by movement of water in a glass capillary tube of bore 0·8 mm. This tube, which is placed between parallel metal plates, forms part of the capacitor. The

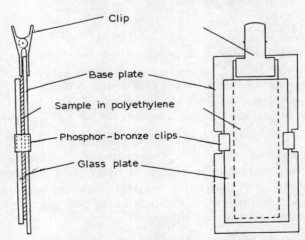

Fig. 7.65. Holder for monomer film used in recording dilatometer.[61]

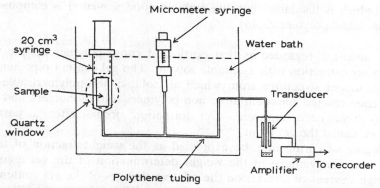

Fig. 7.66. Recording dilatometer for monomer film polymerisation.[61]

change in capacitance of the arrangement is directly proportional to the position of the air–water interface in the capillary tube. The micrometer syringe is used to adjust the position of the water–air interface and to calibrate the system. As the sample in the sachet is contracted during polymerisation, the water is drawn along the capillary tube and a continuous trace of the progress of polymerisation can be recorded.

METHODS FOR THE STUDY OF POLYMERISATION ACCOMPANIED BY CROSS-LINKING SOL–GEL ANALYSIS

Various models have been proposed for the kinetics of network formation.[84,100,117,189,206,271–273,288,289] However, differences of opinion exist on the applicability of these theories to various polymerising systems.

The application of kinetic models requires a knowledge of the following kinetic parameters:

- initiation efficiency;
- propagation rate constant with monomeric double bonds;
- reactivity ratio;
- propagation rate constant with pendant double bonds;
- cyclisation rate constants;
- bimolecular radical termination constants for sol–sol, gel–gel and sol–gel reactions.

A full estimation of these rate constants appears very difficult.

Extension of classic kinetic analysis to a free-radical polymerisation with cross-linking is not straightforward. Polymer network formation seriously complicates the kinetic analysis, particularly in the post-gelation period where two macromolecular systems exist together:[184]

1. sol (which is soluble in good solvents) consists of linear and branched polymer chains; and

2. gel (which is insoluble, but swellable in good solvents) is composed of cross-linked polymer chains.

The gel content is regarded as that portion of cured polymer sample which remains after extraction with a suitable solvent. The gel content represents the pure cross-linked structures from which all soluble non-polymerisable components (non-reacted photoinitiator, non-polymerisable monomers and other additives) have been removed. Gel formation occurs after a particular conversion, called the gel point.[10,225,246]

The degree of curing may be expressed as the weight fraction of the gel formed. The disadvantage of the weight determination of the gel amount is that at high degrees of conversion the relative changes of the gel content with conversion are small and difficult to determine. With brittle materials solvent cracking may further complicate the gravimetric method. Alternatively, chemical analysis of extracts, obtained with suitable solvents, may give useful information about the sol content and about its chemical composition. The use of HPLC enables the quantitative determination of extractable compounds (photoinitiator, monomers, oligomers, etc.) (see the section entitled 'Determination of polymerisation rate (monomer conversion) by liquid chromatography, p. 366).[168]

Measurements of conversion of monomer to sol and gel are commonly made by gravimetry using Soxhlet extraction.

$$\text{Gel content} = \frac{w - w_g}{w} \times 100 \quad (\%) \tag{82}$$

where w and w_g are weight of cured sample (before extraction) and weight of residues after solvent extraction (i.e. weight of gel).

The gel point is determined from conversion of monomer to sol and gel versus time of reaction (Fig. 7.67). The gel point is very much dependent on the concentration of cross-linking monomer (Fig. 7.68). It can also be determined by viscometric methods, e.g. using a low shear rheometer.

Polymerisation up to the gelation point proceeds, in general, in three stages:

1. formation of linear polymer chains with pendant–internal double bonds;
2. branching via these double bonds; and
3. intermolecular cross-linking leading to gelation.

The gel, once formed, grows very rapidly consuming sol polymers and monomer molecules (Fig. 7.67).

Generally, there is a good relation between the gel content and pendulum hardness (see the section entitled 'Pendulum hardness'—p. 428) (Fig. 7.69). Both curves have a similar form: a rapid increase which is followed by a slower increase which approaches a final value.

Gel melting temperatures can be determined by a sensitive DSC calori-

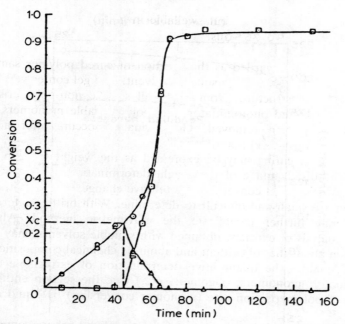

Fig. 7.67. Conversion time plot for conversion of monomer (△) to sol, (□) gel and (○) total for a given polymerisation. X_c = gelation point.[184]

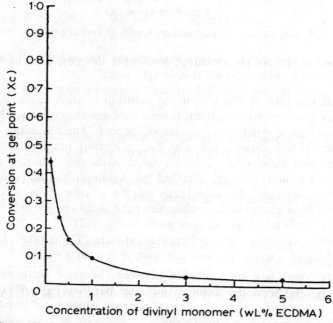

Fig. 7.68. Conversion at gelation point (X_c) versus concentration of cross-linking monomer.[169]

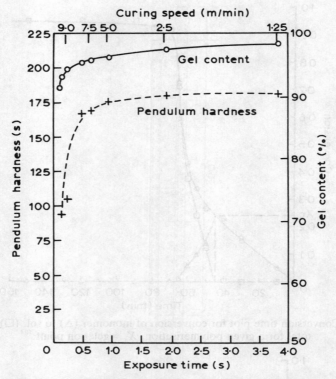

Fig. 7.69. Gel content and pendulum hardness versus exposure time.[225]

meter. However, it can be measured by simple devices, such as those listed below.

- A small test-tube is filled with the solution at high temperatures, then material is quenched to temperatures where gelation occurs. The test-tube is turned upside down and then slowly heated. The temperature at which the material falls down is taken as the gel melting temperature (T_m).[86]
- A U-shaped tube (Fig. 7.70) is filled with the solution at elevated temperature and the gel allowed to form such that there remains difference between the two levels of the U-tube. The upper level is observed by a cathetometer while the tube is slowly heated. The melting temperature is taken as the kink point in Fig. 7.71.[278]
- In another technique, the melting temperature (T_m) is taken by following the sinking of a small ball as a function of temperature.[281]

MEASUREMENTS OF SWELLING OF POLYMER NETWORKS

The swelling can be related to any process in which a cross-linked polymer network is in contact with solvent.[119,240]

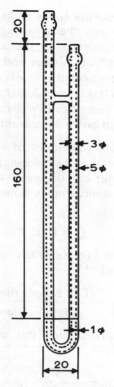

Fig. 7.70. Gel melting temperature apparatus.[278]

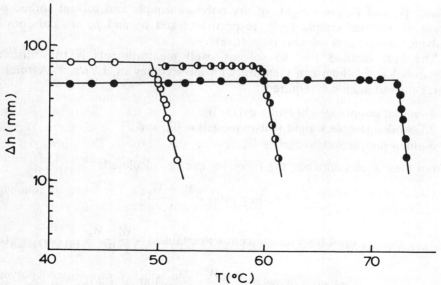

Fig. 7.71. Examples for determining the melting temperature.[278]

- *Saturation swelling.* The network is placed in contact with a large excess of solvent and allowed to reach thermodynamic equilibrium. The volume fractions of solvent and polymer in the swollen gel at saturation are related to the network, type of polymer and type of solvent.
- *Equilibrium swelling.* The network is exposed to limited amounts of solvent. After equilibrium has been reached, the relative magnitude of the different thermodynamic quantities is obtained as a function of deformation, network structure and polymer–solvent pair.

The time evolution of swelling (dynamics of swelling) and deformation of swollen networks (uniaxial and biaxial stretching) can also give information on structure of cross-linked polymer and degree of swelling.

The degree of swelling (Q) at any time is defined as the network volume (V) relative to its dry volume (V_d):

$$Q = V/V_d \tag{83}$$

The volume fraction of polymer in the swollen network (v_2) can be determined from the measured swelling ratio:[132]

$$v_2 = \frac{\text{Volume dry polymer}}{\text{Volume swollen polymer}} = \frac{(\text{Dry sample diameter before swelling})^3}{(\text{Sample diameter after swelling})^3} \tag{84}$$

The measurements are usually carried out under a microscope with an accuracy of ± 0.002 mm.

$$v_2 = \frac{W_p/\rho_p}{W_p/\rho_p + W_s/\rho_s} \tag{85}$$

where W_p and W_s are weights of dry polymer sample and solvent sorbed per gram of polymer sample (gel), respectively; and ρ_p and ρ_s are densities of polymer sample and solvent, respectively.

The best method for the solvent swell measurements is to utilise an electrobalance, which can measure with an accuracy of ± 1 μg.[238] A total of three measurements is required:

1. initial sample weight (sol + gel) = W_i;
2. swollen sample weight (solvent + gel) = W_s; and
3. final sample weight (gel) = W_g.

From these measurements, the following can be calculated:

$$\text{Swelling ratio} = \frac{W_s - W_g}{W_g} \tag{86}$$

$$\text{Sol fraction (percent extractables)} = \frac{W_i - W_g}{W_i} \tag{87}$$

$$\text{Degree of swelling} = \frac{W_s - W_i}{W_s} \times 100 \quad (\%) \tag{88}$$

Solvent swell measurements are long (days), tedious, and require some minimum thickness to avoid erroneous results from excessive surface evaporation. In a conventional measurement of solvent swell, the swollen sample is removed from the solvent, sandwiched between two pieces of similar-sized filter paper and placed in a weighing bottle and weighed. Next, the sample is quickly removed from the weighing bottle, which is again weighed. The swollen sample weight is simply the difference between the two weighings. This method was found to be highly precise. Typically, 2–3 days are required to obtain equilibrium. It is not unusual for the swollen sample weight to continue to change over a period of days, weeks, or months, necessitating an extrapolation to obtain the zero-time equilibrium swell ratio.

The cross-linking density (or the degree of cross-linking) (Γ) is the number of cross-linked monomeric units per primary chain, and is given by eqn (89):

$$\Gamma = \frac{(\bar{M}_n)_0}{(\bar{M}_n)_c} \tag{89}$$

where $(\bar{M}_n)_0$ is the number-average molecular weight of the primary chain (a primary chain is the linear molecule before cross-linking), and $(\bar{M}_n)_c$ is the number-average molecular weight of a network chain.

For the determination of cross-link density (Γ) the Flory–Rehner equation can be applied, eqn (90), which relates cross-link density to the measured volume fraction of polymer in the swollen network (v_2) sample which has reached its equilibrium swollen state:[101,238]

$$\frac{\Gamma}{V_0} = \frac{-[\ln(1 - v_2) + v_2 + \chi_1 v_2^2]}{V_1\left(v_0^{2/3} v_2^{1/3} - \frac{2v_2}{f}\right)} \tag{90}$$

where χ_1 is the polymer–solvent interaction parameter, V_1 is the molar volume of solvent, V_0 is the volume of the polymer, v_0 is the volume fracture of the polymer in the diluent–polymer mixture at the time of cross-linking, and f is the functionality of cross-links.

GLASS TRANSITION TEMPERATURE (T_g) FOR CHARACTERISATION OF THE DEGREE OF CROSS-LINKING

It has been shown that between the glass transition temperature (T_g) and cross-link density (Γ) is a simple linear relation given by eqn (91):[4,11,103,248]

$$T_g = T_{g(l)} + K\Gamma \tag{91}$$

where $T_{g(l)}$ is the glass transition temperature of the corresponding linear polymer, Γ is the cross-link density, and K is a constant (which depends on the functionality of cross-link polymers). However, eqn (91) cannot be applied to systems of high cross-link density, such as polyesters,[58] because the 'copolymer polyeffect' has an effect on the cross-linking. In this case, T_g is given by eqn

(92):[11,16,183]

$$T_g = \frac{\sum \psi_i T_{g(i)}}{\sum \psi_i} \qquad (92)$$

where ψ_i is the parameter characterising the network composition (segments, chain ends, and cross-link mers), and $T_{g(i)}$ is the component value characterising the contribution of the ith species to T_g. In the literature a number of equations are presented which try to improve eqn (91) for the 'copolymer effect', for a given system studied.

The T_g is usually measured from DSC thermograms,[17,80,182,183] or from dynamic mechanical thermal analysis.[165] IR spectrometry allows the disappearing of the groups which participate in the cross-linking reaction to be followed.

The T_g can be used as an index of chemical conversion during the curing processes.[8,311]

DETERMINATION OF THE PHOTOCROSSLINKING NUMBER

The photocrosslinking number, is always calculated with the assumption that this process has to be accompanied by simultaneous chain scission (photodegradation process), which occurs during UV irradiation of cured formulation. The relation between photocrosslinking number (X) and photodegradation (chain scission) (S) and solubility of irradiated sample (s) is expressed on the basis of Charlesby's radiation theory (see the section entitled 'Absorption of light by photocured formulations'—p. 346) by the following simplified equation with the assumption that molecular weight of a sample is random:

$$s + \sqrt{s} = \frac{S}{X} + \frac{1}{X\overline{DP}} \times \frac{1}{I_a} \qquad (93)$$

where s is the weight fraction of the soluble polymer, s is $1 - g$, where g is the gel fraction, $\overline{DP}$ is the average degree of polymerisation, and I_a is the intensity of light absorbed during time t. The value of $s + \sqrt{s}$ is, generally, plotted as a function $1/I_a$ (Fig. 7.72). The slope of the experimental line gives the X value and the intercept gives the S value.

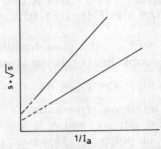

Fig. 7.72. A typical plot of $s + \sqrt{s}$ as a function of the reciprocal of irradiation intensity ($1/I_a$).

APPLICATIONS OF MODERN FTIR SPECTROSCOPY

The modern FTIR spectrometer is interfaced to a minicomputer which allows subtraction, addition, normalisation or integration spectra.[95,121,125,172] In addition more sophisticated programs are available, such as a ratio method,[77,130,173] factor analysis[5,190] and a non-linear optimisation technique.[185] The latter is extremely useful in determining the number of pure components in an unknown formulation.

The high sensitivity of FTIR spectrometers has allowed the development of more powerful techniques such as attenuated total reflectance (ATR) or reflection-absorption (RA). Recently, new spectroscopical techniques such as diffuse reflectance IR FT (DRIFT), surface electromagnetic wave (SEW), photothermal beam deflection (PBD) and emission or photoacoustic spectroscopies (PAS) have been developed; however, these methods require some sort of modifications in order to be applied as an analytical method in a radiation-curing analysis.

ATR IR spectroscopy

ATR IR spectroscopy is a powerful qualitative method for functional-group identification. ATR IR spectroscopy samples the outer 1–2 μm and is markedly insensitive to changes in the outer 100 Å (10 nm).[121,125,308]

In the ATR technique, the sample is pressed against a special crystal, termed an internal reflectance element (IRE). The materials used for the IRE crystal are listed in Table 7.10.

When the angle of incidence of the IR radiation at the sample–IRE crystal interface is greater than or equal to the critical angle, total internal reflection occurs (Fig. 7.73). The radiation existing in the IRE after multiple internal reflections is attenuated by the absorption characteristic of the sample (Fig. 7.74). When the intensity of this reflected radiation is ratioed against a background without sample, a spectrum very similar to a transmission spectrum results. By employing a crystal with a greater refractive index, or increasing the angle of incidence, the depth of penetration is reduced and profiling of the surface is possible.

In the ATR technique it is very important to maintain good contact between the sample and the IRE crystal. Poor contact will drastically diminish the signal-to-noise ratio resulting in poor-quality spectra. One way to improve the signal-to-noise ratio is to increase the effective surface area in contact with the ATR crystal.

A major advantage of ATR technique is its ability to perform surface depth profiling measurements. By varying the incidence angle of the incoming IR beam, it is possible to change the penetration depth into the surface and obtain information from various surface layers.

There are a number of different commercially available ATR accessories (Fig. 7.75). Liquid ATR is a variant of solid ATR. In liquid ATR, the IRE crystal is surrounded by a vessel into which the liquid is poured (Fig. 7.76). In

Table 7.10

Characteristics of materials for internal reflection spectroscopy used in the IR region[240]

Material	Transmission range (μm)	Refractive index (n_1) at 10 μm	Critical angle θ (degrees)	Desirable characteristics	Undesirable characteristics
Germanium (Ge)	2–11.5	4.00	22	It has the highest refractive index of all materials and is completely insoluble	Brittle and fractures easily under pressure
Silicon (Si)	0.5–6.2	3.42	26	It has the second highest refractive index and is a very hard and inert crystal	Oxidises in air
Thallium-bromide-iodide (TlBr–TlI) (KRS-5)	0.5–35	2.37	39	It does not fracture under high pressure	High toxicity. It should not be used with water solutions
Silver chloride (AgCl)	0.4–23	1.98	49	Reflector plates are easily prepared from the rolled sheet using only a razor blade. It is soft and easily deformed	It should not be left in contact with a metal surface because of corrosion
Zinc selenide (ZnSe) (Irtran-4)	1.0–18	2.4		It is insoluble, and thus ideal for all liquid-sampling applications	Inordinately expensive
Cadmium telluride (CdTe) (Irtran-6)	2.0–2.28	2.8			Inordinately expensive

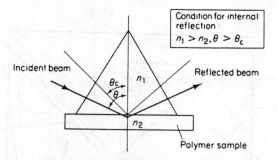

Fig. 7.73. Condition for a single internal reflection.

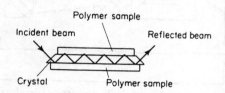

Fig. 7.74. Multiple internal reflection.

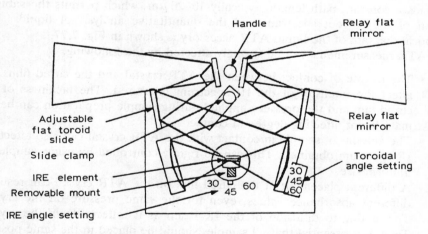

Fig. 7.75. Optical diagram of the ATR accessory (produced by Bio-Rad, USA).

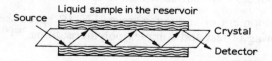

Fig. 7.76. Multiple internal reflection in liquid ATR. The IR beam travels through a crystal, and undergoes multiple reflections at the crystal–liquid interface.

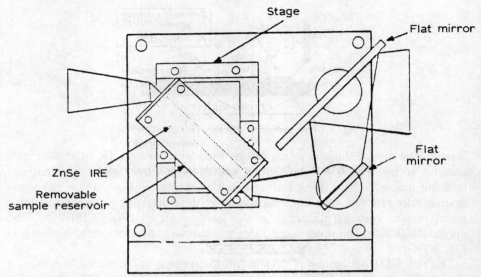

Fig. 7.77. Optical diagram of the liquid ATR accessory (produced by Bio-Rad, USA).

addition, the very reproducible sample contact surface with the IRE crystals yields a constant path length, typically 10–20 μm, which permits the subtraction of the solvent spectrum and the quantitative analysis of liquids. The optical diagram of the liquid ATR accessory is shown in Fig. 7.77.

ATR measurements can be strongly affected by the following.

- The degree of contact between the ATR crystal and the cured films can affect the intensity of the IR spectrum measured. The tackiness of the cured film and the pressure applied during sample preparation can be the main factor affecting degree of contact.
- The amount of sample in contact with the ATR crystal can also affect the IR spectrum obtained. This can be rigidly controlled by using samples of uniform size.
- A different placement of the cured film on the ATR crystal can result in different absorbance values, even for the same pressure on the crystal. This is due to diffusion of the IR beam as it reflects down the crystal. Thus, it is essential that all samples should be placed in the same position for optimum repeatability.
- The position of the ATR crystal in the IR beam is critical. Change in the position produces different absorbances for identical samples.

The recent development of computer-assisted FTIR spectroscopy has enabled accurate monitoring of fast and complex polymerisations.[71,147,150,247,316,317]

RA FTIR spectroscopy

Very thin cured coatings (e.g. monolayers) applied in electronic devices cannot be analysed by the ATR technique because of the bad contact with the IRE

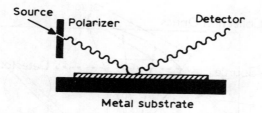

Fig. 7.78. Conditions for RA IR spectroscopy.

crystal. The only technique which can be used to record the IR spectra of such samples is to employ *p*-polarised IR radiation at large angles of incidence (grazing angles). The sensitivity of the measurements is then enhanced dramatically. In this technique a sample is deposited on a highly reflective metal surface such as platinum, gold, silver or nickel (Fig. 7.78).[112,275] The optical diagram of the polarised grazing angle reflectance accessory is shown in Fig. 7.79.

The RA FTIR technique is very useful for transparent samples; however, it is very difficult, if not impossible, to use this technique for the characterisation

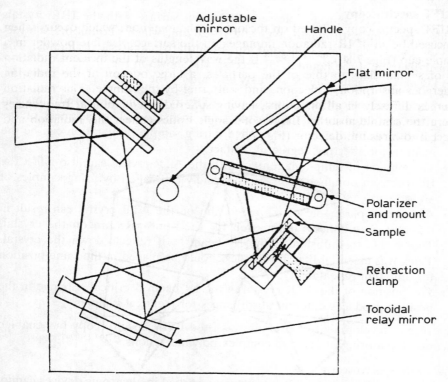

Fig. 7.79. Optical diagram of the polarised grazing angle reflective accessory (produced by Bio-Rad, USA).

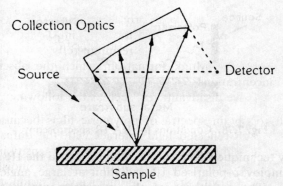

Fig. 7.80. Conditions for diffusion-reflectance IR spectroscopy. In diffuse reflectance, the incident beam undergoes absorption and scattering by the powdered sample and is collected and focused on the detector.

of opaque (pigmented) materials, because of the high absorption of most fillers.[3] However, this technique has a very high sensitivity allowing the detection of even a few monolayers of surface coverage.[21]

DRIFT spectroscopy

DRIFT spectroscopy is based on the optical phenomenon, which occurs when a focused beam of IR radiation impinges on the surface of a fine powder in a sample cup (Fig. 7.80).[104,120,178,318] If the wavelengths of the incident radiation are of similar sizes to that of the particles, a large portion of the radiation undergoes absorption, reflection, and scattering by the sample. This radiation emerges diffusely in all directions, having suffered attenuation at frequencies where the sample absorbs. Large solid angle optics collect this radiation and direct it towards the detector (Fig. 7.81).

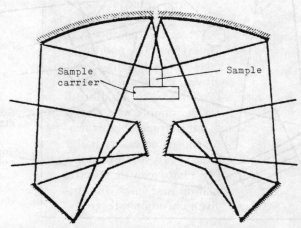

Fig. 7.81. Optical diagram of the diffuse reflectance accessory (produced by Bio-Rad, USA).

Typically, slightly ground potassium bromide (KBr) is used as the reference matrix. The powdered sample is diluted to approximately one part in 10–100 into this KBr, and its spectrum run. The resultant reflectance spectrum is then converted into a Kubelka–Munk format, in which the abscissa is directly proportional to concentrations.

This method does have disadvantages, such as the following:

- it is difficult to obtain spectra of polymeric films because the specular components of the reflected light interact with the diffusively scattered light, which may lead to band distortion or even intensity inversions;
- the KBr powder, which is very hygroscopic, absorbs moisture leading to undesirable reactions with the sample under investigation, thus perturbing the measurements.

However, the DRIFT technique can be used for surface depth profiling studies.[205] By varying the amount of KBr powder overlaying the sample surface, it is possible to enhance the surface selectivity and eliminate the specular light components.

Photoacoustic FTIR spectroscopy

Many solid cross-linked (cured) samples do not lend themselves to routine transmission or reflection techniques. The samples are almost insoluble, difficult to grind into a powder or of irregular shape. PAS is a technique which permits fast, non-destructive analyses of such samples with essentially zero sample preparation.[129,149,177,207,251,297,298]

When IR radiation impinges on a sample it absorbs energy at frequencies characteristic of the vibrational frequencies of its constituent molecules, and this energy is converted into heat. As is shown in Fig. 7.82, this heat will cause the gas at the sample's surface to warm up and expand. In a sealed volume, this creates an increase in pressure—an acoustic or sound signal which can be measured by microphone.

Because the beam in an FTIR spectrometer is modulated, the signal that the microphone measures is an interferogram, and this can be transformed to give a spectrum. The sample spectrum is usually ratioed against a carbon-black spectrum, because this absorbs all of the radiation falling on it. The result is an absorbance-like spectrum (Fig. 7.83). In addition, because the photoacoustic effect occurs in the surface of the sample, it is possible to obtain depth-related information by altering the interferometer mirror velocity.[159,207,298] A typical optical diagram of the photoacoustic accessory is shown in Fig. 7.84.

PAS has been found to be a very useful method for the study of cured polymers.[62,63,253,254,266,292] The photoacoustic spectrum is measured as a function of time at a fixed modulation frequency. In this way, changes are obtained in the concentration of a given chromophore group within a thickness of the sample upon the curing process.

As curing takes place, the chromophore (e.g. double bonds) is consumed and thus the depth of penetration of light will increase as the concentration of

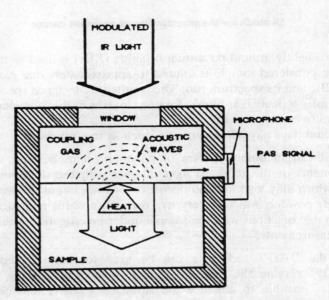

Fig. 7.82. Schematic diagram of the photoacoustic FTIR cell used for condensed samples.[298]

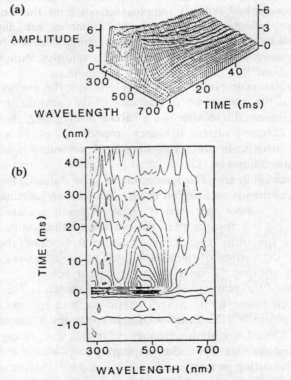

Fig. 7.83. Impulse response photoacoustic spectrum of a bulk polymer film: (a) isomeric projection, and (b) contour map.[159]

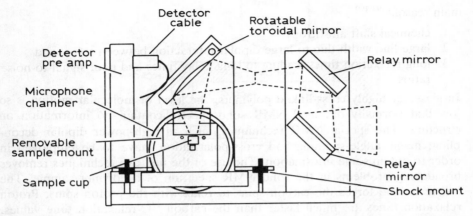

Fig. 7.84. Optical diagram of the photoacoustic accessory (produced by Bio-Rad, USA).

photoactive chromophore decreases. To quantify the extent of cure of a film as a function of depth, the depth profile must be known.

In order to obtain the depth profile of curing, this method requires that a number of physical properties of the cured film are known. The depth profiling experiments require spectra of a sample at a variety of modulation frequencies. From a new generation of cross-correlation photoacoustic spectrometers one can see at a glance which signals are coming from near the surface and which are coming from within or from the back of the film.

NMR SPECTROSCOPY

NMR spectroscopy is almost limited to polymeric materials which contain ^{1}H, ^{13}C, ^{15}N, ^{19}F and ^{31}P atoms. The detailed descriptions of the principles of NMR spectroscopy have been given elsewhere.[24,126,194,233,240,255] The most important information which can be obtained from NMR spectroscopy is as follows:

- chemical shift;
- chemical shift anisotropy;
- dipole–dipole coupling;
- scalar or 'J' coupling; and
- relaxation times (spin–lattice relaxation time and spin–spin relaxation time).

A major advance in NMR spectroscopy arises with the application of FT NMR spectroscopy.[94,95]

Molecular motion or the lack of it plays an important role in NMR spectroscopy. For that reason the most developed NMR spectroscopy deals with polymer solutions. In a non-soluble, cross-linked polymer, lack of motion makes it impossible to obtain high-resolution NMR spectra for three main reasons:[93,199]

1. chemical shift anisotropy;
2. large line width due to large dipolar interaction between spins; and
3. long relaxation times leading to long delay times and poor signal-to-noise ratios.

In glassy or highly cross-linked polymers, the level of molecular motion is so low that normally obtained NMR solid spectra contain no information on structure. The special NMR techniques, such as high-power dipolar decoupling, magic angle-spinning and cross-polarisation, have to be employed in order to obtain some information. The use of the first two techniques removes broadening problems in the solid NMR spectrum of a polymer sample. The last technique forces the carbon spins to relax with the proton spins. Proton relaxation times are much faster than the carbon ^{13}C relaxation time values. This technique also allows improved sensitivity by a factor of four. In spite of these three advanced techniques of the solid NMR, the obtained solid NMR spectra are not as good as liquid NMR spectra.

The solid NMR spectroscopy can be applied in radiation curing to the following:

- identification of unknown materials;
- quantification of cure reactions;[232]
- study of side reactions;
- study of degradation of network structures;
- molecular dynamics studies;
- study of cross-linked structures;[15,144] and
- study of swollen and deformed networks.[56,74,122,191–193,280]

Certain portions of the cured polymer have stronger solvent interactions than other regions, such that the flexible parts of the macromolecule give high-resolution NMR signals, where the rigid portions associated with the cross-link regions give very broad NMR signals which often cannot be clearly observed. In addition, the signals from the solvents themselves are affected by the swollen polymer such that in certain cases significant signal splitting is observed. In most cases these solvents give the best NMR spectra of the polymer.

ESR SPECTROSCOPY

All free-radical processes involved in radiation polymerisation from a theoretical point of view should be directly investigated by ESR spectroscopy. However, without photoinitiators, the ESR signals obtained in direct photopolymerisation of pure monomers are generally too weak for identification of radical species.[243] A number of papers have been devoted to the study of free-radical polymerisation using ESR spectroscopy.[12,154,166,262,320–322]

The concentrations of the unpaired spins of free radicals are very low, sometimes at the threshold sensitivity level of the ESR spectrometer (10^{11} spins/0·1 ml). Modern computerised ESR spectrometers are designed to increase their sensitivity and speed of signal registration of paramagnetic species at their steady-state concentration by spin accumulation in combination with rapid scan techniques in the millisecond range. Spectral resolution can be increased in some cases, when the ESR spectra can be saturated with microwave power by applying combined electron–nuclear double (ENDOR) or triple (TRIPLENDOR) resonance techniques.[155]

Principal disadvantages in the ESR measurements of radiation polymerisations are as follows.

- Measurements must be made at low temperatures, i.e. at liquid-nitrogen temperature (77K). In order to follow polymerisation the sample has to be rapidly quenched by being frozen in liquid nitrogen and then placed in the ESR spectrometer.
- Resultant spectra of formulations, monomer and photoinitiator are often very difficult to interpret.
- Transient intermediates with lifetimes less than about 1 μs may be difficult to observe using ESR spectroscopy. This limitation (in the absence of significant spin polarisation) is imposed by the spin–lattice relaxation times characteristic of free radicals in solution ($\sim 10^{-6}$–10^{-7} s). However, since some spin polarisation is always present, the relaxation criterion is not the primary limitation on the technique in this case.
- Although presently available ESR spectrometers are capable of detecting about 10^{11} spins per gauss of line width, sensitivity considerations are very critical in transient ESR spectroscopy. Since the amplitude of a spin resonance signal is approximately proportional to the inverse square of the peak-to-peak line width, transient radicals with wide lines may be difficult to observe with adequate signal-to-noise, even with extensive signal averaging.
- For free radicals with short lifetimes, the bandwidth of the detection system must be sufficiently broad. However, since the sensitivity is inversely proportional to the square root of the bandwidth (in inverse seconds), an increase in the bandwidth by a factor of 100 will decrease the spectrometer sensitivity by a factor of 10. Some sensitivity can be regained by the use of a computer to average transients; however, some loss of sensitivity for short-lived free radicals still remains.

Some of these problems of detecting free radicals in radiation polymerisation can be solved with the application of stable free radicals (spin traps) added to the irradiated formulations (spin-trapping technique).[166] The most common stable free radicals used are 2-nitroso-2-methyl propane (*t*-butyl-nitroxide), nitrobenzene, or phenyl-*N*-tert-butylnitrone, which react with free radicals (R·) according to the following:[243]

$$R^{\cdot} + CH_3-\underset{\underset{CH_3}{|}}{\overset{\overset{CH_3}{|}}{C}}-N{=}O \longrightarrow CH_3-\underset{\underset{CH_3}{|}}{\overset{\overset{CH_3}{|}}{C}}-\underset{\underset{R}{|}}{\overset{\overset{O^-}{|}}{N^+}} \qquad (94)$$

$$R^{\cdot} + \underset{}{\text{Ph}-N=O} \longrightarrow \underset{R}{\text{Ph}-N^{+}(O^{-})} \quad (95)$$

$$R^{\cdot} + \underset{\text{CH}(CH_3)_2}{\text{Ph}-C(H)=N^{+}(O^{-})} \longrightarrow \underset{\text{CH}(CH_3)_2}{\text{Ph}-C(H)(R)-N^{+}(O^{-})} \quad (96)$$

Nitrogen and proton hyperfine splitting of the spin adducts is a diagnostic parameter for the identification of the radical spin-adduct formed.

Recently, an interesting paper has been presented devoted to the application of ESR spectroscopy to the study of the kinetics of free-radical polymerisation in bulk to high conversion.[105] For the first time, the ESR spectroscopy has been utilised in measuring the concentration of the propagating radicals throughout the course of the bulk polymerisation of methyl methacrylate and thus, in conjunction with conversion measurements, has yielded directly values of kinetic parameters (k_p and k_t).

ESCA FOR THE STUDY OF POLYMER SURFACES

Electron spectroscopy for chemical analysis (ESCA), also called X-ray photoelectron spectroscopy (XPS), is perhaps, the most valuable technique for studying polymeric surfaces, and provides the following information:[27,29,41,53,54,79,107,224,240,260,264,276]

- the elemental composition (except for hydrogen),
- specific details of the chemical structure (bonding state and/or oxidation level of most atoms),
- depth of profiling, and
- surface heterogeneity.

When a soft X-ray beam interacts with core electrons the following processes occur (Fig. 7.85):

- photoionisation, which occurs with the removal of a core electron (emission of a photoelectron);
- shake-up, which is excitation of a valence electron from an occupied to an unoccupied level simultaneously with photoionisation; and
- shake-off, which is ionisation of a valence electron accompanying photoinitiation.

Shake-up and shake-off give rise to satellite peaks on the low kinetic energy side of the main photoionisation peak (Fig. 7.86). Aromatic molecules, in particular, show large probabilities for $\pi \to \pi^*$ shake-up satellites.

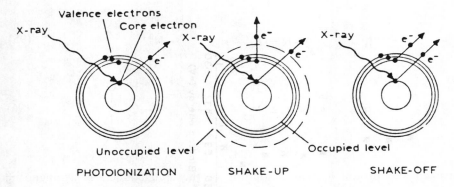

Fig. 7.85. Interaction of soft X-rays with electrons in a molecule.

The kinetic energy of emitted photoelectrons (E_k) is given by eqn (97):

$$E_k = E_{h\nu} - E_b - \phi \quad \text{(eV)} \tag{97}$$

where $E_{h\nu}$ is the energy of the bombarding X-ray, E_b is the binding energy, and ϕ is the work function that is established for each spectrometer.

The binding energy (eV) is a sensitive function of the atomic environment, which is defined by the nature of the atom with which the ejected electron was associated and the atoms bound to that atom. A systematic study of a large number of homopolymers provides a compilation of substituent effects on the C_{1s}, N_{1s}, O_{1s}, F_{1s}, Si_{2p}, P_{2p}, S_{2p} and Cl_{2p} levels.[55] In Fig. 7.87 some of the data pertaining to substituent effects on C_{1s} levels in polymers are shown.

Measurements of the intensity and energy distribution of the photoemitted electrons yields information on the surface properties of an examined material.

- Each peak in an ESCA spectrum (usually covering a 1000 eV range) is indicative of the core-level energy of an element.

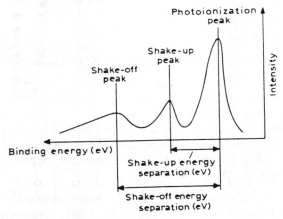

Fig. 7.86. Electron core-level spectra for a given sample.

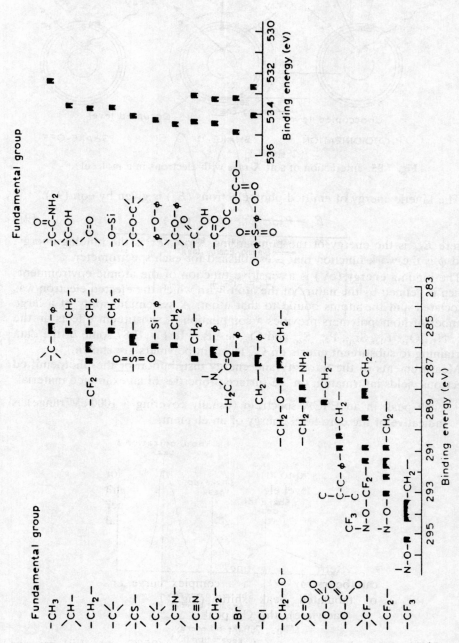

Fig. 7.87. Tabulation of experimental chemical shifts for C_{1s} and O_{1s} levels.[55]

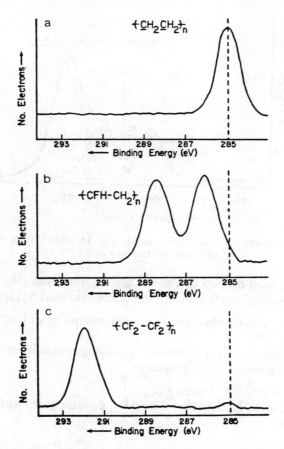

Fig. 7.88. ESCA C_{1s} spectra of (a) polyethylene, (b) poly(vinyl fluoride), and (c) polytetrafluoroethylene.[224]

- Chemical groups attached to an atom undergoing photoionisation alter the binding energy of core-level electrons. At high resolution (a spectral range of typically 20 eV), for a given core level of an element, a number of different sub-peaks may be observed, with each sub-peak representing a different molecular environment. In Fig. 7.88 are shown peak shifts of the C_{1s} spectrum in three analogical polymers: polyethylene, poly(vinyl fluoride) and polytetrafluoroethylene.
- Sub-peaks can be resolved from a complex curve envelope with the assistance of tabulated peak shifts (Fig. 7.87). For example, in poly(methyl methacrylate), the C_{1s} peak spectrum consists of the three overlapping sub-peaks (Fig. 7.89): (i) hydrocarbon-like environment (55%), (ii) carbon singly bonded to oxygen (25%), and (iii) carbons bound to oxygen with both single and double bonds (ester carbons)

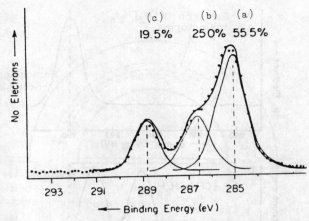

Fig. 7.89. Overlapping of the three sub-peaks in the ESCA C_{1s} spectrum of poly(methyl methacrylate).[224]

(20%). The well-defined doublet of the O_{1s} spectrum (Fig. 7.90) indicates equal portions of single-bonded and double-bonded linkages.

A diagram schematically illustrating the components of an ESCA spectrometer is shown in Fig. 7.91.

ESCA spectroscopy has a number of advantages as a spectroscopic method in the study of radiation cured polymers:

- it is a non-destructive technique;
- study is possible *in situ*, in the working environment, with a minimum of preparation;

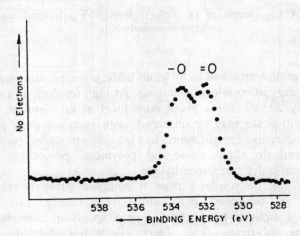

Fig. 7.90. The well-defined doublet of the ESCA O_{1s} spectrum of poly(methyl methacrylate).[224]

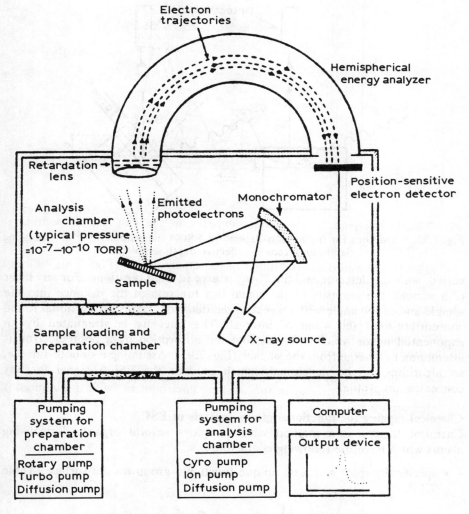

Fig. 7.91. Schematic diagram of the ESCA spectrometer.[224]

- a large amount of information is available from a single measurement;
- the method provides the capability of studying surface, sub-surface and bulk depth profiles; and
- the data are often complementary to those obtained by other spectroscopic methods.

Application of the ESCA for depth profiling

The angular-dependent ESCA technique allows measurement of a depth–concentration profile from the surface down into the bulk.[279] As shown in Fig. 7.92, the photoelectron emission from the sample is measured as the angle is

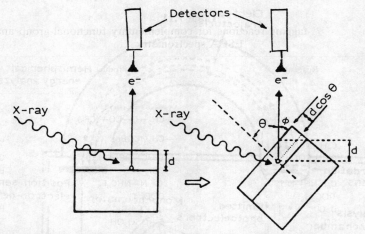

Fig. 7.92. Conditions for the angular-dependent ESCA measurements:[224] d—sampling depth, and d cos θ—effective depth sampled.

varied, with the detector and the X-ray source in fixed positions. For any layer of a sample, the intensity of the signal is a function of the distance into the sample and of the angle with respect to the detector, and is proportional to the concentration of the atom of interest. The intensity is attenuated by an exponential factor which indicates that the absorption by a photoelectron is attempting to emerge from the surface (Fig. 7.93). Assuming a smooth surface, an algorithm can be developed, which predicts the ESCA signal for any concentration profile.[234,244]

Chemical tagging for functional-group analysis in ESCA

Chemical tagging is a method which employs several organic derivatising agents which have the following characteristics:[91]

- specifically react with one unique functional group on the surface (Table 7.11);

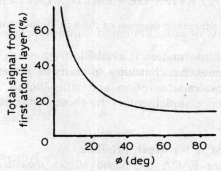

Fig. 7.93. Plot of percentage of the total signal from the first atomic layer as a function of θ (cf. Fig. 7.92).[224]

Table 7.11
Example of possible tagging reactions for complementary functional-group analysis in ESCA spectrometry

Functional group	Reagent	Product	Analysed species
$>C=C<$	Br_2/CCl_4	$>C(Br)-C(Br)<$	Br_{3p}
$-CH_2-OH$	$CF_3(CF_2)_2COCl$	$-CH_2-O-CO-(CF_2)_2CF_3$	F_{1s}
	CBr_3COOH $\}$		
	$C_6H_{11}NCNC_6H_{11}$	$-CH_2OCOBr_3$	$Br_{3p3/2}$
$>C=C-OH$	$ClCH_2COCl$	$>C=C-OCOCH_2Cl$	Cl_{2p}
$-CH_2-C<$	Br_2	$-CBr_2-C<$	Br_{3p}
$>C=O$	$C_6F_5NH-NH_2$	$>C=N-NHC_6F_5$	F_{1s}
$-COOH$	$TiOC_2H_5$	$-COO^-Ti^+$	Ti_{4s}
	$BaCl_2$	$-(COO^-)_2Ba^{2+}$	Ba_{3s}
$-C-OOH$	SO_2	$-C-OSO_2OH$	S_{2p}
$-\underset{O}{\overset{}{C}}-C-$	HCl	$>C(OH)-C(Cl)<$	C_{12p}
$-NH_2$	C_6F_5CHO	$-N=CHC_6F_5$	F_{1s}
	$C_2H_5S-COCF_3$	$-NH-COCF_3$	F_{1s}
	CS_2	$-NHCSSH$	S_{2p}
$-SH$	$AgNO_3$	$-SAg$	Ag_{3d}, S_{2p}
	$(C_6H_3(NO_2)_2(COONa)-S)_2$	$-S-S-C_6H_3(NO_2)_2(COONa)$	S_{2p}, Na_{1s}

- contain atom(s) that are easily detected by its ESCA signal; and
- increase detection sensitivity.

The chemical tagging has found special application to the study of the specific interactions in adhesion.[28]

MEASUREMENTS OF PHYSICAL AND MECHANICAL PROPERTIES OF CURED SAMPLES

The most common tests of radiation-cured samples include examination of the following:

- physical properties such as hardness, polarity of surface (by contact angle measurements), water absorption and adhesion to the cured surfaces, lubrication of surfaces; and
- mechanical properties such as tensile strength, elongation, modulus and impact resistance.[141,174,175]

All physical and mechanical properties of cured polymers depend very much on the following:

- formulations of cured samples;

- method of curing (photocuring or electron-curing) and curing conditions (atmosphere, temperature, time, energy of radiation);
- molecular weight and cross-link density; and
- progress of degradation processes which simultaneously occur during radiation curing.

HARDNESS MEASUREMENTS

Hardness generally correlates both rigidity of cross-linked structures and mechanical strength of a material. Hardness is not mathematically defined. The hardness value is dependent on the measuring instrument used.

Hardness cannot be easily expressed in fundamental units. It is a term, applied to several different qualities of a material, which depends on a combination of several fundamental material properties. The most common methods of hardness measurement employed in cured materials are as follows:[303]

- resistance of a material to indentation by a pointed or spherical indenter;
- resistance of a material to scratching by another material, by a sharp point drawn across a surface, or by rubbing with abrasives;
- energy absorbed from oscillating pendulum; and
- energy absorbed from a dropped object.

Hardness tests are designed to determine the ability of a polymeric material to resist surface deformation or rupture under concentration stresses.

Indentation hardness

The indentation hardness methods are based on the ease with which a penetrator enters the surface and produce a permanent deformation. A hardened steel point, ball, or diamond point of definite shape is forced by a load into the surface of the cured material and the depth of penetration of the size of the remaining impression is measured. The units of penetration are usually arbitrary, but can be converted into millimetres of indentation. Some of the impressions are microscopic, and can be only measured after the removal of the penetrator with a micrometre microscope. The hardness is then calculated from the unrecovered projected area.

There are a number of methods which vary, not in principle, but in the size and shape of the penetrator and in the type of measurements made. The most common methods employed are as follows:

- Rockwell hardness;
- Brinell hardness;
- Vickers hardness; and
- Knoop hardness.

However, many other methods and instruments for the measurements of indentation hardness have been described in the literature.[200]

When a hard penetrator (diamond indenter) is pressed on the cured polymer flat surface, deformation of the surface occurs which is initially purely elastic. But as the load increases a critical point is reached where the elastic limit of the softer material is exceeded and plastic deformation begins. As the load continues to increase, the material around the hard sphere is totally plastic and any further increase in load increases the dimensions of the indentation but produces no change in the main pressure. Because some cured materials are hard, plastic deformation is relatively limited and the sample begins to crack.

Most of measurements are made in the range of 0·05–0·5 N where characteristic Vickers diamond imprint diameters are in the range 3×10^{-6}–50×10^{-6} m. A high-quality microscope with numerical apertures better than 0·6 and a magnification better than 500× must be used. Microscope techniques require highly skilled workers and training to avoid and/or minimise errors.

Since polymeric materials (even cross-linked) creep under load, the measurements must be made in accordance with a strict time programme. In measurements only a very small area of the surface of the material is involved. For that reason, indentation hardness measurement methods are not only used for studying the cure of polymers, but are also widely used for identification and control of production, e.g. of paint coatings.[209,215,216,265]

Scratch hardness
Scratch tests involve drawing a hard object or a precisely ground point over the surface of a material, with resulting tearing of the surface. These tests find a wide application in the coating fields, where they are used in the study of adherence and hardness of films, and in the plastic fields, where the scratching of transparent materials and the removal of scratches is an important problem. The following methods are the most common:

- pencil hardness; and
- ploughing test.

In hardness determination by the scratch method, the results can be obtained in the following ways:

- determination of the weight required to obtain a scratch of specified width or depth (e.g. Martens method);
- determination of the width or depth of a scratch obtained at a specified or standard load on the point; and
- determination of the tangential pressure required to produce a scratch at a specified load on the point.

Calculation of results from scratch measurements for cured materials is difficult, because the stylus develops a shearing force to break bonds between upper and lower layers of material to produce debris.

Pencil hardness
Pencil hardness, because of its simplicity, is a very common method in measuring the hardness of photocured polymers.[174]

Using drawing pencils of different grades is a simple practical method of classifying the scratch resistance of plastics. A set of drawing pencils with different hardness ranging from extremely soft (2B) to extremely hard (9H) are placed firmly against the plastic surface at an angle of 45°, pressure is applied gradually, and position of the pencil is changed until it is perpendicular to the surface. It is then drawn across the polished surface of the plastic. The pencil carbon will be broken if the plastic is appreciably harder than the carbon. A furrow will be ploughed into the surface if the pencil is harder than the plastic. Lead pencils of increasing hardness (2B, B, HB, F, H, 2H, 3H, 4H, 5H, 7H, 8H and 9H) (*cf.* ASTM D 3363 pencil hardness) are used until one is found which first shows a visible mark on the test surface. In the testing of cured coatings, the tearing of the film is taken as the end point. This method is quite reproducible; however, the scale of hardness ratings is arbitrary and has serious disadvantages. The steps in hardness between grades of pencils is not uniform. The hardness depends on the proportions of graphite and clay in each grade. The more graphite and less clay, the softer the pencil. Despite these disadvantages this method is often used in the coating industry. Typical pencil hardness for different polymers is as follows: poly(methyl methacrylate)—9H, polystyrene—2H, cellulose acetate—4H, cellulose nitrate—H, epoxies—H–2B.[23]

Ploughing test
In this method a hard blunt material like diamond is loaded and pulled across a surface to make a groove. The width and depth of the groove are a measure of hardness.

The Martens method consists of finding such a load that when applied to a diamond cone at an angle of 90°, will produce a scratch 10 μm wide. To make a measurement, a number of scratches must be made at different loads, some are wider and some narrower than 10 μm. From these determinations a graph of the scratch width–load relationship is plotted, and then by interpolation the load at which the scratch width equals 10 μm, and which is defined as the Martens hardness number, is read from the graph.

In another method described as a fixed-depth scratch test (Fig. 7.94), the specimen is moved by hand horizontally relative to the stylus.[237] Constant depth conditions are achieved by using the two outer load arms to carry most of the applied load.

Pendulum hardness
The determination of pendulum hardness is a widely used method of measurement which is employed in research, development and production control. This method requires that the coating (i) has a minimum thickness, which in the DIN method must be at least 30 μm, and (ii) is applied to a flat substrate.

Pendulum hardness can be used to provide a spot measurement for a given curing speed.[51,225] It can also be used to obtain a curing curve as a function of exposure time (Fig. 7.95).

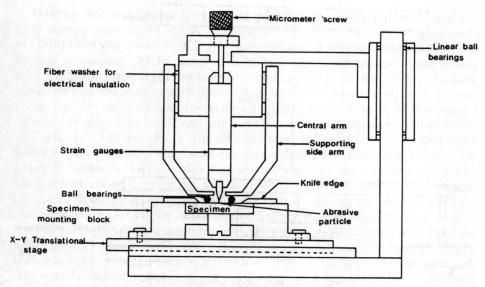

Fig. 7.94. Fixed-depth scratch test apparatus.[200]

In pendulum hardness measurement a very hard pointed pivot is set rocking on the sample surface and the time to decrease the amplitude of the swing by 50% is used as a measure of hardness.

A typical pendulum hardness meter (Fig. 7.96) consists of a horizontal metal beam with a long, very light aluminium wire pointer above the fulcrum which serves as a Vickers diamond pyramid with a 136° angle. Stabilising weights are

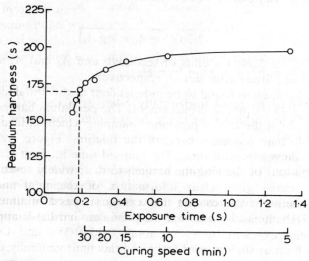

Fig. 7.95. Pendulum hardness versus exposure time (curing speed) for a given formulation.[225]

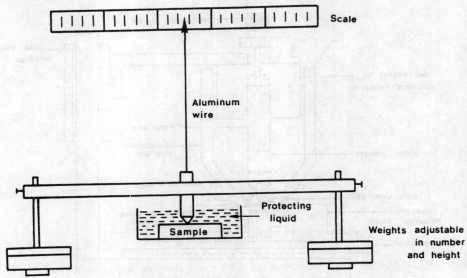

Fig. 7.96. Pendulum sclerometer.[200]

suspended rigidly below the ends of the horizontal beam. The total weight should be adjustable in such a way that the diamond can clearly indent the surface and not just rest on it. Only when a clear indent is obtained will the results become reproducible. The period of oscillation is about 1 s and the amplitude is measured after 60 s. The whole pendulum instrument should be placed in a plastic box to minimise the effects of air currents. A centimetre scale fixed near the top of the box can be used to record the amplitudes A_0 and A_t, and hardness (H_p) can be calculated from eqn (98):

$$H_p = \frac{t}{2\cdot303(\log A_0 - \log A_t)} \quad (98)$$

where H_p is the hardness (in units of seconds); and A_0 and A_t are the initial amplitude and amplitude after time t, respectively.

In practice, hardness is found to be independent of the initial amplitude, A_0, within the range 25–50 mm of displacement. The pendulum hardness (H_p) is an arbitrary measure of the rate of pendulum damping which, in turn, is related to the energy-absorbing processes beneath the fulcrum. Figure 7.97 shows how the amplitude decreases with time. The elapsed time at point C, expressed in seconds, is obtained by drawing the tangent to the curve at $t = 0$.

The curing curve gives clear information of the relationship between pendulum hardness of the coating and exposure speed or time (Fig. 7.95). From these results it can be seen that the pendulum hardness initially increases rapidly between exposure times of about 0·15–0·35 s and then gradually approaches a limit at slightly over 200 s. As this limit generally depends more on the binder system and less on the photoinitiator, the maximum pendulum hardness is not particularly suitable as a criterion of initiator reactivity.

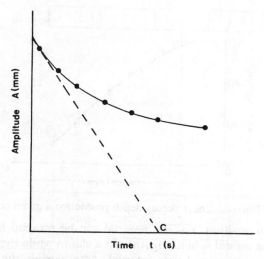

Fig. 7.97. Curve of amplitude decrease with time showing hardness in seconds at point C.[200]

Dynamic methods

Dynamic tests are based on the principle of energy absorption. The impact of a lightweight object dropped on a specimen from a fixed height causes a forced elastic deformation at the surface which disappears over a short period of time. This is essentially tesitng the material at one frequency of vibration.

Cure depth profiles

Cure depth profiles can be obtained by microhardness measurements.[59,302] The specimens prepared for Knoop microhardness measurements can be made in a stainless-steel mould that contains a slot of dimensions $15 \times 4 \times 2$ mm, and a top plate (Fig. 7.98). The mould is filled with composite for curing and the top

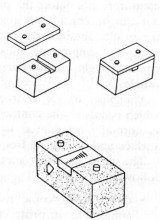

Fig. 7.98. Stainless-steel mould with cover plate, and pattern of increasing Knoop indentations with depth.[302]

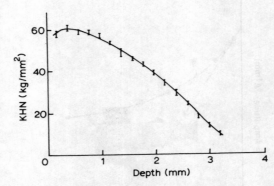

Fig. 7.99. Microhardness versus depth profile for a given composite.[302]

plate pressed into position. Excess material can be scraped from the entrance of the mould. The mould is held together in a clamp while the light enters from one side for a given measured time interval. After curing, the top of the mould and a sample strip is removed and measured as a function of depth of material at 0·2 mm intervals. Microhardness testers usually contain a Knoop diamond indenter and a fixed load (e.g. 200 g). Data are usually calculated as Knoop hardness numbers (KHN) and plotted to give hardness versus depth profiles (Fig. 7.99).

Commercially available hardness testers consist of an anvil, an elevating screw, a loading and indentation measuring unit, and a test data evaluator (Fig. 7.100). The hardness calculator can be programmed for any of three hardness test modalities: Brinell, Vickers, or Knoop.

CONTACT ANGLE MEASUREMENTS

Contact angle measurements monitor the polarity of the outer few Ångströms of a sample and the data can render values for critical surface tension. These measurements detect the presence of, or change in, surface functional groups, but can be considered neither qualitative nor quantitative. Interpretation of the results obtained depends on a number of assumptions. These methods are relatively simple and inexpensive. A number of excellent reviews on the contact angle measurements have been published in books and review articles.[9,87,114,151,221,222,240,324]

The basis of the contact angle techniques is the three-phase equilibrium which occurs at the contact point at the solid–liquid–vapour. This equilibrium is normally considered in terms of the surface and interfacial tensions or surface and interfacial free-energies present.

The most common methods of measuring the contact angle include the following:[222]

- direct microscopic measurements of the three-phase interface with a goniometer or protractor which directly measures the angles;

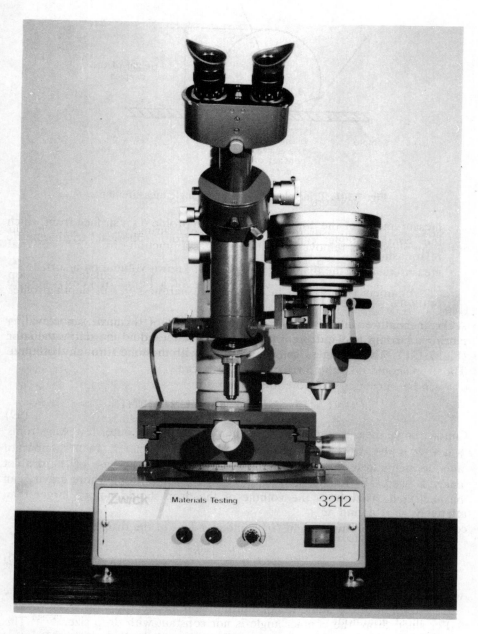

Fig. 7.100. Commercially produced hardness tester (type 3212, Zwick, Germany).

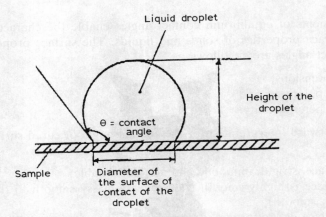

Fig. 7.101. Principle of the contact angle measurements.

- measurements of the dimensions of a drop profile on a surface from which the contact angle can be calculated from spherical trigonometric relationships;[250]
- measurement of the diameter of a drop of known volume on a surface;[143]
- rise in a capillary or on a vertical plate of a liquid of known surface tensions.

A direct measurement of the apparent contact angle (θ) can be performed by placing a tangent to the drop at the point where the liquid intercepts the solid (Fig. 7.101). θ is the angle this tangent forms with the solid through the liquid. The drop should take the form of a spherical cap.

θ can be calculated from the following ratio:

$$\tan\frac{\theta}{2} = \frac{2h}{d} \qquad (99)$$

where h is the drop height, and d is the drop diameter. In a number of measurements instead of drop diameter the length of the base of the droplet image is used. The relationship applies only for small drops where gravity can be neglected. Therefore, the volume of the drop should not be larger than $0.1\,\mu l$.

For large drops, drop height (h) is independent of the drop size:

$$\cos\theta = \frac{1 - \rho g h^2}{2\gamma_{l,v}} \qquad (100)$$

where ρ is the liquid density, g is the gravitational acceleration; and $\gamma_{l,v}$ is the surface tension of the liquid in equilibrium with its vapour.

For small drops the contact angle is not constant with drop size.[106,115] This effect is attributed to the surface heterogeneity and the line tension. For that reason it is advisable to use drops with a diameter greater than 3 mm.

Measurements of equilibrium contact angles enable the characterisation of various surface properties of solids and liquids. The surface properties which affect contact angles are as follows:

- surface tensions;
- surface polarities; and
- surface heterogeneity.

A straightforward interpretation of the contact angle of cured surfaces can be made if

- the surface is rigid, immobile and non-deformable;
- the surface is highly smooth (i.e. appears glassy smooth to the eye or is optically smooth);
- the surface is uniform and homogeneous (polymer blends and block copolymer surfaces are known to have two or more phases present, each of which may have different properties);
- the surface is clean (dirty surfaces are often of a patchy character);
- the liquid must not cause extraction or partitioning of material from the solid phase into the liquid phase; and
- the liquid does not swell the solid surface.

A diagram schematically illustrating the components of an apparatus for measuring the contact angle is shown in Fig. 7.102.

The critical surface tension (γ_c) is a useful parameter obtained by measurements of the contact angle of each of a series of homologous liquids on the test material. A number of methods have been suggested for how to estimate the critical surface tension of a solid surface from contact angle data.[102,113,116,198,223,228,313–315,323–325]

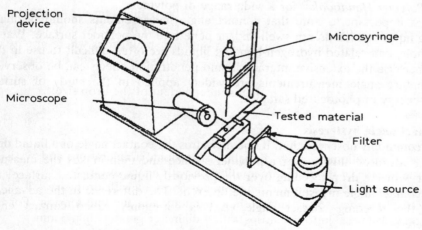

Fig. 7.102. Apparatus for the contact angle measurement.

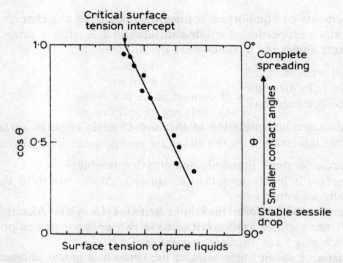

Fig. 7.103. A typical Zisman plot for determination of the critical surface tension (γ_c).[324]

In the simplest way a graphical treatment (Zisman plot) of the contact angle data is used to deduce γ_c (Fig. 7.103). The cosines of the contact angle for a range of pure liquids on a given solid are plotted against the liquid surface tensions. γ_c is given by the intercept at $\cos \theta = 1$ and is defined as the surface tension of that liquid which would only just totally spread on the solid surface. This is an empirical measure related to the surface free-energy of the solid and is called the *critical surface tension for wetting* of that particular solid, and is empirically related to the surface constitution. Even small changes in the outermost atomic layer are reflected in a change of γ_c. Values are tablulated in the *Polymer Handbook*[25] for a wide range of polymers.

It is important to note that contact angle measurements should be made using liquids which do not swell and/or penetrate the polymer surface. Water, alcohols, and related hydrogen-bonding liquids are often difficult to use in this test because the extensive interaction and penetration effects can be observed.

Contact angle measurements are widely applied in the study of surface free-energy of photocured samples.[158]

Contact angle hysteresis

It is commonly observed that, if one measures the contact angle of a liquid drop being advanced slowly over a polymer surface and then makes the measurement with the drop receding over the previously liquid-contacted surface, the two contact angle measurements are different. The difference in the advancing and the receding contact angle (Δ_γ) is commonly called contact angle hysteresis.[151]

Contact angle hysteresis measurements are commonly made by the methods of contact angle measurements. In the most applied method a drop is advanced over a solid surface by inclining the surface, that is, making an incline plane and adjusting the angle of incline until the drop moves slowly over the polymer surface or just begins to move. In this way the advancing angle is on the low end of the plane and the receding angle is on the upper portion of the plane.

One of the major causes of contact angle hysteresis is penetration of liquid into the surface region of the polymer. For example, the water-penetration effect can be very rapid for very thin films. Contact angle hysteresis provides considerable information on the character and properties of polymer surfaces, especially in water environments.

ADHESION OF WATER TO THE SURFACE OF CURED POLYMERS

The adhesion of various materials to the surface of cured polymers plays an important role in industry, i.e. in processes for obtaining coatings and printing on cured surfaces.

The following are the most common methods for determining water adhesion.[148]

- Measurements of the wetting angle (θ). The work of adhesion of water (W) is given by eqn (101).

$$W = \gamma_w(\cos \theta + 1) \tag{101}$$

where γ_w is the surface tension of the water.

- Measurements of the angle of slope of polymeric sheets to a level initiating the displacement of droplets of water deposited on the surface of these sheets (α). The work of adhesion of water (W) is given by eqn (102):

$$W = \frac{\rho V g \sin \alpha}{d} \tag{102}$$

where ρ is the density of water, V is the volume of water droplet, g is the gravitational acceleration, and d is the maximum (transverse) diameter of the plane of contact of the droplet with the surface of the strip.

- Pendulum and rotating disc adhesiometry. The water droplet rolling-off angle measurement principle is shown in Fig. 7.104. Samples are fixed on a smooth, flat-surfaced supporting material such as a slide glass. First, at the horizontal level 0·3 ml distilled water is placed on a sample. The water droplet rolling-off angle is measured when the water drop starts rolling down as the slide glass is gradually inclined. The measurement is normally performed within 30 s of the water drop being placed. Lower 'water droplet rolling-off angle' indicates higher water repellency.[146]

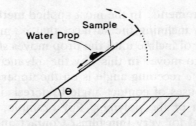

Fig. 7.104. Principle of the measurement of water droplet rolling-off angle.[146]

Water absorption measurements

Water absorption by the various formulations can be determined by the weight difference of films after subjecting them to dry and wet conditions. The cured samples can be easily dried by storing them in a vacuum desiccator over phosphorus pentoxide or calcium sulphate at room temperature for a given time.

Samples subjected to the wet conditions should be placed in the vapour-temperature controlling relative humidity chamber for a given period. The amount of water absorbed per gram of dry film should be calculated for the three samples in order to obtain averaged values.

Many UV-cured acrylates, epoxides and thiol-enes suffer a dramatic loss of tensile strength during exposure to moisture and/or elevated temperatures. Certain formulations are especially sensitive and lose up to 95% of their dry tensile strength in a humid environment.[22]

MEASUREMENTS OF THE LUBRICATION OF SURFACES

Radiation curing is a common method for the surface modification (hydrophobic–hydrophilic properties, lubrication properties, etc.) of polymeric materials.

A coefficient of friction (μ value) is commonly used for evaluating the lubrication of surfaces, but is often difficult to determine, especially for elastic and adhesive polymers because of their complicated hysteresis occurring in a cycle of deformation and recovery during the frictional motion, which is probably associated with adhesion force and displacement force.[30]

μ can be measured in a simple apparatus, as shown in Fig. 7.105. A slider, at the bottom of which a sample film is attached, travels horizontally on the glass plate in the distilled water. The force needed for the slider to travel at a rate of 10 mm/min is converted by a low-friction pulley to a vertical force, which is continuously recorded using a 100 g load cell.

As an example, Fig. 7.106 shows the forces required for a slider attached with films before and after graft photoinduced polymerisation to travel under different loads.

The frictional force can be measured with the aid of a tensile testing

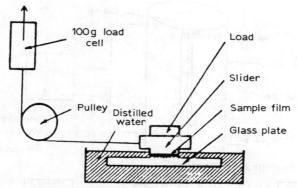

Fig. 7.105. Schematic diagram of the apparatus for the determination of the coefficient of friction (μ).[299]

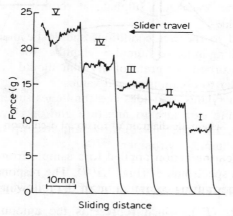

Fig. 7.106. The force required for a slider attached with the polymer film to travel on the glass plate in water.[299]

device.[293] In this method, the frictional force generated on the polymer surface during travelling on a glass plate in aqueous environment is measured.[299]

The frictional profiles can be divided into five types:

1. sharp peaks with steep slopes;
2. sharp peaks with a rapid fall but slow rise;
3. low amplitude at high frequency;
4. relatively smooth curve; and
5. smooth but increasing with time.

DYNAMIC MECHANICAL SPECTROMETRY

Dynamic mechanical spectrometry[217,245,265,284,312] is a technique which determines the dynamic properties of a material by measuring the response to a

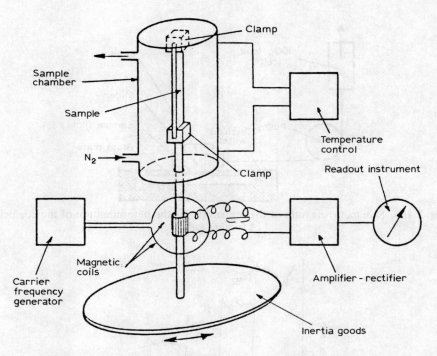

Fig. 7.107. Schematic diagram of torsional oscillation apparatus.

forced sinusoidal shearing action applied to a sample from a servomotor in a rheometric dynamic spectrometer (Fig. 7.107). The response is measured by a torque transducer, permitting computation of the following.

- Elastic modulus (E'), which represents the amount of energy stored elastically upon deformation and recovered per cycle.
- Loss modulus (E'') which represents the energy lost to viscous dissipation, such as heat.
- Complex modulus (E), defined in eqn (103).

$$E = E' = E'' = \frac{h_1}{h_2} \qquad (103)$$

where h_1 and h_2 are the strain and the stress amplitude heights, respectively (cf. Fig. 7.108).

- Loss tangent, or damping factor ($\tan \delta$), which in a cyclic deformation is the ratio of energy lost to energy stored:

$$\tan \delta = \frac{E'}{E''} \qquad (104)$$

Viscoelastic properties of a polymeric material are revealed as a sinusoidal variation of stress and strain (Fig. 7.108).

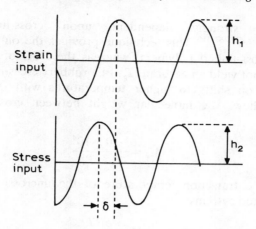

Fig. 7.108. Stress and strain response in the linear viscoelastic region.

E'' and the loss tangent are sensitive enough to measure molecular motions within a polymeric material and also to measure various transitions, relaxation processes, and the structure and morphology.

The experimental data are commonly presented as the dynamic tensile storage modulus (E') and loss modulus (E'') plotted as a function of temperature. The temperature which corresponds to the maximum in the α peak, given by the log E'' curve, $T(E''_{max})$ is defined as T_g (Fig. 7.109).

This technique is especially useful in studying cured cross-linked polymeric systems, as the dynamical properties and loss behaviour of cross-linked

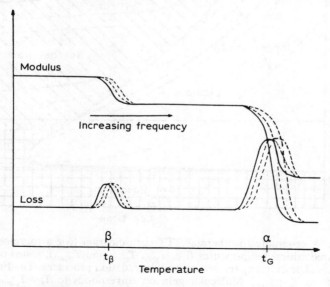

Fig. 7.109. Dynamic behaviour in a transition region.

materials are strongly dependent upon cross-link density and architecture.[98,269,277,287,310] This technique provides the only accurate value of T_g of highly cross-linked polymers, whereas the DSC of highly cross-linked polymers does not yield an accurate T_g. In a rubbery cross-linked structure, the T_g or α-transition shifts to higher temperatures with increasing cross-link density, and allows the molecular weight between cross-links (M_c) to be calculated:[217]

$$M_c = \frac{3 \cdot 10 \times 10^4}{T_g - T_{g0}} \tag{105}$$

where T_{g0} is the transition temperature of the uncrosslinked polymer. For glassy cross-linked systems:

$$M_c = \frac{\rho RT}{E} \tag{106}$$

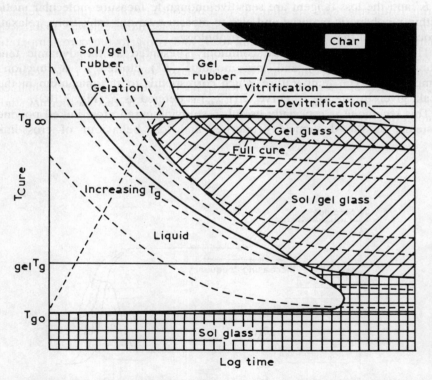

Fig. 7.110. A generalised isothermal TTT cure diagram (for a thermosetting system) showing three critical temperatures (i.e. $T_{g(0)}$, $T_{g(\infty)}$, and $T_{g(gel)}$), states of the material, and contours characterising the setting and degradation processes. The full cure contour corresponds to $T_g = T_{g(\infty)}$. Molecular gelation corresponds to $T_g = T_{g(gel)}$. Other iso-T_g contours are included (-----).[311]

where ρ is the density of a cross-linked sample, R is the gas constant, and E is the relating modulus.

Dynamic mechanical behaviour is significantly affected by the stoichiometric balance of reactants and the cure conditions.[312]

- If a sinusoidal strain is applied to a perfectly elastic material, the resulting stress is in phase with the applied strain.
- If the sinusoidal strain is applied to a pure viscous material, the stress is 90° out of phase with the strain.
- If the phase angle (δ) is defined as the difference between the applied strain and resulting stress, then for the case of the perfectly elastic solid, $\delta = 0$. For the case of the purely viscous material, $\delta = 90°$. Viscoelastic cured materials possess an intermediate response, i.e. $0 < \delta < 90°$.

If the radiation curing is accompanied or followed by thermal curing (backing) it is very useful to study changes which occur during isothermal cure as a function of time.[89,108–110,311] The various changes occurring in the material during isothermal cure are characterised by contours of the times to reach the events. Relevant contours could include molecular gelation, macroscopic gelation, vitrification corresponding to T_g and rising to the cure temperature (T_{cure}), devitrification corresponding to T_g decreasing to T_{cure} because of thermal degradation, and char formation corresponding to T_g increasing to T_{cure} because of thermal degradation. The progress in the isothermal cure process and the state of materials are shown in the isothermal time–temperature–transformation (TTT) cure diagram shown in Fig. 7.110.

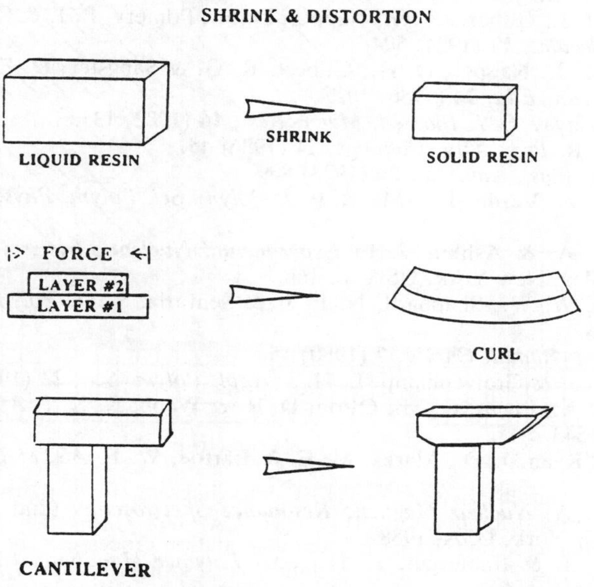

Fig. 7.111. Shrink and distortion of sample shapes observed during radiation curing.

DISTORTIONS OF RADIATION-CURED SAMPLES

During radiation curing distortions of a final sample are often observed. Distortion can be a result not only of shrinkage on curing, but also of the specific structural feature being examined, such as cantilevers (Fig. 7.111). A uniform shrinkage will result in a uniform reduction in size. If it is done in layers, the result is 'curl'. Curl is most pronounced in parts having an unsupported extension, or cantilever. This is what often happens with multilayers of photocurable resins, i.e. in three-dimensional photopolymerisation or photofabrication.[176]

REFERENCES

1. Acitelli, M. A., Prime, R. B. & Sacher, E. *Polymer*, **12** (1971) 333.
2. Alberda van Ekenstein, G. O. R. & Tan, Y. Y. *Euro. Polym. J.*, **24** (1988) 1073.
3. Allara, D. L. In *Characterization of Metal and Polymer Surfaces* (Vol. 2), ed. L. H. Lee. Academic Press, New York, USA, 1977.
4. Andrady, A. L. & Sefcik, M. D. *J. Polym. Sci. Polym. Phys. Edn*, **21** (1983) 2453.
5. Antoon, M. K., D'Esposito, L. & Koenig, J. L. *Appl. Spectr.*, **33** (1979) 349.
6. Appelt, B. K. *Polym. Eng. Sci.*, **23** (1983) 367.
7. Appelt, B. K. & Abadie, M. L. *Polym. Eng. Sci.*, **23** (1983) 125.
8. Aronhime, M. T. & Gillham, J. K. *J. Coatings Technol.*, **56** (1984) 35.
9. Aveyard, R. & Haydon, D. A. *Introduction to Principles of Surface Chemistry*. CUP, London, UK, 1973.
10. Baeumer, W., Koehler, M. & Ohngemach, J. In Conf. Proceed. on RADCURE'86, 4/43.
11. Bank, L. & Ellis, B. *Polymer*, **23** (1982) 1466.
12. Ballard, M. J., Gilbert, R. G., Napper, D. H., Pomery, P. J. & O'Donnell, J. H. *Macromolecules*, **17** (1984) 504.
13. Ballard, M. J., Napper, D. H., Gilbert, R. G. & Sangster, D. F. *J. Polym. Sci. Polym. Chem. Edn*, **24** (1986) 1027.
14. Bandyopadhyay, S. *J. Biomed. Mater. Res.*, **16** (1982) 135.
15. Bauer, D. R. *Prog. Org. Coatings*, **14** (1986) 45.
16. Becker, R. *Plast. Kautsch.*, **20** (1973) 809.
17. Bellenger, V., Verdu, J. & Morel, E. *J. Polym. Sci. Polym. Phys. Edn*, **25** (1987) 1219.
18. Bethe, H. A. & Ashkin, J. In *Experimental Nuclear Physics* (Vol. 1), ed. E. Segre. Wiley, New York, USA, p. 166.
19. Bindstrup, W. W., Shappard, N. F. Jr & Senturia, S. D. *Polym. Eng. Sci.*, **26** (1986) 358.
20. Blank, E. *J. Radiat. Curing*, **7** (1980) 15.
21. Boerio, F. J. & Greivenkamp, L. H. *J. Appl. Polym. Sci.*, **22** (1978) 203.
22. Bolon, D. A., Lucas, G. M., Olson, D. R. & Webb, K. K. *J. Appl. Polym. Sci.*, **25** (1980) 543.
23. Boor, L., Ryan, J. D., Marks, M. E. & Bartoe, W. F. *ASTM Bull.*, **145** (1947) 68.
24. Bovey, F. A. *Nuclear Magnetic Resonance Spectroscopy* (2nd edn). Academic Press, New York, USA, 1988.
25. Brandrup, J. & Immergut, E. H. (eds) *Polymer Handbook* (3rd edn). Wiley, 1990.
26. Bräuchle, C. *Polym. Photochem.*, **5** (1984) 21.

27. Briggs, D. *Polymer,* **25** (1985) 1379.
28. Briggs, D. & Kendall, C. R. *J. Adhesion Adhesives,* (1982) 13.
29. Briggs, D. & Seah, M. P. (eds) *Practical Surface Analysis.* Wiley, Chichester, UK, 1983.
30. Briscoe, B. J. & Tabor, D. In *Polymer Surfaces,* ed. D. T. Clark & J. F. Feast. Wiley, New York, USA, 1978, p. 1.
31. Buback, M. *Makromol. Chem.,* **191** (1990) 1575.
32. Burnett, G. M. & Melville, H. W. *Proc. Royal Soc.* (London), **A189** (1947) 456.
33. Bush, R. W., Ketley, A. K., Morgan, C. R. & Whitt, D. G. *J. Radiat. Curing,* **7** (1980) 20.
34. Campa, De La, J. G., Abajo, De, J., Mantecon, A. & Cadiz, V. *Euro. Polym J.,* **23** (1987) 961.
35. Campbell, D. & White, J. R. *Polymer Characterization.* Chapman & Hall, London, UK, 1989.
36. Carlini, C., Ciardelli, F., Rolla, P. A. & Tombari, E. *J. Polym. Sci. Polym. Phys. Edn,* **25** (1987) 1253.
37. Carlini, C., Ciardelli, F., Rolla, P. A. & Tombari, E. *J. Polym. Sci. Polym. Phys. Edn,* **27** (1989) 189.
38. Carlini, C., Ciardelli, F., Rolla, P. A., Tombari, E., Li Bassi, G. & Nicora, E. In Conf. Proceed. RADTECH '89 Europe, 1989, p. 369.
39. Carlini, C., Martinelli, M., Rolla, P. A. & Tombari, E. *J. Polym. Sci. Polym. Lett. Edn,* **23** (1985) 5.
40. Carlini, C., Rolla, P. A. & Tombari, E. *J. Appl. Polym. Sci.,* **41** (1990) 805.
41. Carlson, T. A. *Photoelectron and Auger Spectroscopy.* Plenum Press, New York, USA, 1975.
42. Carre, C., Lougnot, D. J. & Fouassier, J. P. *Macromolecules,* **22** (1989) 791.
43. Carrozzino, S., Levita, G., Rolla, P. A. & Tombari, E. *Polym. Eng. Sci.,* **30** (1990) 366.
44. Chambers, S., Guthrie, J., Otterburn, M. S. & Woods, J. *Polym. Commun.,* **27** (1986) 209.
45. Charlesby, A. *J. Polym. Sci.,* **11** (1953) 513.
46. Charlesby, A. *Proc. Royal Soc.* (London), **A222** (1954) 542.
47. Charlesby, A. *Atomic Radiation and Polymers.* Pergamon Press, Oxford, UK, 1960.
48. Charlesby, A. & Pinner, S. H. *Proc. Royal Soc.* (London), **A249** (1959) 367.
49. Chiu, W. Y., Carratt, G. M. & Soong, D. S. *Macromolecules,* **16** (1983) 348.
50. Chottiner, J., Sanjana, Z. N., Kodani, M. R., Lengel, K. W. & Rosenblatt, G. B. *Polym. Comp.,* **3** (1982) 59.
51. Christensen, J. E., Wooten, W. L. & Whitman, P. J. *J. Radiat. Curing,* **14** (1987) 35.
52. Chu, B. & Lee, D. *Macromolecules,* **17** (1984) 926.
53. Clark, D. T. In *Polymer Surfaces,* ed. D. T. Clark & W. J. Feast. Wiley, Chichester, UK, 1979, p. 309.
54. Clark, D. T. *Pure Appl. Chem.,* **54** (1982) 415.
55. Clark, D. T. & Thomas, H. R. *J. Polym. Sci. Polym. Chem. Edn,* **16** (1978) 791.
56. Cohen-Addad, J. P., Domard, M., Lorentz, G. & Hertz, J. *J. Phys.* (Paris), **45** (1984) 575.
57. Collins, G. L. & Constanza, J. R. *J. Coatings Technol.,* **51** (1979) 57.
58. Cook, W. D. *Euro. Polym. J.,* **14** (1978) 715.
59. Cook, W. D. *J. Dent. Res.,* **59** (1980) 800.
60. Crivello, J. V. & Lam, J. H. W. *J. Polym. Sci. Polym. Lett. Edn,* **17** (1979) 759.
61. Cundal, R. B., Dandiker, Y. M., Dabies, A. K. & Salim, M. S. In *Radiation Curing of Polymers* (Special Publ. No. 64), ed. D. R. Randell. Royal Society of Chemistry, London, UK, 1987, p. 172.

62. Davidson, R. S., Ellis, R., Wilkinson, S. & Summergill, C. *Euro. Polym. J.*, **23** (1987) 105.
63. Davidson, R. S. & Lowe, C. *Euro. Polym. J.*, **25** (1989) 159.
64. Davies, A. K., Cundall, R. B., Bate, N. J. & Simpson, L. A. *J. Radiat. Curing*, **14** (1987) 22.
65. Decker, C. In *Materials for Microlitography* (ACS Symp. Ser. 266), ed. L. F. Thompson, G. G. Wilson & J. M. J. Fréchet, American Chemical Society, USA, 1984, p. 207.
66. Decker, C. In *Radiation Curing of Polymers* (Special Publ. No. 64), ed. D. R. Randell. Royal Society of Chemistry, London, UK, 1987, p. 16.
67. Decker, C. *Macromolecules*, **23** (1990) 5217.
68. Decker, C. & Fizet, M. *Makromol. Chem. Rapid Commun.*, **1** (1980) 637.
69. Decker, C. & Moussa, K. *Makromol. Chem.*, **189** (1988) 2381.
70. Decker, C. & Moussa, K. *Macromolecules*, **22** (1989) 4455.
71. Decker, C. & Moussa, K. *Polym. Mater. Sci. Eng.*, **60** (1989) 547.
72. Decker, C. & Moussa, K. *Makromol. Chem.*, **191** (1990) 963.
73. Decker, C. & Moussa, K. *J. Coatings Technol.*, **62** (1990) 55.
74. Deloche, B. & Samulski, E. T. *Macromolecules*, **14** (1981) 575.
75. Derome, A. E. *Modern NMR Techniques for Chemical Research*. Pergamon Press, New York, 1987.
76. DeSchryver, F. C. & Boens, N. *J. Oil Colour Chem. Assoc.*, **59** (1976) 171.
77. Diem, H. & Krim, S. *Appl. Spectr.*, **35** (1979) 421.
78. Di Giulio, C., Gautier, M. & Jasse, B. *J. Appl. Polym. Sci.*, **29** (1984) 1771.
79. Dilks, A. In *Electron Spectroscopy: Theory, Techniques and Applications* (Vol. 4), ed. A. D. Baker & C. R. Brundle. Academic Press, London, UK, 1981, p. 277.
80. DiMarzio, E. A. *J. Res. Natl. Bur. Stand.*, **68A** (1964) 611.
81. Dionisio, J. M. & O'Driscoll, K. F. *J. Polym. Sci. Polym. Chem. Edn*, **18** (1980) 241.
82. Dole, M. *The Radiation Chemistry of Macromolecules*. Academic Press, New York, USA, 1973.
83. Doornkamp, A. T. & Tan, N. Y. *Polym. Commun.*, **31** (1990) 362.
84. Dušek, K. *Adv. Polym. Sci.*, **78** (1986) 3.
85. Eckhardt, H., Prusik, T. & Chance, R. R. *Macromolecules*, **16** (1983) 732.
86. Eldridge, J. E. & Ferry, J. D. *J. Phys. Chem.*, **58** (1954) 992.
87. Elliot, G. E. P. & Riddiford, A. C. *Rec. Prog. Surface Sci.*, **2** (1964) 111.
88. El-Shimi, A. & Goddard, E. D. *J. Colloid Interface Sci.*, **48** (1974) 242.
89. Enns, J. B. & Gillham, J. K. In *Polymer Characterization* (ACS Advances in Chemistry Series No. 203), ed. C. D. Craver. 1983, p. 27.
90. Evans, A. J., Armstrong, C. & Tolman, R. J. *J. Oil Colour Chem. Assoc.*, **61** (1978) 251.
91. Everhart, D. S. & Reilley, C. N. *Surface Interface Anal.*, **3** (1981) 126.
92. Ewing, G. E. (ed.) *Analytical Instrumental Handbook*. Marcel Dekker, New York, USA, 1990.
93. Fedotov, V. D. & Schneider, H. *Structure and Dynamics of Bulk Polymer by NMR Methods*. Springer Verlag, Berlin, Germany, 1989.
94. Ferrar, T. C. & Becker, E. D. *Pulse and Fourier Transform NMR*. Academic Press, New York, USA, 1971.
95. Ferraro, J. R. & Basile, L. J. (eds) *Fourier Transform IR Spectroscopy, Applications to Chemical Systems* (Vols I–IV). Academic Press, New York, USA, 1979.
96. Fischer, E. & Kunze, W. In Conf. Proceed. RADTECH '89 Europe, 1989, p. 669.
97. Fischer, E., Kunze, W. & Stapp, B. *Analystechn. Berichte*, Heft 60E., Perkin-Elmer, 1988.

98. Fitzgerald, J. J. & Landry, C. J. T. *J. Appl. Polym. Sci.*, **40** (1990) 1727.
99. Fizet, M., Decker, C. & Faure, J. *Euro. Polym. J.*, **21** (1985) 427.
100. Flory, P. J. *J. Amer. Chem. Soc.*, **63** (1941) 3083, 3091, 3096.
101. Flory, P. J. & Rehner, J. Jr *J. Chem. Phys.*, **11** (1943) 521.
102. Fowkes, F. M. *Ind. Chem. Chem.*, **56** (1964) 40.
103. Fox, T. G. & Loshaek, S. *J. Polym. Sci.*, **15** (1955) 371.
104. Fuller, M. P. & Griffits, P. R. *Anal. Chem.*, **50** (1978) 504.
105. Garrett, R. W., Hill, D. J. T., O'Donnell, J. H., Pomery, P. J. & Winzor, C. L. *Polym. Bull.*, **22** (1989) 611.
106. Gaydos, J. & Neumann, A. W. *J. Colloid Interface Sci.*, **120** (1987) 76.
107. Ghosh, P. K. *Introduction to Photoelectron Spectroscopy.* Wiley, New York, USA, 1983.
108. Gillham, J. K. *Polym. Eng. Sci.*, **19** (1979) 670.
109. Gillham, J. K. In *Developments in Polymer Characterization* (Vol. 3), ed. J. E. Dawkins. Applied Science, London, UK, 1982, p. 159.
110. Gillham, J. K. *Polym. Eng. Sci.*, **26** (1986) 1429.
111. Gobran, R. H., Berenbaum, M. B. & Tobolsky, A. V. *J. Polym. Sci.*, **46** (1960) 431.
112. Golden, W. C. In *Fourier Transform IR Spectroscopy, Applications to Chemical Systems* (Vol. 4), ed. J. R. Ferraro & L. J. Basile. Academic Press, New York, USA, 1985, p. 315.
113. Good, R. J. *Adv. Chem. Ser.*, **43** (1964) 74.
114. Good, R. J. *Surface Colloid Sci.*, **11** (1979) 1.
115. Good, R. J. & Koo, M. N. *J. Colloid Interface Sci.*, **71** (1979) 283.
116. Good, R. J. & Kotsidas, J. *Adhesion*, **10** (1979) 17.
117. Gordon, M. *Proc. Royal Soc.* (London), **A268** (1962) 240.
118. Göring, W. *Farbe Lack*, **83** (1977) 270.
119. Gottlieb, M. In *Biological and Synthetic Polymer Networks*, ed. O. Kramer. Elsevier, London, UK, 1988, p. 403.
120. Griffiths, P. R. & Fuller, M. P. In *Mid-Infrared Spectroscopy of Powdered Samples*, ed. R. J. H. Clark & R. E. Hexter. Heyden, London, UK, 1982, p. 63.
121. Griffiths, P. R. & Haseth, J. A. de. *Fourier Transform Infrared Spectrometry.* Wiley, New York, USA, 1986.
122. Gronsky, W., Stadler, G. & Jacobi, M. M. *Macromolecules*, **17** (1984) 74.
123. Guthrie, J., Jeganathan, M. B., Otterburn, M. S. & Woods, J. *Polym. Bull.*, **15** (1986) 51.
124. Gutierriez, A. R. & Cox, R. J. *Polym. Photochem.*, **7** (1986) 517.
125. Harrick, N. J. *Internal Reflection Spectroscopy.* Wiley-Interscience, New York, USA, 1967.
126. Harris, R. K. *Nuclear Magnetic Resonance Spectroscopy.* Longman, New York, USA, 1986.
127. Hatchard, C. G. & Parker, C. A. *Proc. Royal Soc.* (London), **A235** (1956) 518.
128. Heller, H. G. & Langan, J. R. *J. Chem. Soc. Perkin Trans.* **2** (1981) 341.
129. Hess, P. (ed.) *Photoacoustic, Photothermal and Photochemical Processes at Surfaces and Thin Films* (Topics Current Physics, Vol. 47). Springer Verlag, Berlin, Germany, 1989.
130. Hierschfeld, T. H. *Anal. Chem.*, **53** (1981) 2232.
131. Hill, D. J. T. & O'Donnell, J. H. *J. Polym. Sci. Polym. Chem. Edn*, **20** (1982) 241.
132. Hoffman, A. S., Jameson, J. T., Salmon, W. A., Smith, D. E. & Trageser, D. A. *Ind. Eng. Chem. Prod. Res. Dev.*, **9** (1970) 158.
133. Holmes, B. S. & Trask, C. A. *J. Appl. Polym. Sci.*, **35** (1988) 1399.
134. Hoyle, C. H., Cranford, M., Trapp, M., No, Y. G. & Kim, K. J. *Polymer*, **29** (1988) 2033.
135. Hoyle, C. E., Hensel, R. D. & Grubb, M. B. *J. Radiat. Cuting*, **11** (1984) 22.

136. Hoyle, C. E., Hensel, R. D. & Grubb, M. B. *Polym. Photochem.*, **4** (1984) 69.
137. Hoyle, C. E., Hensel, R. D. & Grubb, M. B. *J. Polym. Sci. Polym. Chem. Edn*, **22** (1984) 1965.
138. Hoyle, C. E., Keel, M. & Kim, K. J. *Polymer*, **29** (1988) 18.
139. Hoyle, C. E. & Kim, K. J. *J. Radiat. Curing*, **12** (1985) 9.
140. Hoyle, C. E. & Kim, K. J. *J. Appl. Polym. Sci.*, **33** (1987) 2985.
141. Huemmer, T. F. *J. Radiat. Curing*, **1** (1974) 3.
142. Hult, A., MacDonald, S. A. & Wilson, C. G. *Macromolecules*, **18** (1985) 1801.
143. Hutchinson, H. *J. Oil Colour Chem. Assoc.*, **72** (1989) 265.
144. Hvilsted, S. In *Biological and Synthetic Polymer Networks*, ed. O. Kramer. Elsevier, London, UK, 1988.
145. Ikeda, M., Teramoto, Y. & Yasutake, M. *J. Polym. Sci. Polym. Chem. Edn*, **16** (1978) 1175.
146. Iriyama, Y., Yasuda, T., Cho, D. L. & Yasuda, H. *J. Appl. Polym. Sci.*, **39** (1990) 249.
147. Ishida, H. & Scott, C. *J. Polym. Eng.*, **6** (1986) 201.
148. Janczuk, B. *Int. Polym. Sci. Technol.*, **16** (1989) T81.
149. Jasse, B. *J. Macromol. Sci. Chem.*, **A26** (1989) 43.
150. Jin, J. R. & Meyer, G. E. *Polymer*, **27** (1986) 592.
151. Johnson, R. E. Jr & Dettre, R. In *Surface and Colloid Science* (Vol. 2), ed. E. Matijevic. Wiley, New York, USA, 1963, p. 85.
152. Joshi, M. G. *J. Appl. Polym. Sci.*, **26** (1981) 3915.
153. Joshi, M. G. & Rodriguez, F. *J. Polym. Sci. Polym. Chem. Edn*, **26** (1988) 819.
154. Kamachi, M., Kuwae, Y., Kohno, M. & Nozakura, S. *Polym. J.*, **17** (1985) 541.
155. Kevan, L. & Bowman, M. K. *Modern Pulsed and Continuous-Wave ESR*. Wiley, New York, USA, 1990.
156. Kilb, R. W. *J. Phys. Chem.*, **63** (1959) 1838.
157. Kim, K. J. & Hoyle, C. E. *J. Radiat. Curing*, **12** (1985) 9.
158. Kim, H. C., Song, J. H., Wilkes, G. L., Smith, S. D. & McGrath, J. E. *J. Appl. Polym. Sci.*, **38** (1989) 1515.
159. Kirkbright, G. F., Miller, R. M., Spillane, D. E. M. & Sugitani, Y. *Anal. Chem.*, **56** (1984) 2043.
160. Kloosterboer, J. G. *Adv. Polym. Sci.*, **84** (1988) 1.
161. Kloosterboer, J. G. & Bressers, H. J. L. *Polym. Bull.*, **2** (1980) 205.
162. Kloosterboer, J. G. & Lijten, G. F. C. M. *Polymer*, **28** (1987) 1149.
163. Kloosterboer, J. G. & Lijten, G. F. C. M. *Polym. Commun.*, **28** (1987) 2.
164. Kloosterboer, J. G. & Lijten, G. F. C. M. In *Crosslinked Polymers* (ACS Symp. Ser. 365), eds R. A. Dickie, S. S. Labana & R. S. Bauer. American Chemical Society, USA, 1988, p. 409.
165. Kloosterboer, J. G. & Lijten, G. F. C. M. *Polymer*, **31** (1990) 95.
166. Kloosterboer, J. G., Lijten, G. F. C. M. & Greidanus, F. J. A. M. *Polym. Commun.*, **27** (1986) 268.
167. Kloosterboer, J. G. & Lippits, G. J. M. *J. Radiat. Curing*, **11** (1984) 10.
168. Kloosterboer, J. G., Genuchten, H. P. M. van, Hei, G. M. M. van de, Melis, G. P. & Lippits, G. J. M. *Org. Coat. Plast. Chem.*, **48** (1983) 445.
169. Kloosterboer, J. G., Hei, G. M. M. van de & Boots, H. M. J. *Polym. Commun.*, **25** (1984) 354.
170. Kloosterboer, J. G., Hei, G. M. M. van de, Gossink, R. G. & Dortant, G. C. M. *Polym. Commun.*, **25** (1984) 322.
171. Knight, R. W. In Conf. Proceed. Radtech '91 Europa, 1991, p. 84.
172. Koenig, J. L. In *Optical Techniques to Characterize Polymer Systems*, ed. H. Bässler. Elsevier, Amsterdam, The Netherlands, 1989, p. 1.
173. Koenig, J. L., D'Esposito, L. & Antoon, M. K. *Appl. Spectr.*, **31** (1977) 292.
174. Koleske, J. V. In Proc. Radcure EUROPE '87, 1987, p. 1.

175. Koshiba, T., Huang, K., Foley, S., Yarusso, A. & Cooper, S. *J. Mater. Sci.*, **17** (1982) 1447.
176. Krajewski, J. J. *J. Coatings Technol.*, **62** (1990) 73.
177. Krishnan, K. *Appl. Spectr.*, **35** (1981) 549.
178. Krishnan, K., Hill, S. L. & Brown, R. H. *Amer. Lab.*, **12** (1980) 104.
179. Kunze, W. & Stapp, B. In Proc. Conf. on Radiation Curing ASI, 1988, p. 319.
180. Ledwith, A. *Pure Appl. Chem.*, **49** (1977) 431.
181. Le, D. D. *J. Radiat. Curing*, **12** (1985) 2.
182. Lee, G. & Hartmann, B. *J. Appl. Polym. Sci.*, **28** (1983) 823.
183. Lee, G. & Hartmann, B. *J. Appl. Polym. Sci.*, **29** (1984) 1471.
184. Li, W. H., Hamielec, A. C. & Crowe, C. M. *Polymer*, **30** (1989) 1513.
185. Liu, J. & Koenig, J. L. *Anal. Chem.*, **59** (1987) 2609.
186. Louden, J. D. & Roberts, T. A. *J. Raman Spectr.*, **14** (1983) 365.
187. Lu, C. & Czanderna, A. W. (eds) *Applications of Piezoelectric Quartz Crystal Microbalances*. Elsevier, Amsterdam, The Netherlands, 1984.
188. Lugnot, D., Carre, C. & Fouassier, J. P. *Makromol. Chem. Macromol. Symp.*, **24** (1989) 209.
189. Macosco, C. W. & Miller, D. R. *Macromolecules*, **9** (1976) 199.
190. Malinowski, E. R. & Lowery, D. G. *Factor Analysis in Chemistry*. Wiley, New York, USA, 1980.
191. Marshall, G. L. *Euro. Polym. J.*, **22** (1986) 217 and 231.
192. Marshall, G. L. & Lander, J. A. *Euro. Polym. J.*, **21** (1985) 959.
193. Marshall, G. L. & Wilson, S. L. *Euro. Polym. J.*, **24** (1988) 933 and 939.
194. Martin, M. L., Delpuech, J. J. & Martin, G. J. *Practical NMR Spectroscopy*. Heyden, London, UK, 1980.
195. Martinelli, M., Rolla, P. A. & Tombari, E. *Instr. Measur.*, **34** (1985) 417.
196. Martinelli, M., Rolla, P. A. & Tombari, E. *IEEE Trans. Microwave Theory Technol.*, **33** (1985) 779.
197. Martinelli, M., Rolle, P. A. & Tombari, E. *Microwaves & RF*, **25** (1986) 3.
198. Matsunaga, T. *J. Appl. Polym. Sci.*, **21** (1977) 2847.
199. McBrierty, V. J. *Polymer*, **15** (1974) 503.
200. McColm, I. J. *Ceramic Hardness*. Plenum Press, New York, USA, 1990.
201. McGinnis, V. D. *J. Radiat. Curing*, **2** (1975) 3.
202. McGinnis, V. D. In *Ultrafast Light Induced Reactions in Polymers* (ACS Symp. Ser. 25), ed S. S. Labana. American Chemical Society, USA, 1976, p. 135.
203. McGinnis, V. D. & Dušek, D. M. *J. Paint Technol.*, **46** (1974) 23.
204. McGinnis, V. D. & Ting, V. W. *J. Radiat. Curing*, **2** (1975) 14.
205. McKenzie, M. T. & Koenig, J. L. *Appl. Spectr.*, **39** (1985) 408.
206. Miller, D. R. & Macosco, C. W. *Macromolecules*, **9** (1976) 206.
207. Miller, R. M. In *Photoacoustic, Photothermal and Photochemical Processes at Surfaces and in Thin Films* (Topics Current Physics, Vol. 47), ed. P. Hess. Springer Verlag, Berlin, Germany, 1989, p. 171.
208. Mitchell, J. Jr (ed.) *Applied Polymer Analysis and Characterization: Recent Developments in Techniques, Instrumentation, Problem Solving*. Hanser, Munich, Germany, 1987.
209. Monk, C. J. H. & Wright, T. A. *J. Oil Colour Chem. Assoc.*, **48** (1965) 520.
210. Moore, J. E. In *UV Curing, Science and Technology*, ed. S. P. Pappas. Technology Marketing Corp., Stamford, CT, USA, 1978, p. 133.
211. Moore, J. E., Schroeter, S. H. & Shultz, A. R. *Coat. Plast. Prep.*, **35** (1975) 239.
212. Moore, J. E., Schroeter, S. H., Shultz, A. R. & Stang, L. D. In *Ultra Fast Light Induced Reactions in Polymers*. (ACS Symp. Ser. 25), ed. S. S. Labana. American Chemical Society, USA, 1976, p. 90.
213. Morgan, C. R. & Ketley, A. D. *J. Polym. Sci. Polym. Lett. Edn*, **16** (1978) 75.
214. Morgan, C. R., Magnatta, F. & Ketley, A. D. *J. Polym. Sci. Polym. Chem. Edn*, **15** (1977) 627.

215. Morris, R. L. J. *J. Oil Colour Chem. Assoc.*, **53** (1970) 1.
216. Morris, R. L. J. *J. Oil Colour Chem. Assoc.*, **56** (1973) 555.
217. Murayama, T. *Dynamic Mechanical Analysis of Polymeric Materials.* Elsevier, London, UK, 1978.
218. Nakamichi, T. *Prog. Org. Coatings,* **8** (1980) 19.
219. Nakamichi, T. *Prog. Org. Coatings,* **14** (1986) 23.
220. Nakamichi, T., Yamazaki, L., Otsubo, Y. & Watanabe, K. *J. Jap. Soc. Colour. Mater.,* **59** (1986) 2.
221. Neumann, A. W. *Adv. Colloid Interface Sci.*, **4** (1975) 105.
222. Neuman, A. W. & Good, E. J. In *Surface Colloid Science* (Vol. 2), ed. R. J. Good & R. R. Stromberg. Plenum Press, New York, USA, 1979, p. 31.
223. Neuman, A. W., Good, R. J., Hope, C. J. & Sejpal, M. *J. Colloid Interface Sci.,* **49** (1974) 291.
224. Oechsner, H. *Thin Film and Depth-Profile Analysis* (Topics Current Physics, Vol. 37). Springer Verlag, Berlin, Germany, 1984.
225. Ohngemach, J., Koehler, M. & Wehner, G. In Conf. Proceed. RADTECH '89 Europe, 1989, p. 638.
226. Oster, G. & Yang, N. *Chem. Rev.,* **68** (1968) 125.
227. Otsubo, Y., Amari, T. & Watanabe, K. *J. Appl. Polym. Sci.,* **29** (1984) 4071.
228. Owens, D. K. & Wendt, R. C. *J. Appl. Polym. Sci.,* **13** (1969) 1741.
229. Ozawa, T. *Bull. Chem. Soc. Japan,* **38** (1965) 1881.
230. Ozawa, T. *J. Therm. Anal.,* **2** (1970) 301.
231. Pacansky, J. & Waltman, R. J. *J. Radiat. Curing,* **15** (1988) 12.
232. Paci, M., Del Vecchio, E. & Campana, F. *Polym. Bull.,* **15** (1986) 21.
233. Pasika, W. M. *Carbon-13 NMR in Polymer Science.* Academic Press, New York, USA, 1972.
234. Paynter, R. W., Ratner, B. D. & Thomas, H. R. In *Polymers as Biomaterials,* ed. S. W. Shalaby, A. S. Hoffman, B. D. Ratner & Horbett. Plenum, New York, USA, 1984.
235. Pemberton, D. R. & Johnson, A. F. *Polymer,* **25** (1985) 529.
236. Phillips, R. *Sources and Applications of Ultraviolet Radiation.* Academic Press, New York, 1983.
237. Prasad, S. V. & Kosel, T. H. In *Proceedings of the International Conference on Wear of Materials,* Vancouver B.C. ASME, New York, USA, 1985, p. 59.
238. Prime, R. B. *Thermochim. Acta,* **26** (1978) 165.
239. Quach, A. & Hansen, C. M. *J. Paint Technol.,* **46** (1974) 40.
240. Rabek, J. F. *Experimental Methods in Polymer Chemistry.* Wiley, Chichester, UK, 1980.
241. Rabek, J. F. *Experimental Methods in Photochemistry and Photophysics* (Vol. I–II). Wiley, Chichester, UK, 1982.
242. Rabek, J. F. *Mechanisms of Photophysical Processes and Photochemical Reactions in Polymers.* Wiley, Chichester, UK, 1987.
243. Rånby, B. & Rabek, J. F. *ESR Spectroscopy in Polymer Research.* Springer Verlag, Berlin, Germany, 1977.
244. Ratner, B. D. In *Polymers in Medicine: Biomedical and Pharmaceutical Applications,* ed. E. Chiellini, P. Giusti, C. Migliares & L. Nicolais. Plenum Press, New York, USA, 1986, p. 13.
245. Read, B. E. & Dean, G. D. *The Determination of Dynamic Properties of Polymers and Composites.* Wiley, New York, USA, 1978.
246. Renson, C. In Proc. Conf. on RADCURE ASIA '88, 1988, p. 356.
247. Richter, E. B. & Macosko, C. W. *Polym. Eng. Sci.,* **18** (1978) 1012.
248. Rietsch, D., Davellose, D. & Froelich, D. *Polymer,* **17** (1976) 858.
249. Roffey, C. G. *Photopolymerization of Surface Coatings.* Wiley, Chichester, UK, 1982.

250. Rotenberg, Y., Boruka, L. & Neumann, A. W. *J. Colloid Interface Sci.*, **93** (1983) 169.
251. Rozencwaig, A. *Photoacoustics and Photoacoustic Spectroscopy*. Wiley, New York, USA, 1980.
252. Russell, G. T., Napper, D. H. & Gilbert, R. G. *Macromolecules*, **21** (1988) 2133.
253. Salazar-Rojas, E. M. & Orban, M. W. *Prog. Org. Coatings*, **16** (1988) 371.
254. Salim, M. S., Cundall, R., Davies, A., Dandikar, Y. & Slifkin, M. In Radcure Conference, Basel, FC85-422, 1985.
255. Sanders, J. K. M. & Hunter, B. K. *Modern NMR Spectroscopy*. OUP, Oxford, UK, 1987.
256. Sauerbrunn, S. R. In Conf. Proceed. RADTECH '88 North America, 1988, p. 219.
257. Sawyer, L. C. & Grubb, D. T. *Polymer Microscopy*. Chapman & Hall, London, UK, 1987.
258. Schneider, N. S., Sprouse, J. F., Hagmaner, J. F. & Gillham, J. K. *Polym. Eng. Sci.*, **19** (1979) 304.
259. Schulz, G. V. & Harborth, G. *Makromol. Chem.*, **1** (1947) 106.
260. Seah, M. P. *Surface Interface Anal.*, **2** (1980) 222.
261. Senogoles, E. & Woolf, L. A. *J. Chem. Educ.*, **44** (1967) 157.
262. Shen, J., Tian, Y., Zeng, Y. & Qiu, Z. *Makromol. Chem. Rapid Commun.*, **8** (1987) 615.
263. Shultz, A. R. *J. Polym. Sci. Polym. Symp.*, **25** (1968) 115.
264. Siegban, K. *Science*, **217** (1982) 111.
265. Skrovanek, D. J. & Schoff, C. K. *Prog. Org. Coatings*, **16** (1988) 135.
266. Small, R. D., Ors, J. A. & Royce, B. C. In *Polymers In Electronics* (ACS Symp. Ser. 242), ed. T. Davidson. American Chemical Society, USA, 1984, p. 325.
267. Soh, S. K. & Sunberg, D. C. *J. Polym. Sci. Polym. Chem. Edn*, **20** (1982) 1299, 1315, 1331 and 1345.
268. Spanier, R. F. *Ind. Res.*, (1975) 73.
269. Spathis, G., Konton, E. & Theocaris, P. S. *J. Polym. Sci. Polym. Chem. Edn*, **25** (1987) 1285.
270. Spinks, J. W. T. & Woods, R. J. *An Introduction to Radiation Chemistry*. Wiley-Interscience, New York, USA, 1990.
271. Stockmayer, W. J. *J. Chem. Phys.*, **11** (1943) 45.
272. Stockmayer, W. J. *J. Chem. Phys.*, **12** (1944) 125.
273. Stockmayer, W. J. *J. Chem. Phys.*, **13** (1945) 199.
274. Struik, L. C. E. *Physical Ageing in Amorphous Polymers and Other Materials*. Elsevier, Amsterdam, The Netherlands, 1978.
275. Swalen, J. D. & Rabolt, J. F. In *Fourier Transform IR Spectroscopy, Applications to Chemical Systems* (Vol. 4), ed. J. R. Ferraro & L. J. Basile. Academic Press, New York, USA, 1985, p. 283.
276. Swingle, R. S. & Riggs, W. M. *Crit. Rev. Anal. Chem.*, **5** (1975) 267.
277. Takahama, T. & Geil, P. H. *J. Polym. Sci. Polym. Lett. Edn*, **20** (1982) 453.
278. Takahashi, A., Nakamura, A. & Kagawa, J. *Polym. J.*, **3** (1972) 207.
279. Takahashi, M. *Prog. Org. Coatings*, **14** (1986) 67.
280. Takahashi, S. *J. Appl. Polym. Sci.*, **28** (1983) 2847.
281. Tan, H. M., Moet, A., Hiltner, A. & Baer, E. *Macromolecules*, **16** (1983) 28.
282. Tanny, G. B., Lubelsky, A., Rav-Noy, Z. & Shchon, E. In Radcure Conference, Basel, FC 85-440, 1985.
283. Taylor, J. R. & Foster, H. *J. Oil Colour Chem. Assoc.*, **54** (1971) 1030.
284. Theweleit, E. & Kunze, W. *Kunststoffe*, **77** (1987) 870.
285. Thommes, G. A. & Webers, V. J. *J. Imaging Sci.*, **29** (1985) 112.
286. Thompson, D., Song, J. H. & Wilkes, G. L. *J. Appl. Polym. Sci.*, **34** (1987) 1063.
287. Timm, D. C., Ayorinde, A. J. & Foral, R. F. *Brit. Polym. J.*, **17** (1985) 227.

288. Tobita, H. & Hamielec, A. E. *Makromol. Chem. Macromol. Symp.*, **20/21** (1988) 501.
289. Tobita, H. & Hamielec, A. E. *Macromolecules*, **22** (1989) 3098.
290. Tobolsky, A. V. *J. Amer. Chem. Soc.*, **80** (1958) 5927.
291. Tobolsky, A. V., Rogers, C. E. & Brickman, R. D. *J. Amer. Chem. Soc.*, **82** (1960) 1227.
292. Torres-Filho, A., Perondi, L. F. & Miranda, L. C. M. *J. Appl. Polym. Sci.*, **35** (1988) 103.
293. Triolo, P. M. & Andrade, J. D. *J. Biomed. Mater. Res.*, **17** (1983) 149.
294. Tryson, G. R. & Shultz, A. R. *J. Polym. Sci. Polym. Phys. Edn*, **17** (1979) 2059.
295. Tulig, T. J. & Tirrell, M. *Macromolecules*, **14** (1981) 1501.
296. Turi, E. A. (ed.) *Thermal Characterization of Polymeric Materials*. Academic Press, London, UK, 1981.
297. Urban, M. W. *J. Coatings Technol.*, **59** (1987) 29.
298. Urban, M. W. *Prog. Org. Coatings*, **16** (1988) 321.
299. Uyama, Y., Tadekoro, H. & Ikada, Y. *J. Appl. Polym. Sci.*, **39** (1990) 489.
300. Vanderhoff, J. M. In *Ultra Fast Light Induced Reactions in Polymers* (ACS Symp. Ser. 25), ed. S. S. Labana. American Chemical Society, USA, 1976, p. 162.
301. Watanabe, K., Amari, T. & Otsubo, Y. *J. Appl. Polym. Sci.*, **29** (1984) 57.
302. Watts, D. C., Amer, O. & Combe, E. C. *Brit. Dent. J.*, **156** (1984) 209.
303. Weingarten, H. *Technische Härtemessung*. Carl Hauser Verlag, Munich, Germany, 1972.
304. Wetton, R. E. In *Developments in Polymer Characterization* (Vol. 6), ed. J. E. Dawkins. Applied Science Publisher, Houston, 1985.
305. Wight, F. R. *J. Polym. Sci. Polym. Lett. Edn*, **16** (1978) 121.
306. Wight, F. R. *J. Radiat. Curing*, **8** (1981) 24.
307. Wight, F. R. & Hicks, G. W. *Polym. Eng. Sci.*, **18** (1978) 378.
308. Willis, H. A. & Zichy, V. J. I. In *Polymer Surfaces*, ed. D. T. Clark & W. J. Feast. Wiley, New York, USA, 1978, p. 287.
309. Wilson, J. E. *Radiation Chemistry of Monomers, Polymers and Plastics*. Dekker, New York, 1974.
310. Wingard, C. D. & Beatty, C. L. *J. Appl. Polym. Sci.*, **41** (1990) 2539.
311. Wisanrakkit, G. & Gillham, J. K. *J. Coatings Technol.*, **62** (1990) 35.
312. Wolfe, S. V. & Toda, D. A. *J. Macromol. Sci. Chem.*, **A26** (1989) 249.
313. Wu, S. *J. Polym. Sci. Polym. Symp.*, **34** (1971) 19.
314. Wu, S. *J. Colloid Interface Sci.*, **71** (1979) 605.
315. Wu, S. *Polymer Interface and Adhesion*. Dekker, New York, USA, 1982.
316. Yang, Y. S. & Lee, L. J. *Macromolecules*, **20** (1987) 1490.
317. Yang, Y. S., Lee, L. J., Tom Lo, S. K. & Menardi, P. J. *J. Appl. Polym. Sci.*, **37** (1989) 2313.
318. Yeboah, S. A., Griffith, P. R., Krishnan, K. & Kuehl, D. *Proc. SPIE Int. Opt. Eng.*, **V289** (1981) 105.
319. Zhu, S. & Hamielec, A. E. *Macromolecules*, **22** (1989) 3093.
320. Zhu, S., Tian, Y., Hamielec, A. E. & Eaton, D. R. *Macromolecules*, **23** (1990) 1144.
321. Zhu, S., Tian, Y., Hamielec, A. E. & Eaton, D. R. *Macromolecules*, **31** (1990) 154.
322. Zhu, S., Tian, Y., Hamielec, A. E. & Eaton, D. R. *Polymer*, **31** (1990) 1726.
323. Zisman, W. A. *Ind. Eng. Chem.*, **55** (1963) 19.
324. Zisman, W. A. *Adv. Chem. Ser.*, **43** (1964) 1.
325. Zisman, W. A. *J. Paint Technol.*, **44** (1972) 42.
326. *Lamps for Special Purposes* Catalogue, Philips (1979).
327. *Philips Lighting* Catalogue (1991).
328. *High Intensity Light* Catalogue, Kratos Schoeffel Instrument Gmbh, Germany, 1974.
329. *Optics Guide*, Melles Griot, Holland, 1975–1991.

Chapter 8

UV-Radiation Sources and Radiation Devices for Industrial Production Processes

JAN F. RABEK

Karolinska Institute, Polymer Research Group, Department of Dental Materials and Technology, Alfred Nobels Alle 8, Box 4064, S-14104 Huddinge, Sweden

Introduction 453
UV-Radiation Sources 454
   Mercury vapour high-pressure and medium-pressure lamps 454
   Doped and addictive lamps 457
   Xenon lamps 457
   Pulsed xenon lamps 463
   Microwave-powered lamps 463
Reflectors for UV Lamps 466
Principles of Designing a Suitable UV-Curing Device 473
Industrial Methods of UV Curing 476
Geometrical Arrangements of UV Lamps 479
Spot-Curing Devices 482
Service Life of UV Lamps 484
Operation Maintenance of UV Lamps in UV-Curing Industrial Devices . . . 485
Lamp Replacement 487
Insulation of UV Lamps in UV-Radiation-Curing Industrial Devices 488
Air Cooling Systems for the UV Lamps in Industrial Systems 488
Measurement of UV-Curing Radiation 491
Computer Control in UV-Curing Industries 497
Industrial Line UV Processors 498
Ozone Produced in UV-Curing Industrial Devices 499
Safety Regulations for UV-Curing Devices 500
References 500

INTRODUCTION

In the 1970s the chemical industry started to produce very effective photoinitiators for UV curing. From then on the areas of application of photocuring have rapidly increased and include the following industrial applications:

- printing ink drying in the printing industry;
- screen printing on glass;
- paint curing in the timber and furniture industries;
- adhesive curing and surface hardening;
- application of thin and thick layers; and
- exposure of photopolymer layers in the reprography, microelectronic and silicon chip industries.

In industrial society, the phrase 'UV drying' process is used instead of 'UV-radiation curing'. It is important to use a precise and correct definition of UV-photocuring processes which have nothing to do with 'drying processes'. During UV curing liquid monomer, photosensitiser and additive formulations are polymerised to the cross-linked, non-soluble structures. Most formulations are UV cured without solvents, or with only a very small proportion of organic solvent, and evaporation of them from formulations is a serious problem in any production. Today, the modern photocurable formulations in the chemical industry are very highly developed and advanced. The UV curing of printing inks, paints, lacquers (pigmented and transparent) and adhesives occurs in industrial production processes within a few seconds. The new technological UV-curing processes require very effective UV lamps and technological devices, and of course knowledge and experience of how to use them.

UV-RADIATION SOURCES

Today on the market are available a lot of different UV lamps produced by many manufactures. It is almost impossible to present and describe all of these UV-radiation sources available for industrial production processes. For that reason the author has selected one of many producers of these UV lamps—Philips Lighting—to present one of the most modern UV mercury vapour discharge lamps used widely in UV-curing industries.[1]

Mercury vapour high-pressure and medium-pressure lamps

In everyday conversation, lamps which operate with an internal pressure in excess of 1 bar (10^5 Pa) and up to 10 kbars (1000 kPa) are frequently referred to as 'high-pressure' lamps and the term 'low pressure' is applied to fluorescent lamps as their internal pressure is less than 1 bar (10^5 Pa). The 'medium-pressure' lamps operate at 1 bar (10^5 Pa) or slightly higher.

UV metal halide (MH) high-pressure lamps (type HPA for UV-B and UV-A (300–400 nm) and HPM for UV-A + violet (350–450 nm) Philips), mercury vapour high-pressure lamps and medium-pressure lamps (type HOK and HTQ for UV-C to orange (200–600 nm), Philips) are made from heat-resistant quartz glass.[1] They are able to withstand heat, mechanical stress and some chemical environments when the UV installations are operated. A comparison of different UV-radiation sources is given in Table 8.1.

Sources and devices for industrial processes

Table 8.1
Comparison of different UV-radiation sources[1]

Radiant excitance (W/cm)	UV efficiency to P_{el}[a] (%)	Tube temperature (°C)	Spectrum (Hg + MH)[b] (nm)	Service life (h)
High-pressure/medium-pressure lamps				
30–120	15–25	600–900	200–600	1000–2000 (note decline in radiant power!)
Low-pressure lamps				
max. 10	max. 30	40	various narrow spectra 250–600	up to 5000 (note decline in radiant power!)

[a]P_{el}—power consumption of lamp.
[b]Hg—mercury discharge lamp; MH—metal halide discharge lamp.

The choice of a UV lamp depends on the UV-curing process and its industrial application. Subdivision of the optical radiation spectrum as per DIN 5031 T.7 c.f. Table 8.2:

UV   100–380 nm
       UV-C   100–280 nm
       UV-B   280–315 nm
       UV-A   315–380 nm
Light   380–780 nm
IR   780–1 μm
       IR-A   780–1400 nm
       IR-B   1400–3000 nm
       IR-C   3000–1 μm

For example, many photoinitiators generate significantly more radicals in very short UV-C radiation, than in longwave UV-A radiation. However, recently a number of new photoinitiating systems have been developed for longwave

Table 8.2
Subdivision of the LP radiation spectrum

	X-rays	Vacuum ultra-violet	Near ultra-violet	Visible	Near-infrared		Far-infrared
λ, Å	1	2 000	4 000	7 000	10 000		
nm	0·1	200	400	700	1 000		
μm			0·4	0·7	1	50	500
ν, Hz		1.5×10^{15}	7.5×10^{14}		3×10^{14}	6×10^{12}	6×10^{11}
$\bar{\nu}$, cm^{-1}		50 000	25 000	14 300	10 000	200	20

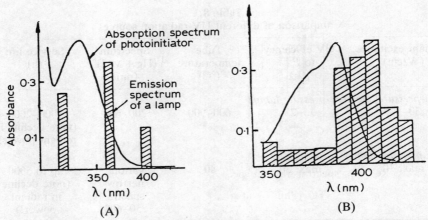

Fig. 8.1. Overlapping of different absorption spectra of photoinitiators. (A) Diethoxyacetophenone and (B) thioxanthone, and different emission spectra of lamps (see Chapter 2).

UV-A and even visible light irradiation. It is very important that the absorption spectrum of the photoinitiator must overlap the emission spectrum of employed UV-radiation source (Fig. 8.1).

The choice of a UV lamp should be a result of compromise with the absorption spectrum of the photoinitiating system. In Figs 8.2 and 8.3 are shown spectra of the mercury vapour high-pressure lamps types HOK and HTQ, respectively.

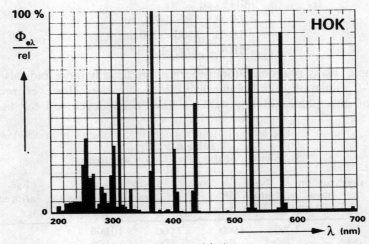

Fig. 8.2. Relative spectral radiant power $\left(\dfrac{\phi_{e\lambda}}{\text{rel}}\right)$ distribution of the high-pressure mercury lamp type HOK (Philips) at nominal rating measured with a bandwidth of 5 nm.[2]

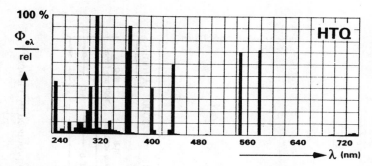

Fig. 8.3. Relative spectral radiant power $\left(\frac{\phi_{e\lambda}}{\text{rel}}\right)$ distribution of the high-pressure mercury lamp type HTQ (Philips) at nominal rating measured with a bandwidth of 5 nm.[2]

In comparison to the high-pressure UV lamps, in Fig. 8.4 are shown spectra of some low-pressure fluorescent lamps.

Most of these lamps, designed for industrial applications, are in the form of long tubes 500–2000 mm long (Fig. 8.5); however, they can also have a much more conventional shape, like a mercury vapour high-pressure lamp with a top reflector type HPR 125 W (Philips) (Fig. 8.6).[2] The emission spectrum of the last lamp is shown in Fig. 8.7. This type of lamp is widely used in industries for printing-plate exposure, curing UV adhesives and paints.

Doped and addictive lamps

These are lamps where MHs have been added to the mercury dose in order to shift the spectral output of the lamp into a more useful part of the UV spectrum. In Figs. 8.8 and 8.9 are shown spectra of MH high-pressure UV lamps types HPA and HPM, respectively.[2]

In general, doped lamps convert shortwave UV radiation to longwave, and are useful where the good penetration of shortwave radiations are useful for coating/interface cure problems such as flexible PVC, or where lack of shortwave radiation can be used to create surface effects such as matting.

Any future increased use of doped lamps for general UV-curing applications probably depends on the formulation of curable coatings which will specifically respond to longwave UV radiations or visible light.

Xenon lamps

Continuous xenon lamps are one of the most intense sources of UV, visible and IR radiation.[3] They have a continuous radiation distribution with a light superimposition of spectral bands, and in the near-IR range two pronounced bands are produced (Fig. 8.10). The colour quality of xenon lamps is similar to that of sunlight, with a correlated colour temperature of about 6000 K.

Xenon lamps operate as an intense point source using xenon at about 20 atm (2020 kPa) pressure. A high-powered discharge in xenon is compressed into a

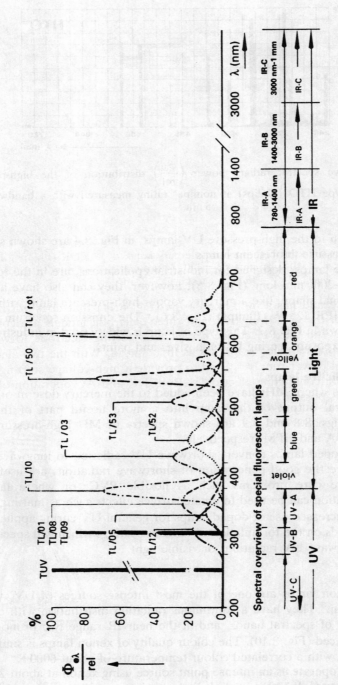

Fig. 8.4. Relative spectral radiant power $\left(\dfrac{\phi_{e\lambda}}{\text{rel}}\right)$ distribution of the different fluorescent lamps produced by Philips.[2]

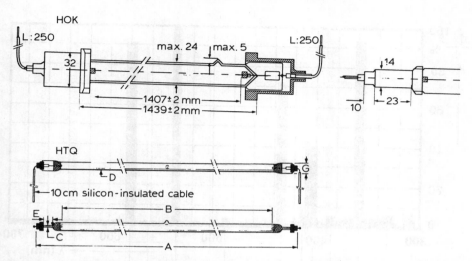

Fig. 8.5. Tubular shape of high-pressure mercury lamps type HOK (Philips) and type HTQ (Philips): A = 510–1485 mm, B = 440–1400 mm; C = 3–4 mm; D = 12–24 mm; E = 18 mm, and G = 15–23 mm.[2]

very small space and has a bulb-like design (Fig. 8.11) that does not require water cooling of the arc. Both two- and three-electrode models are commercially available. The third electrode serves as a starting electrode to which a high-voltage pulse is applied as the arc is first started. With the two-electrode models a special power supply is used to provide a high-voltage starting pulse through the main conducting electrodes. Both DC and AC operation of most of these lamps is possible.

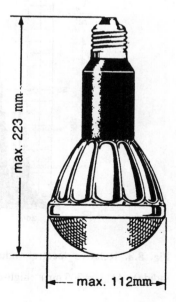

Fig. 8.6. Mercury vapour high-pressure lamp with reflector type HPR 125 W (Philips).[2]

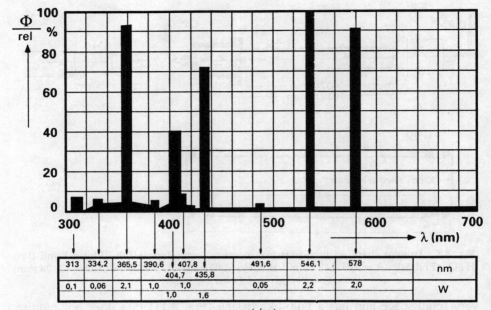

Fig. 8.7. Relative spectral radiant power $\left(\dfrac{\phi_{e\lambda}}{\text{rel}}\right)$ distribution of the mercury vapour high-pressure lamp type HPR 125 W (Philips).[2]

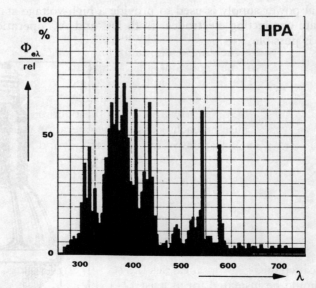

Fig. 8.8. Relative spectral radiant power $\left(\dfrac{\phi_{e\lambda}}{\text{rel}}\right)$ distribution of the MH (iron/cobalt iodide) ($\lambda_{max} = 365$ nm) high-pressure lamp type HPA (Philips) for industrial production.[2]

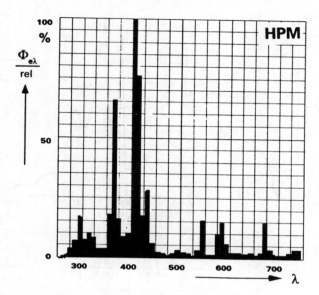

Fig. 8.9. Relative spectral radiant power $\left(\dfrac{\phi_{e\lambda}}{\text{rel}}\right)$ distribution of the MH (gallium/lead iodide) ($\lambda_{max} = 420$ nm) high-pressure lamp type HPM (Philips) for industrial production.[2]

The optimum operation conditions for these lamps are as follows.[3]

1. The lamp should be operated vertically, with the cathode at the top.
2. The ventilation of the lamp house should be controlled. Insufficient ventilation leads to overheating of the lamp and shortened life, whereas too much ventilation may prevent the lamp from heating and the required pressure in the lamp will not be obtained. Housings correctly designed for these lamps are commercially available.
3. It is necessary for these lamps to warm up to the operating temperature as fast as possible, because on each occasion the operation goes through a sputtering phase which is destructive to the electrodes and to the inside of the quartz envelope.
4. The current through the lamp should be adjusted in such a way that the dissipation of the rated wattage is within 5%. Too high a dissipation may cause failure due to either an excessive pressure rise in the envelope or heating of a ribbon in one of the seals. Too low a dissipation, particularly if combined with excessive ventilation, will cause electrode damage and sharply reduce the lamp's life.
5. The electrical connection which carries the starting high-voltage pulses to the lamp should be insulated for at least 12–15 kV.
6. A lamp should be left burning if the period for which it must operate does not exceed 30 min.
7. Power supplies for these lamps are available from the manufacturers.

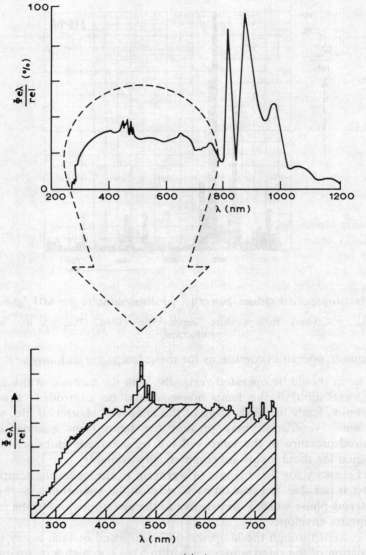

Fig. 8.10. Relative spectral radiant power $\left(\dfrac{\phi_{e\lambda}}{\text{rel}}\right)$ distribution of the typical xenon lamp type CSX 900 W (Philips).[4]

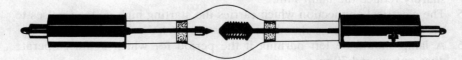

Fig. 8.11. Xenon lamp (vertical burning position) type CSX 900 W (Philips).[4]

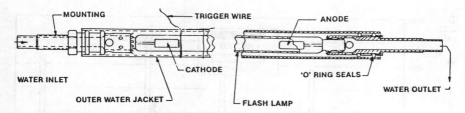

Fig. 8.12. Water-cooled xenon flash lamp, designed for industrial production.[5]

Continuous xenon lamps have very limited applications in UV-curing industries, except by using them in the spot curing devices when the visible range of light radiation is required.

Pulsed xenon lamps
Pulsed (flash) xenon lamps (Fig. 8.12) for photocuring inks and coatings, but mostly in reprography operate at 50–150 pulses/s with average power of 100–300 W/cm.[5] The spectral output of xenon lamps can be changed by increasing the current density and the temperature that the xenon gas is heated to. Higher current densities shift the spectral output toward the shorter wavelength and conversely lower current densities shift the spectral output towards the longer wavelength. Water cooling (5 litres/min) of the flash lamp is necessary. The maximum water-jacket temperature must not exceed 20°C.

The emission spectrum of a typical pulsed xenon lamp is shown in Fig. 8.13.

Microwave-powered lamps
In microwave-powered lamps, the microwave energy generated by the microwave power units is directly transferred to the mercury and MH fillers

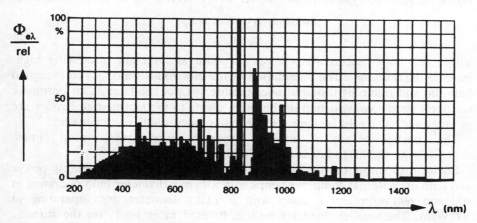

Fig. 8.13. Relative spectral radiant power $\left(\dfrac{\phi_{e\lambda}}{\text{rel}}\right)$ distribution of xenon flash lamp type STX (Starma, Ltd).[6]

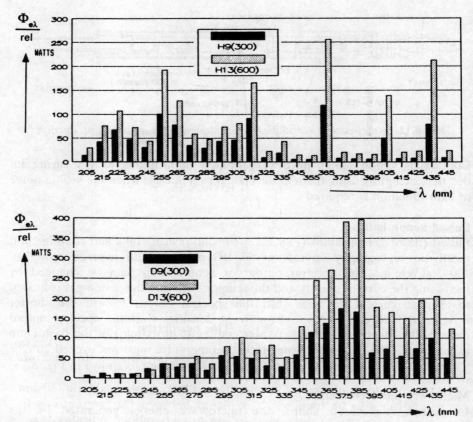

Fig. 8.14. Relative spectral radiant power $\left(\dfrac{\phi_{e\lambda}}{\text{rel}}\right)$ distribution of different types of microwave-powered lamps.[8]

within the bulb heating it to a plasma state and thereby producing high-intensity light energy. The elimination of the electrodes translates to extended bulb life and stable light output over the life span of the lamp. This permits a relatively simple construction of the bulb, owing to the fact that electrodes are not required. The tubular lamp can be made much smaller in diameter.[7–9]

The spectral output of different types of microwave-powered lamps, depending on the MH filler is shown in Fig. 8.14.

Figure 8.15 shows the optical effect of the combination of operating power and bulb diameter. It compares lamps with a 13 mm diameter bulb operating at 400 and 600 W/in and a lamp with a 9 mm diameter bulb operating at 300 W/in. The smaller diameter bulb is focused more finely on the surface. The focused width will be wider for a larger diameter source, and even wider in proportion as the surface becomes less of an approximation of a line.

The two essential elements of the high-power microwave-excited lamp are

Sources and devices for industrial processes

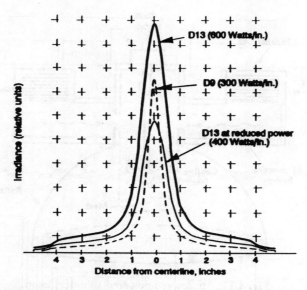

Fig. 8.15. Peak versus width of focus for different types of microwave-powered lamps.[8]

the magnetron and bulb (Fig. 8.16). The magnetron generates microwave energy at 2450 MHz, which is transferred through waveguides to the lamp bulb. The conductive, ionised gas in the bulb is an absorber for microwave energy.

Instead of a long tube bulb (300 mm) (Fig. 8.16), a spherical bulb (30 mm in diameter) can also be used which is placed between two magnetrons (Fig. 8.17). Such a lamp can operate with the quasi-parabolic reflector (see the section entitled 'Reflectors for UV lamps' below).

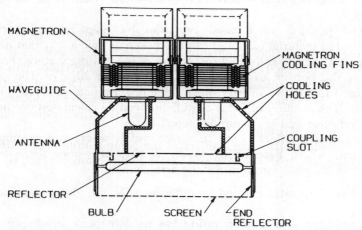

Fig. 8.16. Microwave generator for powering microwave-excited lamps.[8]

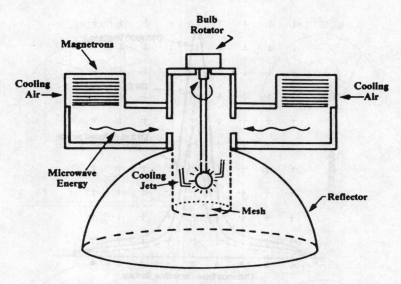

Fig. 8.17. Microwave-powered spherical lamp.[1]

REFLECTORS FOR UV LAMPS

In almost all industrial photocuring devices UV lamps work together with reflectors, which enable the radiation energy to be directed and concentrated to give the best distribution and intensity level to suit the practical application.

Reflectors for UV lamps have a variety of sizes and shapes (Fig. 8.18).[3]

- The elliptical or focusing reflectors give the highest intensity and therefore the most efficient cure effect. They can also be specially designed to create high illumination anywhere between 10 and 70 cm from the lamp (Fig. 8.19).

    The position of focused radiation by elliptical reflectors on the UV-cured material (coating) can be chosen in such a way that illumination (focused energy) is located on the top of the coating surface (Fig. 8.20(a)) or at the bottom of the coating layer (Fig. 8.20(b)).[7] The position of focused radiation can very much influence the final properties of a cured layer. The position of focus of radiation can be calculated on the optics of the reflector and UV lamp diameter.
- Non-focusing reflectors are employed widely where curing temperature is critical, where cure needs to be slower, or where three-dimensional objects are photocured in a static arrangement.

The relative dose rates emitted by different reflectors differ significantly (Fig. 8.21).

Each reflector with a lamp inside has its own peak irradiance and focus width, which depend on the diameter of lamp tube (Fig. 8.22). The smaller

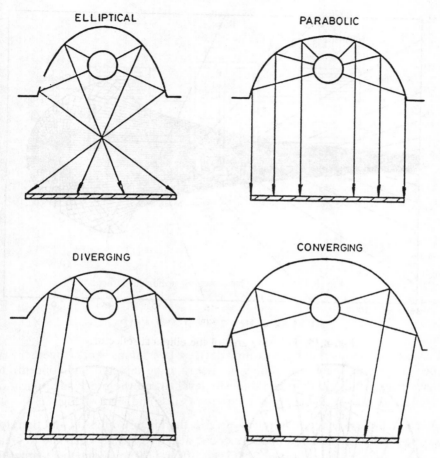

Fig. 8.18. Different shaped optical reflectors, which reflect equal irradiation over a fixed width.[10]

diameter bulb will be focused more finely on the surface. The focused width will be wider for a larger diameter lamp, and even wider in proportion as the source becomes less of an approximation of a line.

For very small lamps bulbs like xenon lamps or microwave-powered lamps quasi-parabolic reflectors can be used (Fig. 8.23).

When photocured formulations are sensitive to the air nitrogen-purged irradiation can be employed (Fig. 8.24). Where curing temperature is critical a water cooling system is employed. Pure water is almost transparent to UV radiation and almost totally non-transparent to IR radiation (heat). The loss in light intensity through about 1 cm of pure water is 20–30% in the UV region, about 20% in the visible region, more than 50% in the IR region (from 700 nm to 1 mm), 80% between 1 and 1·5 mm and total absorption above 1·5 mm IR radiation.

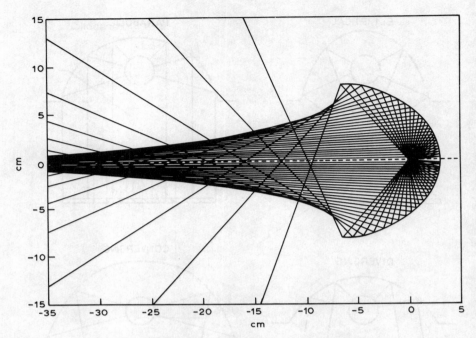

Fig. 8.19. Focusing area of the elliptical reflector.[11]

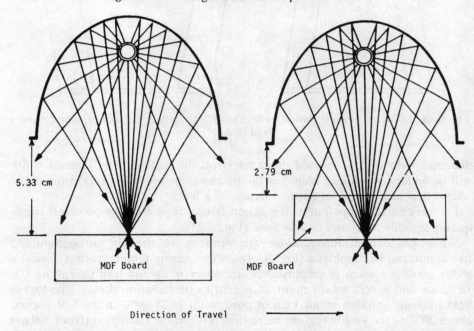

Fig. 8.20. The position of focused radiation by the elliptical reflector on the cured layer.[7]

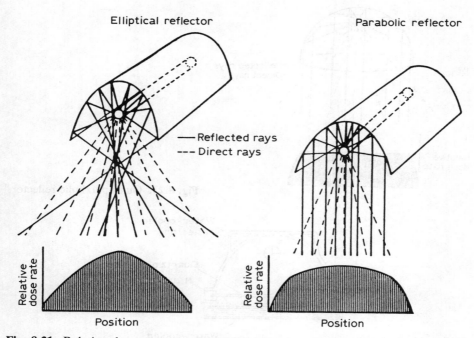

Fig. 8.21. Relative dose rate versus position (cm or inch) for elliptical and parabolic reflectors.[1]

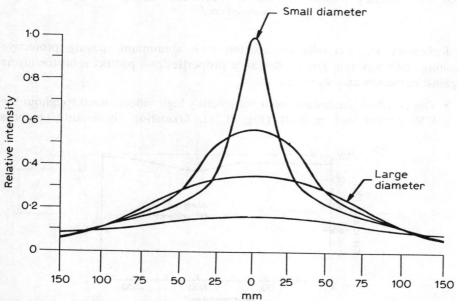

Fig. 8.22. Relative dose versus position for lamps with different diameters of the tube.[12]

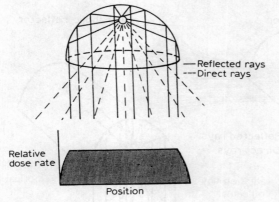

Fig. 8.23. Quasi-parabolic reflector.[1]

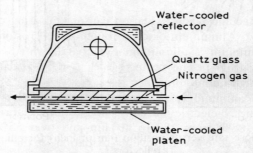

Fig. 8.24. Specially constructed reflector with water cooling system and nitrogen atmosphere.[10]

Reflectors are generally constructed with aluminium, having protective coatings that maintain good reflectance properties and protect reflector layers against corrosion and abrasion.

- The polished aluminium has a consistently high reflectance throughout the UV, visible and near-IR (Fig. 8.25). Oxidation significantly reduces

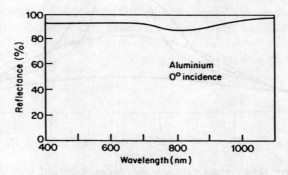

Fig. 8.25. Reflectance of aluminium.[3]

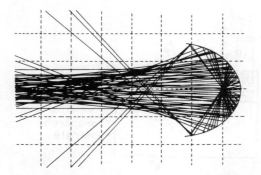

Fig. 8.26. Scattering of reflected radiation by oxidised surface of an elliptical reflector.[12]

aluminium reflectance, and oxidation causes scattering of reflected radiation (Fig. 8.26). Aluminium oxide is tough and corrosion resistant.
- Protected aluminium is aluminium overcoated with a dielectric film or silicon dioxide (SiO_2). The protective film limits oxidation and helps to maintain the high initial abrasions.
- UV-enhanced aluminium is aluminium overcoated with a film of a UV-transmitting dielectric (usually magnesium fluoride). Reflectance averages over 88% from 180 to 400 nm and over 85% throughout the visible. The dielectric layer prevents oxidation of the aluminium surface and provides abrasion resistance.[3]
- Multilayer dielectric coatings have good reflectance to UV radiation and poor reflectance to IR radiation. These multilayer coatings have dichroic properties and reflectors with a such coatings are often referred to as 'dichroic reflectors'. Dichroism is the optical property of a material, usually associated with anisotropic crystals which are birefringent, or double refracting in which light of different wavelengths will be absorbed or reflected by it. The thickness of the film and its index of refraction affect the wavelengths which will be reflected or transmitted.[9]

When specific wavelengths are reflected, interference has occurred in those wavelengths. When the film thickness is a multiple of the quarter-wavelength in the film, that wavelength will be reflected. More exactly, reflected and non-reflected (transmitted) wavelengths are related to the film thickness and its index of refraction:[9]

$$\lambda_{\text{reflected}} = \frac{4nl}{\text{'odd'}} \qquad (1)$$

$$\lambda_{\text{transmitted}} = \frac{4nl}{\text{'even'}} \qquad (2)$$

where n is the index of refraction, l is the thickness of layer, and 'odd' and 'even' are integers.

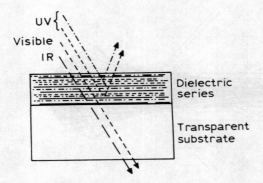

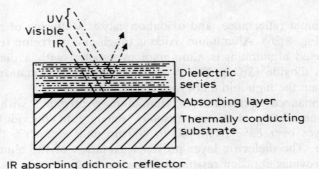

Fig. 8.27. Multilayer dielectric coatings on transparent versus opaque bases.[9]

To produce a reflector with a high spectral range, a large number of thin layers of hard, transparent dielectric materials are deposited one on another. Each layer has a different index of refraction from its adjacent layer, and the optical thickness of each layer is extremely precisely controlled. Durable and reliable multilayer coatings are usually vacuum deposited using various inorganic oxides such as aluminium oxide (Al_2O_3) and silicon dioxide (SiO_2). Wide spectrum dichroic coatings will use several different oxides and a large number of layers (50 or more) to achieve the cumulative interference required. In Fig. 8.27 are shown a multilayer dielectric coating on two different types of base material, transparent and opaque.[9]

In the case of a transparent base, a range of wavelengths is reflected by the dielectric layers, while longer wavelengths (visible and IR) pass through. This type of reflector is constructed by depositing the layers directly onto glass or quartz.

On the opaque base, the multilayer coating reflects UV in the same way, and passes longer wavelengths. The base material has an absorbing (black) coating on it, in which the visible and IR waves are converted to heat. If the

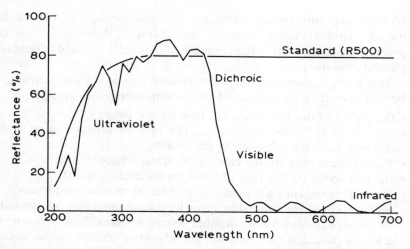

Fig. 8.28. Spectral reflectance by multilayer dielectric coated reflector constructed on an opaque, conductive base.[9]

base material is thermally conductive, the heat is easily removed by cooling. In Fig. 8.28 is shown the spectral reflectance from a multilayer dielectric coated reflector constructed on an opaque, conductive base.[9]

The manufacturers of dichroic reflectors claim an 80% absorption of shortwave IR under ideal conditions and at 80° angle of incidence of reflected light, but in practice there are several factors which reduce the effectiveness considerably:

1. as already stated only 60–75% of the total radiation is reflected;
2. the angle of incidence varies from 0° to 60°;
3. even conventional reflectors absorb approximately 15% of the IR radiations; and
4. there is some unfiltered longwave IR radiation.

Allowing for these factors, the best estimate of IR reduction is approximately 30%, but some tests show an even lower figure of about 20–25% compared with the 50–55% reduction using a water filter.[13]

PRINCIPLES OF DESIGNING A SUITABLE UV-CURING DEVICE

In the designing of a suitable UV-curing device it is necessary to take into consideration the following.[13]

1. Spectral distribution of the emitted radiation from the UV lamp (i.e. type of UV lamp). In principle, the emission spectrum of a UV lamp should overlap the absorption spectrum of the UV photoinitiating system (photoinitiators) in a given UV-cured formulation (see Fig. 8.1).

2. Minimal exposure energy necessary to cause curing, i.e. initiate cross-linking. Excess energy is costly and harmful, because it can cause photodegradation of the cured material[14,15] and necessitate photostabilisation.[16]

This energy, measured in joules (where 1 joule = 1 watt second), can be determined by an analysis of the system, taking into account

- the power of the UV lamp, i.e. type of UV lamp;
- the geometry of the system, i.e. the type of reflector used and the distance from the substrate to the source;
- the exposure time (in the case of the static arrangement);
- the belt speed (in the case of the dynamic arrangement);
- in the dynamic curing system it should be taken into consideration, that highly focused slit irradiance gives increased capability by overcoming threshold exposure value, in other words, a wide diffuse exposure over a longer period of time may not give the same effective cross-linking as a highly focused slit system given the same lamp power (reciprocity failure).

Up to now, the increased lamp power per unit length has been the most practical and successful method of achieving improved curing performance for a given size of UV-curing device.[13]

Increased lamp power has several advantages:

- higher UV intensity level;
- more compact reflector system for a given power output;
- more compact control system for a given output;
- potentially lower cost equipment; and
- more powerful systems can be fitted where available space is limited.

The disadvantages of higher powered lamps are

- higher heat concentration on the substrate;
- potentially greater heat problems in lamp and reflector design;
- shorter lamp life; and
- higher lamp voltages.

An approximate guide to the length–power relationship of lamps currently available is as follows:[13]

- Short lamps (<240 cm) 240 W/cm maximum
- Medium lamps (240–1000 cm) 160 W/cm
- Long lamps (>1000 cm) 120 W/cm

3. Temperature during UV-curing process. Excess temperature can cause evaporation of some volatile constituents from the UV-cured formulation. However, some UV-cured formulations contain initiators of thermal polymerisation, and a high temperature is sometimes required in order to initiate thermal polymerisation (in order to polymerise the remaining monomer which was not cured under UV irradiation).

Development of heat can affect negatively all photocuring process.[11]

- Heat plays a negative role in wood coatings, especially when pine wood is used. An excess temperature makes the resin exude from the knots preventing a good adhesion of the overlaid varnish. An excess of heat spoils the coating of not perfectly seasoned wood (air bubbles) and wood veneers (dimension alternations). In the presence of hot-melt glued edges, overheating can be a serious problem.
- For decorative papers (graphic arts), overheating can be a problem, especially for low-weight paper because the normal humidity contained in the paper is strongly reduced, altering the mechanical behaviour of the printed sheet. It is particularly important in web-offset printing to avoid high temperature and consequently inducing high fragility in the paper.
- Plasticisation of paper and cardboard using UV-curable glue will no longer be critical since the very thermosensitive plastic film can be kept at temperatures lower than 40°C during the process.
- It is also very important to keep the surface temperature of plastics (polymeric films) as low as possible, when curing of UV coatings involves adhesion, or lay-down. The gloss is badly affected by heat.
- The application of UV-curable solder resists on printed circuit boards requires that the substrate not be overheated in order to prevent damage to the underneath layers.
- The most heat resistant are textile materials made of cotton, viscose, nylon and silk.

The removing of heat produced by UV lamps can be very important for the UV-curing process to proceed. It can be done by the following.

- Adjusting the distance between the lamp and the cured product.
- Adjusting the velocity of a web or chain type conveyor.
- Employing UV lamps with dichroic reflectors (see the section entitled 'Reflectors for UV lamps'—p. 466).
- Using water cooling quartz filters between UV lamp and cured surface (see Fig. 8.24). Water filtration has three advantages over dichroic reflectors:[13]

    (i) it is a well tried and tested system with many units in operation world-wide;
    (ii) water filtration filters the IR from all the radiated energy whereas dichroic reflectors only treat the reflected radiations (45–75% depending on reflector shape).
    (iii) dichroic reflectors are expensive and difficult to manufacture in the perfect elliptical shape which is important for good focusing.

- Application of water-cooled plates or water-cooled cylinders, located under the substrate at the irradiation area or point (Fig. 8.29), provided that substrate thickness is not too high or its thermal conductivity is not too small. The heat can even burn the exposed surface in spite of using the cold plates.

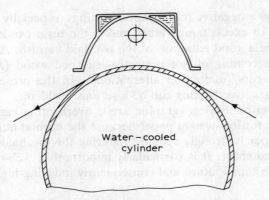

Fig. 8.29. Water-cooled cylinder.[13]

- Air impingement (air cooling systems) (see section entitled 'Air cooling systems for the UV lamps in industrial systems'—p. 488). The use of a cold air blower above and/or under the cured surface has very low cooling efficiency.

In practice the heat problems limit the maximum power that can be generated by UV lamps and this limited the availability of high rated lamps in long (tube) lamp systems. Long lamps are more difficult to cool evenly, and are subject to mechanical bending loads which increase with length.

4. Ventilation in order to remove vapours formed from UV-cured material. Many UV-curable formulations contain volatile components which will attack the lamp jacket and reduce its transparency, making it white and opaque. Most UV lamps produce some amount of ozone (see section entitled 'Ozone produced in UV-curing industrial devices'—p. 499) which can also cause ozonisation of cured formulations. In certain cases it may be necessary to seal the lamp housing completely, using a quartz window below the lamp.
5. Safety regulation (see Vol. IV, Ch. 14).

INDUSTRIAL METHODS OF UV CURING

UV curing may be carried out in one of the following.

1. *Static arrangement.* The cured material is placed in an exposure frame and then exposed to UV irradiation of the appropriate wavelength and intensity in a fixed geometrical position (Figs 8.30 and 8.31). The exposure time is varied to ensure the substance receives sufficient energy to cause a successful cure.

    In Fig. 8.30 is shown a technique used when extremely temperature-sensitive materials which are not particularly photosensitive must be exposed to high levels of UV irradiation. A dichroic mirror (see the

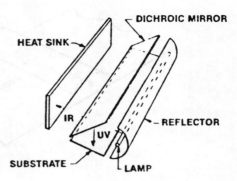

Fig. 8.30. Photocuring arrangements for extremely temperature sensitive materials.[10]

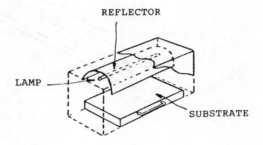

Fig. 8.31. Photocuring chamber for the bath processing.[10]

section entitled 'Reflectors for UV lamps'—p. 466) reflects only the UV radiation onto the surface to be cured, while allowing the visible and IR radiation to pass through the mirror for absorption by a heat sink.

In Fig. 8.31 is shown a curing chamber for bath processing. A parabolic reflector provides uniform radiation over a relatively large area.

2. *Dynamic arrangement.* It is usually carried out with conveyor-type systems using tubular UV lamps (see the section entitled 'UV-radiation sources'—p. 454) with either elliptical or parabolical reflectors (see the section entitled 'Reflectors for UV lamps'—p. 466). In a typical industrial photocuring device a web or chain type conveyor (which transports small objects) travels in the direction of the UV lamps (Fig. 8.32). A system of mirrors is provided from all sides of a web or under the conveyor chain so that a product or small object subjected to photocuring is cured without rotating or indexing a part. In other photocuring devices lamp configuration can differ significantly depending on the cured product.[10]

3. *Rotational arrangement.* The cured material rotates in front of UV lamps. In Fig. 8.33 is shown a technique where many small cured objects rotate on frames in front of microwave-powered lamps.[17]

In Fig. 8.34 is shown a technique for curing on-mandrel or large diameter devices. The cylindrical object is rotated in front of a single

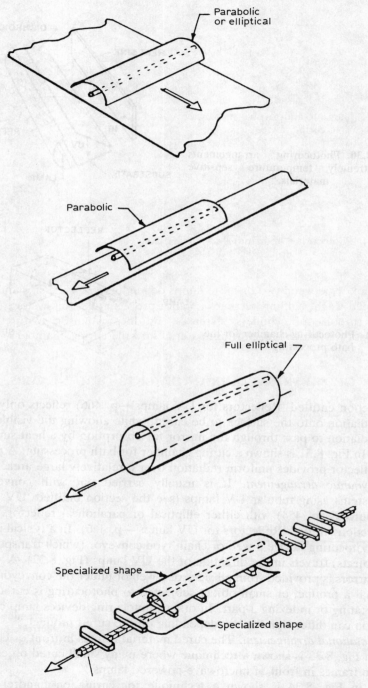

Fig. 8.32. Photocuring devices: web or chain type conveyors.[10]

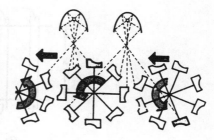

Fig. 8.33. Translate-and-revolve motion photocuring device.[1]

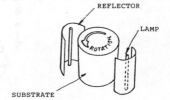

Fig. 8.34. Photocuring on-mandrel or large diameter objects.[10]

lamp or between two or more lamps depending upon the size and surface speed. Light baffling becomes a major problem with this system approach as the size of cylinders increases and the instant-off feature is most important while transporting material into the curing chamber.

GEOMETRICAL ARRANGEMENTS OF UV LAMPS

Linear discharge lamps fall off towards the edges, i.e. where a transversal arrangement is used for a product, there is the possibility of an inadequate curing process if the arc length is not greater than the working lamp width (Fig. 8.35). It is recommended to have the working width + $(2 \times 4\,cm)$ = arc

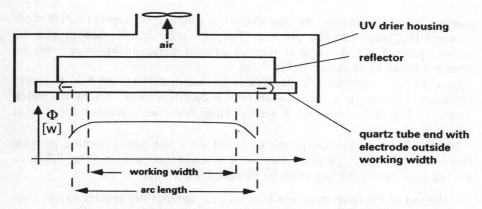

Fig. 8.35. Definition of the working lamp width.[2]

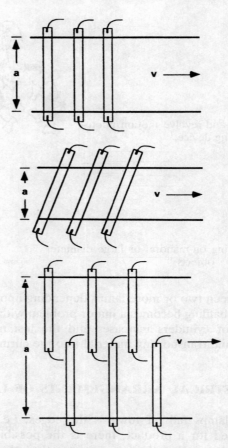

Fig. 8.36. Geometrical arrangement of UV lamps in stages: a—working width of the UV-cured material, and v—feed velocity.[2]

length. There should also be maintained a distance of approximately 5 cm between the lamp and the product. The UV tube lamps used in painted timber, painted metal, painted film or printed paper technologies can be arranged in stages as shown in Fig. 8.36.

In wide line industrial UV processors for width as large as 5 m it is very important to achieve a high linear power density on the whole irradiated width. In Fig. 8.37 is shown a typical lamp bank arrangement for 5 m of substrate.[12]

Most of the tube UV lamps are designed for a horizontal burning position (Fig. 8.38); however, they can also work in a vertical position (Fig. 8.39). In the last case the air cooling must be designed accordingly:[1]

- the top of the lamp must not heat up to a significantly greater extent than the bottom end of the lamp; and

Sources and devices for industrial processes

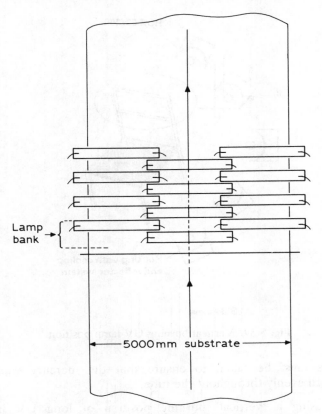

Fig. 8.37. Typical UV lamp bank arrangement for 5000 mm substrate.[12]

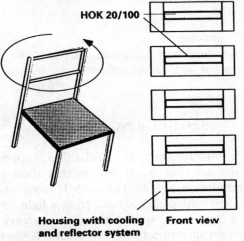

Fig. 8.38. Typical horizontal burning UV lamp position.[1]

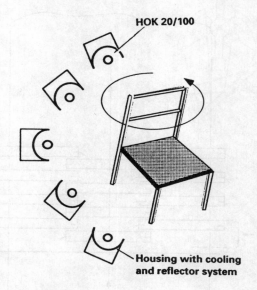

Fig. 8.39. Vertical burning UV lamp position.[1]

- measures must be taken to ensure that the mercury vapour is not distributed evenly throughout the tube.

Instead of using a 'vertical' burning position of long UV lamps it is recommended that several short UV lamps are employed which can be installed horizontally and in parallel above each other (Fig. 8.38). This system has a number of advantages, such as

- lower voltages;
- fewer insulation measures;
- homogeneous distribution of radiation; and
- precise concentration with swivel reflectors.

SPOT-CURING DEVICES

UV spot curing is widely used in mechanical, optical and electronic industries.[18] These devices (Fig. 8.40) require that short-arc xenon (see the section entitled 'Xenon lamps'—p. 457) or arc MH lamps are employed.[19,20] In order to have a high intensity of UV output from a light guide and a constant UV output during a long time, special short-arc mercury lamps have been designed.[19] The emission spectrum from such a lamp is shown in Fig. 8.41.

The spot curing devices with a maximum wavelength of 470 nm (Fig. 8.42)

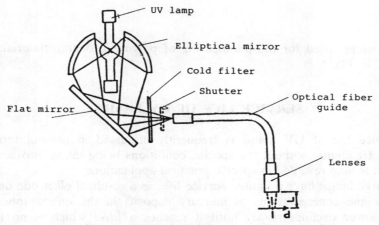

Fig. 8.40. An optical system in a spot cure device.[19]

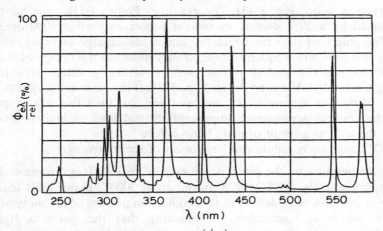

Fig. 8.41. Relative spectral radiant power $\left(\dfrac{\phi_{e\lambda}}{\mathrm{rel}}\right)$ distribution of the short-arc xenon lamp in a spot cure device.[19]

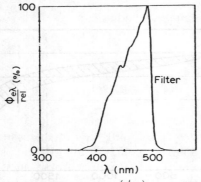

Fig. 8.42. Relative spectral radiant power $\left(\dfrac{\phi_{e\lambda}}{\mathrm{rel}}\right)$ distribution of light source in a dental-cure device.[21]

are also widely used for the photocuring of polymeric dental materials (see Vol. IV Ch. 13).[21]

SERVICE LIFE OF UV LAMPS

The service life of UV lamps is frequently discussed in general terms in industry, frequently without the specific conditions being taken into account. The truth is then revealed in specific practical applications.

Discharge lamps have a limited service life, as a result of electrode damage and very fine concentrations of mercury deposits in the quartz tube. The radiant power declines slowly until it reaches a level which is no longer adequate for ensuring a reliable process. The lifetime curve differs for each specific type of lamp (Fig. 8.43). For example, Philips HOK type lamps are characterised by a 20% decline in radiant power only after 1000 operating hours. The length of time for which the lamps are capable of being operated following this time will frequently depend very much on the design of the UV-curing device and the cooling system (see the section entitled 'Principles of designing a suitable UV-curing device'—p. 473). If the air cooling system is not clean, or if the fan motors which are used are old, then this will frequently reduce the cooling performance imperceptibly and lead to higher operating costs for the user as a result of early lamp failure.[1]

The UV lamps early failure can be a result of the following.[1]

- The temperature on the pinch which may be critical. It is recommended to be approximately 250°C with a maximum of 300°C. If this limit temperature of 300°C is exceeded, then the oxidisation process of the molybdenum plate, which is responsible for ensuring that the pinch is tight, is accelerated. The threshold value is approximately 300–350°C. Negative effects will be experienced even if this value is exceeded for a short time.
- Metal brackets on the bases which are not spaced adequately from the

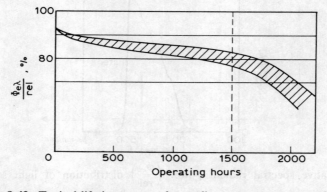

Fig. 8.43. Typical lifetime curve of a medium-pressure mercury lamp.

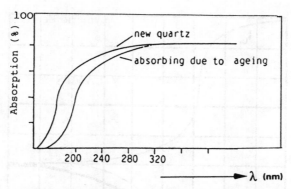

Fig. 8.44. The UV absorption (%) of quartz.[22]

electrodes which feature a voltage of up to 2 kV may cause microscopic holes in the quartz tube as a result of flash-over and thus permit oxygen to enter the tube. The lamp will fail as a result. A metal-free zone directly around the lamp is recommended.
- Excess current will build up in the lamp if appropriate ballasts are not used. Only well-tried and tested ballasts recommended by the manufacturers can be used.
- The destruction of quartz transmittance of UV irradiation (Fig. 8.44).

OPERATION MAINTENANCE OF UV LAMPS IN UV-CURING INDUSTRIAL DEVICES

A typical operation curve of a UV lamp installed in a UV-curing device is shown in Fig. 8.45. The lamp must reach its nominal primary current in 2 min (with reduced cooling) (Fig. 8.46) or 5 min (with 100% cooling). The run-up phase is

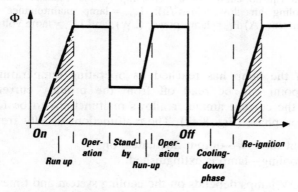

Fig. 8.45. Typical operating curve of a UV lamp.[1]

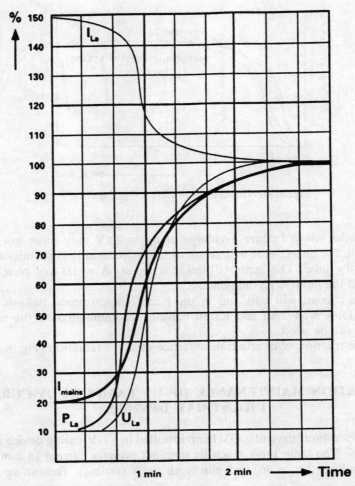

Fig. 8.46. Typical run-up curves of a UV lamp (HOK 140/120 Philips) (at room temperature, cooling air flow 500 m³/h): I_{main} = lamp maintenance current (A); I_{La} = lamp current (A) P_{La} = lamp power (kW), and U_{La} = lamp voltage (V).[1]

complete when the lamp has reached its operating temperature and lamp current. This point can be read off from the primary current ammeter. Application of the current ammeter allows malfunctions to be found quickly during the run-up phase (Fig. 8.47). The malfunctions usually are:[1]

- lamp current too low—output fluctuates; or
- excessive cooling—lamp is extinguished.

Reignition for UV lamps depends on the cooling system and times vary from 3 to 30 min.

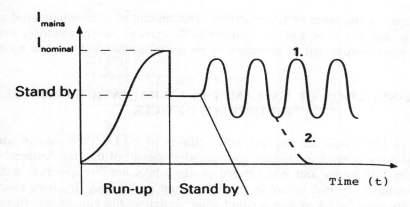

Fig. 8.47. Typical malfunction curve (U_{mains} = constant): 1. Lamp current too low—output fluctuates. 2. Excessive cooling—lamp is extinguished.[1]

Lamp power can be controlled by the simple technique of increasing or decreasing the capacitance required to electrically stabilise the lamp arc. However, application of different capacitors has limited range ratings of only 100, 85, 62·5 and 50%. Recent developments in electronic control have made possible fully variable power selection from as low as 20% up to 100%.[1]

Variable power has several advantages:[13]

1. power selected can be more closely matched to power required and so reduce energy wastage of overheating;
2. lower minimum power means less heat when shuttered, idling or at very low running speeds such as those required during set-up; and
3. power selected for individual lamps can be modified over lamp life to compensate for ageing effects on the lamp and thus potentially increase lamp life.

The disadvantage of variable lamp power is that it is important that the operator should be experienced and knowledgeable in the effects on the cure rate of variables such as colour, speed, coating weight, substrate material, etc. Also the careful monitoring of lamp life, together with routine checks and maintenance of the system is very important. UV output monitors (see the section entitled 'Measurement of UV-curing radiation'—p. 491) are now available to assist with these requirements, and are recommended particularly where variable lamp power selection is available.[13]

LAMP REPLACEMENT

If UV lamps are inserted or replaced with bare hands, then finger prints will be left on the quartz. The grease which is thus applied to the surface of the quartz is burnt into the quartz when the lamp is operated, and leads to the destruction of quartz transmittance of UV radiation. The amount of radiation emitted may decline, and the tube may be completely destroyed. In consequence, finger prints must be removed, if necessary by means of alcohol and distilled water.[1]

INSULATION OF UV LAMPS IN UV-RADIATION-CURING INDUSTRIAL DEVICES

Most of UV lamps are applied with voltages of <1 kV. Voltages of up to 5·2 kV occur in the lamp circuit, the secondary circuit of the transformer.[1]

When UV paints and water-based printing inks are being cured, a high relative humidity may occur in the UV-curing device and this high relative humidity must be taken into account when designing the equipment. In these cases, it is recommended that the ballast and capacitors be accommodated in separate junction boxes whereby the length of cable connecting the lamps should not be too large (antenna effect due to the high voltages involved). The insulation resistance of quartz is inversely proportional to temperature. When the lamp is being ignited—high current in the lamp pinch—high field strengths are encountered at the electrodes, and this high current may at the same time be accompanied by a reduced insulation resistance of the quartz in the area of the pinch and the base. In these situations, unfavourable design of metal brackets may be responsible for flash-over. It is very time-consuming and expensive to detect such flash-over but the result is that the lamp is immediately put out of action.[1]

It is recommended to use ceramic, steatite or other suitable insulation materials. Metal brackets ought to contain adequate air clearance and leakage paths. Brackets should be installed in an elastic manner in order to prevent a situation whereby torsional forces and axial expansion are transmitted to the lamp during the heating-up process.[1]

AIR COOLING SYSTEMS FOR THE UV LAMPS IN INDUSTRIAL SYSTEMS

The most common system employed in industry is the air cooling system. The type of air cooling system to be employed will depend on the application process and on the amount of space available:[1]

- air induction above the product from the room;
- inducted air feed from outside;
- air exhaust above;
- air exhaust to the inside.

The process of inducting air above the lamp is more advantageous than forced ventilation. Discharge lamps require a stable zone of air around the lamp (Fig. 8.48) if they are to be able to generate an arc which is stable in terms of

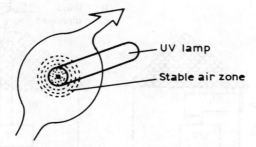

Fig. 8.48. Stable zone of air around a UV lamp.[1]

spectrum and radiant power. This zone is maintained if an induction process is used, but it is very quickly disturbed if forced ventilation is used (Fig. 8.49). In consequence, the service life of the lamp is reduced as a result of blackening. In the case of ozone-generating UV lamps, the cooling air must always be exhausted.

If an arrangement of central ducts for cooling air is used, then this might create problems which would have a considerable negative effect on the service life of the lamps.[1]

- The flow velocities in the ducts will vary whenever any UV-curing device is switched on or off.
- Butterfly valves for devices which operated before the new UV-curing machine was installed, will also affect the cooling performance.
- Summer/winter operation with different temperatures outside mean that it is necessary to provide some form of temperature regulation in order to be able to calculate and operate successfully throughout the entire year with the same production values. If cold winter air is inducted/exhausted at the same flow velocity as summer air, then this may rapidly cause the ends of the lamps to blacken.
- If a UV-curing device is installed in an arrangement of other machines, then it is frequently also necessary to rearrange the ducts for air induction and air exhaust. This will change the impact pressure values, so that the

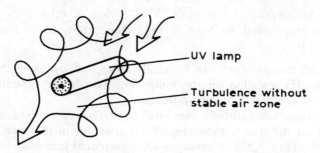

Fig. 8.49. Turbulence around a UV lamp due to the forced ventilation of cooling air.[1]

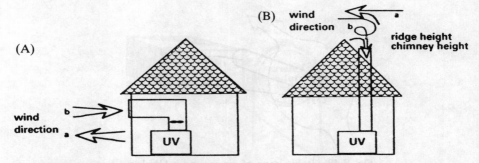

Fig. 8.50. Ventilation systems through (A) the walls and (B) the roof of a building. a—Exhaust air facilitated, and b—exhaust air hindered.[1]

built-in blowers will not be adequate or will be suddenly too powerful. This must be carefully checked.
- Simple induction and exhaust ducts, which are not appropriate in technical terms and which are routed through the walls or roofs of a building (Fig. 8.50) may reduce the service life of the lamps because of blackening (excessive cooling) or may cause the pinch to break (as a result of overheating) if the wind direction is not favourable.
- Individual cooling air systems are recommended for cooling the UV lamps. The possible temperature ranges of the inducted air must be taken into account when designing the industrial operating system.

There are the following consequences of cooling UV lamps:

1. *Overheating of pinch.* Rapid oxidisation of molybdenum plates and penetration of oxygen into the lamp discharge chamber. This causes a very rapid end to service life.
2. *Overheating of quartz tube.* Above approximately 950°C, deformation of quartz tube, sag until internal pressure is at maximum 1·5 bar (150 kPa), lamp is extinguished. The lamp may fracture.
3. *Excessive cooling of quartz tube.* The lamp does not reach its nominal performance, the tube ends blacken.

The following are the temperature limits for the operation of UV lamps: base/pinch temperature maximum 300°C, quartz tube 600–900°C. It is generally recommended to have a full temperature control over the UV operating lamp.[1]

- The pinch temperature can be easily measured by using thermocouples (Ni—Ni—C) installed on the lamp side with lower potentials than the protective conductor or zero potential.
- Quartz tube temperature can also be measured with a thermocouple located on the quartz tube (Fig. 8.51), but not in the area of the warm electrodes (Fig. 8.52). Warning!—Only insulated instruments can be used, otherwise there is a risk of flash-over when the lamp is operated.

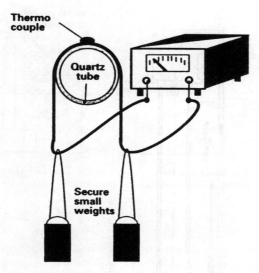

Fig. 8.51. Application of thermocouple to the quartz tube surface temperature measurements.[1]

It is recommended that the temperature and current values be recorded as a function of time either by digital means or by an analogue plotter. This will ensure that a record of data will be available for orientation in the event of any subsequent malfunction. The ideal situation would be to record all the events of a normal UV-curing device operation day. Operating hour meters which have been installed provide the opportunity of monitoring the service life of the lamp and also the opportunity of recognising any problems at an early stage or of indicating that the UV-curing device is functioning in a reliable manner.[1]

MEASUREMENT OF UV-CURING RADIATION

The measurement of the UV radiation in radiometric units (W/cm^2 or W/in^2) falling on the cured surface of a product requires the use of a UV radiometer.[3,22–24]

A radiometer is basically an electronic signal analyser with the electronic signal input coming from the detector probe, which consists of a detector, a filter and an input optical attachment such as a cosine-correction diffuser or fibre-optic assembly. The detectors are vacuum photodiodes with specific photocathode materials sensitive to the relevant region of the UV spectrum to be measured.[3]

Portable UV measuring instruments are widely used to check intensities of UV lamps periodically. However, these devices have a number of disadvantages.[22]

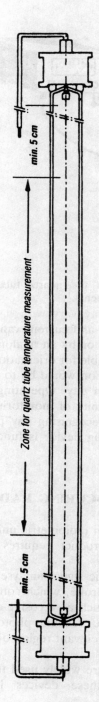

Fig. 8.52. Zone for the quartz tube temperature measurements.[1]

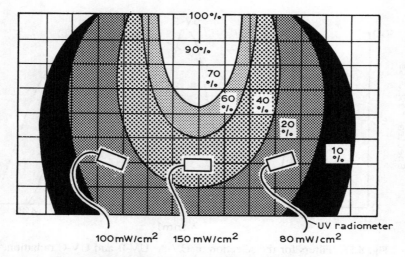

Fig. 8.53. Different readout under the same UV lamp.[22]

1. The values monitored on the display vary with distance, position and angle of the UV sensor with respect to the UV lamp (Fig. 8.53).
2. Short-time intensity variations due to varying voltage cannot be monitored.
3. For safety reasons UV-curing devices are usually shielded. Therefore, access to the lamps is difficult in most cases.
4. Sun-glasses and skin protection are required when working on unshielded systems.

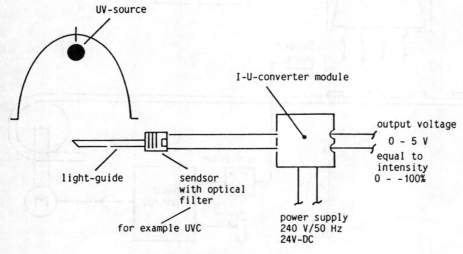

Fig. 8.54. UV radiometer.[22]

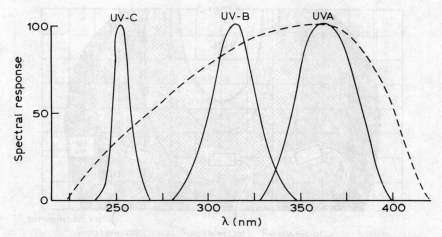

Fig. 8.55. Filters for the selection of UV-A, UV-B and UV-C radiation.[26]

It is recommended to use the continuous control UV radiometers which consist of an array of radiation detection sensors and suitable interferences (I–U converter with amplifiers) (Fig. 8.54). The sensor consists of a light-guide that samples a part of the irradiated light into an optical filter. Only the selected wavelengths are passed into the photocell by this filter (Fig. 8.55).

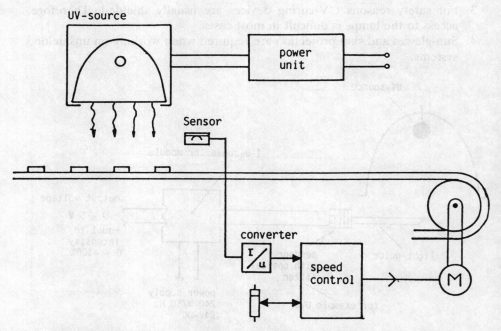

Fig. 8.56. UV-curing production line dose-controlled by a UV radiometer.[22]

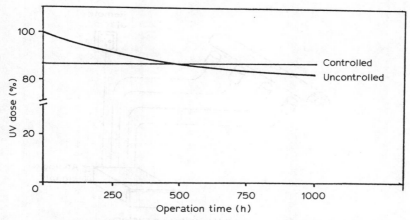

Fig. 8.57. Change of UV dose during operation time.[22]

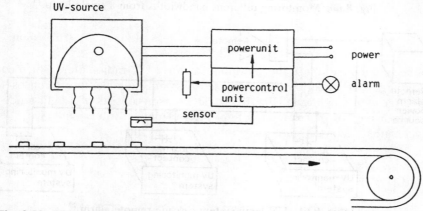

Fig. 8.58. UV-curing production line intensity controlled by a UV radiometer.[22]

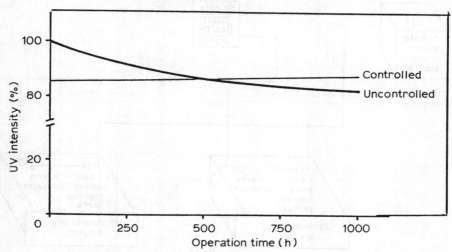

Fig. 8.59. Change of the UV intensity during operation time.[22]

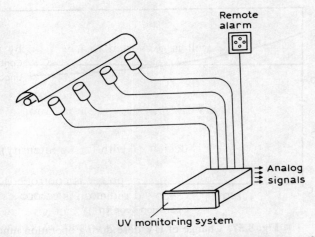

Fig. 8.60. Monitoring different bandwidths from a single lamp.[23]

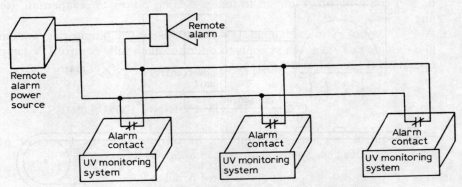

Fig. 8.61. UV lamp system cascade remote alarm.[23]

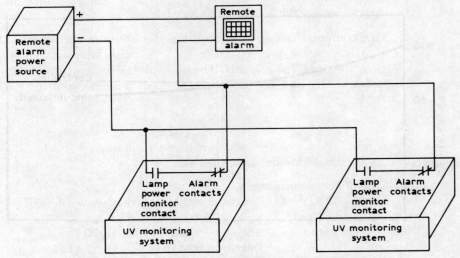

Fig. 8.62. UV lamp power qualified remote alarm.[23]

The current of the photocell is converted to a voltage by the 'convertor-module'. This output voltage (0–5 V) can monitor the relative intensity of emitted radiation of 0–100%. This value can simply be displayed using an analogue or digital instrument that is calibrated in per cent (%).

Such continuous-control UV radiometers can be used for automatic process control.[22,23]

- Production speed (which is decreased with falling intensity) (Fig. 8.56). In this case the UV dose is controlled (Fig. 8.57).
- To achieve constant intensities (lamp power is controlled) (Fig. 8.58). In this case the intensity of UV emitted radiation is controlled (Fig. 8.59).
- Automatic lamp life monitoring (Fig. 8.60) can give an early warning signal of any deterioration in UV lamp output. A connected computer (see the section entitled 'Computer control in UV-curing industries'—below) could be programmed to increase lamp power to compensate for any such reduction.
- Application of a multichannel UV-radiation source monitoring system allows logic remote alarm systems to operate which fully control UV lamp operation (Figs 8.61 and 8.62).

COMPUTER CONTROL IN UV-CURING INDUSTRIES

Computer control has many advantages in complex electrical control of industrial curing devices. Some of the most important are

- very compact and 'user-friendly' control panels,
- a higher level of reliability, and
- the easy incorporation of a large range of additional controls and monitoring functions.

For example, the computer can easily control the very complex sheet-fed card printing device where there may be more than fourteen lamps drying six colours with two in-line coating heads. Drying and UV curing may be by combinations of UV, IR and hot air operated from a single control, with additional features such as variable power selections and UV intensity monitoring.[13]

Computer control can considerably simplify the set-up and application of these complex installations, and in addition it can facilitate many additional control features at little additional cost, such as:[13]

- automatic lamp life monitoring;
- in conjunction with UV monitors the control can give early warning of any deterioration in UV output, and could be programmed to increase lamp power to compensate for any such reduction;
- continuous monitoring of cooling air and water, giving early warning of any loss of efficiency;

- storage of lamp power requirements for individual jobs;
- visual indications of any faults so that minimum time is taken in diagnosis.

INDUSTRIAL LINE UV PROCESSORS

These industrial devices are designed depending on their specific use. For example, processing rates (speed of movement) of cured material differ significantly, e.g. coating for wood can be UV cured with a speed of 8–10 m/min, whereas ink and printing coating on paper or polymeric films can be cured with a speed of 50–100 m/min. The UV-cured material can be as wide as 5 m (paper and polymeric films).

In Fig. 8.63 is shown a typical industrial UV processor for curing coatings on the surface of wood.[7] This system is designed to cure the coatings at a linear rate of speed of nominally 8 m/min. The first 4–5 mils (100–125 μm) of the UV-curable composition is applied by spraying a curtain coating to the medium-density fibre board. After coating, the material enters a flash-off tunnel to remove residual volatile solvents. The residence time is usually 3 min at approximately 50°C in circulating air. The typical solvent concentration ranges from 30 to 40% by weight. To maintain a 3 min dwell in the flash-off tunnel at 8 m/min, a tunnel 24 m long is required.

After the residual solvent is removed from the coating the substrate enters a pre-gel tunnel, which contains microwave-powered lamps (see the section entitled 'Microwave-powered lamps'—p. 463) with emission peaks at 420 and 360 nm. This pre-gelling process also requires a 3 min dwell. Thus, to maintain the rated line speed of 8 m/min a 24-m-long tunnel is required. In order to complete the gelation phase 192 low-pressure microwave-powered lamps are required.

At that point the coating is soft and under-cured at the surface. To impart the hard and scratch-resistant properties required of this coating, a final cure phase is required. This is accomplished using medium-pressure UV lamps operating at 80 W/cm.

To achieve a line speed of approximately 2·5 m/min three lamps are required. The first lamp has an output which peaks at 420 nm while the next

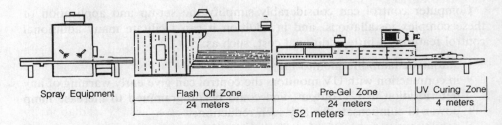

Fig. 8.63. UV-cure station for photocuring coatings on wood.[7]

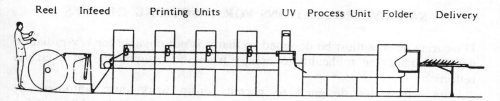

Fig. 8.64. A printing machine for producing high-quality forms incorporating a straight-through web-line and UV curing.[25]

two rows of lamps have an output of 265 nm. This means that in order to reach the desired production rate of 8 m/min a total of nine rows of 80 W/cm lamps are required.

This means that the combined length of all UV-cure stations will be approximately 52 m.

Installation of the lamp bank arrangement in an industrial UV processor for UV-curing coatings on paper or polymeric films can differ depending on the design (Fig. 8.64).

In many UV-curing industrial installations, particularly for printing, a great deal of curing efficiency is sacrificed in order to fit equipment into extremely confined spaces, insufficiently close to the substrate, and often with machine parts shadowing areas of exposure.

OZONE PRODUCED IN UV-CURING INDUSTRIAL DEVICES

The photons emitted in UV radiation are characterised by energy which increases as the wavelength decreases. Below 240 nm, the energy of emitted photons is 5·2 eV, which is greater than the binding energy of the oxygen molecule (O_2) so that the generation of very reactive ozone (O_3) is possible. This compound exist only for a short time and then decays. Ozone is characterised by an intense odour, and is damaging to health and chemically aggressive (especially to rubbers). Odour threshold is 0·01 ml/m^3 (ppm). For these reasons the ozone must be exhausted by means of the lamp cooling air (see the section entitled 'Insulation of UV lamps in UV-radiation-curing industrial devices'—p. 488).

Ozone occurs to a greater extent when the lamp is cold, because the cold quartz tube is still transparent for very shortwave UV radiation. In special UV curing installations, it is possible for the UV zone to be flushed with nitrogen in order to avoid UV absorption of oxygen. This procedure avoids the generation of ozone.

Many manufacturers supply 'ozone-free lamps', which are made of a different type of quartz material, but reduce the overall UV radiation by approximately 10%. In some UV-curing applications, the lower level of UV-C radiation may cause problems.

SAFETY REGULATIONS FOR UV-CURING DEVICES

UV-curing devices must be designed so that it is not possible for UV radiation which is damaging to health to escape from the equipment, not even via reflection.

Flaps which are designed to permit set-up work must automatically switch-off the UV radiation source when they are opened. The UV installations must comply with the relevant regulations for each particular application.

Regulations for UV-curing devices *Safety Regulations for UV Installations in the Printing Industry and in the Paper-Processing Industry* are available from BG Druck- und Papierverarbeitung, 6200 Wisbaden, Germany.

All UV-radiation devices must be marked with a label:

> **Danger: UV radiation!**
> **Protect skin and eyes**
> **against direct and**
> **indirect UV radiation!**

REFERENCES

1. Stowe, R. W. In *Conf. Proceed. Radtech '89 Europa,* (1989) p. 587.
2. *Philips Lighting* Catalogue, 1991.
3. Rabek, J. F. *Experimental Methods in Photochemistry and Photophysics* (Vol. 1–2). Wiley, Chichester, UK, 1982.
4. *Lamps for Special Purpose Philips* Catalogue, 1979.
5. Blanck, E. *J. Radiat. Curing,* **7** (1980) 15.
6. *Specialist Lighting Division Starna Ltd* Catalogue, 1989.
7. Schaeffer, W. R. In *Conf. Proceed. Radtech '91 Europa,* (1991) p. 583.
8. Stowe, R. W. In *Conf. Proceed. Radtech '90 North America,* (1990) p. 165.
9. Stowe, R. W. In *Conf. Proceed. Radtech '91 Europa,* (1991) p. 104.
10. Knight, R. E. In *Conf. Proceed. Radtech '91 Europa,* (1991) p. 84.
11. Gringeri, F. In *Conf. Proceed. Radtech '91 Europa,* (1991) p. 547.
12. Ramler, W. J. In *Conf. Proceed. Radtech '91 Asia,* (1991) p. 470.
13. Efsen, M. *Conf. Proceed. Radtech '91 Europe,* (1991) p. 114.
14. Rabek, J. F. *Mechanisms of Photophysical Processes and Photochemical Reaction in Polymers.* Wiley, Chichester, UK, 1985.
15. Rånby, B. & Rabek, J. F. *Photodegradation, Photooxidation and Photostabilization of Polymers.* Wiley, London, UK, 1975.
16. Rabek, J. F. *Photostabilization of Polymers.* Elsevier, London, UK, 1990.
17. Schaeffer, W. R. In *Conf. Proceed. Radtech '89 Europa,* (1989) p. 131.
18. Hood, R. In *Conf. Proceed. Radtech '90 North America,* (1990) p. 159.
19. Nakaima, Y., Ohnishi, Y. & Hiramoto, T. In *Conf. Proceed. Radtech '91 Asia,* (1991) p. 492.

20. Osawa, O. In *Illum. Eng. Inst. Japan Report Res.* (1987) LS-87-12.
21. Janda, R. In *Conf. Proceed. Radtech '91 Europa,* (1990) p. 720.
22. Müller, R. In *Conf. Proceed. Radtech '91 Europa,* (1991) p. 221.
23. Benedick, D. R. In *Conf. Proceed. Radtech '90 North America,* (1990) p. 341.
24. May, J. T. In *Conf. Proceed. Radtech '91 Europa,* (1991) p. 763.
25. Jefferis, J. In *Conf. Proceed. Radtech '91 Europa,* (1991) p. 181.
26. *Ultra-Violet Product Range Macam Photometers Ltd* Catalogue, 1991.

Chapter 9

Electron-Beam Processing Machinery

SAM V. NABLO

Energy Sciences Inc., 42 Industrial Way, Wilmington,
MA 01887, USA

Introduction 504
    Processor geometries 505
Electron Processor Penetration Performance 511
    General 511
    Scattering effects 512
    Charge deposition effects 514
Electron-Beam Generation Techniques 515
    Introduction 515
    Distributed and scanned processors 516
    Cold cathode systems 517
        Ion-induced secondary emission 517
        Field emission 518
    High-temperature emitters 520
Processor Control Systems 521
    General 521
    Process quality control 522
Exit Window Technology 523
    Introduction 523
    Foil considerations 524
    Foil support structures 527
Shielding Considerations 528
    General 528
Product Handling Geometries 532
    General 532
    Air path heating considerations 534
    Web handling 534
    Sheet handling 536
    Fibers, filaments and wire 538
    Three-dimensional and molded products 541
Inerting and Ozone/Nitrogen Dioxide/Homopolymer Control 542

General 542
Analytical techniques for assessment of inerting efficacy 544
Hybrid inerting techniques 545
Performance Monitoring Procedures/Techniques 547
Passive monitoring 547
Active monitoring 550
References 551

INTRODUCTION

The 60 year history[1] of electron processors has been dominated by three phases of development: the pioneering work,[2-4] primarily using pulsed machines in the 1930s and 1940s; the advent of the medium to high energy DC and pulsed machines[5] of good high-voltage insulation integrity of the 1950s and 1960s; and the introduction of the compact, self-shielded low-energy (<500 kV) DC machines[6] of the last two decades. This last phase has been particularly dramatic in that it provided well-engineered self-shielded designs that incorporated the radiation safety and environmental control (inerting) features into a free-standing package that could be easily integrated into new or existing production lines. The approach eliminated the costly dedicated facility designs involved with the earlier, higher energy systems, where the fabrication of a concrete enclosure to house the processor and its often awkward underbeam product handling machinery was mandatory. Along with this 'room size' enclosure aspect of the earlier scanned equipment was the hazard of 'whole body' operator access to the source, which made the safety requirements and redundant interlock protection all the more critical.

The advent of compact self-shielded processors, both of the extended DC curtain type and of scanned and pulsed form, has provided integrated designs of much greater intrinsic safety and reliability. This has been a most important aspect of the acceptance of these energy sources by basic industry, in particular for graphics, paper and film manufacture, where most of these 700 industrial units (c. 1990) are now applied.

The main interest of the polymer chemists/chemical engineers who use these electron sources for material converting, is in the low-energy regime (i.e. treatment thicknesses of 500 g/m^2 or less). For this reason, we will focus on processors with operating voltages of 300 kV or less. This is, in effect, a good line of demarcation between low- and medium-energy equipment. It represents the point at which self-shielding is still practicable (e.g. 2 cm of lead for high-power systems), and at which single gap accelerators are still reliable. This last point is an important consideration in processor design and is illustrated in Fig. 9.1. Clearly the electrons must be accelerated in vacuum which means that the exciting voltage for acceleration must somehow be reliably sustained in the vacuum system. What is shown in Fig. 9.1 is the behavior of voltage insulation in a vacuum-insulated gap. The ability to hold-off or insulate voltage is found to vary as the square root of gap length;

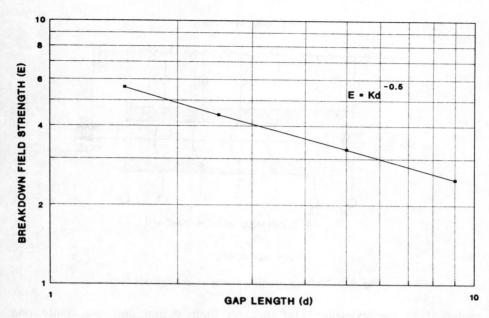

Fig. 9.1. Vacuum insulation strength as a function of gap length.

i.e. $V = kd^{1/2}$,[7] so that tripling the gap (resulting in a 10-fold increase in the machine cross-section) only has the effect of increasing the operating voltage by 70% (e.g. $3^{1/2}$). As a consequence, one reaches a point of diminishing returns at 300 kV in a single gap design with the large number of machine size dependent, design parameters that must be taken into account; e.g. shielding weight, vacuum envelope volume, pumping system capacity, high-voltage cable, connector and bushing design, etc.

For the higher energy equipment, this vacuum-insulation design limitation is eliminated by the use of voltage grading, shown schematically in Fig. 9.2. Here, the accelerating or total voltage used in the electron accelerator is divided into n gaps using a resistive divider which now uniformly establishes the operating potential of each electrode in the vacuum tube from $-V$ to ground, usually in 50–100 kV steps. Ceramic or glass rings bonded to stainless electrodes are typically used with the grading resistors adjacent to the tube in an insulating medium such as high-pressure gas (SF_6 or $CO_2 + N_2$). The reader is referred to the extensive literature for more detail.[8,9]

Processor geometries

The most widely used electron processor configurations employing single acceleration gaps ($V \leq 300$ kV), multiple, linear filaments and integrated shielding are shown schematically in Figs 9.3 and 9.4. The more complex, higher voltage, graded-gap accelerators are typically vault shielded and do not lend themselves readily to integration into a production line. The reader is

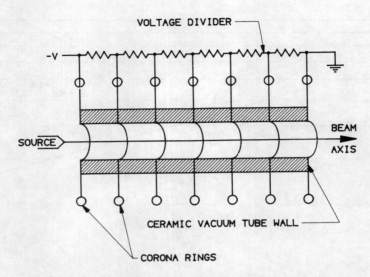

Fig. 9.2. Voltage grading along an accelerator tube.

referred to the extensive literature on their design and use. Only one manufacturer of scanned equipment[10] has successfully developed low-energy scanned units of compact geometry, as illustrated in the self-shielded processor of Fig. 9.5. Several promising designs employing scanned electron accelerators excited by compact pulsed transformer supplies have been employed, particularly by investigators in the former Soviet Union.[11]

In the adaptation of a processor to a converting line (such as a coater or printer), it is necessary to provide a convoluted path for the product between the energy source and the 'outside' world occupied by the system operators. This indirect path must be so designed that it collimates and absorbs the penetrating radiation generated by the electrons as they stop in matter. If the processor accelerates electrons through a potential of n kilovolts (kV), the penetrating radiation (bremsstrahlung) will have a maximum energy of n kiloelectron volts (keV). Shielding of the electrons themselves is a much simpler challenge to the designer, as their penetration capability due to their high stopping power (rate of energy loss) in matter, is very limited. A simple comparison of electron and photon (X-ray, bremsstrahlung) useful penetration depths is shown in Fig. 9.6.[12] The important considerations for shield design are discussed later in this chapter.

Because many (indeed most) of the high-volume applications of electron processors involve coatings on thermolabile materials such as paper and film, the processor geometry must accommodate contactless handling of the treated surface before, and for some time after treatment. Clearly the time required after passage through the beam will be that required for the electron initiated polymerization to go to a 'tack-free' or non-blocking condition. Present systems utilize rollers, drums, or air bars in the Selfshield® for this purpose

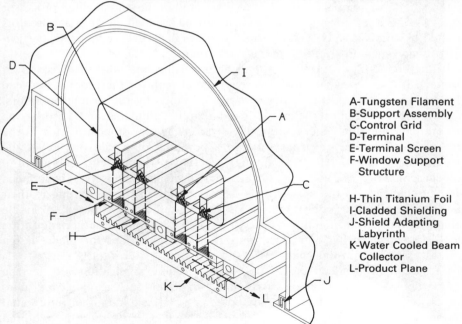

A-Tungsten Filament
B-Support Assembly
C-Control Grid
D-Terminal
E-Terminal Screen
F-Window Support Structure

H-Thin Titanium Foil
I-Cladded Shielding
J-Shield Adapting Labyrinth
K-Water Cooled Beam Collector
L-Product Plane

Fig. 9.3. A 200 kV × 1·65 m Electrocurtain® rated at 1520 Mrad m/min (courtesy Energy Sciences Inc., Wilmington, MA, USA). The cutaway view shows the (four) parallel gun configuration used in these units.

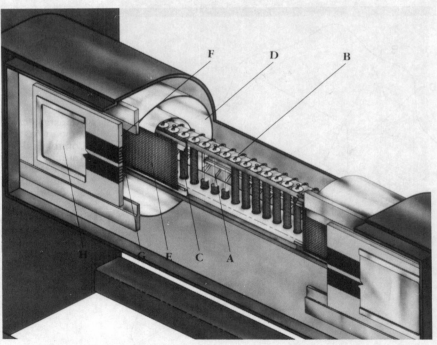

Fig. 9.4. A 200 kV × 1·30 m Broadbeam® rated at 915 Mrad m/min (courtesy RPC Industries, Hayward, CA, USA). The cutaway view shows the multiple filament–control grid assemblies, which have been used in units up to 3·3 m in width.

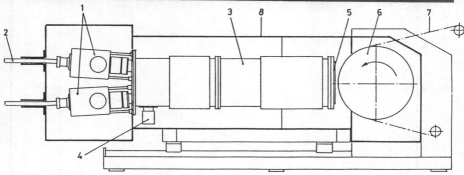

High-performance Electron-Beam Equipment

Acceleration voltage max. 250/280 kV,
penetration of electrons max. 320 g/m^2 at an ionisation of 80 %,
dose capacity 800 m/min, 10 kGy,
dose accuracy over the working width better $\pm$ 3 %.

1. Accelerator with two cathodes,
2. High-voltage cable connection,
3. Scanning system,
4. Pumping system,
5. Electron-beam exit window, inertisation zone, disconnection point for maintenance work with locking system,
6. Drum for material supply,
7. Material inlet / outlet,
8. Radiation shielding.

Fig. 9.5. A 280 kV × 1·0 m high-performance electron-beam system rated at 800 Mrad m/min (courtesy Polymer-Physik GmbH and Company, Tubingen, Germany). The schematic shows the dual cathode scanned geometry used for uniform illumination of a cooled drum supported product.

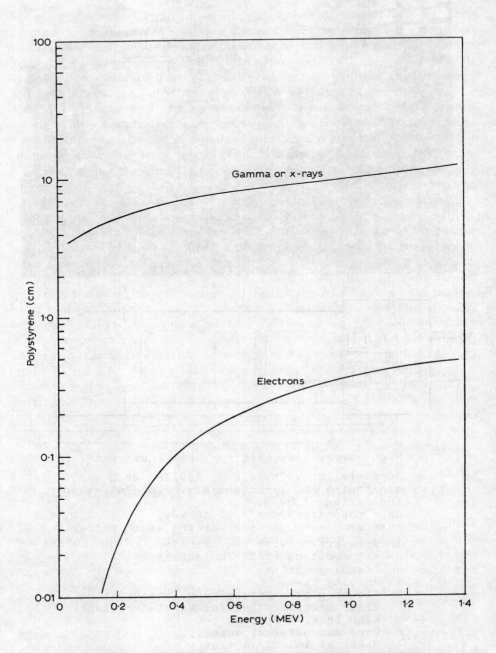

Fig. 9.6. Effective working depths for various radiation sources.

with the treated surface facing the processor. Several geometries are in widespread use which permit horizontal entrance to the processor with an oblique path in the process zone to break the 'line of sight' from the beam to the outside world. The more convoluted the path, the better the shielding efficacy. Panel handling systems have been designed for the wood products industry with gentle angles or with convoluted 180° reversal. Most of the present designs use collimated drum assemblies, which provide good thermal and physical stabilization of the product during treatment, small angles (approximately 10°) with relatively long collimators and larger angles (30–90°) with short collimators. Clearly the shield design must reflect the flexuring capability of the product and is discussed in the product handling section of this chapter (see p. 532).

Energetic electrons are capable of generating hazardous gases such as ozone, nitrous oxide or nitric acid vapor, or hydrogen gas when irradiating a hydrocarbon such as polyethylene. There are well-established limits for these species in the work area, as regulated, for example, by OSHA[13] in the US. As a consequence, it is necessary to eliminate the source or limit them to the process zone. Typically, the former route is chosen with the use of purging the process zone with an inert gas (at say 100 ppm of oxygen) so that it is greatly reduced from the 210 000 ppm of ambient oxygen. Because the important electron-induced free-radical initiated addition polymerisation process is inhibited by oxygen, it is also important for cure quality assurance to provide such a purged environment. How this is incorporated into the self-shielded processors is discussed in the section on inerting in this chapter (see p. 542).

The evolution of the modern electron processor has been truly a multidisciplinary effort. It has required the coupled knowledge of the accelerator engineer, the radiation physicist and chemist, the software designer, the chemical engineer, and the converting equipment designer/operator to reach its present advanced but 'not yet mature' state of development.

ELECTRON PROCESSOR PENETRATION PERFORMANCE

General

One of the most difficult to understand characteristics of electron processing is that of depth and uniformity of penetration. Because the equipment is used by technicians usually skilled in the thermal or diffusion limited processes which radiation curing is replacing, the dramatically differing physics of energy absorption in matter for an energetic electron needs, at least, a phenomenological understanding by the end-user.

The absorption of the kinetic energy of an incoming electron in the target (product) is almost totally based upon electron–electron scattering,[14] where the primary or incident electron 'sees' only the orbital electrons present in the absorbing target. As a consequence, the rate of energy loss by the primary electron is dependent upon the atomic number (or electron density) of the absorber, as well as upon the velocity of the primary electron...the slower the electron or the lower its energy, the higher its rate of loss of energy in this target.

As a consequence of this multiple scattering energy absorption process, low-energy electrons are well suited to efficient energy transfer to the thin coatings or films (50 g/m^2 or less) utilized in most converting applications such as printing and coating. Some insights into the features of energy delivery from industrial processors can be obtained from a study of their energy deposition profiles in matter.[15]

Scattering effects

It is possible to determine energy deposition profiles or depth dose curves with good precision using laminates of thin (10–50 g/m^2) radiochromic dosimeter films.[16,17] Hence the dose profile with depth can be determined for an equivalent bulk hydrocarbon, and good predictions made for higher Z materials. The major disturbance to the ideal irradiation geometry is the multiple scattering of the electron in the processor window, and during its flight path in air. These effects are illustrated for a 200 kV processor in Fig. 9.7. Curve A (continuous) shows the absorption profile in radiochromic film (nylon or cellophane) under the 'normal' irradiation conditions: a 12-μm thick titanium window (54 g/m^2) and a 2 cm drift path (window–product separation), with the dosimeter laminate or stack, i.e. the simulated product, drawn

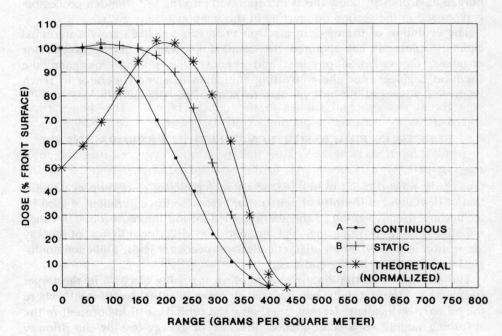

Fig. 9.7. 200 kV depth–dose performance.

continuously through the distributed electron beam. It should be noted that the effect of the air is to add 24 g/m^2 of additional scattering absorption, because air at NTP has a thickness of 12·3 g/m^2 per cm. As shown from the experimental curve taken with a laminate of 35 g/m^2 dosimeters, the maximum depth of penetration of the electrons under these conditions is 400 g/m^2.

Curve B (static) shows a similar depth–dose measurement, but in this case with a static irradiation of the laminate. In this case the laminate was treated at the center of the electron distribution from the window, so did not receive the large-angle scattered electrons seen in curve A. As a consequence, the depth–dose distribution is much steeper—as shown, the half-dose point has increased by 25% while the end-product remains unchanged.

Curve C (theoretical) shows the dose profile predicted[16] for monoenergetic 200 keV electrons with no consideration of spectral modification due to scatter before incidence on the laminate (product).

What can be immediately observed by a comparison of curves A, B, and C is a depression of the target penetration by the 60–70 g/m^2 of intervening material (window foil and air path) present in the processor geometry. Of greater significance to electron processing is the 'flattening' of the energy deposition profile near the surface due to the (now) relatively large entrance angles of the electrons in the target due to their 70 g/m^2 of multiple scattering, so that the absorption near the surface is markedly elevated. This is a very important effect, somewhat fortuitous for the surface processing application, in that, for example, in the case shown (curve A in Fig. 9.7), the variation in treatment is less than ±7% for the first 150 g/m^2 of penetration (curve A) compared with the ±30% predicted (curve C) for the ideal case of orthogonal incidence. As explained earlier, confinement of irradiation to the central region of the beam extends this ±7% region to over 200 g/m^2. A similar analysis at 350 keV has been reported in Ref. 17.

Figure 9.8 illustrates the spectral effects of multiple scattering taken from some recent experimental studies in this energy range.[19] The data shown utilized a 200 kV accelerator potential (just as in curve A of Fig. 9.7) in a machine using a 25 μm Mylar window, so that the effective beam energy at the air–window interface is 190 keV. How this spectrum is degraded from the near monoenergetic primary beam (A), to the 10% spread (FWHM) at 2·5 cm air path (B) to the 25% spread at 20 cm air path (C) is shown from these spectra measured in air with silicon surface barrier detector pulse height analysis. Although these distributions are only indicative of the spectra at the surface of a product because of the effects of the detector itself, they demonstrate the spectral softening and low-energy tail which assists in the important flattening effects, shown in the experimental curves A and B of Fig. 9.7.

A measure of the optical/performance quality of an electron processor is given by the slope or steepness of these depth–dose curves. For example, a deposition profile taken with the badly degraded spectrum of Fig. 9.8 at 20 cm (240 g/m^2 of air) would show no plateau at shallow depths, so that the energy deposited or dose would decline monotonically with depth. Several studies of

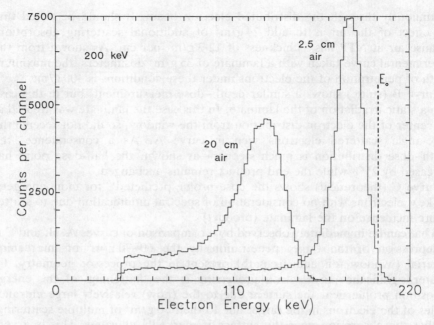

Fig. 9.8. 200 keV spectral degradation with a 20 cm air path.

these effects across the energy regime of interest for industrial application are available in the literature.[20,21]

Charge deposition effects

It is clear from earlier work[22–24] that the penetration profile for electrons in insulators (dielectrics) can be seriously affected by electrostatic deceleration due to charge trapping. Since many of the materials used in electron processing are insulators (film, paper, paperboard, protective varnishes, etc.) the effects of charge build-up must be considered. For the electron energies and doses used in converting applications, the charge density delivered to the product is only a few $\mu C/cm^2$, so that the development of electric field gradients in the product sufficient to affect the beam energy are unlikely. Of greater concern is discharge 'puncture' of the product[25] (where the field build-up exceeds the dielectric strength of the target), or where the charge is delivered to a sandwiched conductor so that a (spiral) capacitor is formed in the rewind roll.[26] Processing at elevated temperatures or reduced dose rates can be used to reduce the first problem, while the second, typically encountered with a barrier metal foil sandwiched in a film or paper laminate, can be eliminated by grounding of the conductor to the rewind roll shaft (which must, in turn, be adequately grounded).

A more subtle problem in charge trapping arises from the inability of a shielded or protected surface to attract a neutralizing image charge. Such a

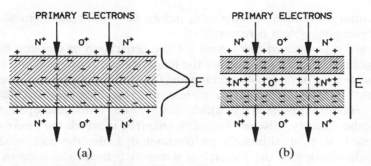

Fig. 9.9. Charge trapping effects in a folded dielectric in air.

situation can be encountered in the electron irradiation (cross-linking) of a folded extrudate, as in Fig. 9.9. Here the product can trap charge (electrons) while moving through the process zone; since the process zone is filled with a highly ionized plasma, typically of nitrogen, there is an abundance of free charge to neutralize the surfaces. This does occur, and upon emergence from the processor the product is electrically neutral. But now the film may be opened for rewind and the unneutralized inner surfaces can provide great problems in product handling. Indeed, the effect may render the film difficult to handle and almost impossible to separate.

This type of electrostatic difficulty can be easily managed by 'opening' the folded film in the process zone so that the inner surfaces now have access to neutralization. The film can then be managed outside the process zone without difficulty if no further triboelectrification (frictional charging) is allowed to occur. How this can be done is shown schematically in Fig. 9.9.

ELECTRON-BEAM GENERATION TECHNIQUES

Introduction

The electron-beam generators used in radiation processing are, in principle, very high voltage vacuum tubes with a 'transparent' anode or window, so that the accelerated electrons transmitted through the window can be applied directly to the product under ambient conditions if desired, or at least at 1 bar (10^5 Pa) environmental pressure in the process zone. Because of the need for a long mean free-path between electron–electron scattering events in the accelerator system, quite good vacuum conditions are required, for example, 10^{-5} mmHg ($1 \cdot 33 \times 10^{-3}$ Pa) or better, in order to provide electron optical performance of high quality. Early accelerators utilized continuous oil or mercury diffusion pumped systems (typically mercury in order to eliminate problems of insulator or cathode contamination) while later systems now utilize turbomolecular or cryogenic pumping.[27] These systems have been developed to deliver very high reliability with window fault tolerant performance. Service periods well in excess of 10 000 h are provided by these pumps, given normal preventive maintenance, before any major maintenance of the dynamic pumping system is required.

There are a number of techniques for the generation of electron beams at high power levels which can provide the longevity and reproducibility required for the processing of materials in commercial quantities. In general, they range from 'cold' or room temperature cathodes to high temperature, low work function emitters ($T > 800°C$) which have been well developed for the vacuum-tube industry.[28] Regardless of the manner in which the electron beam is generated, it must ultimately be distributed across the exit window for uniform illumination of the product as it moves beneath the window. So, in general, processors have taken the form of two-dimensional planar or strip geometry sources through which the product is moved for processing.[29] Before we examine the types of electron sources (cathodes/filaments) in common use, we will first look at the basic differences of the distributed or 'static' elongated source and the sport type or 'dynamically' scanned source.

Distributed and scanned processors

In spite of the fact that one of the earliest commercial electron processors[29] utilized a distributed source (large-area cathode), subsequent developments[30] were largely with spot-type beams, accelerated and then scanned typically in a two-dimensional raster over the window plane. In effect, these scanned processors resulted from the addition of a scanning assembly (electrostatic or electromagnetic) to the end of the acceleration tube of a spot-beam high-voltage accelerator, some of the earliest of which were of Van de Graaff's design. Because of the 'flared' geometry of these systems in which the full-energy scanned electrons drift to the elongated window as a continuously scanned spot, this portion of the processor is referred to as the 'horn'.

There are certain consequences of this approach which are worth noting. The first of these is the relatively high current density in the spot beam which results in a high dose rate experienced by the product. The current density in the scanned spot will typically be >100 times that calculated on the basis of the machine's output current divided by the beam (raster) area at the window plane. This is usually not important (for cross-linking, for example) but may be a factor in electron-initiated polymerization processes where radical–radical recombination loss[31] due to the high radical concentration in the spot, may lead to a significant diminution in dose efficacy, for example, where degree of cure via addition polymerization is important to processed product quality.

The second characteristic of the scanning geometry which should be understood, is the consequence of the varying angle of incidence from the orthogonal unscanned central ray ($\theta = 0°$) to the extreme (typically, $\theta = 30°$) at the edge of the scan. As discussed in the section entitled 'Exit Window Technology'—p. 523, the mean scattering angle of an incident electron of given energy is determined by its path length in the window.

Clearly its scattering path length is increasing as it moves toward the ends of its longitudinal scan, hence the emergent electron has a *diminished energy* (penetration capability) as well as a much *higher* emergent half angle *of scatter* and consequently reduced current density or dose rate. This effect is much more pronounced in low-energy (<500 kV) systems and has been (partially) compensated for by increasing the scan time proportionally to the angle of incidence on the window.[32] This can compensate for the reduced dose rate effects of scatter by increasing the beam residence time, but it cannot compensate for the reduced penetration capability with longitudinal scan distance, which may or may not be important to the product quality. For surface curing, it clearly is not; where 'edge-to-edge' bulk treatment uniformity is critical to product quality, it must be taken into account. Detailed studies of these effects for industrial processors have been reported in the literature.[21]

Because of the relatively wide window elements used in electron processors, (5–10 cm between longitudinal heat sinks or coolant lines) some transverse support structure is necessary. Clearly this structure, be it a series of closely spaced fins of good thermal conductivity, or a closely packed array of holes in foil sandwiching plates (see 'Exit Window Technology', Fig. 9.17), will display an angular dependence on electron absorption; the larger the solid angle the support structure subtends in the beam, the higher the beam loss due to absorption in the structure. Various techniques have been used successfully to reduce this loss in scanned structures, including matching the support fin angle to the angle of electron incidence (Hinsch, J. E., private communication, 1976) or longitudinal magnetic scanners to maintain orthogonal electron incidence ($\theta = 0°$) across the full window width.[33]

Cold cathode systems

The use of cold cathodes in electron processors has been pursued for decades.[29,34] Their passive nature (e.g. no internal heating, no sensitive coatings, etc.) would appear to provide an electron source of superior characteristics: low cost, power efficient, long lifetime, simple maintenance. They are of two types.

Ion-induced secondary emission

The principle of this system is shown in Fig. 9.10,[35,36] in which an inert gas discharge (typically helium) is established at the ground plane or vacuum envelope of the processor.[37,38] The cylindrical (or, in some cases, shaped) cathode is insulated above it and held at high negative voltage so that as (helium) ions from the discharge strike the cathode (usually molybdenum), secondary electrons are ejected from its surface and accelerated back through the plasma-filled discharge cavity to the window. The entire vacuum envelope is held at relatively high pressure (10^{-3} mmHg (0·133 Pa)) to accommodate the requirements of the discharge.

A number of geometries incorporating this principle have been described in the literature,[37–40] and the approach has been widely used in the pulsed mode for the cavity excitation of gas discharge lasers. A major problem in their use lies in control and monitoring of the electron beam, i.e. dose rate delivered.

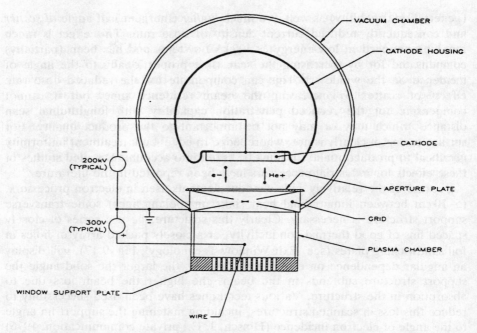

Fig. 9.10. Schematic for a 200 kV ion-induced secondary emission cathode processor.[40,41]

Pulse width modulation of the helium discharge has recently been used for this purpose, i.e. to vary the accelerator duty cycle proportional to product speed.[41] Processor output is difficult to control via current monitoring alone, that is, through the return electron current to the power supply. The power supply cannot differentiate between an ion reaching it at high voltage (assuming singly charged ions!), and an electron leaving it. The output is most effectively monitored by some non-electrical scheme, such as the type of real-time radiation monitor discussed later in the monitoring section of this chapter or in Refs 41 and 42.

Field emission[43]

The field emission cathode operates on the well-understood principle that electrons can be extracted directly from a conducting surface by the application of a sufficiently intense electric field (e.g. 50 MV/cm). The emitters usually take the form of needle-point arrays, foil laminates on edge, carbon felt, etc. With a sufficiently steep negative voltage pulse applied to this cathode array, a plasma of good uniformity is quickly generated over the cathode surface and it will serve as a high brilliance electron source for extended periods, e.g. 10s of μs, before resistive electrical heating of the emitter begins to disrupt the emitter integrity.

Such systems have been studied under repetitively pulsed conditions

(100–1000 Hz) for lower power industrial applications, largely for sterilization (Refs 35 and 36, and Mesyats, G. A., private communication, 1989), where the high dose rate they offer has been shown to provide distinct process advantages. Various schemes have been applied to enhance cathode lifetime, since cathode erosion has proven to be the main limitation of these devices. Because the repetitively pulsed power supplies offering output voltages in the 0·2–1·0 MV range are very compact, fully integrated self-shielded in-line units of <0·5 m³ volume have been built. One such system for the bilateral sterilization of web[35,36] is shown in Fig. 9.11. Due to the significant dose rate effects experienced in most free-radical initiated reactions, these high dose rate systems will continue to be developed largely for 'non-curing' use.

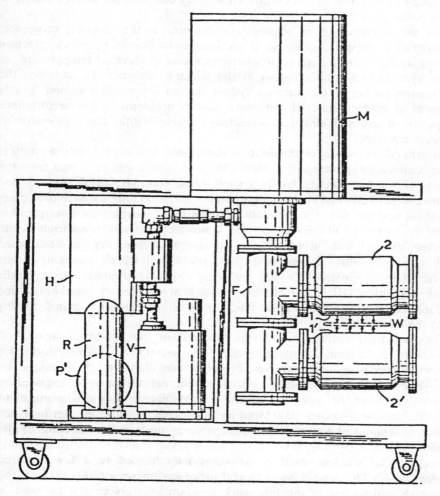

Fig. 9.11. 200 kV, 100 Hz, 0·1 μs cold cathode bilateral processor.[36]

High-temperature emitters

A wide range of electron emitters are available to the processor designer once the complications of supplying power to the emitter within the high-voltage terminal have been accepted. Their operating temperatures are determined by the current density required to satisfy the processor performance and by the difficulty of maintaining thermionic emission. This degree of difficulty is characterized by the electron work function of the emitter surface—high work function refractory metals such as tantalum or tungsten require high temperatures (1900 and 2200°C respectively); other rare earth compounds, such as LaB_6 (lanthanum hexaboride) have lower work functions and can operate well at reduced temperatures (1500°C), while low work function surfaces such as the alkali earth oxides (barium and strontium) can work at surface temperatures as low as 800°C.

For the selection of an appropriate emitter for the electron processor, consideration must be given to its resistance to poisoning (work function elevation) due to accidental atmospheric exposure at elevated temperature, or from hydrocarbon contamination in the vacuum chamber. In addition, the self-contamination of the vacuum system due to evaporation or ion sputter removal of emitter material can cause serious problems in the performance longevity of exposed electrical insulating surfaces within the accelerator or vacuum envelope.

In general, the more 'contamination vulnerable' low work function emitters have been used in scanned processors—they are rather well protected from the effects of window damage by their small area, good structural integrity and relative ease of thermal reconditioning. Very short cathode maintenance times for these systems are reported and lanthanum hexaboride, for example, has been shown to recover fully from the atmospheric exposure associated with window failure.[1] Since window failure in modern processors can be avoided with good preventive maintenance, this problem (cathode survival) is less severe than originally thought. Indirectly heated cathodes are typically employed with a DC heater immersed in a heat-shielded cylinder whose end surface is coated with the emitter. In other cases, the emitter is coated directly onto the heater surface.

For the larger emitter geometries involved in the planar or unscanned processors, a different approach is used. For example, a high performance 2-m-wide processor may involve several linear filaments (longitudinal or transverse) so that large areas (e.g. $5 cm^2$) of emitter may be required to achieve the 5 mA or more of current per centimeter of width required to deliver dose rates above 1000 Mrad m/min. That is, a lower performance emitter is employed to provide the large area, uniform illumination of the window.

Because the window itself is now thermally limited to a few hundred $\mu A/cm^2$ under DC conditions at typical processor energies (200–300 kV), the very high emissivities of the low work function emitters cannot be used as effectively as is possible in the 'pulsed' window illumination of the scanned systems. Hence, present industrial designs utilize directly heated tungsten, thoriated tungsten, or tantalum filament arrays which typically utilize some form of reflective heat shield to reduce radiative heat loss from the now incandescent emitter surface, and employ a spring-type tensioning of the filament to accommodate thermal expansion, metal creep and to control gravitational sag if arrayed in the horizontal plane. This geometry will continue to prevail until sealed, low-voltage electron-processor 'tubes' are developed for low-energy (≤ 100 kV) applications. A return to the more efficient, low work function emitters will then be practicable.

It is worth noting that ferromagnetic materials are rarely used in the construction of electron vacuum tubes/processors and they offer no shielding to the electron beam from the effects of internal or external magnetic fields on the electron beam optics. The effects of internal fields (for example, from DC heater currents) are considered by the manufacturer, but the proximity of the processor to unshielded electrical conductors in the industrial setting must be given some attention in order to ensure that the beam optics of the processor remain undisturbed.

PROCESSOR CONTROL SYSTEMS

General

Modern electron processors can be provided with a broad range of controllers,[44-47] depending upon the process requirements. These systems have evolved from the early simple manual analog controllers, to dedicated digital controllers, programmable controllers, and dedicated microprocessors. As the applications have grown in complexity and sophistication, the real-time computer control of the system has aided in providing the required process quality assurance. In particular, the incorporation of expert operator knowledge into the software permits many of the tedious and error-prone tasks involved in the operation of such high voltage, vacuum insulated equipment to proceed completely automatically.

These controllers, first introduced in commercial equipment in 1979,[44] have eliminated many of the barriers to broader commercial use of electron processing. In particular, they facilitated their incorporation into high-speed converting lines where high 'up-time' reliability was mandatory, and eliminated the need for a high level of expertise 'round-the-clock' from a skilled operator.

Well-designed controllers now provide several levels of intelligence.

- Built-in protection against harmful commands. Such commands may be invoked by the operator because of either a mistake or lack of training, and the software is designed to reject such commands.
- Control of all sequences necessary to start-up and operate the processor. Once these sequences are initiated, the software provides self-diagnosis and protection if interlocks are violated or if component failures occur.

- Optimum sequencing which is dependent upon the accelerator's physical condition. For example, these vacuum-insulated accelerators must be voltage conditioned after a shut-down period, i.e. where no electrical stress was present in the accelerator section. For such automated conditioning, the software is designed to do the sequence decision making until the system reaches 'ready' status in an optimum manner.

Because these processors are typically slaved to track the product speed and deliver a fixed dose over a very broad dynamic range, the microcomputer controller is usually connected via a serial data communication link (e.g. RS-232 type) to a host computer used for system control. For the processor controller, the host computer simply appears to be another terminal—all messages sent to the controller terminal screen are transmitted to the host computer, and, in turn, the host computer can send keyboard equivalent entry signals to the microcomputer controller. Hence, the host computer completely controls the electron beam processor, while allowing the operator to monitor and override the host control from the controller terminal, if desired.

In such a system, the only link to the host computer is the standard serial link, usually with optical coupling for noise immunity and electrical isolation. The host can now be any type of computer from a PC to a mainframe, offering great flexibility in system design, integration of peripheral processing equipment and data logging.

Process quality control

These controllers can be set to provide an audible and visible error warning when operating in the automatic processing mode. For example, if operated in the 'fixed dose' mode, the processor maintains a fixed ratio of beam current to product velocity. If the combination of selected dose and measured product speed requires a beam current greater than that available from the processor, a terminal screen warning and audible alarm are provided until corrective action is taken.

Continuous data logging is conducted via the RS 232 serial link so that all processor and machinery status parameters can be sampled and logged at appropriate rates—several times per second if desired or at the same rate at which the terminal screen is updated. Further details of performance monitoring procedures used with these processors are presented in the section entitled 'Performance monitoring procedures/techniques'—p. 547.

Figure 9.12 illustrates some data taken over a modest operating range (X6) from a microprocessor controlled electron sterilizer,[47] for which the dose set points were checked with the use of calibrated radiochromic film dosimetry. Agreement well within the accuracy of the dosimetry (±3%) is demonstrated for this test for which a transport speed of 40 m/min was selected. Equally precise control is shown for tests conducted over a variable speed range where the controller has been instructed to maintain constant treatment (dose) to the product.

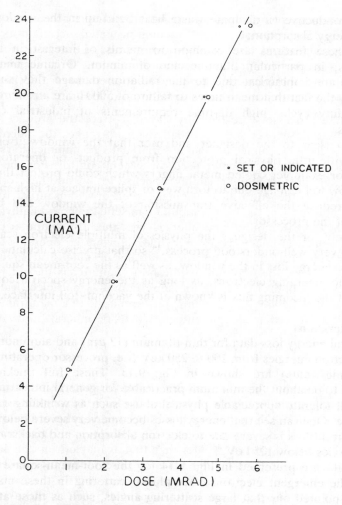

Fig. 9.12. Current as a function of dose for a fixed speed of 40 m/min at 195 kV.

EXIT WINDOW TECHNOLOGY

Introduction
The critical interface between the vacuum envelope in which electrons can be accelerated, and the world in which the industrial converter wishes to apply electron processing, is the window 'plane' of the energy source. The window must be, at the same time, reasonably transparent to the electrons to be directed to the product, vacuum-tight and resistant to chemical attack from chemicals generated in the process zone, of adequate strength in tension at elevated temperatures to provide leak-free performance, and sufficiently

thermally conductive to dissipate waste heat build-up in the window due to electron energy absorption.

All of these features are common to metals of interest in aerospace applications, in particular titanium and aluminum. Organic materials or composites are impractical due to the radiation damage they suffer, emphasized by the fact that mean times to failure of 5000 h are a requirement for the high duty cycle, high up-time requirements of industrial converting applications.

It is also clear to the designer and user that the window itself can be provided with little physical protection from product- or operator-inflicted damage. For example, cooled metal fingers which could protect the window from window foil damage due to torn web or splice impact at high speeds, will inevitably reduce the effective transmission of the window and hence the efficiency of the processor.

Fortunately for the designer the physics of multiple electron scattering in matter is a very well-understood process,[14] so that precise calculations can be made of the energy loss in the window, as well as the root-mean-square scatter angle of the emerging electrons, as long as the energy spectral content and direction of the incoming flux is known at the vacuum–foil interface.

Foil considerations

Some typical energy loss data for thin titanium (13 μm) and aluminum (25 μm) foils at electron energies from 100 to 250 keV (i.e. processor operating voltages of the same value) are shown in Fig. 9.13. These foil thicknesses are considered to be about the minimum practicable for general industrial use, and neither will tolerate appreciable physical abuse such as wrinkling or abrasion or puncture. One can see that energy losses become very severe below 100 keV and the current loss is severe due to electron absorption and backscatter in the foil at energies below 125 keV.[48]

Some data are presented in Fig. 9.14 for the root-mean-square scattering angle of the emergent electrons after plural scattering in these materials. It should be pointed out that large scattering angles, such as those at 100 keV, result in enhanced surface treatment of the product so that poor depth–dose profiles are the end result of high window (or air path) scatter. Indeed, this may be desirable for certain surface-curing applications, as pointed out previously, but in general, the degraded emergent electron spectrum which results from excessive foil scatter leads to undesirable effects for electron processing and the designer usually seeks to minimize the effect where possible. Very thin foils of metallized organic films, and of low Z metals such as beryllium, lithium and carbon have been evaluated with little success for industrial application.

A property of foils which can have very serious implications is that of gas porosity or pin-holing. Air (or N_2) leakage into the vacuum system can lead to immediate ionization by the electron beam, so that positive ion bombardment of the gun/high-voltage terminal assembly of the processor results. This

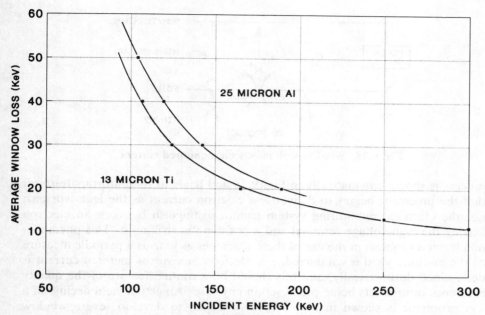

Fig. 9.13. Electron energy loss in a foil as a function of energy.

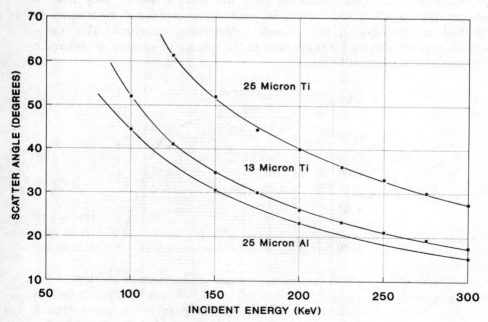

Fig. 9.14. Root-mean-square scatter angle of electrons in a foil.

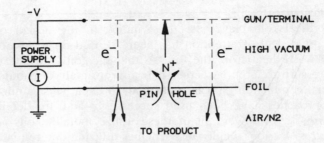

Fig. 9.15. Window leak effects on monitored current.

process is shown schematically in Fig. 9.15 and leads to the unfortunate result that the processor begins to deliver less electron current as the leak worsens; i.e. the electrical monitoring system cannot distinguish between an electron leaving the high-voltage terminal and a positive ion striking it. This phenomenon is not a problem in the use of these machines as long as a periodic measure of the machine yield is conducted, i.e. electron flux versus metered current is determined dosimetrically, and this should be a routine feature of the quality assurance of products being processed in any case. An actual yield decline of a 2 m processor is shown in Fig. 9.16 as it began to develop severe window leakage. 'Severe' in this case may not necessarily mean that the processor pressure has reached a warning plateau. Consider the process shown in Fig. 9.15. The fate of the ions is controlled by the electric field gradients in the processor so that upon ionization near the window plane, they have no opportunity to affect the ionization gage or spectrometer monitoring the residual gas pressure in the vacuum (accelerator) envelope. The ionized species may be trapped for some time in the volume of vacuum dominated by the accelerated electron beam.

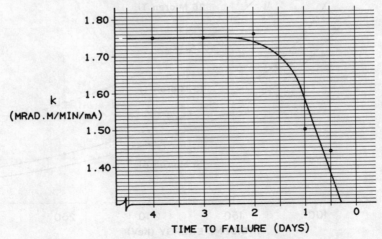

Fig. 9.16. Yield deterioration with window pin-holing.

Foil support structures

Due to the critical role played by the window in the design of electron processors, a great deal of attention has been focused on techniques for the elimination of waste heat from the foil, so that the foil load due to the one atmosphere pressure differential across it is appropriately distributed, and so that the thermal conduction path in the foil to the heat sink is minimized. A number of approaches have been used successfully and are detailed in the patent literature,[49-53] however, all of these utilize metallic foils and cooled support structures. Various windowless schemes utilizing differentially pumped apertures or slots have been studied and developed,[1,54] but are considered to be too cumbersome and expensive for practical electron processors, especially for wide products.

Clearly, the opacity offered by any support frame is critical, as frame absorbed electrons affect the processor efficiency—the maximum energy investment has been made in the electrons which reach the window (ground) plane.

Figure 9.17 shows the most common approaches to foil support, all of which are described in detail in the referenced literature. Figure 9.17(A) shows an arcuate, unsupported foil which will maintain the 1 bar (10^5 Pa) pressure over a 2–4 cm width—since the tensile strength of the foil is now crucial, this approach is limited to titanium foil in the 50 μm range and above, and therefore is used mostly at higher energies (250 kV and above).

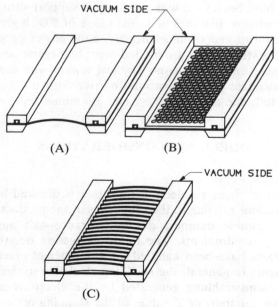

Fig. 9.17. Foil mounting geometries.

The drilled hole window frame of Fig. 9.17(B) offers fairly high transparency (75%) in practical grids, but very poor thermal conduction to the edge heat sinks. As a result, this approach[53] has been used on large area, low current density designs where the maximum foil–frame temperature must be restricted to provide acceptable operating lifetime. Clearly, some reinforcement of the grid assembly is required to prevent buckling of the structure under 1 bar (10^5 Pa) at elevated (operating) temperatures, so that, in practice, widths of under 10 cm are used in order to prevent build-up of the grid thickness which seriously enhances absorption of the emergent electrons possessing large scattering angles.

The finned window of Fig. 9.17(C) has proven to be the most satisfactory in terms of its trade-off of minimum conduction path to the heat sink and buckling strength.[49] In view of the need to prevent fin buckling in the transverse direction at high current densities, curved fin arrays have been developed[50] which provide a smooth motion of the fin during heat-up across the foil surface, and avoid any longitudinal variation in transparency with temperature as the processor rises to maximum electron current density at the window plane.

Almost all electron processors utilize demountable window foils so that they can be easily serviced in the event of chemical attack or physical failure due to abuse, embrittlement or overheating. Techniques for the chemical preservation of foils using platinum group metals have been quite helpful in extending foil lifetime under corrosive operating conditions, while composite foil techniques to provide improved thermal conductivity–electron transmission–fin compliance matching have been used with the fin type support structure.[50]

Electron beam window lifetimes lie in the range of 5000 h given appropriate electron optical control and cooling integrity. Low power processors used in applications such as sterilization are limited more by organic seal failure at the foil–frame interface, since there is no physical reason why these foils should not provide an adequate pressure–vacuum barrier 'forever' if physical flexure and stress of the foil, due to pressure cycling, are minimized.

SHIELDING CONSIDERATIONS

General

The shielding design[55-57] of an electron processor is dictated by a number of practical considerations relating to the properties of the product: thickness and flexibility, width, atomic number, desired entrance–exit angles, transport speed and inerting requirements. The following section describes how these design considerations have been handled for a number of practical cases with self-shielded designs. In general, the shielding challenge to design arises from the penetrating bremsstrahlung generated by the electrons as they stop in matter. The atomic number, or Z value, of the stopping material is critical in view of the sensitive variation (roughly linear) of bremsstrahlung yield with Z.

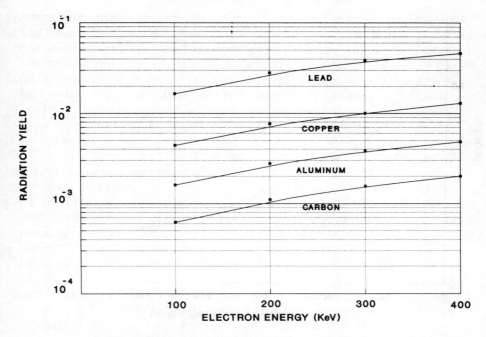

Fig. 9.18. Radiation yield as a function of energy.

Some typical yield values[15] with atomic number are shown in Fig. 9.18 for the 0·1–0·4 MeV electron energy range. The total energy emitted as bremsstrahlung per incident electron varies as ZE^2 where Z is the target atomic number and E is the electron energy. This figure is small for the energy region of interest here; as shown in Fig. 9.18, only 1% of the energy of a 300 keV electron stopped in copper is converted into penetrating radiation.

Of greatest importance for strip or curtain type processors is the spatial geometry of the penetrating bremsstrahlung or continuous radiation sources they generate. These processors provide a slab or two dimensional strip of X-rays, emitted roughly isotropically in the low energy regime (e.g. ≤ 500 keV) and tilted to the forward direction as energy increases in the manner shown in Fig. 9.19. These 'regimes' are illustrated in Fig. 9.20.[58] Of greatest concern is the reinforcement of intensity or lobes of radiation which are a consequence of such anisotropy. For example, the distribution of Fig. 9.20 results in intense radiation lobes at the ends and, of course, along the length of the strip 'source', in the plane defined by the volume wherein the electrons are decelerated from near the velocity of light to rest, that is, by the 'thin' strip in which they lose their energy. Hence, for the self-shielded sources in the low-energy region, the shielding challenges at the ends of the processor, particularly in the product stopping plane, are the most severe. The challenge is, of course, defined by the surface dose rates accepted for continuous human

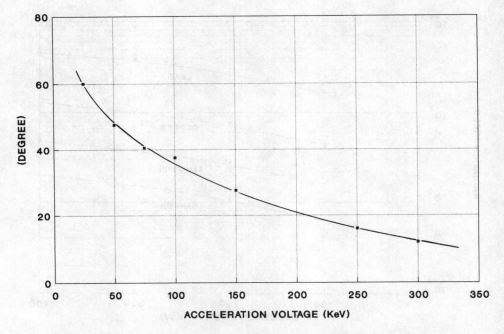

Fig. 9.19. Dependence of angle of maximum intensity on voltage.

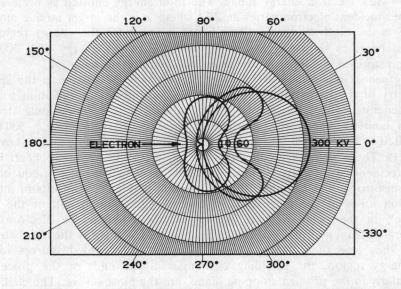

Fig. 9.20. Polar diagram for continuous X-rays generated in a processor.

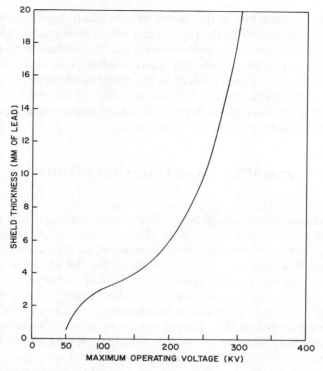

Fig. 9.21. Typical lead thickness as a function of operating voltage.

occupancy. In general, this is accepted to be 0·2 mrem/h (2 μSv/h) or less for regions in which operator access is unrestricted.

It is not possible to lay down precise shielding guidelines because of the many factors affecting the bremsstrahlung source intensity. Since the source strength for the penetrating radiation generated in a given target (product) is determined by electron beam current density and energy, the total beam power is a critical consideration in shield design. Some empirical shielding requirements are shown in Fig. 9.21 for processors operating in the linear current density range of up to 3 mA/cm of machine length, i.e. 600 W/cm for a 200 kV processor.

Lead is the most commonly used shielding material and should be protected from the corrosive materials (such as nitric acid and ozone) typically present in the process zone. This is usually accomplished by stainless-steel sheathing. Since multibounce absorption of the photon via the photoelectric or Compton scattering processes dominates the self-shielded processor range (0–500 kV), care must be taken that the collimator angles and lengths are sufficient to reduce the low-energy secondary electrons and photons to near undetectable levels (i.e. <0·2 mrem/h (2 μSv/h). Because secondary electrons are generated with each photoelectric absorption or Compton scattering event, the presence of a significant electron flux at the mouth of a collimator, for example, may be of much greater concern than the photon flux which produced it. The very high stopping power of matter for soft electrons must be recalled, particularly when selecting a detector offering sufficient discrmination to measure both the soft X-ray and soft electron fluxes present in the shield (collimator) opening at the product exit and entrance points. For this reason, ion chamber type detectors offering a very good low-energy X-ray and electron response are preferred for health physics purposes around electron processors.

PRODUCT HANDLING GEOMETRIES

General

A dramatic change in processor design philosophy was required for the lower energy self-shielded machinery. Up to this time (c. 1972) the much higher energies and product presentation difficulties involved with scanned equipment usually meant that relatively long flight paths for the electron between the window plane and the product were necessary and tolerable. In the case of applications where the (excess) electron beam energy could not be well matched to the penetration needs of the product, the air path was frequently used to degrade or soften the electron spectrum in order to better match its absorption profile to product geometry. Consider, for example, the requirements of the cross-linking of the insulating jackets for hook-up wire. A typical wall thickness of $200 \, g/m^2$ of PVC would be quite uniformly treated by an electron beam energy of 250 keV rather than the 450 keV typically delivered from a 500 kV scanned processor. For example, with an air stopping power of $0.185 \, keV/g \, m^2$, and an NTP air path thickness of $12 \, g/m^2 \, cm$, one can, in effect, 'soften' a 300 keV electron spectrum to 250 keV with the use of an extended (20 cm) drift path between the window and product planes.[19] This result is shown schematically in Fig. 9.22, and, until the advent of the low-energy control offered by the self-shielded processors in the mid-1970s, this technique was widely used for the matching of electron spectral content to the product penetration needs. The importance of this type of optimization for the processing of a low thermal conductivity polymer (jacket) extruded over a low specific heat conductor (such as copper or aluminum) is obvious, especially where high jacket doses are required and the temperature threshold for polymer damage is relatively low.

The example described in Fig. 9.22 is cited as an example of a vault shielded process in which energy matching is critical to a quality product. A more recent example of controlled matching has been described[59,60] in which the deposition profile does not fully penetrate the jacket wall but is adjusted so as to provide a fully processed outer surface and an untreated inner surface, for example, for providing hybrid properties (from fully cross-linked to thermoplastic) across the $100 \, g/m^2$ thickness of a shrink tube or catheter jacket. The depth–dose or penetration profile of a 100 kV beam used for such an

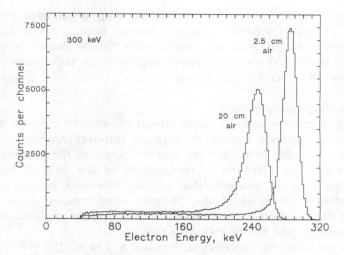

Fig. 9.22. Spectral softening or degradation with air path at 300 kV.

application[61] is shown in Fig. 9.23. The control of the profile 'slope' by product electron illumination geometry, is critical to processes in which a well-defined energy deposition profile is important. A detailed treatment of this problem is available in the literature relating to controlled depth of treatment, for example, for sterilization[62,63] where the process may require significant outer layer penetration for sterility assurance, while keeping the labile product isolated from the primary and secondary electrons. Optimum presentation of the product to a beam of good spectral quality (i.e. minimal scattering) is of major concern in this type of application.

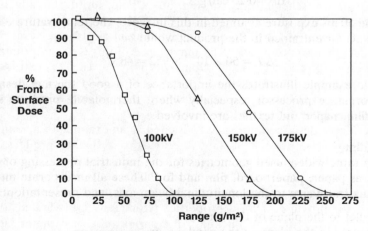

Fig. 9.23. Low-voltage depth–dose profiles.

The development of efficient product handling geometries for electron processing continues as process reaction kinetics become better understood. This important design area has traditionally taken second place to the 'accelerator' itself, and dramatic improvements in the techniques of product presentation will be seen in this decade.

Air path heating considerations

The very high energy delivery rates typical of modern electron processors impose severe demands on air or nitrogen turnover requirements in the process zone in order to reduce the thermal shock to the product. Clearly, some control of the elevation of temperature of the treated layer will be desirable for a given process, due to the increased efficiency of most electron-initiated processes at the elevated temperatures (largely due to enhanced radical mobility). Nevertheless, it is important to eliminate stagnant air in the process zone for the following reason.

Consider an electron processor delivering a 5 Mrad (50 kGy) dose to a product moving at a speed of 240 m/min. For the 20-cm-wide process zone typical of these systems, the average dose rate experienced by the product will be

$$\frac{dD}{dt} = \frac{5 \times 10^6 \text{ rad}}{20 \text{ cm}} \times \frac{2 \cdot 4 \times 10^4 \text{ cm}}{60 \text{ s}}$$
$$= 10^8 \text{ rad/s}$$

This average dose rate of 100 Mrad/s is equivalent to 1000 J/g s or 238 cal/g s.

Now this same dose rate is experienced by the gas surrounding the product, so that if stagnant gas experiences this flux, with a specific heat of 0·25 cal/g °C say, the heating rate will be

$$\frac{dT}{dt} = \frac{238 \text{ cal/g s}}{0 \cdot 25 \text{ cal/g °C}} \quad \text{or} \quad 952 \text{°C/s}$$

So, for the 50 ms exposure assumed in this instance, the temperature elevation of (stagnant) air entrained in the product would be

$$\Delta T = 50 \times 10^{-3} \times 952 = 48 \text{°C}$$

This single example illustrates the importance of a good thermal design in a high-performance processor, especially where thermolabile products such as polymer film, paper and textiles are involved.

Web handling

There are three widely used geometries for the industrial processing of flexible web such as paper, paperboard, film and foil. These all incorporate motion of the product transverse to the longitudinal beam axis, and presentation:

1. parallel to the plane of the window,
2. obliquely to the plane of the window, or
3. circumferentially on the surface of a supporting drum.

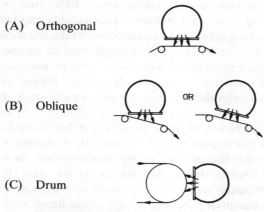

(A) Orthogonal

(B) Oblique

(C) Drum

Fig. 9.24. Web treatment geometries.

In industrial practice, the gap or clearance between the window and its mounting clamp, and the product, must accommodate the practicalities of the following: web flutter, web splicing non-uniformities, relaxation upon web failure or tear, web or coating outgassing, homopolymer volatilization from the coating, hydrogen removal with hydrocarbon irradiation, efficient inerting, electron backscatter from the product and high gas turnover rates to prevent excessive beam heating which would subsequently affect the product, or, indeed, the integrity of the processor itself. Unprotected shielding failure due to the relatively low melting point of lead can be experienced in these systems with inadequate attention of the designer to this important issue of waste heat dissipation and control.

The geometries shown in Fig. 9.24 are the most widely used in the industry. The 'parallel' geometry of Fig. 9.24(A) is the most efficient in terms of energy transfer to the product, in that the scattering air path is minimized and the electron illumination at the product plane is symmetric. This approach can be employed with either a horizontal or vertical draw of product, except that the vertical orientation may offer unacceptable complications to the web path for an in-line system; for example, external web direction reversal with an idler roll may be required if the process will not permit direct roll contact with the active or treated product surface immediately upon emergence from the process zone. Air bars or boundary layer support rolls or 'grater' rolls are often resorted to in such a case. The oblique configuration of Fig. 9.24(B) has been widely used in the graphics application of these machines because it offers a horizontal entrance of the coated web without need for significant angular deflection of the web before it enters the process zone. The main disadvantages of the approach is a greatly enlarged process zone volume with its associated thermal control complications, and a less efficient transfer of energy to the web because of the gradually increasing electron flight path in air as the product moves away from the window plane(s). At 200 kV, for example, these increased losses for a 15° web inclination angle are about 10%—that is, the yield k of the processor is reduced by some 10% due to these increased absorption scattering losses in a processor in which the gap increases from 4 to 9 cm in moving across a split windowed processor (see the section entitled 'Exit Window Technology'—p. 523). Steeper angles will of course, involve greater losses so that 'angled processors' have been used to maintain a minimum air path for non-horizontal (or vertical) web paths. Oblique positioning of the processor[64] has also been used in some designs to eliminate this gap scattering complication.

The drum configuration of Fig. 9.24(C) has been used for at least two decades with high power processors[65–67] in order to provide a firmly supported illumination plane for the product in the process zone, in a geometry which provides a good beam 'dump' for the processor (for those applications involving 'through penetration' of the product or where no product is present for beam energy absorption). Several designs have been employed in which the drum provides integral radiation shielding so that very compact self-shielding geometries are practicable using the intrinsic curved collimation for radiation provided by the drum–shield assembly. For low-energy, high-speed processors employed on surface treatment (curing of inks and coatings, surface modification, etc.), this is a preferred product handling geometry because of its positive (support) product presentation in the beam process zone. Another major advantage lies in its good inerting efficiency in that the air boundary layer at the underside of the web is buried against the drum surface and presents no serious contamination problem in this critical region, especially if nip pinching of the web against the drum surface is employed.

A major disadvantage of the geometry of Fig. 9.24(C) is the increase in electron flight path as the product approaches and leaves the beam axis of symmetry. Clearly the angle of incidence of illumination rises dramatically from 0° as the product traverses the scattered Gaussian distribution at the beam edge. A loss of processor yield results with a significant degradation of the depth dose profile experienced in the product.

Some comparative data are presented in Fig. 9.25 in which a normal, as in Fig. 9.24(A), depth–dose distribution of a 200 kV processor is shown, compared with a profile recorded on a 50 cm diameter drum positioned as in Fig. 9.24(C), with the same minimum air gap at the window axis of symmetry. The deterioration of the electron deposition profile with this modest increase in approach–departure electron flight path length is very significant and can obviously have a significant impact on product treatment uniformity in depth for even the thinnest products, due to the significant elevation in soft or multiply scattered, obliquely incident, electron content of the spectrum at the outer regions along the drum periphery.

Sheet handling

Sheet handling systems for electron processing may be conveniently divided into two categories. Those involved in the wood and metal, panel and ceramic products industries, and those required for plastic sheet and paperboard. The

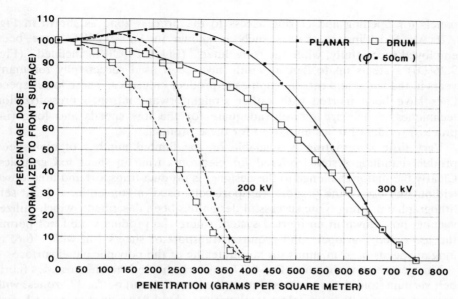

Fig. 9.25. Normal and drum delivered depth–dose profiles.

former involve relatively low speeds with present designs, typically 30–50 m/min, while the latter must accommodate the much higher product speeds of the graphics industry, e.g. above 50 to as high as 400 m/min. Four different geometries currently in use for sheet products are shown in Fig. 9.26.

Gated systems have been used in Japan[68] involving moveable shielding

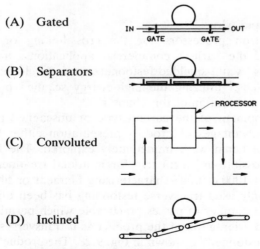

Fig. 9.26. Panel handling configurations.

panels for opening and closing access to the process zone as shown in Fig. 9.26(A). Shielding separators such as shown in Fig. 9.26(B) have been employed successfully, while the convoluted[69] (Fig. 9.26(C)) or inclined[70] (Fig. 9.26(D)) systems up to 2 m in width have been used successfully for many years in the wood and cement composite product industries. These lower speed lines have been inerted with rather straightforward nitrogen pressurization techniques, which are entirely adequate for the low speeds and long (in-process) residence times employed for these products.

Very little information is available on the limited number of high-speed product handling systems offered for electron curing in sheet fed graphics. Clearly the major challenge to the designer is product transport and hold-down without the use of the organic transport belts used in conventional (heat set) lithographic printing. One successful design has been described[71] which utilizes vacuum hold-down in an inerted system where the product would not permit the use of edge-grippers and required transport of sheets 1 m wide × 0·65 m long at speeds of 90 m/min with no scratching of the polymer panel surface.

This geometry is shown in Fig. 9.27 in which the Selfshield® provides fabric belt vacuum hold-down of the sheets in the infeed A and outfeed C zones, with product transfer to a stainless-steel vacuum hold-down in the central high radiation zone B. Inert gas utilization was enhanced by sequencing gas flow from the central process zone to the outer zones in a three-step sequence, with roughly an order magnitude reduction in oxygen concentration at each transition. The sheet handler shown in Fig. 9.27 handles sheets with modest (0·5 cm) edge curl and takes the product directly from the stacker station at the end of the sheet fed offset press.

It is clear that products which have sufficient selvage or unfinished edge margin can be transported by the normal chain gripper or nip roll techniques used in conventional offset printing.

Fibers, filaments and wire

The use of electron processors for the cross-linking of cable jacketing represents one of the earliest commercial applications, internationally, of electron processing. Vault-shielded festooners were developed by several wire and cable companies,[1] utilizing the high-energy scanned processors already referred to in the first section of this chapter.

With the development of the curtain type or unscanned processors in the early 1970s, the advantages of product presentation either longitudinally or transversely to the beam were immediately explored. The long, single pass residence times possible are attractive for multiend treatment with the very high speeds (e.g. > 1000 m/min) characterizing filament or fiber manufacture. The more commonly used transverse festooning has been utilized because of the relatively large number of passes practicable which permits random, or, if desired, controlled orientation of the product as it transverses the beam.

A longitudinal festooner[60] is shown in Fig. 9.28. The product handling system provides a 2 × 10 pass capability with a 30-cm-long beam. Quite precise

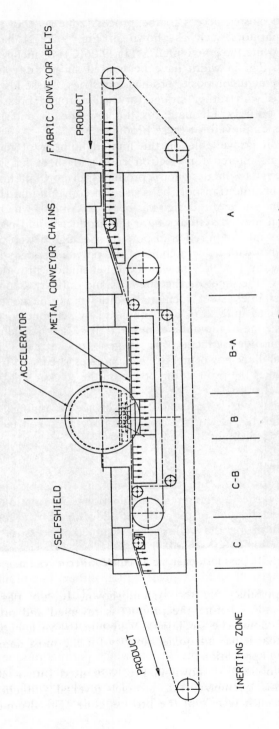

Fig. 9.27. Schematic of sheet stock Selfshield® with vacuum hold-down.

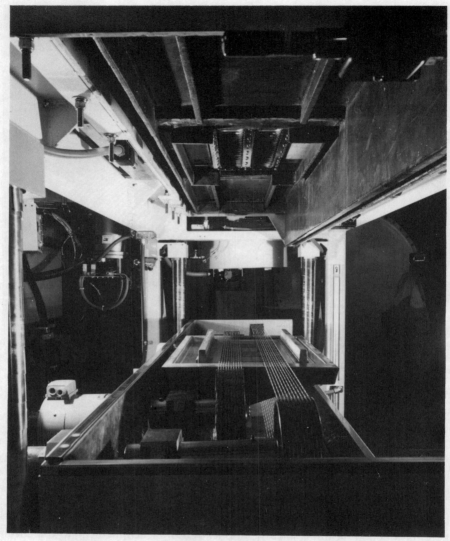

Fig. 9.28. Double 10 pass festooner for a 300 kV × 30 cm system.

product orientation is possible—for the system shown, 10 'up' passes are completed under one window before the product is reversed and presented 'down' to the second window. The small 300 kV system shown is capable of 14 000 Mrad m/min, so that it has adequate capacity for the most demanding polymers for cross-linking applications.

A small reversing festooner[60] is shown in Fig. 9.29 used with a 30 cm × 250 kV processor. In this geometry, full product reversal (direction *and* rotation) is provided on each transit of the process zone. The drum-cooling

Fig. 9.29. Multipass festooner assembly for a 250 kV × 30 cm processor.

techniques employed with these units provide controlled product temperature at high delivered dose, coupled with the ability to restrict energy delivery to a controlled depth in the polymer, in particular to limit heating of the low specific heat conductor about which the polymer (jacket) is extruded. The small 52 pass system shown is capable of 5000 Mrad m/min at 250 kV × 5 kW.

Three-dimensional and molded products
There has been great interest in the application of electron curing to the curing of coatings on formed metal and polymer products. Indeed, the early development of the Electrocure™ systems at Ford Motor Co. was for the automotive dashboard application, using vault shielding and side-fired scanned 300 kV units. The difficulties in this use arise largely from inerting and shielding considerations. Although commercially unsuccessful, the later work of the group at the Volkswagen plant in Wolfsburg, Germany, for automotive wheel rims is significant,[72] while an Electrocure unit has been used for over 15 years in Japan for the finishing of ceramic roof tiles (Nakazato), and more recently for the protection of decorative fiberglass-reinforced gypsum tiles (Achilles Corporation), using rather conventional conveyors.

Fig. 9.30. Electron sterilizer system showing Selfshield® and outfeed.

Great incentive for three-dimensional product handling has been provided by in-line sterilization of molded products, in particular for stored product containers. In this application (for aseptic filling) the product is passed continuously under the sterilizer in either the longitudinal or transverse directions, so that the electron beam defines the 'barrier' between the non-sterile infeed and the interior aseptic tunnel. One configuration for open-mouthed glass, polymer or paperboard containers is shown in Fig. 9.30.[73] Here the 250–500 ml containers are transported on a stainless-steel conveyor in a compact, serpentine Selfshield® which is water washable. Line speeds of 30 m/min are typical for these applications, with the sterilizer output slaved to conveyor speed to maintain a fixed and predetermined dose to provide the requisite sterility assurance level (SAL) for the application.

INERTING AND OZONE/NITROGEN DIOXIDE/HOMOPOLYMER CONTROL

General

The requirements of, and techniques for, 'inerting' of an electron beam processor have been slow to evolve.[74] In fact, the need for providing a low

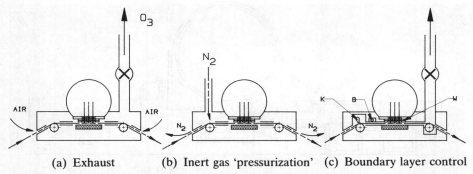

(a) Exhaust (b) Inert gas 'pressurization' (c) Boundary layer control

Fig. 9.31. Techniques for handling gaseous products of irradiation–curing in an electron processor.

oxygen concentration environment in which to accomplish free-radical initiated polymerization efficiently has been a severe impediment in the commercial application of the art, largely due to the absence of reliable, quantitative determinations of the inerting requirements for the addition polymerization chemistry involved. Recent publications have begun to provide this data.[75,76] Nevertheless, the machine designer has been left to his own devices in providing integrated shielding–inert gas handling designs, satisfactory for industrial use.

The general approaches to inerting are illustrated in Fig. 9.31. The first challenge to the product handling 'inerting' design for a continuous processor working under ambient environmental conditions is the elimination of beam-produced ozone and nitrous oxides which can be transported by the product into the work area. Because the tolerable levels of ozone in an occupied area are <0·1 ppm,[13] the gas must be exhausted and incinerated, or diluted with make-up air, or it may be eliminated through self-recombination in the beam-heated process zone. These techniques have been used successfully in self-shielded machinery for curing, sterilization and cross-linking, so that compact processor designs have been possible for ease of integration into continuous production lines. Several geometries[77–79] have been taught in the patent literature with successful commercialization.

As illustrated in Fig. 9.31(a), a successful design which permits high-speed transport of the product without pollution of the workplace, utilizes maintenance of the process zone at negative pressure so that ambient air sweeping into the processor prevents any products of irradiation from reaching the workplace. This technique is typically used for film cross-linking or for drum curing where oxygen inhibition effects are not of concern.

An improved technique is shown in Fig. 9.31(b) where nitrogen or a similar inert gas is used to eliminate the generation of ozone or of nitric acid (from the nitrous oxide) which can result in severe chemical attack of the product handling structure, and undesirable oxide contamination of the process zone (and product surface). This nitrogen 'bleed' or pressurization technique is

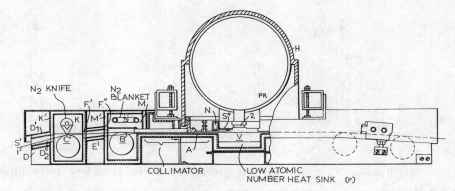

Fig. 9.32. Selfshield® geometry.

effective for elimination of ozone dragged into the work area in low-speed systems, and permits modest temperature elevation of the gas in the beam-affected zone in order to enhance ozone recombination via the $O_3 + O_3 = 3O_2$ reaction. This technique is quite effective for applications involving low product speeds, say to 50 m/min. Continuous external O_3 monitors are typically used for ensuring the efficiency of O_3 control.

For elevated product speeds typical of the bulk of electron processing work, at rates above 50 m/min, more elegant techniques are required as illustrated in Fig. 9.31(c). Here a high speed nitrogen knife (K) is employed to remove or create turbulence in the air boundary layer carried on the product's surface. Other distribution manifolds at B and W (Fig. 31(c)) are employed to maintain a positive pressure in the Selfshield and to provide additional corrective cooling over the window foil at W.

For these industrial systems, the oxygen 'concentration' in the process zone is usually determined by means of a continuous sampling tube located 'somewhere' above the moving product. As shown in the early design[77] of Fig. 9.32, this sampling location A above the web I does not provide an accurate indication of the oxygen concentration at the web surface S, in the web boundary layer, where the electron initiated chemistry of interest in an active curing application is taking place.

Analytical techniques for assessment of inerting efficacy

The second challenge to inerting design is the assurance of an adequate environment with high-speed transport of the product so that an acceptable degree of cure results with modest treatment (dose or specific energy investment). Hence, precise analytical procedures are necessary with which the degree of cure of the product can be determined, not only for its quality assurance but also for optimization of the process. The critical parameters here are dose, dose rate and inert gas control.

A good review of analytical methods for determining the degree of cure in radiation curing has been published.[80] Reference 80 also reports on the use of

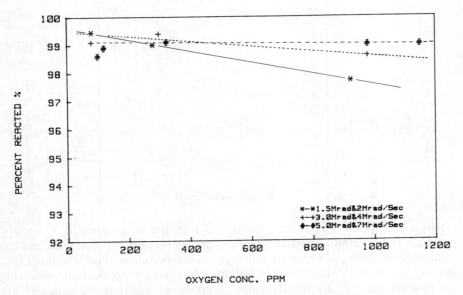

Fig. 9.33. Degree of cure as a function of indicated oxygen concentration.

high-performance liquid chromatography for the assay of unreacted monomer content in printing varnishes. Other work[75] has focused on the use of gas chromatographic procedures on the Soxhlet extracted fractions from the radiation cured coatings, so that a reference base can be established for the process, against which less sophisticated, in-line solvent penetration test techniques can be used for quality assurance purposes. So, for example, a study is first made of degree of cure versus inerting and dose parameters, and these samples which range from 90 to 100% degree of cure, are used as a reference for the simpler droplet penetration test.

Such a sequence is shown in Fig. 9.33 in which dose, dose rate and indicated oxygen concentration in the process zone are all used as independent variables in determining the degree of cure of a five process color/overprint varnish system. Using the geometry of Fig. 9.31(c), the sensitivity of degree of cure on nitrogen purity in the process zone (W) is illustrated in Fig. 9.34. Here the variation in injected gas purity from 50 000 ppm (95%) to 100 ppm (99·99%) resulted in a degree of cure variation from 86% to 99·9%.

Hybrid inerting techniques

It is now possible to use these techniques for the optimization of the sequence of inert gas introduction into the system. In view of the demonstrated need (Fig. 9.34) for low oxygen concentration in the process zone where initiation occurs, an approach has been developed utilizing non-cryogenic gas supplies (such as pressure swing adsorption or molecular sieve generators) in conjunction with high-purity cryogenic supplies. The motivation for this is clearly

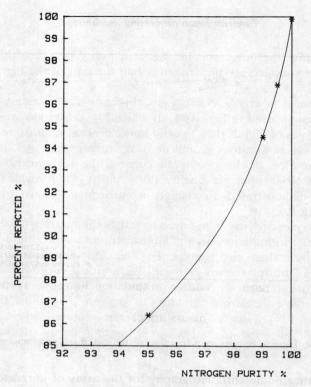

Fig. 9.34. Degree of cure as a function of nitrogen purity.

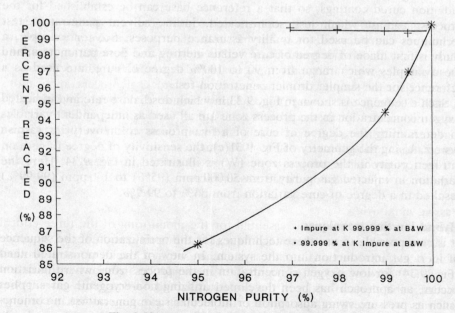

Fig. 9.35. Degree of cure with hybrid inerting.

economic, in that the non-cryogenic gas at a purity of a few thousand ppm oxygen, can be supplied at approximately half the cost of the high-purity LN_2 vaporized gas.

Data are shown in Fig. 9.35 for a gas chromatographic assay of the same ink–varnish system used in Fig. 9.34, in which low purity gas was injected at the infeed and gas of higher (LN_2) purity in the process zone. Referring to Fig. 9.31(c) these are at locations K and at B/W, respectively. As shown in Fig. 9.35, no diminution in the quality of cure of the ink–varnish system was observed with 10 000 ppm gas quality (99% purity) at K, while the varnish system alone demonstrated no change in cure quality down to 30 000 ppm (97% purity) at K.

These techniques are now being used to satisfy the inerting needs of modern radiation-curable formulations by the adjustment and appropriate balancing of the inert gas from these two sources. For the case shown in Fig. 9.35, some 40% of the gas utilizes the lower purity PSA with a concomitant savings of 20% in gas costs. Indeed, as coatings formulations improve which reduce the sensitivity of these 'electron curables' to oxygen inhibition, the high gas purities of the LN_2 systems will no longer be required and non-cryogenic generators will be used to satisfy *both* inerting needs of these electron processors, namely environmental control and degree of cure assurance.

PERFORMANCE MONITORING PROCEDURES/TECHNIQUES

The monitoring of electron processors for product quality assurance as well as for output characterization, is an area still under very active development. Most of the dosimetric techniques well developed for penetrating radiation (X-rays and γ-rays) are totally inadequate for the relatively low penetrations characteristic of the electron energies used in practical process applications (0·10–0·50 MeV). As a consequence, film dosimeters have been developed[81–83] with thicknesses in the range of 10–$50\,g/m^2$, which are suitable for the diagnosis of the electron spectra delivered from these machines (with electron penetration capabilities from 50 to $1000\,g/m^2$).

Electron processor monitoring can be separated into passive and active techniques, which will now be discussed in turn.

Passive monitoring

The use of film dosimetry is essential for the monitoring of the three critical parameters of the processor: machine yield, uniformity, and energy. Machine yield k (Mrad m/min mA) relates machine output current (I) and product speed (v) to dose (D), i.e. $D = kI/v$. It should be invariant with current (because the electron optics of the processor should not vary with current) but will vary with energy simply because the electron stopping power, i.e. energy loss per micrometer of penetration, in the dosimeter or product, varies with

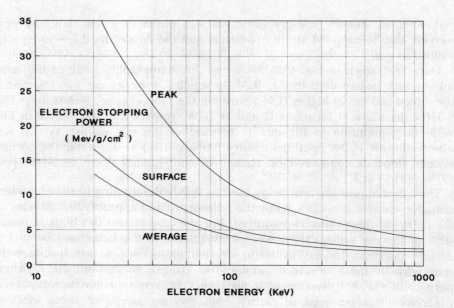

Fig. 9.36. Electron stopping power variation with energy for polystyrene.

energy, as shown in Fig. 9.36. A typical yield variation with energy for a modern processor is shown in Fig. 9.37.[84]

In general, in moving back along the performance curve of Fig. 9.37, one sees the expected yield increase with energy decrease up to the 165 kV accelerating potential at which point the yield curve inflects and begins to

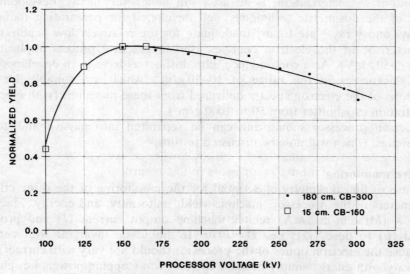

Fig. 9.37. Yield variation with process operating voltage.

decrease. This inflection marks the point where the window absorption losses (discussed in the section entitled 'Exit Window Technology', p. 523) begin to dominate. Further decreases in the incident electron energy (accelerating voltage) dramatically decrease the effective yield, in spite of the continuing increase in stopping power at these reduced energies (see Fig. 9.36)). Obviously, the window foil thickness–air drift path will dramatically affect the inflection point. The case shown in Fig. 9.37 is for a 12·5 μm (56 g/m^2) titanium window and a 2 cm (24 g/m^2) air path.

Care must be taken in providing some degree of electron equilibration for the film dosimeter. For example, a high Z material under the dosimeter during irradiation will overexpose the dosimeter due to the excess backscatter (albedo) the dosimeter receives. If the film is not suitably electron equilibrated, for example, if it is exposed bare, it will provide a conservative indication of energy absorbed since fewer electrons are returned to the film by plural scattering in the surroundings than would occur in the bulk material of the equivalent product (a hydrocarbon, for example). These errors can be quite large (20–30%) for large Z discontinuities in low energy dosimetry.[85] Their impact diminishes with increasing electron energy since surface scattering is less dominant.

The frequency with which yield measurements are conducted will vary with the tolerance of the product for dose variation and its value. For example, in electron sterilization of relatively expensive medical devices,[86] yield data will be recorded and evaluated several times daily, before the process is continued, even with continuous recording of the other parameters (current, speed, voltage). The reasons for this have been discussed in the subsection on foil considerations and relate to the indeterminate nature of simple current monitoring as an indication of processor output. Many less demanding applications, where physical testing of the processed product itself serves as a secondary dosimeter, may involve yield value determinations as infrequently as once a month.

The *uniformity* of the electron fluence delivered by a processor is measured with the use of continuous or segmented dosimeters carried through the process zone at the same plane traversed by the product. This is typically performed at the same operating conditions used in processing (sterilization, curing, or cross-linking, for example) and the strips are then read out with a scanning or static densitometer and the data logged for archival reference. The frequency of this measurement is typically low, once every 4 weeks, in view of the good stability of modern processors in this regard.

The *energy* determination for the processor is accomplished with the use of depth–dose laminates. For example, for a machine operating at 175 kV, the operator is aware that at the product plane, the penetration end-point (extrapolated range) is 280 g/m^2. Clearly, 10 × 35 g/m^2 films will provide an infinitely thick absorber for such a spectrum and the seven films which are exposed will provide a good measure of the penetration profile. Some typical depth–dose determinations for commercial processors reveal the detail available for spectral analysis. In practice, experimental curves are compared with a reference curve for the processor to insure that errors in voltage monitoring (energy setting) due to monitoring resistor deterioration or power supply malfunction (for example, a phase loss) are detected in a timely manner.

Active monitoring

The processor control systems provide ready access to and archival monitoring of the critical operating parameters of this equipment. This normally includes machine current (dose rate), operating voltage (electron energy) and product speed (exposure time). These real-time data, supported by the periodic passive dosimetry measurements reviewed above, normally provide the performance history required for process control.

In order to assure, in real-time, that all of these parameters are sustained in a traceable maner, real-time radiation monitors have been developed[41,42] for the monitoring of the X-ray or bremsstrahlung distribution generated by the electron beam in the window frame–foil support structure of the processor. In this manner the detector cannot be 'fooled' in that the X-ray monitored intensity distribution is directly related to the electron flux and energy spectrum of interest. The monitored flux can, in turn, be directly related to the delivered electron dose by film dosimetry traceable to a standard (or reference radiation source).

The principle of such a real-time monitor is shown in Fig. 9.38, in which a fixed or scannable (as shown) solid state detector is employed to survey the

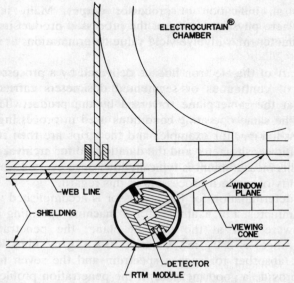

Fig. 9.38. Real-time radiation monitor in Selfshield® system.

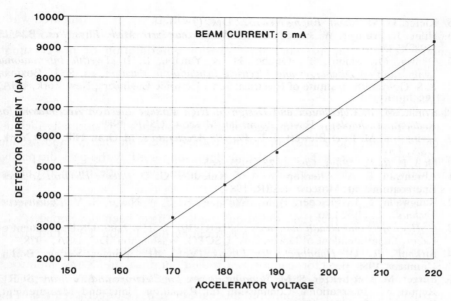

Fig. 9.39. Detector response versus operating voltage at fixed current.

bremsstrahlung field at the window plane, *through* the product plane, if desired. Fixed arrays of these detectors have been developed[42] for the sterilization application and provide an 'all electric' monitor of the electron process which complements the traceable, passive dosimetry techniques described above. The great sensitivity of such a monitor to variations in operating voltage, at fixed current, is illustrated in Fig. 9.39. As shown here, a 1 mm diameter detector[87] provides sufficient sensitivity to spectral variation that control of energy to 1% or less is quite practicable given comparable power supply regulation.

The ability of such active monitoring techniques to provide archival, real-time data on the basic process parameters discussed in the section entitled 'Passive monitoring'—see p. 547, assures their application where the treated product demands an unusual degree of quality assurance. It brings all-electric monitoring to (essentially) an electrical process, and with it a degree of measurement precision not available with the thermal or chemically initiated processes traditionally employed in these industries.

REFERENCES

1. Abramyan, E. A. *Industrial Electron Accelerators.* Energoatomizdat, Moscow, USSR, 1986.
2. Brasch, A., US Patent 1,957,008, 1934.
3. Brasch, A., US Patent 2,429,217, 1947.
4. Brasch, A., US Patent 2,796,545, 1957.

5. Kiefer, D. M. *Chem. Engng News*, 22 Dec. (1969) 31.
6. Hiley, J., Frutiger, W. A. & Nablo, S. V. *Nuclear Instr. Meth. Phys. Res.*, **B24/25** (1987) 985.
7. Okawa, M., Shioiri, T., Okubo, H. & Yanabu, S. In *Twelfth International Symposium on Discharges and Electrical Insulation in Vacuum*, ed. R. L. Boxman & S. Goldsmith. Institute of Electrical and Electronic Engineers, New York, USA, 1986, p. 65.
8. Svinin, M. P. *Calculation and Design of High Voltage Electron Accelerators for Radiation Technology*. Energoatomizdat, Moscow, USSR, 1989.
9. Livingston, M. L. & Blewett, J. P. *Particle Accelerators*. McGraw-Hill, New York, USA, 1962.
10. Holl, P. *Radiat. Phys. Chem.*, **25** (1985) 665.
11. Abramyan, E. A., Alterkop, V. A. & Kuleshov, G. D. *Intense Electron Sources*. Energoatomizdat, Moscow, USSR, 1984.
12. Rangwalla, I. J., Fletcher, P. M., Williams, K. E. & Nablo, S. V. *Pharmaceut. Technol.*, **11** (1985) 36.
13. *OSHA Safety and Health Standards* (29CFR1910), OSHA 2206. US Department of Labor, Superintendent of Documents, USGPO, Washington, DC, USA, 1978.
14. Birkhoff, R. D. *Handbuch der Physik* (Vol. 34). Springer Verlag, Berlin, Germany, 1958, p. 53.
15. Berger, M. J. & Seltzer, S. M. *Stopping Powers for Electrons and Positrons*, ICRU Report 37. International Commission on Radiation Units and Measurement, Bethesda, MD, USA, 1984.
16. Nablo, S. V. & Frutiger, W. A. *Radiat. Phys. Chem.* **18** (1981) 1023.
17. Miller, A. & McLaughlin, W. L. *Radiat. Phys. Chem.* **14** (1979) 525–33.
18. Spencer, L. V. *Energy Dissipation By Fast Electrons*, NBS Monograph 1. Superintendent of Documents, USGPO, Washington, DC, USA, 1959.
19. McLaughlin, W. L., Khan, H. M., Farahani, M., Walker, M. L., Puhl, J. M., Seltzer, S. M., Soares, C. G. & Dick, C. E. *Beta-Gamma*, **2/3** (1991) 20.
20. McLaughlin, W. L., Boyd, A. W., Chadwick, K. H., McDonald, D. C. & Miller, A. *Dosimetry For Radiation Processing*. Taylor & Francis, London, UK, 1989.
21. McLaughlin, W. L., Hjortenberg, P. E. & Batsberg Pedersen, W. *J. Appl. Radiat. Isotopes*, **26** (1975) 95.
22. Lackner, H., Kohlberg, I. & Nablo, S. V. *J. Appl. Phys.* **36** (1965) 2064.
23. Gross, B. & Nablo, S. V. *J. Appl. Phys.*, **38** (1967) 2272.
24. Rosenstein, M., McLaughlin, W. L. & Silverman, J. In *Electron and Ion Beam Science and Technology*, ed. R. Bakish. The Electrochemical Society, New York, USA, 1970, p. 591.
25. Sasaki, T., Hosoi, F., Kasai, N. & Hagiwara, M. *Radiat. Phys. Chem.*, **18** (1981) 847.
26. Frungel, F. B. A. *High Speed Pulse Technology*. Academic Press, New York, USA, 1965, Part A.
27. Roth, A. *Vacuum Technology*. North Holland, Amsterdam, The Netherlands, 1982.
28. Kohl, W. H. *Materials and Techniques for Electron Tubes*. Reinhold, New York, USA, 1960.
29. Schonberg, R. G., Nunan, C. S. & Brown, L. E., US Patent 3,144,552, 1964.
30. Bakish, R., (ed.) *Introduction to Electron Beam Technology*. J. Wiley, New York, USA, 1962, Ch. 1 and Ch. 16.
31. Chapiro, A. *Radiation Chemistry of Polymer Systems*. Interscience, New York, USA, 1962.
32. Karmann, W. Proc. RadTech Europe '89, 1989, p. 595. Florence, Italy, Oct. 9–11.
33. Grishchenko, A. I., Salimov, R. A., Kraizman, A. M. & Fokin, V. S. *RadTech 90*

NA Conference Proceedings (Vol. 1). RadTech International North America, Northbrook, IL, USA, 1990, p. 163.
34. Trump, J. G., US Patent 3,588,565, 1971.
35. Cheever, R. N., US Patents 4,305,000; 4,369,412 and 4,439,686, 1984.
36. Cheever, R. N. *Radiat. Phys. Chem.,* **14** (1979) 267.
37. Danel, F. & Muel, J. *Radiat. Phys. Chem.* **22** (1983) 465.
38. Pigache, D., French Patents 72.38368, 1972; 85.17724, 1985.
39. Wakalopulos, G., US Patent 3,970,892, July 1976.
40. Farrell, S. R. & Smith, R. W., US Patent 4,786,844, 1988.
41. Meskan, D. A. Proc RadTech 90 North America (Vol. 2), 1990, p. 40.
42. Nablo, S. V. *Sterilization of Medical Products* (Vol. II), ed. E. R. L. Gaughran & R. F. Morrissey. Multiscience, Montreal, Quebec, Canada, 1981, p. 210.
43. Gomer, R. *Field Emission and Field Ionization.* Harvard University Press, Cambridge, MA, USA, 1961.
44. Cleland, M. R. *Radiat. Phys. Chem.,* **18** (1981) 301.
45. Wilton, M. S. de. *Radiat. Phys. Chem.,* **25** (1985) 643.
46. Hare, G. E. *Radiat. Phys. Chem.,* **35** (1990) 619.
47. Avnery, T., Cotter, W. & Lawson, P. *Radiat. Phys. Chem.,* **31** (1988) 395.
48. Nablo, S. V., Uglum, J. R. & Quintal, B. S. *Nonpolluting Coatings and Coating Processes,* ed. J. L. Gardon & J. W. Prane. Plenum Press, New York, USA, 1973, p. 179.
49. Colvin, A. D., Nilssen, O. K. & Turner, A. H., US Patent 3,440,466, 1969.
50. Avnery, T., US Patent 4,591,756, 1986.
51. Farrell, S. R., US Patent 4,833,036, 1982.
52. Cruz, G. E. & Edwards, W. F., US Patent 4,785,254, 1988.
53. Okamoto, K. & Matsumoto, T., US Patent 4,829,190, 1989.
54. Schumacher, B. W., Lowry, J. F. & Smith, R. C. Proc. Fourth Int. Elect. Beam Symposium. 1976.
55. Markovic, V. *Radiat. Phys. Chem.,* **18** (1981) 27.
56. Nablo, S. V. *Radiat. Phys. Chem.,* **22** (1983) 369.
57. Chilton, A. B., Shultis, J. K. & Faw, R. E., *Principles of Radiation Shielding.* Prentice-Hall, Englewood Cliffs, NJ, USA, 1984.
58. Flugge, S. (ed.) *Encyclopedia of Physics* (Vol. 30). Springer-Verlag, Berlin, Germany, 1957, p. 356.
59. Cook, P. M., US Patent 3,455,337, 1969.
60. Rangwalla, I. J., Williams, K. E. & Nablo, S. V. *Nuclear Instr. Meth. Phys. Res.,* **B40/41** (1989) 1146.
61. Nablo, S. V. *Nuclear Instr. Meth. Phys. Res.,* **B56/57** (1991) 1232.
62. Cleghorn, D. & Nablo, S. V. *Radiat. Phys. Chem.,* **35** (1990) 352.
63. Nablo, S. V. US Patent 4,652,763, 1987.
64. Kosterka, D. W., US Patent 4,410,560, 1983.
65. Albertinsky, V. I., Svinin, M. P. & Tsepakin, S. G. *Proceedings of the Fourth All Union Conference on the Industrial Application of Accelerators,* NIIEFA, 1982, p. 186.
66. Holl, P. & Foll, E. *Radiat. Phys. Chem.,* **35** (1990) 653.
67. Nablo, S. V. & Tripp, E. P., US Patent 4,521,445, 1985.
68. Mizusawa, K., Ejiri, H., Kimura, T., Hoshi, Y. & Suzuki, H. In Proc. Conf. on Radiation Curing Asia, CRCA, Tokyo, Japan, 1988, p. 287.
69. Czvikovszky, T., Alpar, T., Cazjlik, I. & Takacs, E. *Radiat. Phys. Chem.,* **31** (1988) 639.
70. French, D., Quintal, B. S. & Nablo, S. V., *Radiat. Phys. Chem.,* **18** (1981) 879.
71. Nablo, S. V. *Radiat. Phys. Chem.,* **35** (1990) 46.
72. Holl, P. & Reuter, G. *Radiat. Phys. Chem.,* **22** (1983) 811.
73. Aaronson, J. N. & Nablo, S. V. *Radiat. Phys. Chem.,* **31** (1988) 711.

74. Huemmer, T. F. & Ungurait, J. R. *Electron and Ion Beam Science and Technology*, ed. R. Bakish. The Electrochemical Society, New York, USA, 1970, p. 629.
75. Rangwalla, I. J. & Nablo, S. V. In RadTech 90 N.A. Conference Proceedings, RadTech International, Northbrook, IL, USA, 1990, p. 18.
76. Seng, H. P. *Beta-Gamma*, **3** (1989) 10; **4** (1989) 25.
77. Nablo, S. V., US Patent 4,252,413, 1978.
78. Coleman, G. E., US Patents 3,654,459, 1972; 3,676,673, 1972; 3,790,801, 1974.
79. Troue, H. H., US Patent 3,807,052, 1974.
80. Seng, H. P., Proceedings RadTech Europe, RadTech Europe, Fribourg, Switzerland, 1989, p. 163.
81. Ueno, K. *Radiat. Phys. Chem.*, **31** (1988) 467.
82. McLaughlin, W. L., Miller, A., Abdel-Rahim, F. & Preisinger, T. *Radiat. Phys. Chem.*, **25** (1985) 729.
83. Thalacker, V. P., Simpson, J. T. & Postma, N. B. *Radiat. Phys. Chem.*, **31** (1988) 473.
84. Frutiger, W. A. & Nablo, S. V. *Radiat. Phys. Chem.*, **22** (1983) 431.
85. Seltzer, S. M. & Berger, M. J., *National Bureau of Standards Report NBSIR 84-2931*. National Institute of Standards and Technology, Gaithersburg, MD, USA, 1984.
86. *Guideline for Electron Beam Radiation Sterilization of Medical Devices*, ANSI/AAMI ST31. Association for the Advancement of Medical Instrumentation, Arlington, VA, USA, 1990.
87. Kneeland, D. R., Trygon, Inc., Manchester, MA 01944, Private Communication.

Index

Aberochrome, 540, 342
Abrasive products, 220
Absorption spectrum, 55
Acid-catalyzed chemical amplification, 164–6
Acid-catalyzed cross-linking, 129
Acid-catalyzed depolymerisation, 132
Acid generators, 125, 170
Acrylated oils, 228
Acrylic acid-derived polymers, 150
Acrylic resin impregnated carbon composites, 208
Acrylic systems, 8–9
Addictive lamps, 457
Additives, role of, 68
Adhesives, 16, 37–9
  laminating, 38
  potential applications, 42–3
  pressure-sensitive. See PSA
  see also under specific types
Alkoxymethyl benzoguanamine, 175
Aluminium foil, 219
Amine synergists, 240–1
Amino ketones, 79, 92
Application specific integrated circuits (ASIC), 140
Aryl-aryl sulphides, 79–80
Aspect ratio, 136
ATR IR spectroscopy, 407–10

Backscatter effects, 245–7
Banknote paper, 204
Base generators, 128
Beer–Lambert law, 54, 346–50
Benziketals, 77
Benzophenone–Michler's ketone system, 96
Benzoyl oxime esters, 79
Benzoyl phosphine oxides, 79
Bimolecular termination reaction, 353
2,6-bis(4'-azidobenzylidene)-4-methylcyclohexanone-1, 169
Bisphenol A epoxy acrylates, 227–8
Blisters, 39
Building materials, 209–16

Car wheel rims, 217
Carbonyl chromophore, 53
Casting paper, 34, 197
Cationic polymerisation, 129–30
Cationic systems, 8
Caul sheet, 34
Cell projection technique, 140
Cement-based roof-tiles, 215
Ceramic tiles, 215
Charge transfer complexes (CTC), 58
Chemical actinometry, 342
Chemical amplification, 125, 170–9
Chemical tagging, 424–5

555

Chipboard panels, 211, 213
Chromophoric group, 52, 69
Cinnamic acid, 177
Coating machines, 31–2
  see also under specific types
Coatings, 15–18, 32–7, 206–7
  metal, 36
  on metals, 216–21
  photostability of, 252–6
  plastic, 35–6
  post-cured stability, 248–61
  PVC, 35
  static control, 205
  thermal stability of, 249–52
  three-dimensional plastic objects, 207–8
  see also under specific types
Co-initiator, 354
Complex modulus, 440
Computer control in UV curing, 497–8
Condensation, cross-linking by, 130, 172–8
Contact angle
  hysteresis, 436–7
  measurements of, 432–6
Contrast factor, 134
Cross-linkable polymers, 152–9
Cross-linking, 146–59
  acid-catalyzed, 129
  condensation, by, 130, 172–8
  polycondensation, by, 172–8
  characterisation of degree of, 405–6
  intermolecular, 123
  polymer films, of, 204
Cross-linking agents, network formation via, 124–5, 167–70
Cross-linking sol-gel analysis, 399–402
Cure depth profiles, 431–2
Curing technology, 44–5
Curtain coater, 31–2

Damping factor, 440
Decor papers, 198
  lamination, 210–11
Decorated melamine-cement panels, 215
Decorative building panels, 216
Degradable polymers, 146–52
Depolymerisation, 178–9
  acid-catalyzed, 132
Deprotection, 129, 170–1
Depth profiling, ESCA for, 423–4
Dialkoxyacetophenones, 77
Dianol diacrylate (CDDA), 238
3,3′-Diazidodiphenyl sulfone, 168

Diazonaphthoquinone (DNQ) inhibitors, 159–63
Diazonium salts, 103
Difunctional monomers, 238
α-Diketones, 102
Dilatometric measurements of shrinkage, 396–9
Diluents, 237–40
Dipentaerythritylpentaacrylate (DPEPA), 167, 168
Diphenoxybenzophenone, 112
Diphenylmethane diisocyanate (MDI), 229
Dissolution inhibition principle, 124, 159–67
Distortions of radiation-cured samples, 444
Distributed feedback laser diode (DFB-LD), 150
Doped lamps, 457
DRIFT spectroscopy, 412–13
Dry-etch resistance, 136–7
Dry finishing, 34
Drying technology, 44–5
DSC, 368–71
Dyes, 102
Dynamic mechanical spectrometry, 439–43
Dynamic random access memory devices (DRAM), 122
Dynamic tests, 431–7

Elastic modulus, 440
Electromagnetic radiation, 121
Electromagnetic spectrum, 5
Electron-beam accelerators, 6–7, 242
Electron-beam curing, 30
  applications, 193–223
  devices, 343–6
  mechanism of, 247–8
  operational conditions, 243–7
  paper and board, 194–201
  polished drum, 197
  printing, 201–4
  relevance of rafting, 310–20
  technology, 241–8
Electron-beam generation techniques, 515–21
Electron-beam grafting, 265–8
Electron-beam lithography, 140–1
Electron-beam processing machinery, 503–54
  air path heating, 534

Electron-beam processing machinery—*contd.*
  charge deposition effects, 514–15
  cold cathode systems, 517–19
  configurations, 505–11
  control systems, 521–2
  distributed and scanned systems, 516–17
  exit window technology, 523–8
  fibres, filaments and wire handling, 538–41
  field emission cathode, 518–19
  foil support structures, 527–8
  foils, 524–6
  high-temperature electron emitters, 520–1
  history, 504–5
  inerting, 542–7
    active monitoring, 550–1
    analytical techniques for assessment of efficacy, 544–5
    hybrid techniques, 545–7
  ion-induced secondary emission, 517–18
  ozone/nitrogen dioxide/homopolymer control, 542–7
  passive monitoring, 547–50
  penetration performance, 511–15
  performance monitoring procedures/techniques, 547–51
  process quality control, 522
  product handling geometries, 532–42
  scattering effects, 512–14
  sheet handling, 536–8
  shielding design, 528–32
  three-dimensional and molded products, 541–2
  web handling, 534–6
Electron-beam radiation, 121
Electron-beam storage rings, 143
Electron-curtain accelerators, 242
Electron energy and penetration, 244–5
Electron impact sources, 141
Electron spectroscopy for chemical analysis (ESCA), 418–25
Electron transfer, 76
Electronics, 16
Electrophilic aromatic substitution, 130–1
Ellio-sheet, 218
End-users, 5
Energy losses, 245
Energy transfer processes, 59–61
  applications, 60
EPDM systems, 221

EPDM systems—*contd.*
  photoinitiated cross-linking reaction of, 64
Epoxy acrylates, 226–8
  applications and examples, 228
Equilibrium swelling, 404
Erasable white boards, 206
ESR spectroscopy, 416–18
2-Ethylhexylacrylate (EHA), 239
Europe, 20–1
Excited-state interaction, 99
Excited-state processes, 74
Excited-state reactivity, 57–9
Extrusion, 32

Fibre-board, 213
Fibre-reinforced objects, 208–9
Films
  converting, 204–6
  grafting on, 205
  siliconisation of, 200
Finely focused IB (FIB) exposure technique, 143
First-generation reactive monomers, 237–8
Flexogravure, 29–30
Floppy discs, 207
Formulations, 5
Formulators, 23–4
  future trends, 45
  mergers and acquisitions, 25
  multinationals, 24
Free radicals, 311–15, 317–19
  formation of, 351
FTIR spectroscopy, 407

G value, 344, 345, 346
GaAs-field effect transistors (FETs), 140
Gas plasma, 142
Glass coatings, 36
Glass transition temperature, 405–6
Glassfibre coatings, 36
Glassfibre composites, 208
Glassfibre-reinforced copper-clad prepregs, 208
Grafting
  air temperature, 274–5
  curing, and, 264
  commercial applications, 324–5
  concurrent, 324–5
  definition, 264

Grafting—contd.
  dose and dose rate with ionising radiation, 272–4
  effect of acids and salts-partitioning phenomena, 305–8
  effect of additives on homopolymerisation yields, 291–301
  effect of cationic salts in suppressing homopolymer-partitioning phenomena, 308–10
  effect of commercial additives used in curing, 319–20
  effect of polyfunctional monomers, 308
  effect of structure of monomer and backbone polymer with additives, 301–5
  electron-beam, 265–8
  future developments, 320–5
  homopolymerisation in, 268
  intensity of light with UV, 274
  ionising radiation, 280–1, 283–8, 290
  mechanisms involving additive effects, 305–10
  mechanistic studies, 320–4
  monomers, of, 265–6
  monomers used in curing of,
    cationic systems, 315–17
    free radical systems, 311–15
  oligomers, of, 267
    used in curing—free-radical and cationic initiators, 317–19
  films, on, 205
  partitioning effects—additives, 323–4
  post-irradiation, 268
  reactions, 265–8
  relevance in EB and UV curing, 310–20
  role of ions, 320–3
  role of polyfunctional monomers (PFMs), 280–2
  synergistic effect of acids and salts with polyfunctional monomers, 290–1
  typical systems, 268–72
  UV systems, 266–8, 281–2, 288–91
  mutual technique, with,
    using ionising radiation, 269
    using UV, 270–1
  pre-irradiation technique, with, 271–2
  see also Homopolymerisation
Grafting yield
  effect of additives, 276–91
  physical parameters affecting, 272–6
Graphic art, 14–16, 20
  applications, 25–31

Graphic art—contd.
  potential applications, 42
Gypsum board, 215
Gypsum panels, 215

Halogenated aromatic polymers, 152–5
Hardness measurements, 426–30
Heliogravure, 30–1
Heterolytic bond cleavage, 128
Hexamethoxymethylmelamine (HMMM), 172
Hexamethylene diisocyanate (HMDI), 229
Hexamethylsilazane (HMDS), 137
High-energy radiation, chemical effects of different stopping power, 132–4
High-energy radiation lithography, 121
High-intensity xenon lamp, spectral energy distribution, 338
Holographic method, 382–5
Homolytic bond cleavage, 128
Homopolymerisation
  grafting, in, 268
  effect of additives, 291–301
  presence of acid and polyfunctional monomers, in,
    ionising radiation, 293–5
    UV systems, 296–8
  suppression using co-monomer techniques, 299–301
Hot-melt packaging paper, 199
HPLC, 366
Hydrogen-abstraction reaction, 76
Hydroxyalkylphenyl ketones, 77
α-Hydroxylmethyl benzoin, 86

Indentation hardness, 426
Ink formulators, 23
Inorganic salts in grafting, 283–9
Intaglio inks, 30–1
Intaglio printing, 204
Intermolecular cross-linking, 123
Ion beam (IB) lithography, 143–5, 179–83
Ion beam (IB) radiation, 121
Isophorone diisocyanate (IPDI), 229
Isothermal time-temperature-transformation (TTT) cure diagram, 443

Japan, 20–1
Jet inks, 31

Ketocoumarins, 99
k_p rate constant, determination of, 381–2
k_t rate constant, determination of, 381–2

Labels, release systems for, 34
Laminating adhesives, 38
Lamination of paper and board, 198
Langmuir–Blodgett (LB) films, 146
Laser-induced polymerisation, 105–7
Laser plasma sources, 142
Legislation, 43–4
Light, photochemistry, 51
Light absorption, 51–5
 by photocured formulations, 346–50
Light-curable formulations, 67–73
Light-induced polymerisation, 67–112
Light-induced reactions, 61–6
Light-intensity effects, 73
Line profile, 136
Lithographic inks, 25–7
Lithographic process, 120
 technical aspects, 140–5
Loss modulus, 440
Loss tangent, 440
Lubrication of surfaces, 438

Magnetic coatings, 207
Main-chain degradation, 123–4, 146–59
Main-chain scission, 123, 153
Maleic acid, 167
Market evolution, 12–14
Market share of radiation-curing, 16
Market volume estimation, 12–20
 distribution by application field, 14
 distribution by chemical families, 14
Maskless ion implantations, 143
MD fibre-board panels, 213
Mechanical properties of cured samples, 425–37
Melamine-impregnated panels, 215
Mercury vapour high-pressure and medium-pressure lamps, 333–6, 454–7
Metal coatings, 36
Metallisation base coat, 195
Metals, coatings on, 216–21
Methacrylic systems, 8–9
3-Methylindenecarboxylic acid, 177
Michler's ketone, 96
Microelectronic devices, 122
Microstructure development, 139–40
 dry, 139, 168–70

Microstruture development—contd.
 self-development, 140
 thermal development, 140
 wet, 139
Microstructure generation, 120–2, 146–79
 basic principle of, 123
 chemical strategies for, 123–32
Microwave dielectrometry, 385–8
Microwave-powered lamps, 463–5
Mineral acids in grafting, 283–9
Mini-conveyor UV curing units, 336
Molecular absorption coefficient, 69
Molecular extinction coefficient, 54
Monofunctional monomers, 237–8
Monomers, 5, 237–40
 containing polar groups, 239
 grafting of, 265–6
Morpholino ketones, 79
Moulded products, 541–2

Negatively acting resists, 135, 145
Network formation
 kinetics of, 399
 via cross-linking agents, 124–5, 167–70
Newspapers, high-speed printing, 202
NMR spectroscopy, 415–16
Non-reactive monomers, 240
Novolak epoxide, 130

Objects, 39–40
 coatings, 207–8
Offset machines, 28
Oligomers, 5, 67–8
 grafting, 267, 317–19
Onium salts, 103
Optical density (OD), 54, 74–5
Optical glassfibres, 36
OPV, 27–8
Organo-metallic compounds, 105
Oscillating-plate rheometer, 389
Overprinting vanish. See OPV
Oxygen quenching, 71
Ozone production in UV-curing, 499

Packaging materials, printing of, 201, 202
Paper
 impregnation, 198
 potential applications, 42
Paper and board
 EBC applications, 194–201
 printing with EBC, 201–4

Paper to asbestos-cement coating and lamination, 215
Paper upgrading, 34
Pattern profile, 136
Pencil hardness, 427–8
Pendulum hardness, 428–30
Pentacosa-diynoic acid, 158
Peroxohetero(carbon)poly-tungstic acid, 159
Perrin–Jablonski's diagram, 56
PETP film, 216, 218
Photoacoustic FTIR spectroscopy, 413–15
Photocalorimetry, 371–81
Photochemistry
  basic principles, 51–61
  basic processes in light-induced reactions, 61–6
Photocrosslinking number, 406
Photocrosslinking process, 63–5
Photocuring, 330–3
  bath processing, 477
  industrial applications, 453–4
  temperature sensitive materials, 477
Photocycloaddition reactions, 64–5
Photodegradation under outdoor exposure, 71
Photoinduced polymerisation, direct and sensitised, 61–3
Photoinitiated cross-linking reaction of EPDM systems, 64
Photoinitiated polymerisation, 330–3
  definition, 351
  kinetics of, 351–85
Photoinitiators, 67, 73–112, 354
  α-cleavage, 98
  cationic, 103–5, 315–19
  characterisation, 354
  concentration of, 351
  decomposition, 351
    determination of, 354–6
  efficiency, 355–6
  exhibiting particular properties, 85–94
  free-radical, 317–19
  long alkyl chain, 89
  metal atom, 100
  miscellaneous systems, 94
  pigmented media, 91
  polymeric or copolymerisable, 88
  radical, 73–85
  reactivity, 73–85, 332
  structure-reactivity relationships, 107–12
  uncleavable, 80
  visible, 99–102

Photoinitiators—contd.
  volatility and extractability, 71
  water-soluble, 89
Photolithography, 121, 122
Photomer 4047, 234
Photomer 4116, 241
Photomer 4182, 241
Photomer 7020, 237
Photometric terms and units, 341
Photomodification process, 65–6
Photopolymerisation
  definition, 351
  mechanistic and kinetic studies by holographic method, 382–5
  types of, 351
Photoscission processes, 76
Photosensitisers, 96–112, 354
  structure-reactivity relationships, 107–12
  transition-metal complexes, 101
Photostabilisers, 259–61
Photostability of coatings, 252–61
Physical properties of cured samples, 425–37
Piezoelectric quartz crystal microbalances, 388–9
Pigmented gloss finishing, 213
Pigmented high-gloss coatings, 194
Plasma etching, 168
Plasma sources, 141
Plastic coatings, 35–6
Plastics, potential applications, 42
Playing-card stock lamination, 198
Ploughing test, 428
PMMA, 146–9, 168
  IB resist properties of, 180
  interaction of ions with, 179
  resist films, 150
  solvent-free self-development, 149
Polyaldehydes, 163
Polyaryline imides, 157
Poly(5,7-(bis-1,12-n-butyl-carboxy-methylene-urethane) (P4BCMU), 157
Poly(t-butoxycarbonyloxystyrene) (PBOCST), 129, 170
Poly-t-butylmethacrylate, 150
Poly(t-butyl-p-vinyl benzoate) (PTBVB), 129, 170
Poly(chloromethylstyrene), 154–5, 167
Polycondensation, cross-linking by, 172–8
Polydiacetylenes, 157
Poly(dimethyl glutarimide) (PGMI), 150
Poly(diphenoxyphosphazene) (PDPP), 158

Polyene-thiol-silicone systems, 236–7
Polyester acrylates, 231–3
  applications and examples, 232
Polyether acrylates, 232–3
  preparation, 232–3
Polyether urethane acrylates, 229–31
Polyfunctional monomers (PFMs) in grafting, 280–2
Poly(p-hydroxystyrene), 168
Polyimides, 155–7
Polymer surfaces, study of, 418
Polymeric dissolution inhibitors, 163–7
Polymerisation accompanied by cross-linking sol-gel analysis, 399
Polymerisation rate
  determination of, 356–61
    DSC measurements, by, 368–71
    liquid chromatography, by, 366
    spectroscopical methods, by, 361–6
Polymerisation reaction, characterisation of, 331
Poly(methacryl anhydride) (PMAH), 150
Poly(methacrylonitrile) (PMCN), 151
Poly(methylisopropenyl ketone) (PMIPK), 169
Poly(methylmethacrylate-co-tri(n-butyl)tin methacrylate, 149
Poly(2-methyl-1-pentene sulfone) (PMPS), 163
Poly(olefin sulfones), 163, 164
Poly(p-vinyl phenol), 168
Polyphenylsilsesquioxane, 183
Polyphthalaldehydes end-capped by acylation (PPA), 166
Polysiloxane, 178
Poly(trifluoroethyl-α-chloroacrylate-co-tetrafluoropropyl-α-chloroacrylate), 151
Poly(2,2,2-trifluoroethyl methacrylate) (PTFEM), 151
Poly(trimethylgermylmethyl methacrylate-co-chloromethylstyrene), 155
Poly(trimethylsilylstyrene$_{90}$-co-chloromethylstyrene$_{10}$), 180
Poly(p-vinyl cinnamic acid), 177
Poly(p-vinyl phenol), 172, 175
Positively acting resists, 135, 145
Post-exposure bake (PEB), 138
PPA, 179
Pressure-sensitive adhesives. See PSA
Printing, EBC, 201–4
Products, 21–2
  future trends, 46

Propagation reaction, 353
Protective coatings, 206
Protective films, 200, 205–6
Protective varnish, 198
PSA, 16, 38–9
  removable, 39
  repositionable, 39
Pulsed xenon lamps, 463
Purity analysis of monomers, solvents, diluents and initiators, 350–1
PVC coatings, 35
PVC film holograms, 220

Quantum yield, 55, 57–9, 75
Quinones, 102

RA FTIR spectroscopy, 410–12
Radiation curing
  analytical methods, 332
  appearance, 9–10
  applications, 2–4
  chemistries, 7–9
  consumers' requirements, 43
  ecology issues, 10
  economy issues, 10
  energy savings, 10
  environment, 43–4
  evolution, 10–11
  future trends, 43–6
  growth, 46
  history of technology, 9–11
  investigation methods, 329–452
  limitations, 40–1
  manufacturing and distribution chain, 4–5
  potential applications, 41–3
  sources of, 5–7
  state-of-the-art and market trends, 1–47
Radiation dose, 243–4
Radiation grafting. See Grafting
Radiation-sensitive base generator, 176–7
Radiation sensitivity, 134
Radiometers, 337, 340, 491–7
  continuous-control, 494–7
Radiometric terms and units, 341
Rare earth layer, 206
Raw materials, 5, 21–3, 25
Reactive diluents, 67
Reactive monomers, 237
Reflectors for UV lamps, 466–73
Release papers, 34
Release systems for labels, 34

Resist films
  morphology of, 137–8
  thickness of, 136
Resist patterns, stability of, 137
Resist requirements, 122
Resist systems
  classification, 145
  costs, 139
  inorganic, 159
  one-layer versus multilayer, 145
  positive versus negative, 145
  process compatibility, 138
  requirements of, 134–9
  self-development, 183
  two-level, 182
  wet development, 180
Resolution, 136
Roller coaters, 28, 31
Rotary screen, 32
RTIR spectroscopy, 362–6

Sample thickness of cured thin polymer films, 391–3
Saturation swelling, 404
Scanned-beam accelerators, 242
Scratch hardness, 427
Second-generation monomers, 238
Selective metal-transfer process, 195
Separation sheets, 197–8, 211
Shrinkage measurements, 393–9
Silicone release coatings, 199–201
Siliconisation
  of films, 200
  with EBC, 199
Silk-screen printing, 28–9
Sol-gel analysis, 399–402
Solvent swell measurements, 404–5
Solvents in grafting reactions, 277–80
SOR-lithography, 147, 150
Spectral absorption range, 70
Spectral radiometric units, 341
Spot-curing devices, 482–4
Spraying, 32
Stabilisers, interactions with, 72–3
Static control coating, 205
Steel sheets, 218
Stereo-photo lithography, 37
Stern–Volmer coefficient, 59
Stern–Volmer relationship, 59
Substrates, 39–40
Sulphonyl ketones, 80
Sulphoxides, 80

Suppliers, 22–3
  concentration of companies, 24–5
  future trends, 45
Swelling of polymer networks, 402–25
Swelling ratio, 404
Synchrotron orbital radiation (SOR), 143

Telephone credit cards, 206
TESASEAL, 201
Tetramethylammonium hydroxide (TMAH), 139, 168
Thermal development, 178
Thermal stability of coatings, 249–52
Thioether group, 90–1
Thiol-ene reaction, 65
Thiol-ene systems, 8
Thiol/polyene systems, 236
Three-dimensional objects, 39–40
Three-dimensional plastic objects, coatings, 207–8
Three-dimensional products, 541–2
Tolylene diisocyanate (TDI), 229
Top-surface-imaging processes, 139
1,3,5-Triazin-2,4,6-triamine, 130
ω-Tricosenoic acid, 183
ω-Tricosynoic acid, 158
Trifunctional monomers, 238, 239
Trimethylolpropanetrimethacrylate (TPTM), 167
Tripropyleneglycol diacrylate (TPGDA), 238
Tripropyleneglycol diglycidylether (TPG), 151
Tris(4-*tert*-butoxycarbonyloxy-phenyl)sulfonium triflate, 171
Tunnel walls, 218–19

Unsaturated polyesters, 235–6
  applications, 235–6
  styrene (UPE), in, 7–9
Urethane acrylates, 228–31
  applications, 231
USA, 20–1
UV curing
  basic principles, 49–117
  computer control in, 497–8
  dynamic arrangement, 477
  industrial methods, 476–9
  mini-conveyor units, 336
  ozone production in, 499
  radiation measurement, 491–7
  relevance of grafting, 310–20

UV curing—*contd.*
   rotational arrangement, 477–9
   safety regulations, 500
   static arrangement, 476
   viscoelastic measurements during, 391
UV curing devices
   design principles, 473–6
   on-mandrel or large diameter objects, 477–9
   translate-and-revolve motion, 479
   web or chain type conveyors, 478
UV deblockable acid-releasing systems, 85–8
UV filter sets, 339
UV inks, 25–7, 29
UV lamps, 6, 7, 9, 26, 454–99
   air-cooling systems, 488–91
   geometrical arrangements, 479–82
   insulation, 488
   measurements of, 341
   operation maintenance, 485–7
   reflectors for, 466–73
   replacement, 487–8
   service life, 484–5
   spectral energy distribution, 334, 336, 337
UV light, 121
UV ovens, 7
UV processors, 498–9
UV radiation
   curing terminology, 454
   intensity of, 337
UV radiation sources, 454–65
UV spot curing, 482–4
UV technology, 26
UV-vis lamps, 333

Vacuum curtain, 32
Vacuum-metallisation, 195

Very large scale integration (VLSI)-chip production, 142–3
Vinyl acetate, 167
Vinyl/acrylic pre-polymers, 234–5
Vinyl chloride, 167
Vinyl functional monomers, 240
Viscoelastic measurements during UV curing, 391
Viscoelastic properties, 440
Viscosity measurements during curing, 389–91

Water absorption measurements, 438
Water adhesion, 437–44
   methods for determining, 437
Water-miscible pre-polymers, 233–4
Wet finishing, 33
White-boards, 218
Wire enamelling, 219–20
Wood-based materials, 209–16
Wood-cement panels, 212–14
Wood-coating lines, 215
Wood elements for TV sets, 214
Wood finishing, 33
   potential applications, 42
Wood panel finishing, 214
Wooden base-boards, 214

Xenon lamps, 337, 338, 457–63, 482
X-ray lithography, 122, 141–2
   radiation sources, 142
X-ray masks, 122
   defect repair, 143
X-ray photoelectron spectroscopy (XPS), 418–25
X-ray sources, 141